D0778940

ULTRA MICRO WEIGHT DETERMINATION IN CONTROLLED ENVIRONMENTS

Ultra Micro Weight Determination in Controlled Environments

EDITED BY

S. P. WOLSKY and E. J. ZDANUK

P. R. Mallory & Co., Inc.
Burlington, Massachusetts

INTERSCIENCE PUBLISHERS

A DIVISION OF JOHN WILEY & SONS
NEW YORK · LONDON · SYDNEY · TORONTO

UNIVERSITY OF VICTORIA
LIBRARY
Victoria, B. C.

The paper used in this book has pH of 6.5 or higher.
It has been used because the best information now available indicates that this will contribute to its longevity.

COPYRIGHT © 1969 BY JOHN WILEY & SONS, INC.

All Rights Reserved. No part of this book may be reproduced by any means, nor transmitted, nor translated into a machine language without the written permission of the publisher.

LIBRARY OF CONGRESS CATALOG CARD NUMBER 69-14293
SBN 471 95964 2

PRINTED IN THE UNITED STATES OF AMERICA

Preface

This book, which may be said to be much ado about almost nothing, has been inspired by the steady rising interest in ultra micro mass determination. The microbalance field has developed dynamically in recent years. The concept of micro weighing has been widened markedly by the recognition of the potential of the quartz crystal oscillator. Microbalance utilization has in general been very imaginative and varied. Because the microbalance has been applied so broadly, details of theory, design, and utilization are scattered throughout many publications. It is the prime objective of this volume, therefore, to provide a single effective source of information on the theory, design, applications, and commercial availability of conventional and quartz oscillator microbalances. Sufficient information is included to allow the reader to practice these techniques in a "do-it-yourself manner."

We take this opportunity to acknowledge the encouragement and assistance of many of our colleagues in the conception and preparation of this book. We are particularly indebted to the participants of the Conferences on Vacuum Microbalance Techniques, initiated in 1960, who have helped to destroy the notion that this area of activity was only of limited interest and application. Special recognition must be given to Dr. K. H. Behrndt, who organized the first conference and also introduced the idea of this book to the editors, as well as to Dr. E. A. Gulbransen, who has been a pioneer and also the most consistent contributor in this field.

We wish also to acknowledge the contributions by many individuals to our own microbalance effort since its inception in 1954 while the authors were at the Research Division, Raytheon Company. Among this group are Mr. R. Chase, Dr. W. C. Dunlap, Dr. A. B. Fowler, Mr. G. Freedman, Dr. L. Guildner, Dr. T. H. Johnson, Dr. J. Hawkes, and Dr. H. A. Papazian. We would be remiss in failing to note that our experimental work was made possible through the assistance of Mr. L. Rubin, who solved many of our instrumental problems, Mr. J. Gage, who skillfully fabricated our microbalance systems, Mr. J. Silva, who handled the machine shop chores and Mrs. P. Rodriguez, our most loyal and conscientious assistant.

Finally we must acknowledge those who assisted us in the preparation of this book. We are appreciative of the enthusiastic and cooperative efforts of the authors of the various chapters, the support of Dr. D. G. Wilson, Vice President of R & E, and P. R. Mallory & Co. Inc., and especially of the efforts of Mrs. M. Dorandi, who handled the many secretarial chores involved in the widespread correspondence and actual preparation of the manuscripts.

S. P. Wolsky
E. J. Zdanuk

January, 1969

Contributors

C. L. Angell, *Union Carbide Research Institute, Tarrytown, New York*

F. A. Brassart, *Westinghouse Research & Development Center, Pittsburgh, Pennsylvania*

A. W. Czanderna, *Clarkson College, Potsdam, New York*

D. Fox, *U.S. Army Signal Research & Development Laboratory, The Institute for Exploratory Research, Fort Monmouth, New Jersey*

Theodor Gast, *Technische Universität, Berlin 15, Kurfürstendamm, 195, II*

E. A. Gulbransen, *Westinghouse Research & Development Center, Pittsburgh, Pennsylvania*

D. Hillecke, *Department of Physics, Technische Hochschule, Clausthal, Germany*

M. Katz, *U.S. Army Signal Research & Development Laboratory, The Institute for Exploratory Research, Fort Monmouth, New Jersey*

W. Kreisman, *GCA Corporation, Bedford, Massachusetts*

C. H. Massen, *Department of Physics, Technological University, Eindhoven, Netherlands*

R. Niedermayer, *Department of Physics, Technische Hochschule, Clausthal, Germany*

J. A. Poulis, *Department of Physics, Technological University, Eindhoven, Netherlands*

R. L. Schwoebel, *Sandia Corporation, Division 5123, Sandia Base, Albuquerque, New Mexico*

A. W. Warner, *Bell Telephone Laboratories, Murray Hill, New Jersey*

S. P. Wolsky, *P. R. Mallory & Co. Inc., Laboratory for Physical Science, Burlington, Massachusetts*

E. J. Zdanuk, *P. R. Mallory & Co. Inc., Laboratory for Physical Science, Burlington, Massachusetts*

Contents

PART I

Theory and Design

Chapter 1

Introduction

S. P. Wolsky

Laboratory for Physical Science, P. R. Mallory & Co. Inc., Burlington, Massachusetts.

The progress of many important physical processes can be followed through observation of associated weight changes. Adsorption, desorption, oxidation, reduction, and evaporation are only a few of the more common phenomena which fall into this category. Although these processes can, in many instances, be studied through measurement of other corollary effects, the simple, direct, and absolute nature of gravimetric measurements is very attractive.

This book is concerned with ultra micro weight determination by means of conventional spring and beam balances or a quartz crystal oscillator. Mass is measured directly with the conventional balance and indirectly through its effect on oscillator frequency with the quartz crystal. The relationship between mass and frequency for the quartz crystal and the limits of the method are well defined. Each method possesses its own peculiar advantages and problems. The methods generally do not compete with each other for utility and, in fact, can be used in a complementary fashion. The oscillator is particularly attractive in that with strict environmental control its potential sensitivity is several orders of magnitude greater than that of conventional microbalances. In the remainder of the discussion in this chapter the term "microbalance" is intended to include both weight determination methods.

The increased interest in the microbalance has resulted, at least in part, from the recent rapid technological progress in scientific instrumentation. From a current view it is difficult to believe that only 10–15 years ago the ranks of the microbalance workers were limited to those few who had the resources to fabricate their own instruments. The quartz crystal microbalance was then still unknown. Today highly sophisticated automatic recording microbalance apparatus adequate to satisfy most requirements is available from a large number of commercial sources. In addition parallel progress in vacuum technology has provided a means for the careful environmental control required for many important microbalance experiments.

For some time general microbalance information has been available primarily from papers presented at the informal Vacuum Microbalance Conferences initiated in 1960 by a handful of enthusiasts. The interest in the subject has grown to the point where these conferences are now held regularly on an international basis. It is the prime purpose of this book to support this interest by providing a single reference source for a detailed discussion of the important aspects and applications of the microbalance. The contributors to this volume are drawn from an international group. Each is well recognized for specific contributions in his respective field and many have pioneered some aspect of the development and application of the microbalance.

The book has been designed to be highly practical. Very detailed discussions are deliberately included on design, fabrication, theory, operation, application, and sources of commercial equipment. This information is sufficient to allow the reader to evaluate the microbalance for his purposes and to initiate his own effort. For the reader's convenience the book has been subdivided into three parts, namely, Part I, Theory and Design; Part II, Applications; and Part III, Commercial Equipment. Each chapter in this book is, in itself, a separate entity and can be read as such by the knowledgeable worker. However, each contribution adds another segment of the information necessary for a complete understanding of ultra micro gravimetry with the microbalance. Cross-referencing throughout is also used to integrate the overall effort.

In Section I, a complete review of the microbalance literature provides a current assessment of the field. Details of microbalance design and fabrication will assist those who wish to build their own apparatus or to modify commercially available equipment. The discussion of microbalance theory and of factors influencing balance behavior provides an understanding of the important aspects of balance design and operation. Micro weighing with the quartz crystal oscillator is given separate consideration. Finally, the chapter on vacuum technology offers basic and practical information on the design and construction of a controlled environment microbalance system.

The applications in Part II were chosen as indicative of the types of problems currently being investigated with the microbalance. Surface phenomena, films, high temperature reactions, and sputtering are all topics of current interest. The reader's imagination should be stimulated by the discussion of some of the new microbalance applications such as the determination of the concentration and grain size distribution of dust and the investigation with the quartz crystal oscillator of the physical characteristics of various gases. A wide variety of balance designs are employed by the different authors and interesting new experimental approaches such as the simultaneous measurement of mass and optical spectra and the complementary use of the quartz and conventional microbalances are outlined. A deliberate effort has been made to remove from the description of experimental apparatus and methods all redundancy other than that necessary to make the chapters coherent or to emphasize an important point.

Part III gives a brief description of commercially available equipment and lists potential industrial sources. It is recognized that equipment designs, specifications, and prices will change with time. However, it is considered that this section will be of continuing value in that it will be indicative of available equipment and simplify the reader's task of seeking additional information.

There has been one major deliberate omission in this book, namely a discussion of thermogravimetry. The use of the balance for thermogravimetric studies, however, has been well covered in other texts, e.g., *Thermal Methods of Analysis* by Wendlandt (Interscience, 1964). Also thermogravimetry in the ultra micro range has not been widely employed and may be considered, therefore, outside the general scope of this book.

The field of ultra micro gravimetry in controlled environments still presents many challenges. The ultimate performance of present microbalance methods has not yet been obtained. As illustrated in Part II, new applications of the microbalance are continually appearing. If the advances of the past, however, may be considered as a precursor of the future, great progress can be expected in further broadening the usefulness of this very interesting experimental method.

Chapter 2

Ultramicrobalance Review

A. W. CZANDERNA

Department of Physics and Institute of Colloid and Surface Science
Clarkson College of Technology
Potsdam, New York

I. INTRODUCTION

The subject matter in this chapter will be discussed briefly to provide a broad introductory review for this volume. Detailed discussions of the theory, design, and application of ultramicrobalances will be presented in other chapters. The intent of this chapter is to summarize the historical development of the various types of balances, to cite references to the significant complexities that may be introduced in the measurements, and to reference the most important groups of microbalance applications. It is hoped that this discussion will aid the new host of microbalance experimentalists in avoiding needless repetitious studies of the undesirable mass and force changes that attend to the respective application categories.

The detection of changes in weight is one of the fundamental measurements in the physical sciences. As a result, considerable effort has been expended in the design, construction, and use of ultrasensitive weighing apparatus. The advent of the simple precision microbalance is generally attributed to Warburg and Ihmori in 1886 (1). The steady progress in this field has been discussed in review articles (2–10) and the published proceedings of a series of conferences on vacuum microbalance techniques (11–16). Since extensive references are available in these articles and volumes, no attempt will be made to produce a complete bibliography of the microbalance literature in this chapter. Rather, references will be used for exemplary purposes and to provide a more detailed account of some important topics that were left somewhat incomplete in the most recent review (10). These topics will include classification of microbalances based on the response to an applied force, on the method used for monitoring the response, and on the method of automatic operation. Then, an account of various types of auxiliary equipment required to operate a microbalance will be presented. This will be followed by a discussion of the undesirable disturbances encountered in vacuum microgravimetry. Here, the radiometric forces that result from the thermomolecular flow of gases will be reviewed in some detail because these represent the greatest constraint to ultra precise measurement of mass changes uncovered in the last decade. Finally, the use of different types of microbalances for various applications will be cited.

Only brief mention will be made of the materials used for the construction of microbalances and of the techniques of calibration since these will be covered in detail by Schwoebel in Chapter 3. Most microweighing to date has been carried out using cantilever, spring, and beam microbalances rather than the recently developed quartz crystal oscillator. Hence, the history, definition of terms, etc. that are unique to the quartz crystal oscillator will be provided in the detailed presentation by Warner in Chapter 5.

The versatility of using microbalance techniques for the measurement of small force or mass changes of samples *in situ* has been recognized (17). However, all ultra micro weighing techniques are constrained to monitoring *only* the change in mass of the specimen. The challenge, when an ultrasensitive weighing device has been developed, then, is to be unequivocally certain that the mass change observed *is actually occurring on the specimen under investigation* because of controlled or specified changes in the temperature, pressure, gas composition, etc. This very considerable challenge will be apparent in the remaining sections of this book.

Some confusion may arise in reading the literature concerning microgravimetric results reported as weight or mass changes. The weight, $\mathbf{W}$, of a body is the gravitational force exerted on it by the earth. From Newton's second law, the mass, m, of the body is given by $m = \mathbf{W}/\mathbf{g}$ where $\mathbf{g}$ is the acceleration due to gravity. In a uniform gravitational field, an equal arm balance can be used to determine an unknown mass from a known mass by establishing the point of equivalence of weight. Most calibrations are referred to a mass standard, directly or indirectly. Hence, it is proper to report microgravimetric results as mass changes. It is also obvious that results reported as weight changes should indeed be labeled in units of force (dynes, newtons, or pounds) rather than in units of mass (μg, slugs). Arbitrary conversion of weight to mass without the knowledge of the gravitational constant could result in an error of up to 0.4%.

II. HISTORY

Microbalances have been used for a long time to study such diverse topics as adsorption, desorption, oxidation, reduction, solubilities, stoichiometry, phase change, surface tension, gas densities, magnetic susceptibilities, and long-range forces of molecular attraction. In most of the work prior to 1945, the emphasis of each researcher was simply to develop a suitable method for weighing microgram and submicrogram amounts of material. Consequently, the result was a proliferation of custom-built balances which were delicate in nature, invariably unmanageable in high vacuum, and required unusual craftsmanship to build and operate.

Progress in precision microgravimetry, which follows somewhat the progress in instrumentation, can be divided into three periods of development. During the first period, from 1886 to about 1945, the number of workers was few and the difficulties were great. Emich and Donau (7) made outstanding contributions during this era of custom-made balances. In the next period, from 1945 to 1960, the number of workers in the field increased considerably. Techniques were developed for building microbalances in quantity with a minimum amount of craftmanship. Toward the end of this era, the pioneering work on various types of automation for microbalances was carried out. In the third period, from 1960 to date, the number of scientists and technicians using microbalances has probably increased tenfold primarily because of the advent of commercially available automatic recording apparatus. This has resulted in a shift from an emphasis of designing microbalances to obtain specific results to one of mass production of microgravimetric data. The latter has not necessarily carried with it a concomitant increase in the understanding of the physical systems under investigation because of a lack of detailed understanding of the uncertainties and spurious effects that accompany the techniques used.

III. DEFINITIONS

The definitions encountered in microgravimetry are easier to understand if they are separated into two broad categories. The following discussion is concerned, therefore, first with the definitions that are characteristic properties of the instrument itself and then with those unique to the *use* of the instrument in a micro weighing system. Further discussion of the definitions is provided in Chapter 3.

A *microbalance* is a highly responsive instrument capable of detecting very small changes in the mass of a specimen or an equivalent force. The lower limit of detectability of the microbalance is not well defined and may range from 10^{-10} to 10^{-5} g or an equivalent force range from 10^{-7} to 10^{-2}

dynes. The possible range of mass or force has been narrowed by the use of the term *ultramicrobalance* to indicate a detectability of at least 10^{-7} g or 10^{-4} dynes (18). The *sensitivity* is the magnitude of the reversible or elastic displacement per unit variation in mass or weight. The sensitivity is an inherent property of the instrument; its particular value is specified by the design and construction of the balance although obviously some measurement is necessary to determine its magnitude. The *response* is the reciprocal of the sensitivity and is expressed in micrograms per unit of arc (or per unit of displacement of the pointer at a given distance from the fulcrum). Values of response are often given in the literature in mass units, that is, micrograms. While these are not reciprocal sensitivities as defined here, enough information is usually available in any given article to identify the true value of the response. The *period* is the time required to complete one oscillatory or vibrational cycle of motion of the balance. The *capacity* is the maximum load that can be suspended and placed on the balance without injury to the balance or its operation. In the case of beam balances, the capacity must also include the weight of the beam.

The *precision* of a balance is the minimum variation in mass that can be observed experimentally in a reproducible manner in the absence of any effects from the microbalance system. The *sensibility* indicates the minimum variation in mass that can be observed experimentally in a reproducible manner *in a micro weighing system*. Sensibility is a more practical term than either sensitivity or precision since it is a characteristic of the complete weighing system, that is, its value depends on the response of the instruments and/or operators used to detect the balance deflection and on effects of thermal, pressure, electrical, and mechanical instabilities. The ultimate sensibility obtained when the errors associated with the system are negligible will, by definition, be equivalent to the precision of the balance. *Resolution* or *mass resolution* refer to the minimum mass change that is detectable using the microbalance. These two terms should be used with care since they may represent a statement of either precision or sensibility. The *range*, generally stated for *in situ* measurement, is the maximum variation in mass change that can be measured with the balance at a given load. The *zero-point stability* or *long-term stability* of the balance refers to any time-dependent variation of the precision of the balance or the sensibility of the system. The stability is an especially important balance characteristic when a particular application requires monitoring the total mass of submicrogram quantities for days, weeks, or months. A systematic drift in the zero point for any reason will result in a decrease of the precision of measurement.

Before discussion of the various classifications of microbalances, it is important to provide a useful term that relates the ratio of the sample load to the minimum detectable mass change in any planned microbalance application. The load to precision ratio (LPR) which is favored by the author, and the ratio of load to sample mass resolution, used in Chapter 3 are equivalent for properly describing the necessary balance characteristic. The load to sensitivity ratio (LSR) which is often encountered in the literature is of no value for either (*1*) planning applications of the balance or (*2*) describing one of the fundamental characteristics of the balance. The criticism of LSR for (*1*) is evident from the definition of terms in which both precision and mass resolution are derived from an experimental capability while sensitivity is a particular property of the balance design and/or material. For (*1*), Schwoebel has shown the uselessness of LSR for evaluating the true characteristics of widely differing balances (19). High values of LSR could indicate either large loads or low sensitivities which invalidates its use as an index. However, he has found that the load–sensitivity product (LSP) provides a proper quantitative basis for comparing the true performance characteristics of various types of microbalances (Chapter 3, Tables 3.1–3.5). The adoption of LSP, which was developed during the preparation of Chapter 3, is strongly endorsed as a future standard.

The LPR is of special importance in applications where small mass changes of large samples occur. For example, if we hypothetically consider single crystal wafers of silver, 0.01 cm thick with an area of 1 cm^2 per side, then 100 wafers would weigh 10.49 g, have an area of 200 cm^2 but would adsorb only 5 μg of oxygen at monolayer coverage. A balance with a LPR of 10^8 would detect changes of 0.105 μg, or only 2% of an adsorbed monolayer. Wafers 0.01 cm thick can be prepared from a single crystal (19a), but there are numerous other problems that arise in the measurement of the adsorption of submonolayer quantities of gas using microbalance techniques. The reader may find it enlightening to perform similar calculations for measurements of desorption, solubilities, thin film stoichiometry, iron impurity concentration in magnetic susceptibility measurements, etc. It will be shown that design and operation of a balance with a LPR of 10^8 is a creditable accomplishment and there will be considerable focus on this problem in this book.

IV. CLASSIFICATIONS OF VARIOUS TYPES OF MICROBALANCES

A. Introduction: The Ideal Microbalance

The ideal microbalance, which is most closely approached by the beam microbalance, should optimize a number of characteristics. It should have a large capacity and range, zero point stability, and a precision that is invariant with load; it should exhibit minimal pressure and temperature coefficients, be free from electrostatic and vibrational disturbances, be simple to calibrate, and be stable. It should be possible to subject the chemically inert balance material, frame and housing to a high temperature bakeout for producing ultrahigh vacuum without altering the operation or characteristics of the balance. Finally, the balance should be at least capable of semiautomatic or automatic operation. In the author's opinion, fused silica coated with gold (20) comes closest to representing the ideal choice of balance material although satisfactory results have been obtained by employing fused silica (1–16,21), gold-coated aluminium (22), and even alloys of aluminium (23,24) (see Section IV.E for further discussion of beam materials).

B. Types of Instruments

It is of interest to compare the various types of microbalances by their mode of operation from a pseudohistorical approach to demonstrate the evolution of the microbalance to the currently held view of the ideal instrument. The various designs can be classified as cantilever, spring, beam, or quartz crystal. In general, this classification is based on the response of the instrument to a change in sample mass.

1. Cantilever Microbalances

The cantilever microbalance, which is the simplest type of balance, depends on the measurement of the deflection of a thin fiber to determine the change in weight. In its simplest form, as shown in Figure 2.1, one end of

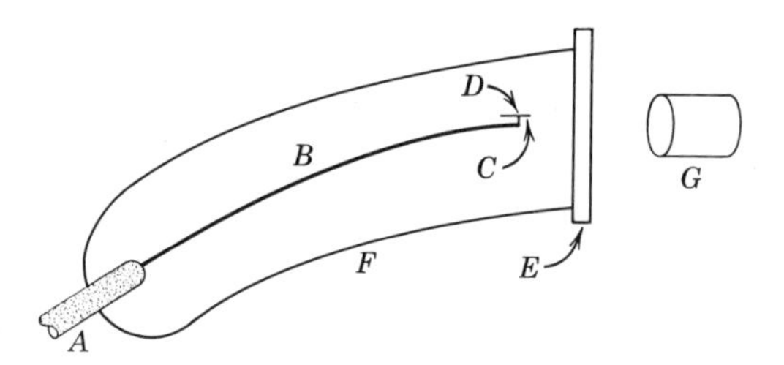

Figure 2.1. Cantilever balance. *A*, glass cane; *B*, quartz fiber; *C*, glass pan; *D*, fiducial fiber; *E*, optical flat; *F*, glass housing; *G*, microscope. (After J. Skelly, *Rev. Sci. Instr.*, **38**, 985 (1967) American Institute of Physics.)

the fiber is fixed and the other end is loaded. An increase in the mass will result in the downward displacement of the free end. It is customary to measure the displacement by an optical technique because of the severe limitation of the capacity of the balance. Characteristics of typical cantilever balances have been discussed (7). Historically, the cantilever balance

Table 2.1. Some Important Properties and Characteristics of Cantilever, Spring, and Beam Microbalances

	Balance type				
Consideration	Cantilever	Spring	Beam–knife edge	Beam–torsional	Beam–pivotal
Response to small change in weight	Bending of a beam	Stretching of a spring	Rotation of beam at primary fulcrum	Rotation of beam at primary fulcrum	Rotation of beam at primary fulcrum
Capacity	10^{-2}–10^{-4} g	Variable depending on helix size; 0.01–10 g, but see LPR	200 g	Less than 1 g in general but see ref. 66	Up to 20 g
Minimum Mass Resolution	~0.01–0.1 μg	~0.1 μg	1 μg	~0.001 μg	~0.01 μg
Range of Load to precision ratio (LPR)	10^3–10^5	10^4–10^6	~10^8	2 × 10^8 or less	2 × 10^8 or less
Typical operation					
Manual	Optical lever	Optical lever for vertical displacement	Riders in a bell jar	Optical lever; torsion head	Optical lever
Automatic[a]	Capacitance	P, T, and C	P, T, and C	P, T, and C	P, T, and C
Usual calibration method	Standard weights	Standard weights	Standard weights	Standard weights, buoyancy	Buoyancy, standard weights
Range	Variable; limited to beam deflection; typical 1000–10,000 μg	Variable; limited by size of helix; typical 1000–10,000 μg	Largest of all balances, 200 g	Deflection: 10–500 μg Automatic > 10,000 μg	Deflection: 200–1000 μg Automatic > 10,000 μg

Virtues	Simple; adaptable to unusual temperatures and pressures; relatively free from vibrations	Simplest in design, operation and for incorporation into a vacuum system; readily outgassed	Large LPR; rugged; easy to use; large range	Large LPR; simple symmetrical design; versatile; adaptable to vacuum and ultrahigh vacuum operation; excellent sensibility over extended pressures and temperatures; can be operated using manual or automatic null techniques; large range; superior zero point stability	
				Superior deflection sensitivity	Extremely rugged, doesn't yaw during use; insensitive to vibrations; good deflection sensitivity
Disadvantages	Susceptible to air currents and electrostatic charge; sensibility limited by incremental resolution at total load; critically limited in capacity and range; fragile	Total load must be weighed to the full accuracy of the helix; limited LPR and range; susceptible to vibrations, especially in vacuum; magnetic damping reduces useful load, must be carefully thermostatted to achieve < 1 μg precision; fragile	Absolute accuracy limited by rider technique; difficult to construct and adapt for ultrahigh vacuum operation; lowest sensitivity about 1 μg; high initial cost	Must be thermostatted, costly to buy attendant apparatus for manual and/or automatic operation; limited range on use as a deflection instrument; requires some manual skills to build; costly jigs are almost essential; zero shifts may result from beam banging arrest or from lateral impacts on balance support stand	
				Fragile; requires manual skills to maintain[b]; sensitive to vibrations; yaws during use	

[a] P: photoelectric; T: transducer; C: capacitance.
[b] These disadvantages are negated when suitable instruments are available commercially.

can be traced to Salvioni (25) in 1901; papers continue to be published (26–28) in which this type of design is used. Details of construction of cantilever balances have been presented in a recent paper (26). The virtues of these balances include simplicity, relative freedom from vibration, and adaptability. They are particularly susceptible to air currents, temperature inhomogeneities, and electrostatic charge. However, the recent design by Kessler and Moore (27), which is automated by a capacitance technique, indicates that balances of this type can be made very convenient by employing sophisticated electronics. Some of the important properties of the cantilever and other types of microbalances are summarized in Table 2.1.

2. *Spring Microbalances*

The spring microbalance consists of a helical spring, usually made of fused silica,* suspended vertically in a tube, as shown in Figure 2.2 (from reference 29). The extension of the spring is essentially proportional to the load where Hooke's law holds and the helix remains essentially circular. The load to precision ratio (LPR) of this type microbalance is its most serious limitation (Table 2.1). The spring balance, which was originated by Emich (30), is sometimes called a McBain-Bakr (31) balance because of their extensive use of it for sorption studies. Kirk and Schaffer (32) have written an excellent review of the construction and application of quartz helix balances. Rand (33) has described a method to prevent the reference arm from tangling with the helix. Spring balances are sensitive to vibration, especially in high vacuum, notoriously fragile, and should be thermostatted to prevent serious reading errors. However, Madorsky (34) has shown that tungsten springs give dependable performance for years and do not have the usual disadvantages. Others have used tungsten helices (35–37) but most devices described employ fused silica (29,38–46). The balances are amenable to magnetic compensation (29,39,42,45,46) and to automatic operation using capacitance (43), transducer (40,46), and photoelectric sensors (39,41,44). They are simple, commercially available, and adaptable to many types of studies. Dell (38), for example, has made use of silica springs in a flow system and discussed buoyancy effects. Moreau (46) and Harrison and Delgrosso (47) have described complete systems although the latter is not of microbalance sensitivity.

3. *Beam Microbalances*

All beam microbalances use the principle of center point balancing in which the primary fulcrum serves as an axis of rotation for equal clockwise

* Editors' Note: The terms "quartz," "fused quartz," and "fused silica" are often used interchangeably to indicate a material rather than the crystalline state of the material.

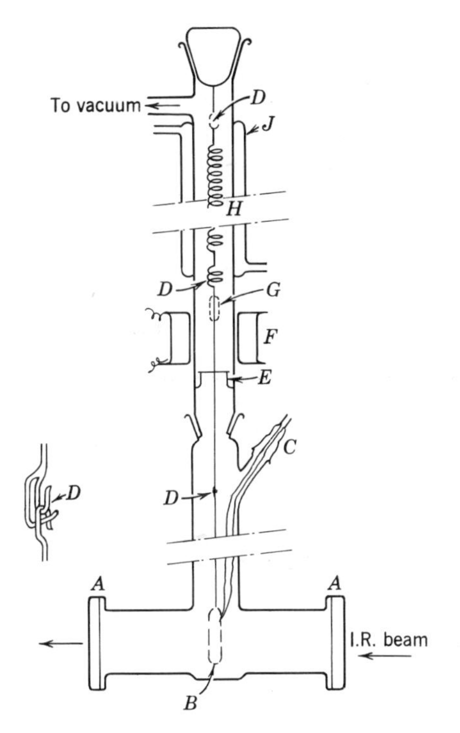

Figure 2.2. Spring microbalance. *A*, infrared transparent windows; *B*, sample holder with sample; *C*, thermocouple; *D*, nonrotating hooks (shown enlarged on the left); *E*, cross bar and stop; *F*, coil; *G*, magnet; *H*, quartz spring; *J*, thermal jacket. (After D. Seanor and C. Amberg, *Rev. Sci. Instr.*, **34**, 917 (1963). American Institute of Physics.)

and counterclockwise moments of force. The moments arise from the action of a force at the secondary fulcrums or yokes where the distance from the primary to the secondary fulcrum is the moment arm as shown schematically in Figure 2.3. It is customary to classify beam balances according to the type of primary fulcrum used. The three categories include the knife edge, torsional, and pivotal balances. Most ultramicrobalances in use today are variants of one of the last two types.

a. Knife Edge. The knife edge balance is of historical importance because it is the prototype of the modern precision vacuum microbalance. It was used first by Warburg and Ihmori (1) and developed considerably by Steele and Grant (48). The conventional silica or agate knife edge bearing on a flat plate constitutes the primary fulcrum of this balance and is also used for the secondary fulcrum in many models. The limitations of this type balance for vacuum microgravimetry have been reviewed (7) and discussed in technical detail (49). Of these, the problems of detecting and compensating mass changes into the submicrogram mass region and obtaining very high vacuum on the large volume of the vacuum enclosure stand out. However, in recent years, the use of the knife edge balance has increased considerably because of the commercial availability of automatic recording instruments (50) and the improvements in vacuum technique, such as feed through collars, base plates, and high speed pumps (51) applicable to systems with a large volume.

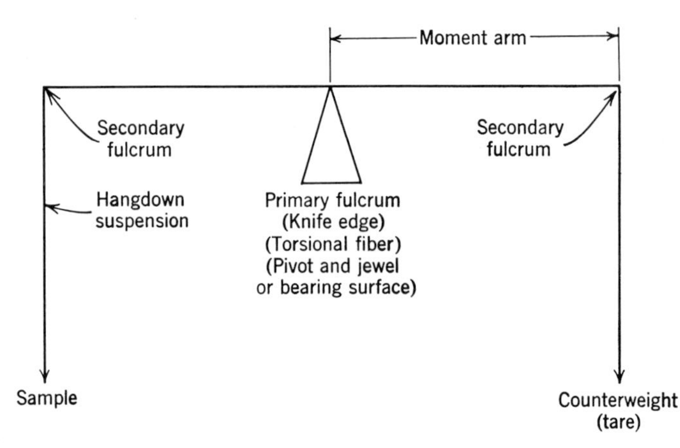

Figure 2.3. Beam microbalance.
Schematic showing principle of center point balancing.

b. Torsional Horizontal Suspension. The most widely used balance is the torsional suspension balance in which the primary fulcrum is a wire, usually of quartz or tungsten, that is attached to the beam. In this type of balance, the torsional moment of the wire is made extremely small compared with the moment exerted by the sample and counterweight suspended from the secondary fulcrums. The early designs of Weber (52), Nernst and Riesenfeld (53), and Pettersson (2) were improved by Emich and Donau (54,55). The development of the Donau version by Gulbransen (56) and Rhodin (57) ushered in the second period of development of vacuum microbalance techniques. It is therefore a proper tribute that many torsional balances are designated Gulbransen-type balances. A lightweight trussed beam was used by Kirk et al. (58) and others (59) to obtain rigidity. The Jennings and Gray microbalance (60) is another widely used trussed beam torsional microbalance; it was one of the first models to be automated using photoelectric techniques. The operation of a torsional balance is extraordinarily reliable and offers the numerous advantages given in Table 2.1 that allow it to approach the ideal microbalance. The major disadvantages include their delicacy, tendency for sidewise yawing, susceptivity to vibration, and in some cases variation of sensitivity with load. However, these may not be serious problems (Chapter 3). The variation of sensitivity with load has been used advantageously for some experiments by Wolsky and Zdanuk (61).

Descriptions of torsional balances that continue to appear in the literature can usually be classified as Gulbransen type (17,24,62–69) or Jennings and Gray type (67–70). Rodder's patented ultrasensitive instrument (71) which has long torsional fibers and a small beam mass, has desirable characteristics. Balances made from Invar (66) and Dural aluminium (24) are not only rugged but can be designed to accept high loads (66). Bakeable variants of the time honored designs that permit operation in ultrahigh vacuum have been described by Wolsky (62), Schwoebel (68,69), and others (70–74). Gouault (75) has discussed the theory of the problems of operation, sensitivity, and stability of torsional balances using tungsten at the primary fulcrum. A general discussion of these problems is included in Chapter 3. Many recent descriptions of torsional balances have been concerned with the mode of automation, i.e. by photoelectric (65,68–70,72, 76–84), capacitance (64,73,85), or inductive (86–88) null point sensing techniques. Some form of magnetic compensation (65,67–70,80–84) is usually employed although Moret and Louwerix (74) managed to use a complicated rider adjustment mechanism. Two widely used commercial instruments employ a torsion type primary support and magnetic compensation; Cahn (81–84) prefers a photoelectric sensor while a magnetic inductive sensor is used in the Sartorius (86–88).

c. Torsional Vertical Suspension. In this technique, a torque is produced on a device suspended by a vertical torsion fiber. The technique, which was introduced by Volmer (89), is frequently referred to as the torsion effusion method. As shown in Figure 2.4, the device is designed so that the rate of change of momentum of some particulate matter produces a torque or a force couple acting about the vertical axis of rotation. The extent of rotation of the fiber is measured frequently by optical techniques. However, it is amenable to compensation at a null position using torsion head (90,91), magnetic (92), capacitance (93), and photoelectric (94) techniques. The steady progress in the use of this technique has increased in the last few years as indicated by the sophisticated descriptions (95–103), which includes a bakeable design (100), and discussion of associated problems (103). Extensive references on applications of the device to problems of chemical interest including vapor pressure measurements, sputtering yields, and thin film studies have been cited in the review by Thomas and Williams (10) and the paper by Peleg and Alcock (102). Freeman and Gwinup (103) have cited references extensively in discussing disturbing effects that result from thermomolecular flow (see Section VII.F) on vertically suspended torsion fibers.

This technique has been combined with the use of beam balances, usually of the pivotal (103,104) or torsion (94) design, by a number of investigators (94,96,103–105) to gain information on both the rate and total mass loss.

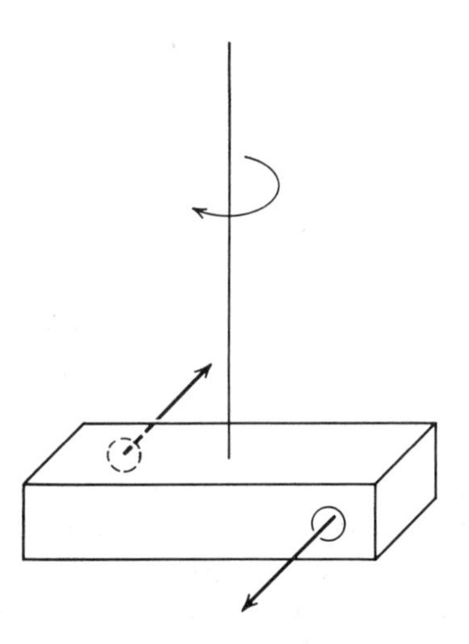

Figure 2.4. Torsional beam microbalance: vertical suspension.

d. Pivotal. In the pivot balance, two sharp points bearing on a surface serve as the primary fulcrum. The choice of the point (pivot) and bearing (jewel) materials has led to an apparent large number of modifications. However, the gas density balance of Stock and Ritter (106,107) is the basic design which has been modified by others (104,108–125). The development of the pivotal balance that has led to its increased use was led by Honig and coworkers (111–114,118). Czanderna (120–122), with the ideal balance as a goal, continued the development of the tungsten–quartz pivot-bearing balance (120,121). He improved the transducer method of automation that was developed by Cochran (115,116) and designed an all quartz bakeable beam (121). The latter was the prototype of the patented trussed beam design (122,126) that can operate satisfactorily after high temperature bakeout at a LPR of $> 10^8$ and at loads of at least 20 g. According to van Lier (127), the balance described in reference 122 will withstand extensive and repeated baking at temperatures exceeding 375° and is amenable to automation by a photoelectric technique. Others (128–130) have found its performance satisfactory even when fabricated from Pyrex (130). The plastic flow experienced by tungsten (122), however, appears to be an important asset to the remarkable operation achieved thus far (111–113, 122). Simons et al. (110) automated their balance with a capacitance technique.

Poulis and coworkers (131,132) have engaged in greatly needed theoretical and experimental work concerning the pivot which included studies of choice of materials used for the pivot and jewel surface. They have shown that LPR'S of 10^8 are possible and have tested balances operable at 25 g loads. Additional exploration on the limitations of the pivot systems described (104,108,115,116,123–125,131–133) will be useful.

The chief disadvantage of the current designs of pivotal balances is that a deflection sensitivity that is comparable with the most sensitive torsional models has not been achieved. However, the LSP is comparable (Chapter 3, Tables 3.3 and 3.4). This has been circumvented to a considerable extent by the use of techniques developed recently to detect very small changes in the deflection (115,116,120,121).

e. Other. In 1944, Vieweg and Gast (133) designed a recording microbalance with a mass resolution of 10 μg that operates on the principle of magnetic coupling between an energized coil mounted on the beam and stationary ones mounted on each side. The current induced in the outer coils by movement of the beam was amplified and adjusted to restore balance. Gast continued this pioneering development until his magnetic suspension balance (134) is now dependable to about 1–2 μg (135), viz. on the fringe of the ultramicrogravimetric range. A balance based on the same principle of operation has been described by Beams (136,137). The outstanding advantage of the magnetic suspension which is applicable to any type of high load balance is that it permits the balance and the sample suspension to be in separate chambers.

The Worden balance (138), which could be classified as a vertically suspended pivot balance, is an intriguing design that has been used extensively by Wade and coworkers (139). The balance is all quartz, compact, inexpensive, and exhibits a very high deflection sensitivity.

Attempts have been made by Behrndt (140) and others (141) to combine the advantages of beam and spring microbalances to achieve bakeable operation and/or greater sensitivity. While these authors have achieved their stated purpose, their designs are extremely intricate. With the modern developments in electronics (see Section IV.D), it is clear that even specialized balances can be simplified to broaden their potential scope of use.

Some of the important properties and characteristics of cantilever, spring, and beam microbalances are compared in Table 2.1. The disadvantages cited may depend markedly on the particular experimental system.

4. *Quartz Crystal Microbalances*

In this form of micro weighing, the change in mass of a specimen is detected as a change in the resonant frequency of a quartz crystal. The effect on the resonant frequency of a piezoelectric crystal resulting from adding mass to the crystal has been known for decades. A detailed study of the effect has been carried out by Sauerbrey (142) while advanced stages of development of micro weighing using this technique have been achieved by the efforts of Stockbridge and Warner (143–148). In a "crystal balance," commercially available quartz crystals fitted with a gold plate are an integral component of an oscillating circuit. The change in frequency of the resonating crystals that results from mass gain or loss is monitored with an electronic counter. The remarkable achievement is that mass changes of the order of picograms have been detected with these devices under carefully controlled conditions. Others (149–157) have used crystal microbalances for studies of thin films (see Chapter 8). The importance of this technique to ultra micro weighing is underscored further by devoting Chapters 5 and 7 to a detailed discussion of the subject.

C. Methods of Monitoring Mass Changes

It would be proper to classify microbalances by the method used for monitoring mass changes. These would include deflection and null operation which are both amenable to automatic operation.

1. *Deflection Operation, Manual or Automatic*

The measurement of the elongation or the deflection of the balance by the use of optical levers or cathetometers has been used for all but the quartz crystal balance. In this type of measurement, the mass changes that can be monitored are limited to the free swing of the beam which restricts the range of the instrument. This method of monitoring, which is exemplified by the Gulbransen gravity balance, is still in use today.

Satisfactory automation has been achieved using capacitance, transducer (linear variable differential transformer), or photoelectric techniques. However, even in the automated form, deflection techniques are restrictive and, in general, most balances are monitored with some form of null technique.

2. *Null Operation, Manual or Automatic Compensation*

The most effective technique for either manual or automatic operation of any balance is the null method, viz. the change in weight of the sample is compensated by the adjustment of some other calibrated force in the system. Compensating forces that have been applied for null techniques include the adjustment of riders, of a buoyancy force, of the torsional moment at the primary fulcrum, and of electrical and magnetic forces.

a. Compensation with Riders. The limitations on the use of adjustment by riders have been discussed (7,114). These include uncertainties that are introduced because it is difficult to reposition the rider and achieve submicrogram precision, and it is not easy to adapt the balance to vacuum operation.

b. Buoyancy Compensation. In the use of buoyancy effects (48), mass changes are monitored by careful measurement of changes in pressure. This technique does not allow changes in weight to be measured in a vacuum or when the samples under study are affected by the pressure of the ambient gas.

c. Torsion Drum Compensation. The application of a torsional moment (58,158) to the supporting torsion fiber can also be used to restore the balance to the null position. However, care must be exercised to avoid nonlinearities, drift, and breakage of the torsion wheel. This procedure also presents considerable difficulties when used in vacuum.

d. Electrostatic Compensation. A capacitance technique (27,43,64,73,85, 110) has been devised in which the force between two charged parallel plates is varied by the amount of charge on the plates. This technique is amenable to high vacuum operation but considerable difficulties are encountered because of the different dielectric properties of gases that might be used in any given experiment.

e. Electromagnetic Compensation. One of the first applications of electromagnetic compensation was made by Angstrom (159) in which 10^{-6} g of material was weighed. The technique was refined by Emich to permit changes of less than 10^{-7} g to be detected. Since then, some form of magnetic coupling has become the most widely used scheme for balancing by a null technique. In the basic electromagnetic compensation technique, a permanent magnet is attached or enclosed on a hangdown suspension fiber. One end of the magnet is placed at the center of a solenoid; the position of the other end is either outside the solenoid or at least in a weaker magnetic field to establish the requisite magnetic field gradient. The compensating force then varies linearly with the current in the solenoid, according to the basic equations for the field of a solenoid (160). Typical modifications of this type have been reported in recent papers (29,45,46,67–70,111–118). This form of electromagnetic compensation will be discussed in more detail in Chapter 3, Section V.B. The alternate method is to place a coil on the beam in a uniform magnetic field; the restoring torque, which rotates the beam about the primary fulcrum, is obtained by changing the current in the coil. This method has been used by Vieweg and Gast (133), Cahn et al. (81–84), and in the Sartorius balance (86–88,161). This alternate method is difficult to adapt to ultrahigh vacuum operation while the basic method must be modified for use in the presence of variable stray magnetic fields.

In principle, all of the above methods of compensation to use a balance as a null instrument can be used manually or for automatic operation. Again, some form of electromagnetic compensation has been used in nearly every scheme developed for automatic or semiautomatic elimination of an error signal as will be discussed later in this article.

D. Methods for Automatic Sensing of Movement from the Null Point

1. Introduction: Automatic Recording Microbalances

Some of the early attempts to automate vacuum microbalances have been summarized by Cochran (115,116). It is interesting that at the First Informal Conference on Microbalance Techniques (11) there was not unanimity of agreement concerning the value of automatic operation compared with manual techniques (17,63,162 versus 114,116,163). There is no doubt today that automatic operation using some type of null balancing with a potentiometric read out offers tremendous advantages for both kinetic and equilibrium studies. It is not infrequent that a significant advance in technique, which is represented by the development of the automatic recording beam microbalance, is the precursor to significant scientific discovery. This is simply a manifestation of providing the observer with an abundance of dependable experimental data—and for increasing the time he has for critical analytical evaluation. There are, of course, experimental situations where manual or deflection techniques should be used but in the last decade the question, "should I use automatic equipment?" has shifted to "why shouldn't I use automatic equipment?" The overwhelming case for automatic operation can be deduced from a rational inspection of Table 2.2.

Automatic operation provides continuous monitoring of all mass changes, more reliable data, an extensive range of compensation, and releases the operator for better use of his time. Deflection methods and manual methods of compensation using a null technique are in many cases simpler to use for short-term experiments which seems to be their only important asset.

The two main disadvantages of automation are cost and difficulty for

Table 2.2. Some Considerations of Automatic Operation of Beam Microbalances Using Magnetic Compensation and Manual Operation Employing a Deflection Method

Considerations	Automatic operation, null magnetic compensation (AONMC)	Manual operation, deflection or null methods (MO)
Adaptability	Can be used effectively and efficiently for all beam balances	Can be used for all types of microbalances except quartz crystal oscillator
Capacity	No effect; probes, mirrors, magnets, etc. are used as tare	Same as AONMC
Range	Large range provided by magnetic compensation	Deflection operation is limited to free swing of beam; null operation same as AONMC
Variation in sensitivity with load and effect on calibration	Precision may vary with load but not the calibration factor	Sensitivity and hence calibration factor vary with load
Zero point stability	Balance stability unaltered but null point sensor may have a significant temperature dependence	Zero point has very small temperature dependence
Zero point detectability	Electronic sensors of deflection from null are superior to manual methods	The best methods possible must be used to detect changes in beam position
Data	Continuous data. Other parameters, viz., T,P, gas composition, etc. simultaneously recorded; zero shifts from any source automatically recorded	Intermittent readings; quality of data may vary from tired or different operators; small zero shifts may be missed

(*continued*)

Table 2.2—*continued*

Consideration	Automatic operation, null magnetic compensation (AONMC)	Manual operation, deflection or null methods (MO)
Temperature control	Necessary for some types of automation; otherwise, as for MO	Necessary for data of highest precision
Vibrational disturbances	Low level steady vibrations are handled without difficulty	Vibrations may be annoying for observation of beam position
Pressure and temperature	Temperature control necessary for null point sensor eliminates further concern	Temperature coefficient of beam *may* require thermostatting as carefully as for automatic operation with temp. sensitive sensor
High temperature bakeout	Operation not possible during bakeout; repeated bakeout cycles may alter calibration by by demagnetizing compensation magnet	Operation not possible during bakeout; readily adaptable to repeated bakeout without altering calibration operation with temp. sensitive sensor
Cost, dollars	Initial investment: several thousand dollars	Initial investment: one to two thousand dollars
Cost, personnel	Continuous attention by technical personnel not necessary	Continuous attention necessary. Continuous twenty-four-hour operation very costly
Adaptation to computer facilities	Digital methods—may be operated directly from voltage output and evaluation of data completed by computer program	Data may be evaluated by computer after manual recording of data

high temperature bakeout. The initial cost of about $15,000 is easily justifiable relative to the advantages gained. The difficulty of maintaining automatic operation during bakeout to 300–400° has not been resolved; however, Addiss' and then Schwoebel's successful use of iron single crystals (67–70) has made a very significant contribution to eliminating this troublesome materials problem. According to van Lier (164), Cunife slowly loses its magnetization on repeated bakeout at 400°, thus, altering the balance calibration.

A general scheme for automation of a microbalance is shown in the block diagram in Figure 2.5. It is customary to operate the balance at a

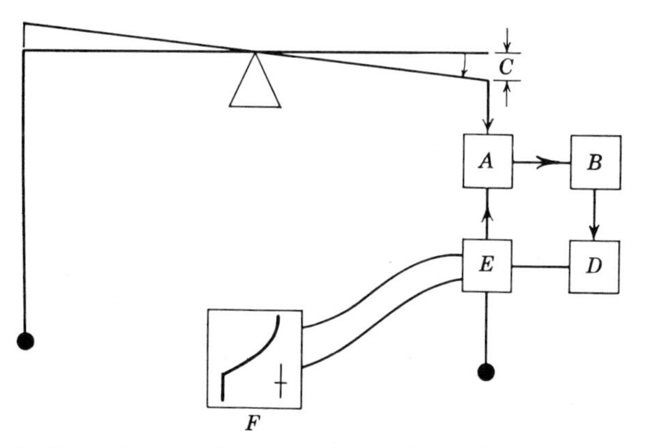

Figure 2.5. Block diagram showing general scheme for automation of a beam microbalance. *A*, sensing device; *B*, error signal because of movement of beam; *C*, mechanical movement of beam; *D*, servo feedback circuit, *E*, restoring force on hangdown suspension; *F*, recorder.

null position. A sensing device (*A*), will provide an error signal (*B*), from the mechanical movement (*C*), of the beam away from the null. The error signal is presented most advantageously as a voltage to a servo feedback circuit (*D*). The latter alters a force acting on the balance (*E*), usually by changing the current in a magnetic compensation circuit, to eliminate the error signal. The changes in the current in the compensation circuit (*F*), are then used to provide a voltage that is related to mass change by some calibration factor. A detailed presentation of a typical automated system is presented in Chapter 3, Section V.C.

The most widely used sensing devices depend on the change in capacitance, magnetic induction, or intensity of light falling on a photocell because of the mechanical movement of the beam. Typical examples of these are shown in Figure 2.6.

2. *Capacitance*

In the capacitance technique (Figure 2.6*a*), one plate of the capacitor moves because of the deflection of the balance. For a parallel plate capacitor, $C = Ke_0A/d$ where C is the capacitance, A the area of the plates, d the distance of separation, e_0 is a constant, K the dielectric coefficient, and Ke_0 the permittivity of the dielectric between the plates. Since very small changes in C can be detected by beat frequency techniques, it can be

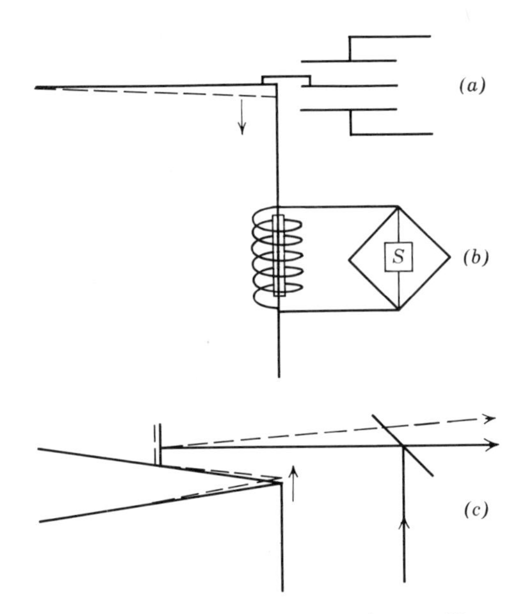

Figure. 2.6. Schematics of sensing devices. (*a*) capacitance; (*b*) magnetic induction; (*c*) photoelectric.

seen that small values of d can also be sensed with this method even when the requisite leads are attached to the balance. Unfortunately, the value of K ($K = 1$ in vacuum) depends on the gas and the pressure of the gas in the system. While this sensing technique might not seem to be desirable for microbalance systems where pressure dependent studies are planned, a reference capacitor could be placed in the system to minimize the ambient effects.

3. *Magnetic Induction*

In the magnetic induction technique (Figure 2.6*b*), the movement of the microbalance hangdown produces inductive effects in a bridged circuit.

Again, the imbalance in the bridge can be used to detect very small changes in displacement. The geometry of the devices used for inducing the current imbalance provides difficulties. The method that depends on the movement of a probe in the core of a transducer or differential transformer is difficult to adapt to ultrahigh vacuum operation. It is also extremely sensitive to temperature changes and gradients which requires elaborate thermostatting.

The inverted method of using the movement of a conductor in a uniform magnetic field (81–84,86–88,133–135) requires the attachment of a current loop to the beam of the balance. This method is not only difficult to adapt to ultrahigh vacuum operation but also has an adverse influence on the deflection sensitivity of the balance and the inertness of the beam, especially to organic gases.

4. Photoelectric

In the photoelectric method (Figure 2.6*c*), the movement of the balance produces either a change in the path of light incident on a mirror attached to the beam or a change in the position of an opaque flag in a light beam. Alternatively, a beam splitter may be used (19). In all cases, the result is a change in intensity of light incident on a photosensitive detector. The change in photovoltage is then a measure of the movement of the balance. This technique not only provides adequate sensitivity to deflection but is readily adaptable to ultrahigh vacuum and is insensitive to changes in pressure and temperature. Of the three methods of sensing deflection described, the photoelectric method may become the most widely used because of the inherent advantages and lack of disadvantages.

E. Materials for Microbalance Fabrication

A critical discussion and comparison of the various materials used in the construction of microbalances will be presented in Chapter 3. It is sufficient, therefore, to note that quartz has been used for fabrication of an overwhelming number of balances because of its inherent advantages. It is chemically inert, has a low density, high tensile strength, and a low coefficient of thermal expansion; it can be easily manipulated, drawn, and fused; it has a high purity and homogeneity and can be readily cleaned and outgassed. The principal disadvantages include a low thermal and electrical conductivity. However, these latter factors are sufficiently important to have led to the investigation of other suitable beam materials (21–24). Aluminium and its alloys have been shown to be as good as quartz as a beam material. Characteristics that are superior to either quartz or aluminum and its alloys have been obtained with a gold-plated quartz beam (20) because the metallic coating provides high thermal and electrical conductivity in addition to the principal advantages of quartz. Since the platinized beam, preferred by Rhodin (7), was used in a thermostatted enclosure, it remains to be established what metal will provide the ideal coating for a quartz beam. Both studies (7,20) were carried out on Gulbransen-type beams.

F. Conclusion: The Ultimate Microbalance

In Section IV.A, an attempt was made to define an ideal microbalance. A better term might have been the ultimate instrument for microgravimetric studies. It is rather obvious that not every investigator will require the ultimate instrument for a given study. In this case, compromises based on information presented in detail in Chapter 3 will be made until a suitable instrument design is reached. However, from the knowledge currently available, the ultimate instrument should be a torsion or pivot beam microbalance made of fused quartz with a metallic (gold) coating. It should operate automatically as a null instrument using a photoelectric sensor and an electronic servo feedback to a magnetic compensation scheme. For the latter, the suspension of a permanent magnet in the field of a solenoid is preferred. The entire system should be designed to permit the attainment of ultrahigh vacuum operation. The pivotal model (122) is preferred by many experimentalists because it is rugged, versatile, and has a load capacity that is usable for a great breadth of studies. The only apparent disadvantage is that it is less sensitive than the torsional model. This loses its prominence, however, because for many studies, a host of undesirable effects, discussed in Section VII, can be considerably greater than the precision of the instrument. Thus, the challenge of being assured that the mass change is actually occurring on the sample under investigation is again noted.

All the other choices made for the ideal balance are practically dictated when a balance is built to operate in ultrahigh vacuum (UHV). When UHV is not needed, other schemes for automation and magnetic compensation have significant merit and should be seriously considered. At the present time, the automatic bakeable torsion balance described by Schwoebel (69), represents the closest approach to the ultimate design that has been described in the literature. A comparable balance of the pivotal design (122) has been perfected by van Lier (164).

Finally, the goal of attaining a LPR of $> 10^8$ has been reached with both torsion and pivotal balances. According to Poulis and Thomas (165), attaining greater sensitivities may be fundamentally limited by Brownian motion (see Chapter 4). The relevance of fluctuation theory to this problem (165) and to the determination of micromagnetic susceptibilities has been discussed (118,166).

The reader should also evaluate the quartz crystal balance as another possible approach to the ideal balance (see Chapters 5, 7, and 8).

V. CALIBRATION TECHNIQUES

Since this subject is extensive no attempt will be made here to assess the development of the various direct and indirect techniques used for calibration of a balance. It is obvious that calibration is a prerequisite for accurate measurements and the method of calibration used will be dependent on the anticipated design and use of the instrument. The direct calibration technique, in which mass standards are added to one hangdown suspension of the balance, is probably the most widely used method because of its simplicity and accuracy. However, the technique of using two buoyancy bulbs of the same mass but different volumes, which evolved from an interdisciplinary collaboration (112,118) and was described by Czanderna and Honig (111,112,114,117), is especially desirable. With this technique, it is possible to establish the relation between mass change and the voltage from the compensation circuit and the sensibility of the balance over wide ranges of pressure. It is possible to obtain over 40 data points with either buoyancy bulb in a normal working day. By judicious choice of the volume of the bulbs, data can be obtained which are insensitive to reading errors in the pressure or the null position of the balance. The data can then be subjected to analysis by the method of least squares (167) to obtain the standard deviation, confidence limits, etc. including differentials (111,112, 114,122) for the precision of the balance *and* the pressure measuring device in the system. Thus, the buoyancy method provides a simple test of the linearity, precision, and accuracy of any gauge that might be used in the operation of the balance at pressures exceeding a few torr.

The papers by Macurdy (168) and Fennell and Webb (169) are interesting contributions. Methods for calibration and assessment of a quartz torsion balance are described (169); a standard deviation of 0.08 μg was obtained for submicrogram ranges.

Calibration procedures are discussed in the Appendix, Chapter 3.

VI. AUXILIARY EQUIPMENT FOR OPERATION OF A VACUUM MICROBALANCE

A. Introduction

In Section IV, the state of development of ultra micro weighing devices was the primary concern rather than their operation. The auxiliary equipment discussed in this section will be restricted to the needs for "classic operation." For this, the balance is operated in a vacuum system; attendant facilities are provided that permit the temperature, pressure, and gas composition to be varied about a sample suspended from the balance. In Section VII.B, undesirable disturbances that arise in classic operation will be reviewed. The measurement of mass change and additional parameters, simultaneously, will be treated in Section VIII.B.

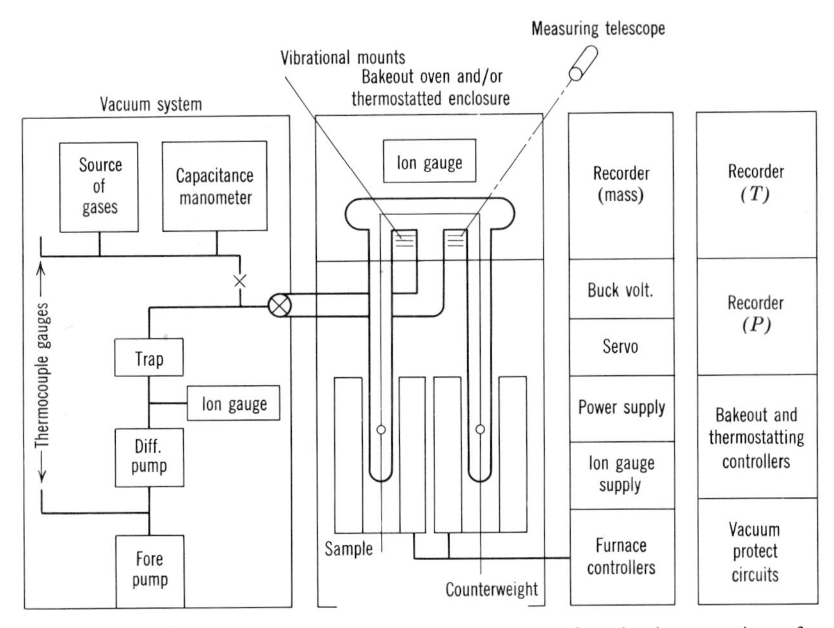

Figure 2.7. Typical arrangement of auxiliary apparatus for classic operation of a beam microbalance.

The cost of fabrication of the balances described in Section IV may be as little as ten dollars. However, when special jigs are used for construction of a balance similar to those described by Podgurski (17), Wolsky and Zdanuk (62), Czanderna (122), and Boggs (70), the cost of fabrication can run well into four figures for the first balance. The cost of the auxiliary equipment for the classic operation of a precision microbalance can easily reach $15,000. Typical needs include a vacuum system and attendant instrumentation, a sturdy vibration resistant mount, furnaces and cryostats, equipment for automatic and/or manual operation and recording, and apparatus for thermostatting an enclosure about the balance and/or hangdown tubes. A typical arrangement of the auxiliary equipment that will be discussed for classic operation is drawn in Figure 2.7.

B. Vacuum Operation

The essential components required for vacuum operation include a chamber for the balance (balance housing), hangdown tubes to enclose the sample and counterweight suspensions, a gas handling system used for adjusting the gas composition and pressure about the sample, and a pumping system.

1. Balance Housing

The chamber for housing the vacuum microbalance is usually of Pyrex glass, quartz, brass, or stainless steel with suitable windows, ports, and flanges attached. The detailed geometry of the arrangement of the balance in the housing, access to pumps and gas handling facilities, and hangdown tubes for beam and vertical torsion balances varies widely with the intended use of the balance. The extensive references cited in Section VIII.A, B, provide a considerable body of source material for any type of needed application.

2. Hangdown Tubes

A system designed for precision studies should employ a symmetrical design of the housing and hangdown tubes (Figure 2.8) as introduced by Rhodin (7,57). The magnitude of the importance of this contribution will be appreciated more when undesirable forces that result from buoyancy and thermomolecular flow are considered (Section VII.E, F). The size of the tubes should be chosen to provide the desired pumping speed and to minimize thermomolecular effects. Hangdown tubes are frequently coated with a conductive material and grounded to eliminate forces from static charges (Section VII.C). Pressure gauges, traps, getters, thermocouple wells or lead throughs, and radiation baffles are frequently attached to hangdown tubes. Since the final design is governed by the anticipated use, systems employing interchangeable metal flanges with gold gaskets permit maximum flexibility.

3. Sample Suspension Fibers

The nature of the suspension fibers used depends on the method of preparation of the sample and insertion into the vacuum system. Rhodin (7) and Gulbransen (8,17) have reviewed the importance of careful sample preparation for precision studies. It is unfortunate that little detailed information is available on suspension fibers. Silica drawn into thin fibers has been used by many investigators for experimental work up to about

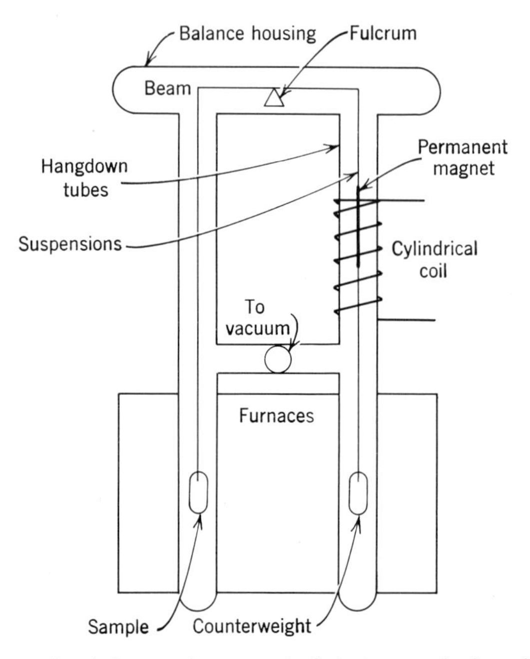

Figure 2.8. Beam microbalance of symmetrical design employing electromagnetic compensation.

1100°. The small manual by Faeth (117) contains simple detailed instructions which serves as a good starting point. For low temperature operation, nichrome, Pyrex, and other metal wires have been used. For temperatures approaching 1600°, Gulbransen employed alumina, platinum, and platinum–rhodium (17). Walker (63), after encountering difficulties with platinum, iridium, rhodium, and tungsten suspensions while employing an induction heating technique, successfully used a chain of sapphire single crystal rods to 1800°. Tripp et al. (170) employed alumina rods for oxidation studies up to 1800°.

A double hook suspension was used by Singer (171) and Soule et al. (172) in anisotropy studies. A copper tube with ultra thin walls cemented to the double hook provided the necessary completely rigid suspension (172). Teflon monofilaments, less than 1 μ in diameter also have been used effectively (172).

4. Vacuum Technique

The advancement in vacuum technology has been phenomenal in the last decade. Excellent books have been published on vacuum technique (173) and ultrahigh vacuum and its applications (174). Both books (173,

174) have extensive references. A general list of references to ultrahigh vacuum, general vacuum, and pertinent vacuum journals is contained in the appendix of reference (174). Therefore, no attempt will be made to review vacuum science further but the importance of utilizing the remarkable techniques for the production and measurement of vacuum cannot be over-emphasized. Microbalance systems that employ good design and excellent vacuum technique will be found in the applications chapters of this book. A composite vacuum system is blocked out in Figure 2.9 for illustrative purposes. For further detail on vacuum theory and design, see Chapter 6.

It is worth noting that pressure measurements made at room temperature *when the sample is at a different temperature* must be corrected for thermal transpiration as discussed in reference 173, p. 59. The empirical correction formulas of Liang (175) are especially helpful.

C. Vibrational Mounts

Special mounts to eliminate undesirable vibrational effects have been used by numerous investigators (7,23,62,68,111,112,115,116,121,122,176) as a general practice. These range from use of concrete pillars for the microbalance base (68) to simply isolating the balance from sources of vibration (62) and can be very unusual (177). Pivotal microbalances have been used effectively without any vibrational mounts (115); the balance housing was supported by an angle aluminium table bolted to the basement floor of a four-story building. Vibrational effects can be serious, especially at the upper levels of multistory buildings. It appears from a concensus of the literature that the safest way to deal with vibrational problems is to erect the microbalance support system on the ground floor of a building and isolate the support from the walls. The housing should be supported by a rigid framework. According to Kissa (178), further attempts to minimize or eliminate vibrational effects should be minimized since the indiscriminate use of resilient materials may magnify vibrations. Potential sources of vibration must be considered, however, in locating a balance (23,176) despite claims to the contrary (81–84) because of the effect on the ultimate sensibility. Requirements for the mounting of seismic helix balances have been discussed (179). Further discussion of vibration problems may be found in Chapters 3 and 4.

D. Thermostatic Operation

Thermostatted enclosures (7,57,62,115,121) have been employed to maintain temperatures to within ± 0.03 to $\pm 0.1°$. In the simplest form, air is drawn over a water cooled heat exchanger from the enclosure with a

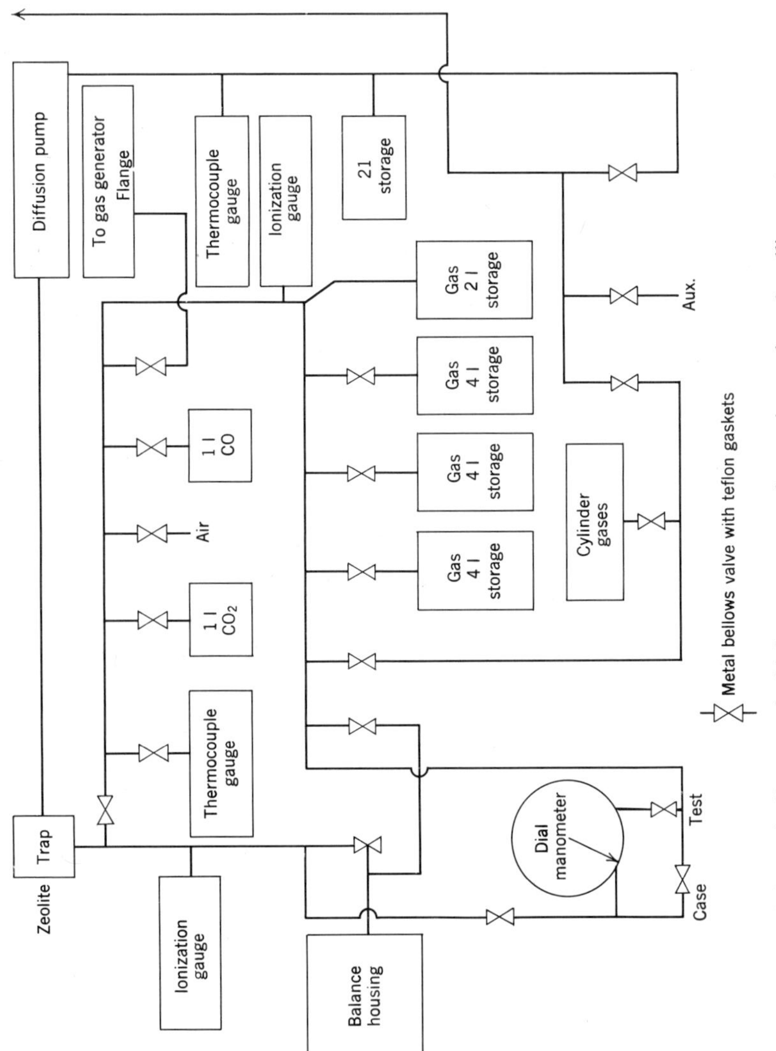

Figure 2.9. Schematic diagram of a high vacuum pumping station and gas handling system.

centrifugal blower and returned to it through two heating elements. The power input of the heaters is regulated by an on–off controller and powerstats. When bakeout ovens are employed, they may also serve as the enclosure for thermostatting. It is convenient to raise and lower ovens and thermostats with a hoist.

E. Automatic Operation

The apparatus required for automatic operation includes the items shown in Figure 2.5. Multipen recorders are especially valuable for displaying simultaneously the change in mass of the sample, its temperature, and the pressure in the system. Other physical parameters measured simultaneously with mass change, which will be reviewed later, also can be conveniently recorded.

F. Manual Operation

Even though the era of automation has descended on vacuum micro weighing, many investigators find numerous advantages in alternate manual operation (17,62,68,172,180). Good results can be obtained using a cathetometer with a resolution of ± 1 μ and some form of magnetic compensation. Satisfactory results can be obtained with a simple potentiometric circuit as described by Czanderna (111), Honig (114), and Faeth (117).

G. High Temperature Operation

The furnace used to heat the sample may present difficult materials problems when very high temperatures are desired. Gulbransen (8,17,66, 181) has pioneered studies of the performance of materials for high temperature furnace tubes and high vacuum operation. Measurements of pressure, apparent pressure, and leak rates in vacuum and in various gases were carried out prior to studies of high temperature reactions. In Chapter 10, Gulbransen discusses high temperature reactions in detail. The measurement and control of temperature can be achieved by the usual engineering techniques except at very high temperatures (8,17,63,66,73,170 181,182). Control can be achieved with intermittent controllers or by maintaining a constant voltage input to the furnace (112). Gradientless furnaces (183) and radiation shields (184) are helpful where large samples are used.

Temperatures can be determined with thermocouples or pyrometers. The thermocouples either can be mounted in the reaction chamber as close to the sample as possible or cemented to the walls of the tube furnace outside the vacuum system. In the latter case, Faeth (117) has shown that a calibration curve (Figure 2.10) relating the thermocouple temperature to the temperature of the sample must be obtained because of radiation loss. Special care must be exercised to correct the brightness of the surface

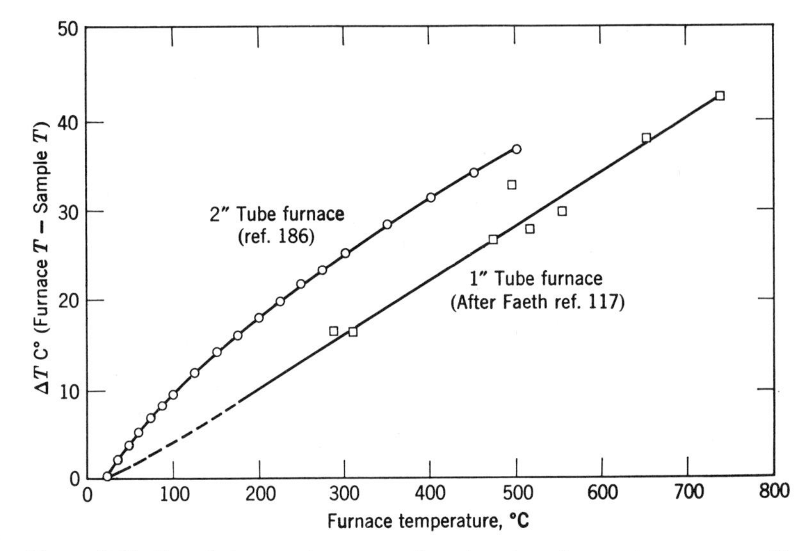

Figure. 2.10. Sample temperature correction at various furnace temperatures. The sample temperature is always lower for the tube furnaces because of radiation losses. (After Faeth, reference 117.)

which is measured with an optical pyrometer for the emissivity of the surface (181). For samples enclosed in opaque furnace tubes, the Stephan–Boltzmann radiation law has been found useful for estimating the surface temperature of a sample (181).

Some problems in induction heating to achieve temperatures of the order of 2000° have been described (73,182).

H. Low Temperature Operation

For low temperature operation, the hangdown tubes can be immersed in liquid–gas or solid–gas refrigerants contained in dewars to maintain a constant low temperature; it is critical to maintain the level of the refrigerant for precision studies. For intermediate temperatures, standard cryogenic techniques can be employed involving controlled heat leaks to refrigerated chambers (185). Again it is important that both sample and dummy suspensions be subjected to the same temperature conditions. For precision studies of the surface area, the temperature of the refrigerant must be determined because of the critical temperature dependence of the monolayer coverage. This eliminates the need to trust the purity of the liquid refrigerants or the fluctuation of its boiling point with change in the atmospheric pressure.

VII. UNDESIRABLE DISTURBANCES OR FORCES

In carrying out measurements with a microbalance, a number of undesirable forces and mass changes are encountered. These disturbances are sometimes referred to as spurious because they mask the measurement desired but in actual fact are painfully real. The undesirable disturbances arise because of buoyancy, convection currents, thermomolecular flow (TMF), static electricity, temperature fluctuations about the sample and the balance case, and adsorption and desorption. These topics will be discussed briefly in the subsequent paragraphs; Poulis will discuss some of the more important effects in depth in Chapter 4.

A. Adsorption and Desorption Effects

Adsorption and desorption effects from the beam of a balance have been used by some (112,136) to account for unknown measured quantities. However, these can be dismissed for most studies by a simple calculation. The *total* surface area of a typical Gulbransen-type torsion balance (cf. reference 17) is about 0.58 cm^2; the area of oversize hangdown suspensions 40 cm long and 0.25 mm in diameter is only 0.02 cm^2. Even with an unrealistically large roughness factor of 5, the total available surface is only 3 cm^2. Using the accepted value of 10.6 $Å^2$ per molecule of water for adsorption on an oxide surface (117), the mass gain at monolayer coverage would be 2.87×10^{-8} g/cm^2 or a total of about 0.09 μg. When the symmetry of the entire system is now taken into account and it is realized that the balance will only detect changes because of a *differential* surface area, it can be seen that monolayer adsorption effects can in general, be ignored. However, adsorption and desorption is important when materials are used in the system that could condense or react with the beam. Thus, the attachment of plastics, glue, etc. to the microbalance is not acceptable technique because multilayer adsorption may occur. It will be shown in Chapter 3 that part of the choice of materials is based on preventing condensation or reaction with the beam. Thus, care should be exercised to prevent mercury vapor from reaching silver globules on the beam of the balance or organic vapors from sorbing on polymethylmethacrylate. It is apparent that a balance will register mass changes from adsorption and desorption when proper precautions and great cleanliness are not observed.

B. Temperature Fluctuations

Errors that could result from temperature fluctuations about the balance can be minimized by using thermostatted enclosures. These errors may be serious for cantilever and spring balances but can be minimized by use of a metallic coating on a fused silica beam for beam balances without thermostatting (120). No report of a study of the effect of metallized cantilevers or springs was found in this literature search. The plot Figure 2.11 shows the need for thermostatting to attain submicrogram operation of observed mass

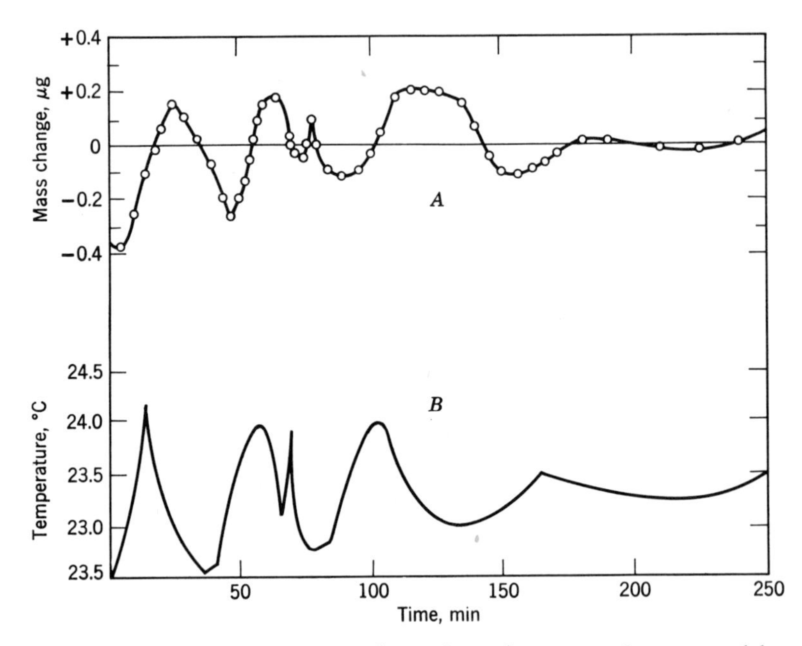

Figure 2.11. Effect of temperature fluctuations about an unthermostatted beam microbalance. The need for temperature control to minimize the 0.3 μg/° mass variation is evident. *A*, "mass" change relative to "zero" at 23.75°. *B*, temperature of the balance housing in an air conditioned room.

change versus temperature in an all quartz pivot balance of the type described by Czanderna and Honig (111,112). These data (186) were obtained with the balance, housing, hangdown tubes, and vacuum system situated in an air conditioned room. A vacuum of 2×10^{-7} torr was maintained while the temperature outside the balance case was monitored with a 40-junction iron constantan thermocouple. The balance was operated manually as a null instrument with magnetic compensation. The quartz sample and dummy bulbs were empty. As can be seen, the temperature coefficient of the balance was 0.3 μg/°.

C. Static Charge Forces

Disturbances that arise from static electricity may be avoided by use of grounded conductive coatings such as tin oxide on the balance housing and hangdown tubes. A grounded metallic coating on the balance beam or the use of a metal beam is the preferred technique for dissipating static charge from the beam (7,23,57,62,63,81–84). Others have found that heating (184), radioactive sources (186), and other ion-forming techniques (187) are useful to dissipate static charge. The use of a grounded silver hangdown tube with an ungrounded hangdown suspension has also been effective (120). Mitsui and Yoshikawa (188), however, were able to use a microbalance to measure the static charge on a polyethylene rod.

D. Convection Currents

At pressures exceeding 100–350 torr, convection currents arise in a heated hangdown tube that produce continuous oscillation of an undamped balance. No detailed study of this phenomenon has been reported. For a 1 in. diam tube, Czanderna (111) found that motion set in at 650 torr at a sample temperature of 500°; at 900°, the motion began at 350 torr. Thomas and Williams (23) noted effects of convection at 100 torr and, apparently, at temperatures below 0°. The effect of the convection current is to decrease the precision of the measurements. Automatic operation, some form of eddy current damping, the use of baffles (208), or all three significantly decrease the annoyance of convection currents. The lowest pressure where convection currents become noticeable depends on the temperature, geometry, and the gas in the system (23,111) and possibly on the nature of an adsorbent surface (23). Further study of this topic will be necessitated when microbalances are used extensively at both high pressures and high temperatures.

E. Buoyancy Forces

Unless a microbalance is being used explicitly to determine the density of gases (106,107,109,110,124), raw data must be corrected for the buoyancy of the gas. A buoyant force, $\mathbf{F}_y$, will exist if an unequal volume, ΔV, is occupied by the beam, suspensions, sample containers, etc. on either side of the fulcrum. The magnitude of $\mathbf{F}_y$ equals the weight ($m\mathbf{g}$) of the displaced fluid which in most work is a gas. For a gas obeying the equation of state $PV = nRT$, the mass, m, of the displaced fluid is given by

$$\mathbf{F}_y/\mathbf{g} = m = PM\,\Delta V/RT \tag{2.1}$$

where P is the pressure in torr, M is the gram molecular weight of the gas, ΔV is in cm^3, T in °K, and R is 62, 364 cm^3 torr mol^{-1} °K^{-1}. This equation shows that the buoyancy effect on the mass is a function of four variables viz. the pressure, temperature, gas, and net buoyancy volume. The advantage of a symmetric system is now evident since ΔV is the only parameter which can be designed to be zero, thus eliminating the correction for buoyancy. Arranging ΔV to be zero initially is a necessary but not complete solution; the *sample, counterweight, and heated or cooled* parts of the suspensions also must have the same volume to eliminate a temperature dependence of ΔV. Even then, Pierotti (189) has shown by an elegant statistical mechanical treatment that second order effects arise from a sample of large surface area. Thus, if the surface area of the sample and counterweight are different and if the surface area of the sample is greater than 1–100 m^2/g, a mass defect will be recorded. Pierotti (189) has shown that the mass defect arises because of the range of van der Waals interaction between the surface and an ambient gas.

It is also evident that the first order buoyancy correction, given by equation 2.1, is based on ideal gas behavior. When real gases are employed, care must be taken to use the equation of state that is appropriate for the gas, temperature, and pressure conditions of the experiment.

F. Radiometric Forces that Result from the Thermomolecular Flow of Gases

When a temperature gradient exists along a microbalance suspension fiber, sample, or counterweight, radiometric forces will be generated because of the thermomolecular flow (TMF) of gases at pressures which generally range from 10^{-3} to 20 torr. The gas species arriving at a unit surface from regions of different temperatures, and hence with different momenta, produce a net force on the fiber in the direction of decreasing temperature. The magnitude of the force depends on the pressure, gas, temperature, temperature gradient, and geometry of the system. The force may be minimized by employing identical suspension fibers, hangdown tubes, and temperature gradients about an identical sample and counterweight. With this ideal arrangement depicted in Figure 2.12, TMF will produce compensating forces in opposite directions about the fulcrum of the balance. The net magnitude of the resultant "spurious" effect, then, is zero as a function of pressure. In practice, it is difficult to obtain ideal compensation and, therefore, there is a resultant force from TMF that is not a simple function of pressure. If all variables remain the same, it is reproducible and may be calibrated. The magnitude of the effect may not become submicrogram until the pressure is as low as 10^{-5} or as high as 200 torr if the geometry of the suspension fiber, sample, and hangdown tube is poor. Thus, Czanderna (190) identified the origin of TMF and suggested an empirical scheme to correct for it. This qualitative explanation is simply an adaptation of the theory of the Knudsen or radiometer type pressure gauge (191). The similarity between the composite plot of radiometric forces measured with a microbalance, shown in Figure 2.13*a*, and the data shown in Figure 2.13*b* taken from Figure 5.36 of reference 173 is striking. The latter data were taken from work by Bruche and Littwin (192).

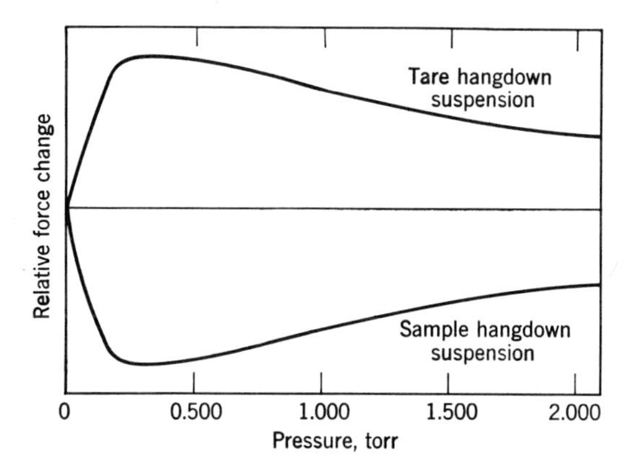

Figure 2.12. Scheme for perfect compensation for the effect of thermomolecular flow of gases on a symmetrical balance hangdown, suspension, and sample geometry.

Historically, the first identification of radiometric effects in thermogravimetry seems to be that of Eyraud and Goton (193) who detected forces equivalent to 5.0 mg at a pressure of 1 torr and 500°. The effect of TMF was defined as a pressure effect by Gulbransen (8). While the presence of TMF makes precision microgravimetry more difficult, it was probably the major thorn that led to Rhodin's development of the symmetrical housing and hangdown system (7,57) to minimize thermal eddy effects. Katz and Gulbransen (184) included the effects of TMF in a general empirical correction to the balance deflection because of pressure while others (136) accounted for it as an adsorption and desorption effect. Czanderna (111) and Czanderna and Honig (112) ascribed the pressure effect to thermomolecular flow and showed that the magnitude of the force depends on the gas, pressure, and temperature gradient of the system and described techniques to correct for it (111,117,194).

The effect of thermomolecular flow was one of the three major topics of discussion at the First Conference on Microbalance Techniques at Fort Monmouth, New Jersey in January 1960 (11). As a result of four papers (184,190,195,196), a discussion of TMF was initiated that involved practically every participant at the conference.

Poulis and Thomas (197,198), in pioneering the theoretical analysis of

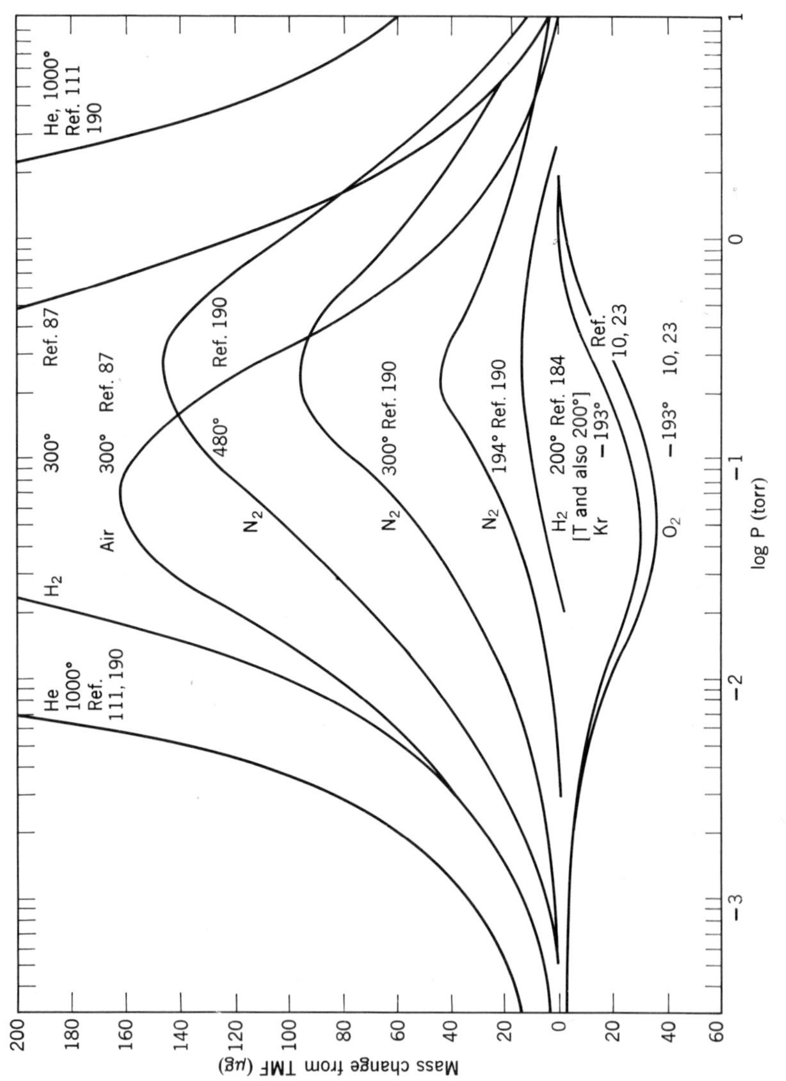

Figure 2.13. (*a*) Composite plot of thermomolecular flow effects published by a number of different workers.

the problem, showed the forces obtained experimentally by others (184,190) could be calculated approximately by modification and extension of Knudsen's theory, thus substantiating the qualitative description (190). About the same time, Krupp et al. (199) reported the results of a detailed experimental study of the dependence of the magnitude of the radiometric force

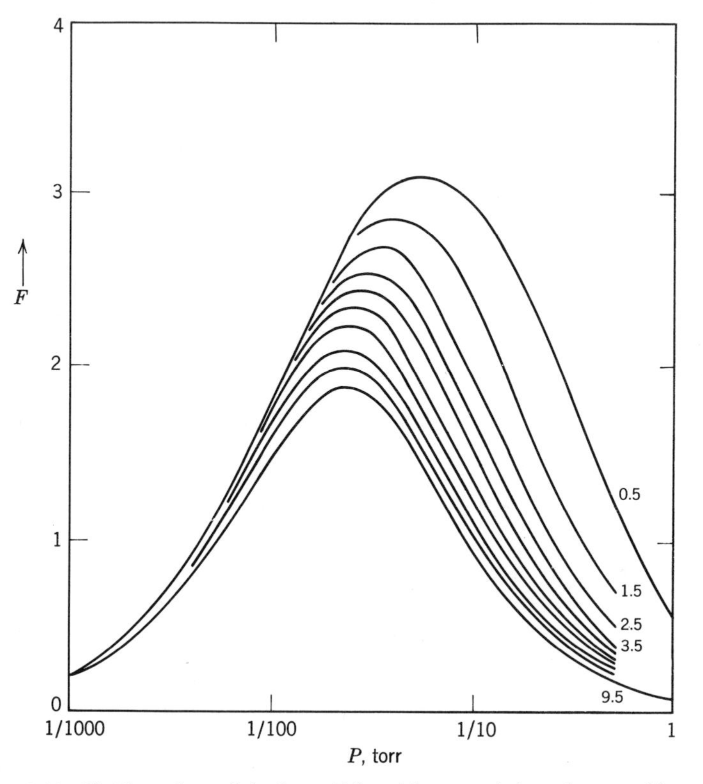

Figure 2.13. (*b*) The values of the force F (in arbitrary units) on the movable vane in a Knudsen gauge, plotted against log P (torr) for a series of values of the distance between the heated surface and the vane. (From observations by Bruche and Littwin, reference 192.) (After Dushman and Lafferty, 2nd Ed., *Scientific Foundations of Vacuum Technique*, 1962, Wiley.)

acting on one leg of a microbalance suspension. Using the methods described (11), they determined the pressure dependence of the force as a function of the temperature and geometry of the specimen and the ambient gas. They used modified quantitative expressions based on theories of Knudsen (173) and Hettner (200) to demonstrate that satisfactory agreement between calculated and measured values could be obtained in the high and low pressure regions (Figure 2.13). Unfortunately, neither theoretical treatment (197–199), has been satisfactory in the transition pressure region, viz. the region of maximum force. Poulis, Massen, and Thomas (201–204) and others (205–206) have not been able to develop a satisfactory equation, even a semi-empirical one (206), that will permit one to compute corrections for the force encountered in the transition pressure region. This approach is similar to that used by Liang (175) in dealing with thermal transpiration. It is indeed unfortunate that, after the most recent attempt (204), it was concluded that the best way to deal with the forces encountered in this region is to calibrate it experimentally. Thus, for equilibrium experiments requiring the greatest precision, the technique developed by Czanderna (207) using partial pressures of active gases may become increasingly important even though it is laborious, time consuming, and limited to balances of excellent long-term stability. For more routine work the methods described (111,112,117,186,190,196, 208) may deserve further study.

Since TMF is dependent on the presence of a temperature gradient, it is obvious the effect will be encountered with a sample at temperatures below room temperature as well as above. This was reported in 1957 (111), 1962 (117,118), 1964 (23), and again in 1967 (208).

Numerous encounters with TMF have been reported in the last several years by a host of investigators (23,29,103,111,112,117,118,120,176,186, 188,190,194–196,199,207–213) in different types of pressure dependent studies. The observations made by Seanor and Amberg (29) and Freeman and Gwinup (103) deserve special mention because the TMF was noted using a quartz spring and vertical torsion balance, respectively, where no opportunity for at least partial compensation existed. All other reports were made by experimentalists using beam balances.

In a study of porous carbons, Thomas and Williams (23) have detected a disturbance they attributed to a high pressure TMF effect although further study seems warranted. These effects, which are attributed to temperature gradients as small as 0.01°/cm at the gas–solid interface, have received further attention in the form of cavity forces (214,215). Finally, new designs for furnaces have been postulated (216) to minimize the force from TMF by directing it orthogonally to the vertical hangdown suspension. These methods, however, will require not only considerable ingenuity to use in practice but also are of doubtful value because vertically directed temperature gradients have not been eliminated.

VIII. APPLICATIONS

In this section, the applications of microbalances for measuring mass changes attending kinetic and equilibrium processes will be arbitrarily categorized according to classic, simultaneous, and future use of the instruments. The bounds for *classic operation* include the use of the device to monitor a change in mass of a sample that results from altering the temperature and/or pressure of various gases about the sample, the measurement of a mass change of the sample such as in evaporation and sputtering, or the measurement of a force exerted on a sample or device suspended from a balance. *Simultaneous* use includes the measurement of the mass and at least one additional physical parameter simultaneously to enhance the analysis of the physical or chemical occurrences in or on the sample or specimen. Changes in the composition but not the temperature or pressure, of the surrounding gases would be considered as an additional parameter. Obviously, there will be cases that are not clearly either "classic" or "simultaneous." *Future* applications will be suggested that appear to have potential; they represent areas of research where correlating mass changes in a specimen with other important physical parameters are considered to be of special importance.

A. Classic Operation

A mass change or a force change may be measured in classic operation of a microbalance. Mass changes that result from an increase or decrease in the pressure about a specimen occur in oxidation, reduction, adsorption, desorption, thermodesorption, photodesorption, surface area, deviations from stoichiometry, decomposition or degradation, and absorption studies. References in these categories in which mass changes have been measured by ultra micro weighing techniques are summarized in Table 2.3. Most of these have been published in the last decade and include the addition of references 217–255. Of somewhat unusual interest are five papers that employ a microbalance to determine mass changes at pressures of several or more atmospheres (124,256–260). Force changes that result during use of the balance to measure evaporation rates, gas densities, gas–solid, liquid–solid, solid–solid, and gas–liquid interactions, and magnetic susceptibilities are also included in Table 2.3. References 261 through 280, which are included in these latter classifications, have been published in recent years. Surface tension, sedimentation, van der Waals interactions, etc. are included under the heading relating to bistate interactions.

The categories listed in Table 2.3 constitute the major fraction of classical use of balances. The diverse applicability of microgravimetric measurements are illustrated by reports of the determination of dust in flue gases (281,282), the measurement of liquid density (137,283), and the determination of molecular weight (284). In the latter case (284), a gas density balance was substituted for a gas chromatographic detector in a

Table 2.3. Selected References: Applications of Ultramicroweighing Techniques to Classical Studies

Applications	References
Oxidation	7, 8, 17, 32, 44, 46, 62, 63, 65, 67, 68, 70, 73, 111, 115, 116, 117, 157, 170, 181, 182, 211, 217, 218, 219, 220, 221, 222, 223, 224, 225, 226, 227, 251, 252, 253, 254, 255, 296
Reduction	7, 17, 184, 228
Adsorption	7, 23, 29, 31, 57, 60, 108, 121, 141, 149, 176, 189, 207, 210, 213, 229 through 240, 298
Desorption	176, 213, 229, 231, 232, 234, 236, 237, 240, 298
Thermodesorption and photodesorption	121, 234, 236, 241, 242, 243
Surface area	7, 23, 46, 87, 111, 117, 176, 194, 208, 213, 228, 233, 234, 237, 239, 244, 298
Deviations from stoichiometry	111, 194, 228, 245, 246, 247, 255
Decomposition or degradation	34, 36, 80, 111, 117, 194, 209, 248, 249, 250
Absorption	249
Superatmospheric pressure	124, 256, 257, 258, 259, 260
Evaporation rates (for vapor pressure)	73, 89, 94, 95, 100, 103, 104, 105, 150, 249, 261, 262, 263
Evaporation rates (sputtering yields, thin film studies)	36, 61, 62, 69, 72, 93, 96, 97, 98, 99, 101, 140, 150, 152, 153, 154, 155–157, 253, 264, 265, 266, 267, 268
Gas densities	38, 106, 107, 109, 111, 112, 114, 117, 124, 269
Gas–solid, liquid–solid, solid–solid, gas–liquid interactions	40, 137, 265, 270, 271, 272, 273
Magnetic susceptibility	27, 60, 64, 77, 91, 92, 118, 123, 171, 172, 180, 247, 274, 275, 276, 277, 278, 279, 280

flow system. The static charge on a polyethylene rod has been measured (188) and the use of a cantilever balance for mass determinations in biological studies was reported (26). Scott and Harrison (285) applied the principal of thermal gravimetric analysis and thermodesorption in a study of the "thermo" oxidation of several UO_2 mixtures. Finally, the impulse of a force has been measured to evaluate the microthrust of a dynamic electric engine (286).

As might be expected a number of valuable *auxiliary* studies can be performed on microgravimetric specimens after classical treatment on the microbalance. These could include electron diffraction and electron microscopic investigations of the surface layers after kinetic and/or equilibrium studies have been carried out. Recent representative studies that fall in this category have been reported by Boggs (287) and Schwoebel (288). Boggs concluded from his studies of the oxidation of polycrystalline tin and its alloys that oxide nuclei are formed preferentially at dislocations on the surface. Schwoebel used a bakeable ultrahigh vacuum ultramicrobalance to study the oxidation of magnesium single crystals. In this elegant study, he has shown that the kinetics of oxide film formation on the basal and prismatic planes are consistent with the Wagner-Hauffe theory of oxide growth by the diffusion of lattice defects.

B. Simultaneous Measurement of Mass and Other Physical Parameters

The use of a horizontal beam balance combined with a vertically suspended torsion balance (94,96,103–105) was mentioned in Section IV.B.5. Balances of this type are frequently used with a mass spectrometer. This type of data has been included in the classic operation of balances. Although the use of a mass spectrometer and a microbalance in Knudsen effusion experiments has enhanced significantly the interpretation of vapor pressure determinations, the simultaneous measurement of mass and other physical parameters should be of greatest value in the study of surfaces of solids and thin films. The facility to resolve a mass change of one part in 10^8, discussed in Section III, in itself is suggestive of extensive application of ultramicrobalances to the fundamental investigation of gas–solid interactions. The integration of combined measurements including mass is a natural extension of the advance in microbalance techniques (120). Interpretation of the dynamic nature of micromass changes will be enhanced considerably if additional parameters are measured simultaneously. Some possibilities include the measurement of the ambient gas composition in ultrahigh and high vacuum, optical absorption and transmission, electron and optical micrography, magnetic susceptibility, magnetic resonance, electrical conductivity, and thermoelectric power. Although some of these will involve considerable experimental difficulty, it is surprising that only a limited number of attempts have been made to date.

Wolsky and Zdanuk (62) used an omegatron mass spectrometer during ultrahigh vacuum operation of a bakeable microbalance. Thus, simultaneous information was generated during outgassing and oxidation of germanium and silicon single crystals. Subsequent use of the instrument, which has been related to the investigation of ion bombardment and thin film phenomena, has produced a steady flow of valuable information (61, 229,231,232,266,268). Gulbransen (181) reiterated the value of combining gas analyzer or mass spectrometer techniques with microbalance measurements for high temperature oxidation studies.

Czanderna and Wieder combined the measurement of the optical transmission and mass change in thin films (120) to discover a gross defect composition of cuprous oxide, $CuO_{0.67}$ (228). The optical data, obtained simultaneously with mass, were then used to relate the auxiliary measurements of density, optical transmittance, electron diffraction and microscopy (228), and magnetic susceptibility (247) to the composition of the (partially) oxidized films. The combined measurement has been used effectively in the study of the reduction (289) and mobility (290) of thin films. It also enabled the optical constants of the gross defect composition to be determined (291).

Amberg and Seanor (29, 292) combined a quartz spring balance with an infrared spectrometer to study the adsorption of carbon monoxide on doped zinc oxide surfaces. It was possible for them to show that the integrated band intensities thus obtained were consistent with a proposed dipolar interaction between the weakly adsorbed CO species and the adsorbent. Angell (293) has devised a similar combined measurement scheme employing an automatic recording balance and spectrometer for the study of gases adsorbed on zeolites. Ward (294) has carried out a similar study though the balance was not in the ultramicro range.

Soule et al. (172) constructed an ultrasensitive Faraday susceptibility apparatus with the capability of studying magnetic susceptibility changes of a paramagnetic gas absorbed on a diamagnetic solid. Thus, Czanderna (186) was able to relate quantitatively the changes in susceptibility of oxygen adsorbed on a silver surface with the amount of oxygen adsorbed.

The use of an automatic recording balance in a flow environment by Gulbransen et al. (217,295,296) could be classified as a simultaneous measurement because supplementary analyses of the reaction products are carried out. The flow system is effective in minimizing diffusion effects during fast reactions. Thus, the authors can more reasonably expect the rate determining step to be adsorption, reaction on the surface, or desorption.

C. Future

The extensive application of precision ultra micro weighing to classical problems will undoubtedly increase markedly because of the existence of commercially available equipment and the fundamental nature of the measurement. Some of the most elegant studies should take place in surface science. For example, microbalances surely will be valuable in determining the mechanism of nucleation and growth of thin films. Little work has been done to study the physical properties of metal oxide thin films. Again, micro weighing techniques will be used to obtain precise information.

It would be anticipated that the combined measurement of mass and additional parameters will increase by orders of magnitude. Many exciting unexploited possibilities exist in addition to those described in Section VIII.B. For example, significant challenges exist simply to develop techniques for combining the measurement of mass and electrical conductivity, thermoelectric power, magnetic spin resonance, and low energy electron diffraction. For studies of heterogeneous catalysts during reaction conditions (299), a microbalance and a gas chromatograph might be combined for useful *in situ* studies.

Perhaps one of the greatest unexplored areas of gas–solid interactions is to determine the nature of a gas atom or molecule adsorbed on a well defined surface. For this, a combined study of adsorption, reaction on the surface, and desorption could be carried out by simultaneously employing a microbalance, infrared (or magnetic resonance) spectroscopy, and a gas analyzer. Further developments in the theory of thermodesorption and photodesorption will provide additional analytical parameters.

In the area of solid state science, the simultaneous measurement of mass change, conductivity, thermoelectric power, and magnetic susceptibility could be accomplished with appropriate pulsing techniques. Thus, detailed correlation of the deviation from stoichiometry and carrier concentration could be achieved.

Finally, in all applications, more sophisticated data analysis and data reduction techniques should be considered. For example, computer programs have been used in TGA analysis (297). The conversion of strip chart records to digitalized data for analysis will attain greater significance as ultra micro weighing techniques are combined with several different parameters for simultaneous measurement.

D. Thermogravimetry

The fields of thermogravimetric analysis (TGA) and differential thermal analysis (DTA) generally make use of micro weighing apparatus rather than *ultra* micro weighing equipment. Since entire volumes have been written on the subject (297) only brief mention will be made in this chapter. Quite obviously many of the techniques developed and employed in TGA and DTA can be of value to the ultramicrogravimetric scientist. The primary difference is that many extraordinarily complex problems that arise in DTA and TGA can be greatly simplified for studies that carry into the sub-microgram region of mass change as has been discussed for the case of thermodesorption of gases (242).

ACKNOWLEDGEMENTS

The author is indebted to Dr. S. P. Wolsky, Dr. R. L. Schwoebel, Dr. R. Tisinger, and Dr. R. Alire for their helpful comments. Partial support of this work by the Division of Air Pollution, Bureau of State Services, Public Health Service, under Research Grant 1 R01 AP 00552-01 is gratefully acknowledged. Acknowledgement is made to the donors of the Petroleum Research Fund, administered by the American Chemical Society, for partial support of this work. Thanks are due my wife, Lucile, for her painstaking proofreading of the manuscripts.

REFERENCES

1. A. Warburg and E. Ihmori, *Ann. Physik. Chem.*, **27**, 481 (1886); **31**, 145 (1887).
2. H. Pettersson, *Proc. Phys. Soc. (London)*, **32**, 209 (1920); Dissertation, University of Göteburg, 1914.
3. F. Emich, *Abderhalden*, **9**, 55 (1919).
4. G. Gorbach, *Mikrochemie*, **20**, 254 (1936).
5. G. Ingraham, *Metallurgia*, **39**, 232 (1949); B. B. Cunningham, *Nucleonics*, **5**, 62 (1949).
6. P. L. Kirk, *Quantitative Ultramicroanalysis*, Wiley, New York, 1950.
7. T. N. Rhodin, *Advan. Catalysis*, **5**, 39 (1953).
8. E. A. Gulbransen, *Advan. Catalysis*, **5**, 119 (1953).
9. K. Behrndt, *Z. Angew. Phys.*, **8**, 453 (1956).
10. J. M. Thomas and B. R. Williams, *Quart. Rev. (London)*, **19**, 231 (1965).
11. M. J. Katz, Ed., *Vacuum Microbalance Techniques*, Vol. I, Plenum Press, New York, 1961.
12. R. F. Walker, Ed., *Vacuum Microbalance Techniques*, Vol. II, Plenum Press, New York, 1963.
13. K. H. Behrndt, Ed., *Vacuum Microbalance Techniques*, Vol. III, Plenum Press, New York, 1963.
14. P. M. Waters, Ed., *Vacuum Microbalance Techniques*, Vol. IV, Plenum Press, New York, 1965.
15. K. H. Behrndt, Ed., *Vacuum Microbalance Techniques*, Vol. V, Plenum Press, New York, 1966.
16. A. W. Czanderna, Ed., *Vacuum Microbalance Techniques*, Vol. VI, Plenum Press, New York, 1967.

17. E. A. Gulbransen and K. F. Andrew, *Vacuum Microbalance Tech.*, **1**, 1 (1961).
18. H. S. Peiser, *Vacuum Microbalance Tech.*, **2**, 1 (1962).
19. R. L. Schwoebel, private communication.
19a. G. Deamoley and D. S. Northrup, *Semiconductor Surfaces for Nuclear Reactions*, Spon Ltd., London, 1966, p. 166.
20. C. H. Massen, J. A. Poulis, S. P. Wolsky, and E. J. Zdanuk, *Vacuum Microbalance Tech.*, **6**, 37 (1967).
21. C. H. Massen, J. A. Poulis, and J. M. Thomas, *Vacuum Microbalance Tech.*, **4**, 35 (1965).
22. Cahn Instrument Co., UHV model RG.
23. J. M. Thomas and B. R. Williams, *Vacuum Microbalance Tech.*, **4**, 209 (1965).
24. B. L. Evans, J. M. Thomas, and B. R. Williams, *J. Sci. Instr.*, **43**, 263 (1966).
25. E. Salvioni, Dissertation, University of Messina, 1901.
26. S. L. Bonting and B. R. Mayron, *Microchem. J.*, **5**, 31 (1961).
27. J. O. Kessler and A. R. Moore, *Rev. Sci. Instr.*, **33**, 478 (1962).
28. J. Skelly, *Rev. Sci. Instr.*, **38**, 985 (1967).
29. D. A. Seanor and C. H. Amberg, *Rev. Sci. Instr.*, **34**, 917 (1963).
30. F. Emich, *Monatsh*, **36**, 407 (1915); through *Chem. Abstr.*, **9**, 2997 (1915).
31. J. McBain and A. Bakr, *J. Am. Chem. Soc.*, **48**, 690 (1926).
32. P. L. Kirk and F. L. Schaffer, *Rev. Sci. Instr.*, **19**, 785 (1948).
33. M. J. Rand, *Rev. Sci. Instr.*, **32**, 991 (1961).
34. S. L. Madorsky, *Vacuum Microbalance Tech.*, **2**, 47 (1962).
35. S. S. Leshchenko, V. L. Karpov, I. K. Karpovich, V. N. Katyshev, and Zh. K. Krumin, *Zavodsk. Lab.*, **29**, 1016 (1963); through *Chem. Abstr.*, **59**, 11679a (1963).
36. M. M. Freundlich, *Vacuum*, **14**, 293 (1964).
37. H. Kambe and S. Igarashi, *Rept., Aeron. Res. Inst. Univ. Tokyo*, **28**, 145 (1963); *Chem. Abstr.*, **60**, 1891g (1964).
38. R. M. Dell and V. J. Wheeler, U.K. Atomic Energy Research Estab. (Gt. Brit.), R 3424 (1960); *Chem. Abstr.*, **55**, 5046 (1961).
39. J. Duchene, *Vortaege Originalfassung Intern. Kongr. Grenzflaechenaktive Stoffe*, **3**, Cologne, 1960, **3**, 1961; through *Chem. Abstr.*, **57**, 8385a (1962).
40. R. M. Joshi, *J. Polymer Sci.*, **35**, 271 (1959).
41. P. Barrett, *Bull. Soc. Chim.* (*France*), **1958**, 376; *Chem. Abstr.*, **54**, 2828 (1960).
42. M. M. Vetykov, R. G. Chuvilyaev, and S. N. Shkolnikov, *Zh. Fiz. Khim.*, **33**, 2370 (1959); *Chem. Abstr.*, **54**, 12671h (1960).
43. J. Westmoreland, *Chem. Ind.* (*London*), **1965**, 2000.
44. C. T. Fujii, C. D. Carpenter, and R. A. Meussner, *Rev. Sci. Instr.*, **33**, 362 (1962).
45. J. Hooley, *Can. J. Chem.*, **35**, 374 (1957).
46. C. Moreau, *Vacuum Microbalance Tech.*, **4**, 21 (1965).
47. R. W. Harrison and E. J. Delgrosso, *J. Sci. Instr.*, **41**, 222 (1964).
48. B. D. Steele and K. Grant, *Proc. Roy. Soc.* (*London*), **A82**, 580 (1909).
49. W. Felgenträger, *Feine Waagen, Wägungen und Gewichte*, 2nd ed., Springer, Berlin, 1933; *Nature*, **132**, 730 (1950).
50. Catalog, Wm. Ainsworth and Sons, Denver, Colo. 80205. Model 15, for example.
51. For examples, see articles in *J. Vacuum Sci. Technol.*, **1**ff, 1964–1964ff.
52. W. Weber, *Werke* (*Berlin*), **1**, 497 (1892).
53. W. Nernst and E. A. Riesenfeld, *Ber*, **39**, 381 (1906).
54. F. Emich and J. Donau, *Handb. Biol. Arbeitsmeth.*, **1**, 183 (1921).

55. J. Donau, *Mikrochemi Emich Fest*, **39**, 10 (1930); **3**, 1 (1931); *Mikrochemie*, **9**, 1 1931); **13**, 155 (1933).
56. E. A. Gulbransen, *Rev. Sci. Instr.*, **15**, 201 (1944).
57. T. N. Rhodin, Jr., *J. Am. Chem. Soc.*, **72**, 4343 (1950).
58. P. L. Kirk, R. E. Craig, J. E. Gullberg, and R. Q. Boyer, *Anal. Chem.*, **19**, 427 (1947).
59. F. Edwards and R. Baldwin, *Anal. Chem.*, **23**, 357 (1951).
60. T. J. Jennings, in *The Defect Solid State*, T. J. Gray, Ed., Interscience, New York, 1957, p. 487.
61. S. P. Wolsky and E. J. Zdanuk, *Vacuum Microbalance Tech.*, **2**, 37 (1962).
62. S. P. Wolsky and E. J. Zdanuk, *Vacuum Microbalance Tech.*, **1**, 35 (1961).
63. R. F. Walker, *Vacuum Microbalance Tech.*, **1**, 87 (1961).
64. R. J. Kolenkow and P. W. Zitzewitz, *Vacuum Microbalance Tech.*, **4**, 195 (1965).
65. J. E. Whittle, *J. Sci. Instr.*, **43**, 150 (1966).
66. E. A. Gulbransen and K. F. Andrew, *Vacuum Microbalance Tech.*, **2**, 129 (1962).
67. R. Addiss, Ph.D. Thesis, Dept. of Engr. Physics, Cornell University, Ithaca, N.Y., 1958.
68. R. L. Schwoebel, Ph.D. Thesis, Cornell University, Ithaca, N.Y., 1962.
69. R. L. Schwoebel, *Surface Sci.*, **2**, 356 (1964).
70. W. E. Boggs, *Vacuum Microbalance Tech.*, **6**, 45 (1967).
71. J. A. Rodder, Brit., 851, 913. *Chem. Abstr.*, **55**, P 8965d (1961).
72. H. Mayer, R. Niedermayer, W. Schroen, D. Stuenkel, and H. Goehre, *Vacuum Microbalance Tech.*, **3**, 75 (1963).
73. N. J. Carrera and R. F. Walker, *Vacuum Microbalance Tech.*, **3**, 153 (1963).
74. H. Moret and E. Louwerix, *Vacuum Microbalance Tech.*, **5**, 59 (1966).
75. J. Gouault, *Compt. Rend.*, **256**, 378 (1963).
76. V. P. Vetrov, N. I. Kokin, and A. N. Klassen, *Nauchn. Soobshch. Aziatsk. Fil. Gos. Vses. Nauchn. Issled. Inst. Tsmentn. Prom.*, **1963**, 66; *Chem. Abstr.*, **63**, 17942h (1965).
77. J. Kaczer, *Czech. J. Phys.*, **13**, 386 (1963).
78. R. Niedermayer and W. Schroen, *Vakuum-Tech.* **11**, 36 (1962).
79. P. Vast, *Bull. Soc. Chim. France*, **1965**, 359.
80. E. Ericson and W. J. Kirsten, *Microchem. J.*, **9**, 82 (1965).
81. L. Cahn and H. R. Schultz, *Vacuum Microbalance Tech.*, **2**, 7 (1962).
82. L. Cahn, *Dechema Monograph.*, **44**, 45 (1962); *Instr. Control Systems*, **35**, 107 (1962).
83. L. Cahn, H. Schultz, and P. Gaskins, *Microchem. J. Symp. Ser.*, **2**, 1027 (1962).
84. L. Cahn and H. R. Schultz, *Vacuum Microbalance Tech.*, **3**, 29 (1963).
85. J. Gouault, *Compt. Rend.*, **256**, 1455 (1963).
86. G. Gorbach, *Anal. Chim. Acta*, **29**, 453 (1963).
87. H. L. Gruber and C. S. Shipley, *Vacuum Microbalance Tech.*, **3**, 131 (1963).
88. H. Meiners, *Dechema Monograph.*, **54**, 131 (1965).
89. M. Volmer, *Z. Physik. Chem. Bodenstein Festband*, **1931**, 836.
90. M. C. Day, L. D. Hulett, and D. E. Willis, *Rev. Sci. Instr.*, **31**, 1142 (1960).
91. C. W. Fleischmann and A. G. Turner, *Rev. Sci. Instr.*, **37**, 73 (1966).
92. J. A. Poulis, C. H. Massen, and P. van der Leeden, *Appl. Sci. Res. B*, **9**, 133 (1961).
93. W. T. Siegle and W. R. Beam, *Rev. Sci. Instr.*, **35**, 1173 (1964).
94. A. E. Wilson, J. H. Kim, and A. Cosgarea, Jr., *Rev. Sci. Instr.*, **36**, 1428 (1965).
95. E. K. Rideal and P. M. Wiggins, *Proc. Roy. Soc.* (*London*), **A210**, 291 (1951).
96. R. S. Bradley and T. G. Cleasby, *J. Chem. Soc.*, **1953**, 1681.

97. D. W. Bassett and A. J. B. Robertson, *Brit. J. Appl. Phys.*, **10**, 534 (1959); *Chem. Abstr.*, **54**, 11585e (1960).
98. K. A. Becker, H. J. Forth, and I. N. Stranski, *Z. Electrochem.*, **64**, 373 (1960).
99. F. B. Humphrey and A. R. Johnson, *Rev. Sci. Instr.*, **34**, 348 (1963).
100. S. Pearson and N. J. Wadsworth, *J. Sci. Instr.*, **42**, 150 (1965).
101. A. R. Beavitt, *J. Sci. Instr.*, **43**, 182 (1966).
102. M. Peleg and C. B. Alcock, *J. Sci. Instr.*, **43**, 558 (1966).
103. R. D. Freeman and P. D. Gwinup, *Rev. Sci. Instr.*, **37**, 773 (1966).
104. R. D. Freeman, Tech. Doc. Rpt. No. ASD TDR 63–754, Part II, W-PAFB, Ohio, Jan. 1965, p. 36.
105. G. Wessel, *Z. Physik*, **130**, 539 (1951).
106. A. Stock and G. Ritter, *Z. Physik Chem.*, **119**, 333 (1926); **124**, 204 (1926); **126**, 172 (1927).
107. A. Stock and G. Ritter, *Z. Angew. Chem.*, **39**, 1463 (1926).
108. S. J. Gregg, *J. Chem. Soc.*, **1946**, 561; **1955**, 1438.
109. E. W. Johnson and L. K. Nash, *Rev. Sci. Instr.*, **22**, 240 (1951).
110. J. H. Simons, C. L. Scheirer, Jr., and H. L. Ritter, *Rev. Sci. Instr.*, **24**, 36 (1953).
111. A. W. Czanderna, Ph.D. Thesis, Purdue University, 1957, W. Lafayette, Ind.
112. A. W. Czanderna and J. M. Honig, *Anal. Chem.*, **29**, 1206 (1957).
113. J. Richlin, M.S. Thesis, Purdue University, 1957, W. Lafayette, Ind.
114. J. M. Honig, *Vacuum Microbalance Tech.*, **1**, 55 (1961).
115. C. N. Cochran, *Rev. Sci. Instr.*, **29**, 1135 (1958).
116. C. N. Cochran, *Vacuum Microbalance Tech.*, **1**, 23 (1961).
117. P. A. Faeth, "Vacuum and Adsorption Technique," Inst. Sci. Tech. Press, Univ. of Michigan, TR-66100-2-X (1962), Ann Arbor, Michigan.
118. A. N. Gerritsen and D. H. Damon, *Rev. Sci. Instr.*, **33**, 301 (1962).
119. A. Langer, *Vacuum Microbalance Tech.*, **4**, 231 (1965).
120. A. W. Czanderna and H. Wieder, *Vacuum Microbalance Tech.*, **2**, 147 (1962).
121. A. W. Czanderna, *Vacuum Microbalance Tech.*, **4**, 57 (1965).
122. A. W. Czanderna, *Vacuum Microbalance Tech.*, **4**, 175 (1965).
123. D. Das, *Indian J. Phys.*, **37**, 582 (1963).
124. R. M. Tisinger, *Rev. Sci. Instr.*, **38**, 547 (1967).
125. A. F. Grigor and W. A. Steele, *Rev. Sci. Instr.*, **37**, 51 (1966).
126. A. W. Czanderna, U.S. 3,224,521 (to Union Carbide) 12-21-1965.
127. J. A. van Lier, private communication.
128. P. A. Faeth, private communication.
129. J. Genco, private communication.
130. J. E. Wilson, private communication.
131. J. A. Poulis, W. Dekker, and P. J. Meeusen, *Vacuum Microbalance Tech.*, **5**, 49 (1966).
132. J. A. Poulis, P. J. Meeusen, W. Dekker, and J. P. de Mey, *Vacuum Microbalance Tech.*, **6**, 27 (1967).
133. R. Vieweg and Th. Gast, *Kunststoffe*, **34**, 117 (1944).
134. Th. Gast, *Vacuum Microbalance Tech.*, **3**, 45 (1963).
135. Th. Gast, *Vacuum Microbalance Tech.*, **6**, 59 (1967).
136. J. W. Beams, C. W. Hurlburt, W. E. Lotz, and R. M. Montague, *Rev. Sci. Instr.*, **26**, 1181 (1955).
137. J. W. Beams and A. M. Clarke, *Rev. Sci. Instr.*, **33**, 750 (1962).
138. New Instruments, *Rev. Sci. Instr.*, **36**, 1906 (1965).
139. W. H. Wade and D. E. Meyer, private communication.

140. K. H. Behrndt, *Vacuum Microbalance Tech.*, **1**, 69 (1961).
141. V. E. Vasserberg, *Kinetika i Kataliz*, **3**, 556 (1962).
142. G. Sauerbrey, *Z. Physik*, **155**, 206 (1959).
143. A. W. Warner and C. D. Stockbridge, *Vacuum Microbalance Tech.*, **2**, 71 (1962).
144. C. D. Stockbridge and A. W. Warner, *Vacuum Microbalance Tech.*, **2**, 93 (1962).
145. A. W. Warner and C. D. Stockbridge, *Vacuum Microbalance Tech.*, **3**, 55 (1963).
146. C. D. Stockbridge, *Vacuum Microbalance Tech.*, **5**, 147 (1966).
147. C. D. Stockbridge, *Vacuum Microbalance Tech.*, **5**, 179 (1966).
148. C. D. Stockbridge, *Vacuum Microbalance Tech.*, **5**, 193 (1966).
149. W. H. Wade and L. J. Slutsky, *Vacuum Microbalance Tech.*, **2**, 115 (1962).
150. J. E. Johnson, *Vacuum Microbalance Tech.*, **4**, 81 (1965).
151. R. P. Riegert, *Vacuum Microbalance Tech.*, **4**, 99 (1965).
152. H. L. Eschbach and E. W. Kruidhof, *Vacuum Microbalance Tech.*, **5**, 207 (1966).
153. R. Niedermayer, N. Gladkich, and D. Hillecke, *Vacuum Microbalance Tech.*, **5**, 217 (1966).
154. A. Langer and J. T. Patton, *Vacuum Microbalance Tech.*, **5**, 231 (1966).
155. W. H. Lawson, *J. Sci. Instr.*, **44**, 917 (1967).
156. R. J. Whitefield and J. J. Brady, *Rev. Sci. Instr.*, **38**, 1670 (1967).
157. C. T. Kirk and E. E. Huber, *Surface Sci.*, **9**, 217 (1968).
158. F. A. Chappell, *Vacuum Microbalance Tech.*, **2**, 19 (1962).
159. K. Angstrom, *Oefversigt Kongl. Vetenskap Akad.*, 643 (1895).
160. F. W. Sears and M. W. Zemansky, *University Physics*, 2nd ed., Addison-Wesley, Reading, Mass., 1955, p. 597.
161. Sartorius Electrona Recording Balance, *Rev. Sci. Instr.*, **28**, 744 (1957).
162. T. N. Rhodin, *Vacuum Microbalance Tech.*, **1**, ix (1961).
163. A. W. Czanderna, unpublished comments; the automatic balances described in references 120 and 121 had been in operation for over a year for studies of the interaction of oxygen with silver and copper.
164. J. A. van Lier, unpublished, private communication.
165. J. A. Poulis and J. M. Thomas, *Vacuum Microbalance Tech.*, **3**, 1 (1963).
166. J. A. Poulis, *Proc. Phys. Soc.* (*London*), **80**, 918 (1962).
167. E. E. Pugh and G. H. Winslow, *The Analysis of Physical Measurements*, Addison-Wesley, Reading, Mass., 1966.
168. L. B. Macurdy, *Vacuum Microbalance Tech.*, **2**, 165 (1962).
169. T. R. F. W. Fennell and J. R. Webb, *Microchem. J. Symp. Ser.*, **2**, 1003 (1962).
170. W. C. Tripp, R. W. Vest, and N. M. Tallan, *Vacuum Microbalance Tech.*, **4**, 141 (1965).
171. J. R. Singer, *Rev. Sci. Instr.*, **30**, 1123 (1959).
172. D. E. Soule, C. W. Nezbeda, and A. W. Czanderna, *Rev. Sci. Instr.*, **35**, 1504 (1964).
173. S. Dushman and J. M. Lafferty, *Scientific Foundation of Vacuum Technique*, 2nd ed., Wiley, New York, 1962.
174. R. W. Roberts and T. A. Vanderslice, *Ultrahigh Vacuum and Its Applications*, Prentice-Hall, Englewood Cliffs, N.J., 1963.
175. S. C. Liang, *Can. J. Chem.*, **33**, 279 (1955).
176. E. L. Fuller, H. F. Holmes, and C. H. Secoy, *Vacuum Microbalance Tech.*, **4**, 109 (1965).
177. W. Schoeniger, *Mikrochim. Acta*, **3**, 382 (1959); through *Chem. Abstr.*, **56**, 8498a (1962).
178. E. Kissa, *Microchem. J.*, **4**, 89 (1960); *Chem. Abstr.*, **54**, 8162f (1960).

179. J. A. Macinante and J. Waldersee, *J. Sci. Instr.*, **40**, 77 (1963).
180. A. W. Czanderna, *Vacuum Microbalance Tech.*, **4**, 159 (1965).
181. E. A. Gulbransen, *Vacuum Microbalance Tech.*, **4**, xi (1965).
182. H. C. Graham and W. C. Tripp, *Vacuum Microbalance Tech.*, **6**, 63 (1967).
183. M. J. Laubitz, *Can. J. Phys.*, **37**, 1114 (1959).
184. O. M. Katz and E. A. Gulbransen, *Vacuum Microbalance Tech.*, **1**, 111 (1961).
185. See *Cryogenics*, Vols. 1– (1960–) for numerous cryostat designs.
186. A. W. Czanderna, unpublished results.
187. S. P. Wolsky, in discussion of ref. 293.
188. T. Mitsui and K. Yoshikawa, *Microchim. Acta*, **1961**, 527.
189. R. A. Pierotti, *Vacuum Microbalance Tech.*, **6**, 1 (1967).
190. A. W. Czanderna, *Vacuum Microbalance Tech.*, **1**, 129 (1961).
191. See, for example, pp. 277–278 and Figure 5.36 of ref. 173.
192. E. Bruche and W. Littwin, *Z. Physik*, **52**, 318 (1928).
193. C. Eyraud and R. Goton, *Bull. Soc. Chim. France*, **1953**, 1009; through *Chem. Abstr.*, **48**, 4898g (1954).
194. A. W. Czanderna and J. M. Honig, *J. Phys. Chem.*, **63**, 620 (1959).
195. S. P. Wolsky, *Vacuum Microbalance Tech.*, **1**, 143 (1961).
196. W. E. Boggs, *Vacuum Microbalance Tech.*, **1**, 145 (1961).
197. J. M. Thomas and J. A. Poulis, *Vacuum Microbalance Tech.*, **3**, 15 (1963).
198. J. A. Poulis and J. M. Thomas, *J. Sci. Instr.*, **40**, 95 (1963).
199. H. Krupp, E. Robens, G. Sandstede, and G. Walter, *Vacuum*, **13**, 297 (1963).
200. C. Hettner, *Ergebn. Exakt. Naturwiss.*, **7**, 209 (1928).
201. J. A. Poulis, B. Pelupessy, C. H. Massen, and J. M. Thomas, *J. Sci. Instr.*, **41**, 295 (1964).
202. J. A. Poulis, *Appl. Sci. Res.*, **A14**, 98 (1965).
203. C. H. Massen, B. Pelupessy, J. M. Thomas, and J. A. Poulis, *Vacuum Microbalance Tech.*, **5**, 1 (1966).
204. J. A. Poulis, C. H. Massen, and J. M. Thomas, *J. Sci. Instr.*, **43**, 234 (1966).
205. T. Steensland and K. S. Førland, *Vacuum Microbalance Tech.*, **5**, 17 (1966).
206. K. H. Behrndt, C. H. Massen, J. A. Poulis, and T. Steensland, *Vacuum Microbalance Tech.*, **5**, 33 (1966).
207. A. W. Czanderna, *Vacuum Microbalance Tech.*, **4**, 69 (1965).
208. J. D. Ferchak, *Rev. Sci. Instr.*, **38**, 273 (N), (1967).
209. S. D. Bruck, *Vacuum Microbalance Tech.*, **4**, 247 (1965).
210. H. L. Gruber, *Monatsh. Chem.*, **95**, 1017 (1964); *Chem. Abstr.*, **62**, 3389c (1965).
211. P. Kofstad and P. B. Anderson, *J. Phys. Chem. Solids*, **21**, 280 (1961).
212. G. Sandstede and E. Robens, *Chem. Ing. Tech.*, **32**, 413 (1960).
213. G. Sandstede and E. Robens, *Chem. Ing. Tech.*, **34**, 708 (1962).
214. C. H. Massen and J. A. Poulis, *Vacuum Microbalance Tech.*, **6**, 17 (1967).
215. J. A. Poulis and C. H. Massen, *J. Sci. Instr.*, **44**, 275 (1967).
216. J. A. Poulis, C. H. Massen, and B. Pelupessy, *Vacuum Microbalance Tech.*, **4**, 41 (1965).
217. E. A. Gulbransen, K. F. Andrew, and F. A. Brassart, *Vacuum Microbalance Tech.*, **4**, 127 (1965).
218. P. E. Blackburn, J. Weissbart, and E. A. Gulbransen, *J. Phys. Chem.*, **62**, 902 (1958).
219. I. Rusznak, D. Levai, and M. Toth, *Vysokomolekul., Soedin.*, **5**, 449 (1963); *Chem. Abstr.*, **59**, 1831c (1963).

220. E. A. Gulbransen and K. F. Andrew, *J. Electrochem. Soc.*, **110**, 476 (1963) and references cited.
221. P. A. Faeth and A. F. Clifford, U.S. At. Energy Comm. Conf-20-16, 22 pp. (1963); *Chem. Abstr.*, **61**, 10302d (1964).
222. R. W. Harrison and E. J. Delgrosso, *J. Sci. Instr.*, **41**, 222 (1964).
223. R. J. Sorenson, U.S. At. Energy Comm., HW 79141 (1963); *Chem. Abstr.*, **60**, 11601g (1964).
224. J. M. Thomas and K. M. Jones, *J. Nucl. Mater.*, **11**, 236 (1964).
225. I. Stamenkovic and I. Blagojevic, *Chem. Abstr.*, **62**, 15742e (1965).
226. E. A. Gulbransen, K. F. Andrew, and F. A. Brassart, *J. Electrochem. Soc.*, **113**, 1311 (1966).
227. W. C. Tripp, R. W. Vest, and H. C. Graham, *Vacuum Microbalance Tech.*, **6**, 107 (1967).
228. H. Wieder and A. W. Czanderna, *J. Phys. Chem.*, **66**, 816 (1962).
229. S. P. Wolsky and A. B. Fowler, *Semiconductor Surface Physics*, Univ. of Pennsylvania Press, 1957, Philadelphia, Pa.
230. P. L. Cannon, *Rev. Sci. Instr.*, **29**, 1115 (1958).
231. S. P. Wolsky and E. J. Zdanuk, *Vacuum*, **10**, 13 (1960).
232. S. P. Wolsky and E. J. Zdanuk, *6th Natl. Symp. Vacuum Technology Transact.*, Pergamon Press, Oxford, 1960; S. P. Wolsky and A. B. Fowler, p. 401.
233. G. Sandstede and E. Robens, *Chem. Ingr. Tech.*, **32**, 413 (1960); *Chem. Abstr.*, **54**, 19081i (1960).
234. A. W. Czanderna, *J. Phys. Chem.*, **68**, 2765 (1964).
235. M. R. Harris, *J. Sci. Instr.*, **41**, 163 (1964).
236. A. W. Czanderna, *J. Coll. Inter. Sci.*, **22**, 482 (1966).
237. A. W. Czanderna, *J. Phys. Chem.*, **70**, 2120 (1966).
238. J. Barto, J. L. Durham, V. F. Baston, and W. H. Wade, *J. Coll. Inter. Sci.*, **22**, 491 (1966).
239. R. A. Pierotti and R. E. Smallwood, *J. Coll. Inter. Sci.*, **22**, 469 (1966).
240. Th. Gast, *Archiv. Technisches Messen*, **361**, R31 (1966).
241. A. Winkel, *Staub*, **22**, 77 (1962); through *Chem. Abstr.*, **57**, 4488a (1962).
242. A. W. Czanderna, *Vacuum Microbalance Tech.*, **6**, 129 (1967).
243. R. Jongepier and G. C. A. Schuit, *J. Catalysis*, **3**, 464 (1964).
244. J. M. Thomas, E. E. G. Hughes, and B. R. Williams, *Nature*, **189**, 134 (1961).
245. A. W. Czanderna and J. M. Honig, *J. Phys. Chem. Solids*, **6**, 96 (1958).
246. J. M. Honig, A. F. Clifford, and P. A. Faeth, *Inorg. Chem.*, **2**, 791 (1963).
247. A. W. Czanderna and H. Wieder, *J. Chem. Phys.*, **39**, 489 (1963).
248. P. D. Garn, *Anal. Chem.*, **33**, 1247 (1961).
249. E. Robens, G. Robens, and G. Sandstede, *Vacuum*, **13**, 303 (1963).
250. P. L. Waters, *Anal. Chem.*, **32**, 852 (1960).
251. B. E. Deal and H. J. Svec, *J. Electrochem. Soc.*, **103**, 421 (1956).
252. H. J. Svec and T. J. Rider, *J. Less Common Metals*, **14**, 103 (1968).
253. R. E. Pawel, *J. Electrochem. Soc.*, **114**, 1222 (1967).
254. J. E. Antill and J. B. Wharburton, *J. Electrochem. Soc.*, **114**, 1215 (1967).
255. M. G. Hapase, M. K. Gharpurey, and A. P. Biswas, *Surface Sci.*, **9**, 87 (1968).
256. W. Bierman and M. Heinrichs, *Can. J. Chem.*, **40**, 1361 (1962).
257. B. Boehlen, W. Hausmann, and A. Guyer, *Helv. Chim. Acta*, **47**, 1821 (1964).
258. B. Boehlen and A. Guyer, *Helv. Chim. Acta*, **47**, 1815 (1964).
259. K. M. Laing, *Vacuum Microbalance Tech.*, **6**, 149 (1967).
260. P. E. Blackburn, *J. Phys. Chem.*, **62**, 897 (1958).

261. N. E. Heyerdahl, *Vacuum Microbalance Tech.*, **5**, 121 (1966).
262. A. Block-Bolten, *Arch. Hutnictwa*, **8**, 81 (1963); through *Chem. Abstr.*, **60**, 2387f (1964).
263. P. D. Zavitisanos, *Rev. Sci. Instr.*, **35**, 1061 (1964).
264. H. Mayer, R. Niedermayer, W. Schroen, D. Stuenkel, and H. Goehre, *Vacuum Microbalance Tech.*, **3**, 87 (1963).
265. I. Haller and P. White, *Rev. Sci. Instr.*, **34**, 677 (1963).
266. E. J. Zdanuk and S. P. Wolsky, *J. Appl. Phys.*, **36**, 1683 (1965).
267. H. H. A. Bath, *J. Sci. Instr.*, **43**, 374 (1966).
268. E. J. Zdanuk and S. P. Wolsky, *Vacuum Microbalance Tech.*, **5**, 111 (1966).
269. E. A. Johnson, D. G. Childs, and G. H. Beaven, *J. Chromatog.*, **4**, 429 (1960).
270. P. Schatzberg, *Vacuum Microbalance Tech.*, **6**, 89 (1967).
271. B. V. Derjaguin, I. I. Abikosova, and E. M. Lifshitz, *Quart. Rev.*, **10**, 295 (1956).
272. J. A. Mann, Jr. and R. S. Hansen, *Rev. Sci. Instr.*, **34**, 702 (1963).
273. K. Edelmann and K. Wulf, *Faserforsch. U. Textiltech*, **12**, 461 (1961); 466 (1961).
274. K. N. Korovkin, N. A. Oks, E. A. Bylyna, and V. B. Evdokomov, *Zh. Fiz. Khim.*, **35**, 677 (1961); through *Chem. Abstr.*, **55**, 17109g (1961).
275. E. Wachtel, *Mem. Sci. Rev. Met.*, **59**, 416 (1962); through *Chem. Abstr.*, **57**, 10889a (1962).
276. A. Blaise and M. A. Peuch, *Comm. Energie At.* (*France*), Rappt. No. CEA-2187 (1962); through *Chem. Abstr.*, **58**, 8506c (1963).
277. R. Havemann, *Z. Chem.*, **4**, 121 (1964).
278. R. Kohlhaas and H. Lange, *Z. Angew. Phys.*, **17**, 448 (1964).
279. L. N. Mulay and L. K. Keys, *Anal. Chem.*, **36**, 2383 (1964).
280. A. Van den Bosch, *Vacuum Microbalance Tech.*, **5**, 77 (1966).
281. Th. Gast, *Dechema Monograph*, **31**, 9 (1959); *Chem. Abstr.*, **54**, 9375c (1960).
282. Th. Gast, *Staub*, **20**, 266 (1960); *Chem. Abstr.*, **55**, 25113a (1961).
283. C. W. Hargens, *Rev. Sci. Instr.*, **28**, 921 (1957).
284. C. S. G. Phillips and P. L. Timms, *J. Chromatog.*, **5**, 131 (1961).
285. K. T. Scott and K. T. Harrison, *J. Nucl. Mater.*, **8**, 307 (1963); *Chem. Abstr.*, **60**, 10153 (1964).
286. P. Malherbe, T. Vogt, and C. Boebel, *Vacuum Microbalance Tech.*, **5**, 97 (1966).
287. W. E. Boggs, R. H. Kachik, and G. E. Pellissier, *J. Electrochem. Soc.*, **108**, 6 (1961); **110**, 4 (1963).
288. R. L. Schwoebel, *J. Appl. Phys.*, **34**, 2776 (1963); **34**, 2784 (1963).
289. H. Wieder and A. W. Czanderna, *J. Chem. Phys.*, **35**, 2259 (1961).
290. A. W. Czanderna, *Vacuum Microbalance Tech.*, **5**, 135 (1966).
291. H. Wieder and A. W. Czanderna, *J. Appl. Phys.*, **37**, 184 (1966).
292. C. H. Amberg and D. A. Seanor, *Proc. Third Intern. Congr. Catal.*, North-Holland, Amsterdam, 1964, p. 122.
293. C. L. Angell, *Vacuum Microbalance Tech.*, **6**, 77 (1967).
294. J. T. Ward, *J. Catalysis*, **9**, 225 (1967).
295. E. A. Gulbransen, K. F. Andrew, and F. A. Brassart, *Carbon*, **1**, 413 (1964).
296. E. A. Gulbransen and F. A. Brassart, *J. Less Common Metals*, **14**, 217 (1968).
297. W. W. Wendlandt, *Anal. Chem.*, **34**, 1726 (1962).
298. B. Evans and T. E. White, *Vacuum Microbalance Tech.*, **6**, 157 (1967).
299. A. W. Czanderna, *J. Colloid Interface Sci.*, **24**, 500 (1967).

Chapter 3

Microbalance Theory and Design

RICHARD L. SCHWOEBEL

Sandia Corporation
Albuquerque, New Mexico

I. INTRODUCTION

In Chapter 2 Dr. Czanderna discussed many important features of microbalances and associated apparatus in addition to several applications of these instruments. A number of important details concerning the theory, fabrication, and operation of microbalances, excluding the quartz oscillator type (Chapter 5), will now be considered. The three types of microbalances described in Section 4, Chapter 2, will be dealt with in detail: the spring, cantilever, and beam balances. First, a theoretical framework for microbalance characteristics will be established. This formulation will be rather complete so those unfamiliar with these instruments can recognize the important parameters determining the static and dynamic characteristics of microbalances. This treatment will include the development of fundamental relations between important microbalance parameters, such as period and deflection sensitivity, and the relation between periods of loaded and unloaded balances. The important design criteria will be established and related to these parameters. Typical values will be provided so characteristics of a specific microbalance may be readily determined from the pertinent equations. In addition, several tables will be used to illustrate comparable features of frequently employed microbalances. Secondly, a discussion and comparison of suitable microbalance materials and their properties will be presented. The important properties of these materials will be compiled and related to various advantages and disadvantages of use. Both conventional and some as yet unused materials will be considered. The third major section will deal with the selection and fabrication of a microbalance following consideration of the pertinent environmental and experimental conditions. In this section several of the construction details critical to the operation of microbalances will be described. This will include a description of fiber fabrication and attachment, period and sensitivity adjustments, and useful aids in construction. Finally, techniques of manual or automatic operation of the completed microbalance will be discussed for various compensation schemes. This discussion will include examples of operational modes such as those combining manual and automatic techniques in which automatic damping may be utilized. The entire system will then be considered in terms of the characteristic system errors and the resultant sensibility or resolution.

II. MICROBALANCE THEORY

An important prerequisite for proper microbalance design is an understanding of the operational principles, limits, and dynamic characteristics of microbalances. In this section the formal relations for deflection sensitivity, period, and variations in sensitivity and stability of the three microbalance systems will be derived.

A. Spring Microbalance

A spring microbalance may be considered essentially equivalent to a straight rod under torsion. This is an excellent approximation for most balances constructed in this geometry operating under the conditions stated below. From the elementary theory for helical springs, it is concluded that the torsion, T', required to rotate a rod of radius R and length l_0 through an angle θ, is

$$T' = \frac{\pi\theta G R^4}{2l_0} \tag{3.1}$$

where G is the shear modulus of the rod material. A helical spring consisting of n turns of radius r may be considered equivalent to a rod of length $2\pi rn$ under a torque given by the mass, M, and a lever arm r, Figure 3.1.

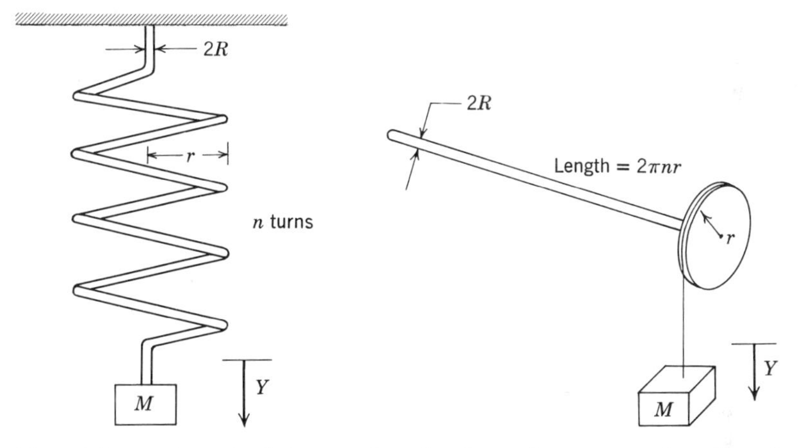

Figure 3.1. A spring microbalance and a rod under torsion are approximately equivalent systems when the pitch of the coil is nearly zero and the fiber diameter and deflection per coil are much less than the spring diameter.

Since the angular rotation of the rod forming the spring will be equal to the deflection y divided by r (i.e. y/rn per turn in Figure 3.1), the deflection will be, using equation 3.1,

$$y = \frac{4Lr^3 ng}{GR^4} \tag{3.2}$$

where g is the acceleration of gravity and L is the suspended load. The deflection sensitivity is

$$\frac{dy}{dL} = \frac{4r^3 ng}{GR^4} \tag{3.3}$$

The essential assumptions used in arriving at equations 3.2 and 3.3 are (*1*) that $r/R \gg 1$, that is, the length of the inside and outside of the fiber is assumed to be very nearly equal; (*2*) the deflections are small in comparison to nr; and (3) the initial pitch angle is assumed to very nearly zero. Deviations of spring systems from the behavior expressed in equations 3.2 have been considered in some detail by a number of authors, including Röever (1), Wahl (2), and Goehner (3). They have shown that a more correct form of equation 3.2 would be

$$y = \frac{4Lr^3ng}{GR^4}\left[\frac{\cos\alpha}{1 + 3/16[\cos^4\alpha/(c^2-1)]} + \frac{2G}{E}\sin\alpha\tan\alpha\right] \tag{3.4}$$

where c is the spring index (equal to r/R), α is the pitch angle and E is the elastic modulus. The term in brackets will differ only slightly from unity for almost all spring microbalances of interest. For example, if r/R is as small as 3, $\alpha = 5°$ and $G/E = 0.35$, the correction in the deflection given by the bracketed term is approximately 2.1%. Clearly this can be ignored in most cases. Wahl (4) has also considered the case in which the deflections are large and calculated appropriate corrections for equations 3.2 and 3.4. Again the corrections are less than 5% if, for example, the pitch angle is less than 5° and the deflection per turn is less than the initial coil radius. Simple theory, therefore, gives the deflection versus load relation (equation 3.2) to within a few percent for springs used as microbalances.

Beginning with equation 3.2 a number of useful relations can be easily derived, such as the self-deflection of a spring microbalance. The self-deflection is the deflection of the balance due to its own mass. The deflection of the uppermost coil, Δy_1, will be

$$\Delta y_1 = (4r^3ng/GR^4)[M(n-1)/n] \tag{3.5}$$

where M is the spring mass and n is assumed to be large. Similarly, the deflection of each coil in the spring due to the remaining portion of spring suspended below it is

$$\begin{aligned}\Delta y_2 &= (4r^3ng/GR^4)[M(n-2)/n]\\ \Delta y_3 &= (4r^3ng/GR^4)[M(n-3)/n]\\ &\vdots\end{aligned}$$

The total self-deflection of the kth coil is then

$$\begin{aligned} y_{\text{self}}^{(k)} &= \sum_{p=1}^{k} \Delta y_p = \frac{4r^3nMg}{GR^4}\sum_{p=1}^{k}\frac{n-p}{n} \\ &= \frac{4r^3nMg}{GR^4}\left[\frac{n-1}{n} - \frac{(n-k-1)(n-k)}{2n}\right] \quad \text{for} \quad k \le n \end{aligned} \tag{3.6}$$

For the entire spring, $k = n$ and

$$y_{\text{self}} \simeq \frac{2r^3n^2Mg}{GR^4} \quad \text{for} \quad n \gg 1 \tag{3.7}$$

Equation 3.7 gives the total self-deflection for a uniform spring of mass M.

Accordingly, equations 3.2 and 3.3 may also be expressed in terms of y_{self}. Therefore, some of the *important characteristics of microbalances can be ascertained simply by observing the total self-deflection.*

Using equation 3.2, the principal oscillatory period, T, of a spring balance will be,

$$T = 2\pi(4Lnr^3/GR^4)^{1/2} \tag{3.8}$$

Assuming that $L \gg M$.

The maximum permissible load, L_{max}, will be limited by the material properties of the spring and also by a desire to operate the balance within bounds of nearly linear response. From the previous discussion, it can be concluded that in order to operate the balance in the region of nearly linear response, the total deflection should not exceed about 1 spring radius/turn, or, using equations 3.2 and 3.7,

$$y_{total} = y_{self} + \frac{4L_{max}r^3ng}{GR^4} \sim nr \tag{3.9}$$

from which L_{max} can be determined.

Spring microbalances have been used for a variety of applications including quantitative microanalysis (5), gas adsorption (6), various biological studies (7), sulfiding of metals (8), magnetic susceptibility measurements (9), degradation of polymers (10), and various gas–solid reactions (11). These instruments are usually characterized by load-sensitivity products (LSP)* in the range of 1–10^2 cm. Other characteristics of these spring microbalances are listed in Table 3.1.

Table 3.1. Characteristics of Some Helical Spring Microbalances

Rod diameter, cm	Coil diameter, cm	Turns, n	Spring material	Deflection sensitivity, cm/g	Typical load g	LSP load sensitivity, product, cm	Ref.
0.02	1.25	15	Quartz	~9	0.500	4–5	5
0.0035–0.017	1.4	40–110	Quartz	10^2–10^4	0.010	1–100	7
0.0075	1.1	23	Tungsten	55	0.400	20	10
—	—	—	Quartz	10^2	0.500	50	11
0.635 × 0.0076 (band)	1.27	~15	Ni–Span C	0.2	40	8	95
Worden Type H	—	—	Quartz	10^2	0.235	24	e.g., 96

* See Chapter 2 for a discussion of LSP and other terms used to characterize microbalance performance.

B. Cantilever Microbalance

From the simple theory of bending, it is concluded that the radius of curvature R, the bending moment M, the moment of inertia about the neutral axis I, and the modulus of elasticity E, are related by the equation

$$1/R = M/IE \tag{3.10}$$

when bending occurs within the usual limiting assumptions. If the radius of curvature is large, R may be expressed as d^2y/dx^2, in which case the deflection y can be calculated in terms of x, the distance along the beam.

Consider a uniform cantilever beam of length l, mass m per unit length,

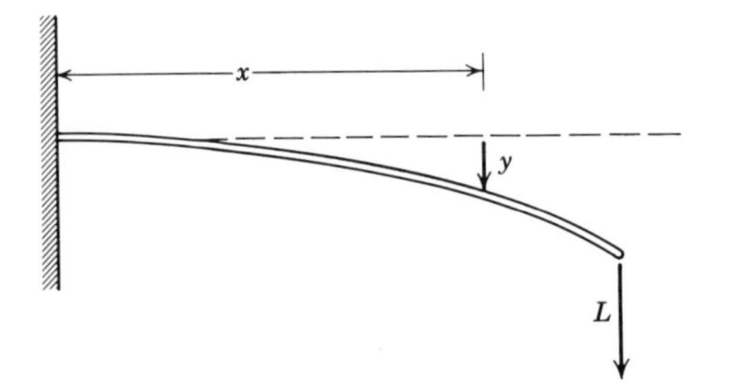

Figure 3.2. A simple cantilever microbalance supporting a load at the extreme end and assumed to be of uniform cross section.

supporting a mass L at its extreme end as illustrated in Figure 3.2. The deflection, y, in terms of x will be given by integrating

$$\frac{d^2y}{dx^2} = \frac{g}{IE}\left[L(l - x) + \frac{m(l - x)^2}{2}\right] \tag{3.11}$$

twice using the boundary conditions that $dy/dx = y(x) = 0$ at $x = 0$. The first term in the brackets of equation 3.11 is due to the load while the second term is due to the mass of the beam itself. The result is that,

$$y(x) = \frac{g}{EI}\left[\frac{Lx^2}{2}\left(l - \frac{x}{3}\right) + \frac{mx^2}{2}\left(\frac{l^2}{2} - \frac{lx}{3} + \frac{x^2}{12}\right)\right] \tag{3.12}$$

If we consider that the beam is of circular cross section, the moment of inertia I, will be equal to $\pi r^4/4$. In this case the deflection at the extreme end of the cantilever will be

$$y(x = l) = \frac{4l^3g}{\pi Er^4}\left\{\frac{L}{3} + \frac{M}{8}\right\} \tag{3.13}$$

where $M = ml$, the mass of the beam. Equation 3.13 gives the total deflection of the end of a uniform cantilever beam due to its own mass M, plus that due to the mass L suspended from its end. The deflection sensitivity is therefore

$$\frac{dy}{dL} = \frac{4l^3g}{3\pi Er^4} \tag{3.14}$$

or, in terms of the self-deflection,

$$\frac{dy}{dL} = \frac{8y_{\text{self}}}{3M} \tag{3.15}$$

where the self-deflection has been determined from equation 3.13 by setting $L \equiv 0$. Again, the deflection sensitivity can be determined simply by observing the self-deflection of the uniform cantilever of (in this case) circular cross section. The primary oscillatory period for the system, assuming $L \gg M$, will be

$$T = 2\pi\left(\frac{4l^3L}{3\pi Er^4}\right)^{1/2} = 2\pi\left(\frac{8Ly_{\text{self}}}{3Mg}\right)^{1/2} \tag{3.16}$$

Cantilever balances have been used for a variety of applications including investigations of surface reactions (12), histochemical (13), and hydration (14) studies. Characteristics of cantilever balances used by some investigators are shown in Table 3.2 as compiled in part by Rhodin (15).

Table 3.2. Characteristics of Some Cantilever Microbalances

Fiber	Fiber diameter, μ	Fiber length, cm	Typical sample mass, mg	Deflection sensitivity, cm/μg	Stated resolution μg	LSP, cm	Ref.
Quartz	—	7 (est)	30 (max)	$\sim 10^{-4}$	~0.01	~3	12
Quartz	130	7	0.5–1.0	4×10^{-5}	~0.1	0.04	17
Quartz	20	5.6	—	~0.2	0.05	—	97
Quartz tube	250 (?)	20	0.2	~0.03	~0.1	~6	13
Phosphor bronze ribbons	0.023 × 0.0051 cm band (typical)	9	0.4	$\sim 10^{-3}$	—	~0.4	14
Steel	0.1 cm band	~10	15–25	—	—	—	18 (16)
Nickel–chrome steel	140	25–40	0.1–0.6	$\sim 5 \times 10^{-4}$	—	~0.3	19

The use of cantilever balances is generally limited to experiments in which the desired load-sensitivity product (LSP) is of the order of 0.1–10 cm. Considerably higher values of LSP can be attained with other microbalance systems; however, cantilever balances are both simple to construct and use. Although cantilevers have been traditionally used as deflection devices, they may be adapted to null deflection compensation techniques.

C. Beam Microbalance

Beam microbalances have received considerably more attention than spring or cantilever balances because of the relatively high values of the load-sensitivity product (LSP) that are readily attainable. Successful use of sensitive beam microbalances usually involves more sophisticated and detailed operational techniques and some experimental skill must be acquired before these delicate instruments can be operated in a routine manner. Robust versions of beam microbalances have been designed and offer obvious advantages in experiments where a greater measure of ruggedness is required. However, these designs sometimes incorporate constructional details which preclude the use of the balance in ultrahigh vacuum or other special environments. Accordingly, several aspects of the experiment should be considered before a specific beam microbalance is selected.

Beam microbalances may be classified either as torsional or gravity balances, depending on the origin of the principle restoring torque operating on the beam. Balances designated as torsional, for example, are those in which the restoring moment is supplied by twisting or bending a suspension fiber. Gravity balances are designed so that any deflection is accompanied by an elevation of the beam and the center of mass of the suspended masses. Beam microbalances have been used in a variety of forms. Of these, three types will be briefly discussed: beam microbalances with horizontal suspensions, vertical suspensions, and pivotal supports. A variety of such microbalances can be discussed in terms of the following schematic diagram, Figure 3.3, where masses L_1 and L_2 are suspended from ends of a beam of length $l_1 + l_2$ supported at point O. The mass of the beam is M, its moment of inertia is I, and the center of mass is located a distance x below the suspension point. The plane containing the points from which the two supported masses are suspended is located a distance d below the suspension point. When the beam is at the equilibrium position,

$$L_2 g(l_2 \cos\theta + d \sin\theta) + Mgx \sin\theta - L_1 g(l_1 \cos\theta - d \sin\theta) + K\theta = 0 \qquad (3.17)$$

where g is the acceleration of gravity and K is the torsion constant of the suspension. When the angular deflection, θ, is very small, we can expand equation 3.17 and to first order:

$$\theta = \frac{L_1 l_1 - L_2 l_2}{(L_1 + L_2)d + Mx + K/g} \tag{3.18}$$

For a symmetrical balance in which l_1 and $l_2 = l$, and L_1 and L_0 are nearly equal, it can be seen that the deflection sensitivity will be

$$\frac{d\theta}{dL} = \frac{l}{2Ld + Mx + K/g} \tag{3.19}$$

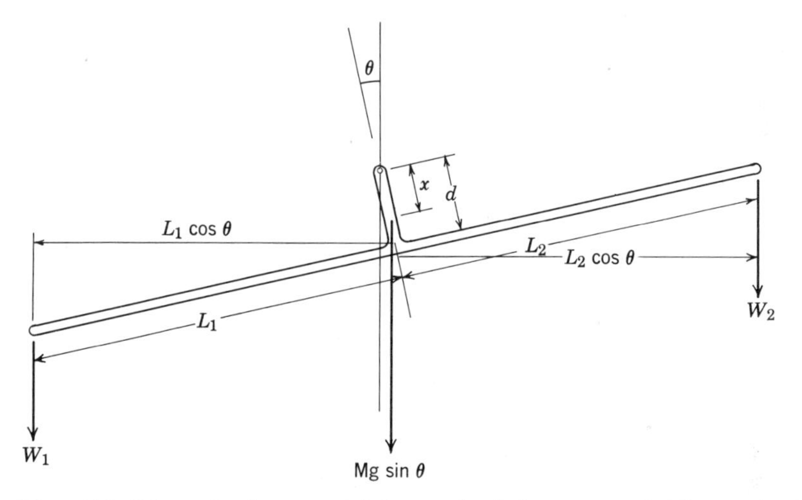

Figure 3.3. Schematic of a symmetric beam microbalance suspended from point *O* and supporting masses at each end. The center of mass of the beam is located a distance x below the suspension point.

for $L_1 = L_2 = L$. When the suspension, *O*, is located above both the beam center of mass and the plane joining the supported masses, the balance will be stable, i.e. there will be a specific value of the deflection angle, θ, for which the net moment on the balance is zero. If the center of rotation (*O* in Figure 3.3) is located below the beam center of mass and the plane joining the supported masses, the angular deflection is given by a similar derivation as

$$\theta^* = \frac{L_1 l_1 - L_2 l_2}{-d(L_1 + L_2) - Mx + K/g} \tag{3.20}$$

In this case, balance stability can be maintained only by providing a sufficiently large torsion constant, *K*, such that the inequality,

$$K/g > d(L_1 + L_2) + Mx \tag{3.21}$$

is satisfied. Balances in which there are other relative positions of the suspension, O, beam center of mass, and the plane containing the points from which the supported masses are suspended, may be considered in an analogous manner. Microbalance deflection sensitivity may be enhanced by designing the beam and suspension system such that the denominator in equation 3.19 for example, is minimized. There will be practical limits, of course, in what reduction can be accomplished.

When the beam is freely oscillating through small angles about the balance position, the approximate equation of motion will be, when x and d are positive,

$$(I + 2Ll^2)\ddot{\theta} + (2Ldg + Mgx + K)\theta = 0 \tag{3.22}$$

and the period will be, solving equation 3.22:

$$T = 2\pi \left[\frac{I + 2Ll^2}{2Ldg + Mxg + K}\right]^{1/2} \tag{3.23}$$

In terms of the deflection sensitivity, equation 3.19, the period is

$$T = 2\pi \left[\frac{(I + 2Ll^2)}{gL} \frac{d\theta}{dL}\right]^{1/2} \tag{3.24}$$

and the period of the loaded balance is directly related to the deflection sensitivity. If the balance is stable when unloaded ($L = 0$), it can be seen from equation 3.23 that the unloaded period T_u, will be

$$T_u = 2\pi \left[\frac{I}{Mxg + K}\right]^{1/2} \tag{3.25}$$

and is, of course, determined by the beam mass, moment of inertia, center of mass position, and the torsion imposed by the suspension system. Since x and K are not usually well known, while M and I can be determined fairly accurately, it is possible to determine approximately the important characteristics of the balance simply by noting its unloaded period (equation 3.25) and relating this to the deflection sensitivity (equation 3.19):

$$\frac{d\theta}{dL} = \frac{l}{2Ld + 4\pi^2 I/T_u^{\,2} g} \tag{3.26}$$

As before, if x and/or d are negative, appropriate sign changes must be made in equations 3.22 through 3.26.

Oscillation in the balance mode (equation 3.23) is, unfortunately, not the only oscillatory mode present in beam microbalances supported by fibers. Two other modes are frequently excited by oscillations in the balance mode. Of principal importance to microbalances utilizing a horizontal suspension is oscillation of the beam about a vertical axis perpendicular to

Table 3.3. Some Characteristics of Beam Microbalances Utilizing Horizontal Suspension Fibers

Beam material	Beam length, cm	Fiber material	Fiber diameter, μ	Typical sample mass, g	Deflection sensitivity, deg/μg	LSP, deg	Ref.
Glass or metal	—	Tungsten or quartz	—	~0.4	—	—	20
Glass	~40	Tungsten	17	0.5	~0.01	~5 × 10^3	21
Quartz	14.5	Tungsten	25	~0.2	~0.008	~1.6 × 10^3	22
Quartz	10	Quartz	23	0.1–0.2	3.0	3–6 × 10^5	23
Quartz	~15	Tungsten	10	~0.5	~0.07	~3.5 × 10^4	24
Quartz	~10	Quartz	~10	0.1–0.15	1.0	1–1.5 × 10^5	25
Quartz	15.2	Quartz	38	0.2	0.2	4 × 10^4	26–28
Glass	16	Tungsten	50	4	0.004	~1.6 × 10^4	29,30
Quartz	14.5	Tungsten	25	0.3	0.002	6 × 10^2	31
Aluminum tubing	6.75 cm, sample to fulcrum	Beryllium–copper ribbon	15 (thickness)	~1	0.023	2.3 × 10^4	32
Glass	~15	Quartz	75	—	~0.0015	—	33
Dural	18	Tungsten	25	0.14	0.016	2.2 × 10^3	34,35
Quartz	10	Tungsten	50	2.5	—	—	36
—	~5 cm, sample to fulcrum	Beryllium–copper ribbon	—	20	~10^{-4}	~2 × 10^3	37
Quartz	~8.4	Tungsten	8.4	~0.2	~0.07	~1.4 × 10^4	38
Quartz (Sartorius)	20	Platinum–iridium	—	1	0.065	6.5 × 10^4	88,89
Dural	18	Tungsten	10	0.14	0.01–0.02	1.4–2.8 × 10^3	91

the suspension. This oscillation promotes motion of the suspended masses on hangdown fibers. If the microbalance is used in low pressure environments (below approximately 10^{-2} torr), the damping coefficients are generally so small that oscillation may continue for several hours. Compensation and automatic damping systems are infrequently designed to deal with this problem; however, some control can be maintained by deflecting the microbalance to an end point and dissipating the vibrational energy through the supporting framework.

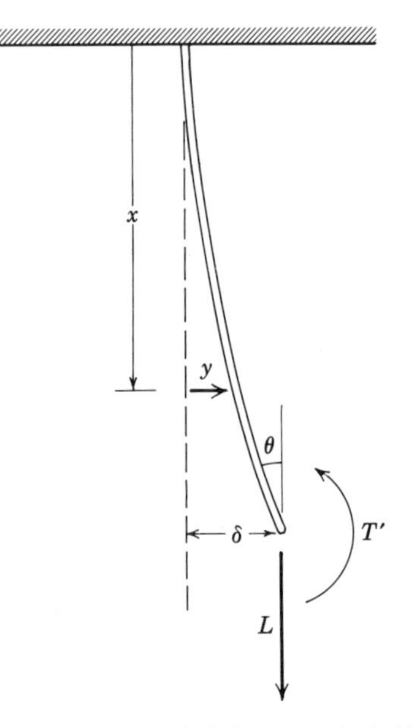

Figure 3.4. Microbalance beams suspended from vertical fibers exert not only a load but also a torque about the end of the fiber on deflection of the beam. The corresponding displacement and effective torsion constant is calculated for uniform fibers.

The elementary theory associated with the operation of beam microbalances with horizontal suspension fibers has been presented earlier in this section. The deflection sensitivity and periods are presented in equations 3.19, 3.24, and 3.25, respectively. In this case the torsion constant, K, is $\pi G R^4/2l_0$ (equation 3.1) as noted previously for fibers in torsion. Some characteristics of a few microbalances using horizontal suspensions are summarized in Table 3.3. The specified deflection sensitivities indicate the mechanical deflection which will occur if there is no compensation.

Beam microbalances utilizing pivotal supports have been used principally by Czanderna and Honig (see Table 3.4). In this case the term in the theoretical development describing the restoring torque due to the suspension (equations 3.17–3.25) must be replaced. If the two elements of the pivotal bearing are perfectly rigid, a frictional force term can be used if it is assumed that effects of plastic flow, bending, etc., are relatively insignificant. Some features of beam microbalances with pivotal supports are noted in Table 3.4.

Beam microbalances utilizing vertical suspensions are relatively infrequently used; however, they offer unique possibilities to experimentalists interested in high deflection sensitivities (see Chapter 2, Section 4). If a vertical suspension is used, the form of the torsion constant, K, in the theoretical development can be deduced from equations similar to those describing fiber bending in the section dealing with cantilever balances. In this case the vertical fiber supports a load L and torque T' is applied to the lower end (Figure 3.4). That is, for small deflections

$$EI\frac{d^2y}{dx^2} = T' - L(\delta - y) \tag{3.27}$$

where δ is the maximum translation of the fiber at $x = l$. Solving for y and using the boundary conditions that $y(0) = 0$, $y(l) = \delta$, and $dy/dx = 0$ at $x = 0$, gives

$$y = \frac{T'}{L \cosh (L/EI)^{1/2}l}[\cosh (L/EI)^{1/2}x - I] \tag{3.28}$$

The effective torsion constant, K, about an axis in the horizontal plane will then be

$$K = (EIL)^{1/2} \cosh (L/EI)^{1/2}l \tag{3.29}$$

in terms of the elastic constant E, fiber length l, axial load L, and the moment of inertia of the fiber cross section, I. Characteristics of some microbalances using vertical suspensions are noted in Table 3.5.

It is interesting to compare the deflection sensitivities of various microbalance systems in current use with those used by Pettersson in 1920 (44) and by Kirk et al. in 1947 (23). With few exceptions, balances in use today have deflection sensitivities one or more orders of magnitude greater than those used by some investigators over two decades ago. The ultimate system sensibility is, of course, directly related to the balance deflection sensitivity for a given deflection detecting system. It would appear that the high system sensibilities currently reported by most investigators reflects on the sophisticated methods used to detect and compensate deflection, rather than the design of microbalances with superior deflection sensitivities.

Table 3.4. Beam Microbalances with Pivotal Supports

Beam material	Beam length, cm	Cups	Tips	Typical sample mass, g	Deflection sensitivity, deg/μg	LSP, deg	Ref.
Quartz	16	Quartz	Tungsten	1–2	0.007	1.4×10^4	39–42
Quartz	18	Quartz	Tungsten	15	0.033	5×10^5	43
Quartz	12 est.	Quartz	Tungsten	Vapor density balance	~0.006	—	80
Pyrex	17	Pyrex	Pyrex	Vapor density balance	—	—	77,81
Quartz	12 est.	Sapphire	Tungsten	1.0	~0.06	6×10^4	82

Table 3.5. Some Characteristics of Beam Microbalances with Vertical Suspensions

Beam material	Beam length, cm	Fiber material	Fiber diameter, μ	Typical sample mass, g	Deflection sensitivity, deg/μg	LSP, deg	Ref.
Hard glass	5	Hard glass	1.5–2 (est.)	0.1–0.2	2.7	$2.7–5.4 \times 10^5$	44
Hard glass	10	Hard glass[a]	6	0.1	0.11	1.1×10^4	45
Quartz	14	Quartz	4	—	0.14	—	46

[a] Torsion fiber used for compensation.

D. Design Criteria for a Beam Microbalance

Some of the important characteristics of microbalance designs may be inferred from the deflection sensitivity and other equations noted in a previous section. The important features of beam microbalances will now be briefly reviewed because of the interest in utilizing this kind of microbalance in a variety of applications.

As noted previously, beam microbalances may be classified either as torsional or gravity balances, depending on the origin of the principle restoring torque operating on the beam. We have considered here that torsional balances include those in which the principle restoring torque is the result of (*1*) twisting a horizontal suspension fiber, or (*2*) bending a vertical suspension fiber. In gravity balances these effects are small in comparison to the restoring torque resulting from elevation of the center-of-mass of the beam system (including the suspended masses) by deflection. In many beam microbalances the center-of-mass of the beam system is located very near the deflection axis or the fulcrum. Consequently, the primary restoring torque is due to (*1*) or (*2*) above, etc. In other microbalance systems (e.g. reference 22) the restoring torque of the horizontal suspension is relatively small and the center-of-mass of the beam system is purposely located slightly below the suspension point so that geometric considerations determine the deflection for a given unbalance. Regardless of the operational mode of the microbalance, some important constructional considerations are as follows:

(*1*) The beam mass should be minimized in accord with maximizing the deflection sensitivity (equation 3.19).

(*2*) The beam should be rigid so the balance is relatively insensitive to the load, i.e. the values of x and d (equation 3.19) should be held nearly constant.

(*3*) The beam should be symmetrical so effects of bending and thermal expansion will have a more clearly defined effect on the deflection sensitivity and the associated balance calibration.

If the balance is to be used as a gravity device, restoring forces due to the suspension can be reduced by use of long, thin fibers of low shear or bending modulus depending on the particular suspension that is employed. If the balance is to be used as a torsional device, constructional procedures must be used to insure that the center-of-mass of the beam system is precisely located on the deflection axis. As previously mentioned, torsion in the suspension may be effectively reduced by purposely designing the beam with negative values of x and/or d (equation 3.19). In this case the geometry of the beam-suspension system is constructed with gravitational and torsional effects in opposition and the deflection sensitivity increased accordingly. If construction of such a balance is contemplated, stability considerations will be important and the designer must evaluate equation 3.18 using appropriate signs for x and/or d to determine the limits of stable balance operation.

Adherence to the idealized design criteria outlined here will result in the construction of microbalances with high deflection sensitivities. Nevertheless, the attainment of high deflection sensitivities is frequently not the only consideration in the use of such instruments. If, for example, the time between measurements in a particular experiment must be less than a certain interval, the microbalance used in that experiment should have a period approximately equal to or less than the desired measurement interval. That is, there will be an upper bound on the desired sensitivity in this case. Similarly, other limitations on sensitivity, load capacity, etc. are frequently imposed by the experimental conditions and these factors should be considered in detail during the preliminary design stages.

III. MICROBALANCE MATERIALS

Experimentalists have frequently attempted to optimize or simplify microbalance construction by an appropriate selection of materials. While some microbalances have been constructed of a single material such as quartz, others have been constructed of a variety of materials with properties matched to the requirements of the individual microbalance component. For example, one might construct the beam of quartz for lightness and rigidity, and use a tungsten fiber suspension for simplicity. The specific material or combination of materials selected for microbalance construction is, of course, contingent on the intended experimental conditions. There have been many different materials used in microbalance construction and no attempt will be made to review the extensive literature in this area. Rather, some of the important properties of both frequently used materials and some materials of potential value will be tabulated. Bonding agents are often required when more than one material is used in microbalance fabrication and some of the important characteristics of these agents will be noted. Similarly, quartz fibers are frequently employed in suspension systems and some of their mechanical properties are shown in Table 3.6.

Materials such as quartz and tungsten have been used extensively in the construction of microbalance beams and suspension systems. These materials possess a number of desirable characteristics for microbalance applications and it is worthwhile to compare their properties to those of several other materials that are coming into increased availability and use.

Table 3.6. Some Mechanical Properties of Quartz Fibers (59)[a]

Diameter, μ	Young's modulus, dynes/cm^2	Shear modulus, dynes/cm^2	Ultimate strength, dynes/cm^2	Strain at failure
1.5	—	—	0.90×10^{11}	—
2.0	—	—	0.80	—
3.0	11.1×10^{11}	6.6×10^{11}	0.65	0.059
4.0	10.3	6.1	0.55	0.054
5.0	9.8	5.8	0.48	0.049
7.0	9.0	5.3	0.39	0.043
10.0	8.5	4.8	0.30	0.035
15.0	7.9	4.2	0.23	0.029
20.0	7.6	3.9	0.17	0.022
30.0	7.1	3.5	0.145	0.020

[a] All are mean values with deviations of $\pm 20\%$ to be expected.

Some of these properties are compared in Table 3.7. Various metals, some of which have been used in balance fabrication, are listed in the upper portion of this table, while for comparison some oxides are listed below. The oxides have been included because of the rapidly developing technology associated with structural ultilization of refractory oxides, and because these materials are representative of those with the highest strength to weight ratios. For example, Anderegg (55) has observed some silica fibers with fracture stresses of up to 3.5 mpsi. Hillig (56) has observed quartz rods with fracture stresses of 1.9 mpsi, and Al_2O_3 and BeO whiskers have been stressed to 2.2 and 2.8 mpsi by Brenner (57) and Ryshkewitch (58), respectively. A number of carbides and borides might also be included in such a compilation of high strength materials. Although the ultimate strength of various oxides is of interest, so are other properties such as the relative thermal conductivity (e.g. compare BeO and the metals), chemical inertness, etc. Oxides are usually poor electrical conductors and several investigators including Walker (29), Moret and Louwerix (36), and Wolsky et al. (87) have successfully metallized quartz microbalances to reduce charging effects, provide for current conduction to the suspended sample, etc. Furthermore, Massen et al. (92) and Wolsky et al. (87) have considered some of the important effects of temperature inhomogeneities along beams and have demonstrated the importance of metalizing insulating beams to reduce these inhomogeneities. The magnitude of this problem depends, of course, on the effective thermal conductivity of the gas surrounding the beam.

If more than one material is used in the construction of a balance, bonding agents are frequently required. There are three materials that have been used to bond metals to nonmetals in the fabrication of microbalances: silver chloride, silver–platinum, and epoxy resin cements. Some of the pertinent properties of these and other possible bonding agents are shown in Table 3.8. The vacuum properties of AgCl have been studied and reviewed in some detail by Martin (83) and this and antecedent publications should be studied if this material is to be used in the microbalance system. The use of metal pastes and epoxy cements has not been fully explored and the technology of cements useful in joining refractory oxides and other materials is undergoing rapid development. Consequently, it is quite possible that a wider selection of materials can be used in microbalance construction than is apparent from the literature.

IV. SELECTION AND FABRICATION OF A MICROBALANCE

The selection of a particular microbalance depends primarily on both environmental and experimental considerations. For example, use of a microbalance in an ultrahigh vacuum environment requires that the instrument be capable of repeated bakeout at temperatures of 200–400°C while maintaining a constant deflection characteristic and calibration. Such a special environmental requirement could be satisfied simply by a proper selection of microbalance materials, but this will not be the only important consideration. Some regard must also be given to the method in which the microbalance will be operated and the particular experiment that is planned. For example, use of an adjustable torsion fiber to restore the microbalance to a null position is a simple procedure in the laboratory environment. However, the adaptation of this compensation scheme for operation in an ultrahigh vacuum environment is a difficult problem. Similarly, the particular experimental application is an important factor. If, for example, specimen mass changes are anticipated to be comparable to the specimen mass, a helical or cantilever balance with the required sensitivity would be a much better choice than a more complicated beam microbalance. Often it will be found that a variety of instruments will satisfy the experimental and environmental conditions of the proposed experiment and the selection of a particular instrument must be based on other factors such as the relative ease of fabrication and repair, operator experience, cost, etc.

In this section we will discuss a few of the significant environmental and experimental considerations and note some of the advantages and disadvantages of various microbalance systems. Finally, we will discuss some methods by which microbalances may be fabricated.

Table 3.7. Properties of Some Materials Used in Microbalance Construction

Material	Maximum density	Ultimate tensile strength, psi $\times 10^{-3}$	Elastic modulus, psi $\times 10^{-6}$	Yield or proof stress psi $\times 10^{-3}$	Torsion modulus, psi $\times 10^{-6}$	Damping capacity, $\times 10^{5}$
Tungsten (W) (drawn)	19.35	590 (48)	52 (47)	—	21.48 (49)	140 (47)
Iridium (Ir)	22.42	—	75 (47)	—	—	—
Molybdenum (Mo)	10.22	400 (49)	42 (47)	—	21.3 (49)	102 (47)
Stainless steel	~7.7	80–160 (47)	28–32 (47)	30–40 (47)	—	see (47)
Invar; 36% Ni–Fe	8.0	~100	~26 (49)	—	—	—
Duraluminum	2.8 (47)	~62 (24°C) ~4 (371°C) (47)	10 (49)	~54 (24°C) ~3 (371°C) (47)	3.98 (49)	1–20 (47)
Quartz (SiO_2) (fused)	2.2 (48)	7.0[a] (48)	10.4 (48)	—	4.5 (48)	—
Diamond	3.51 (48)	—	~170 (54)	—	—	—
Aluminum oxide (Al_2O_3)	3.965 (48)	to 37[b] sintered (51)	76 (54) (xtal) 49 (50) sintered	—	27 (25°C) sapphire	—
Magnesium oxide (MgO)	3.58 (48)	~15 (*RT*-800°C) (53) sintered	35 (54) xtal 30–40 (53) sintered	—	10–16 (53) sintered	
Beryllium oxide (BeO)	3.01 (48)	~15 (53)[b] sintered	~51 (53) (xtal) ~40 (53) sintered	—	~20 (53)	—

[a] See Table 3.6 and also text.
[b] See text.

Minimum creep, %/1000 hr	Coef. of linear expansion, × 10^6	Melting point, °C	Thermal conductivity, cgs units	Resistivity, μohm cm	Previous or (possible) use in microbalance construction
~6 (49) (filament)	—	3380 (47)	0.39 (100°C) (47)	5.5 (47)	Suspension fibers
—	5.7 (49)	2443 (47)	0.014 (100°C) (47)	5.3 (47)	Suspension fibers
—	4.9–5.5 (49)	2600 (47)	0.34 (100°C) (47)	5.17 (47)	(Suspension fibers)
—	9.6–12.1 (49)	—	0.045–0.065 (47)	70–100 (47)	Fibers
—	0.9 (49)	—	—	~80 (47)	Beam
1% at 18,000 psi, 190°C	22.5 (47)	—	0.35–0.38 (100°C)	5.0–5.3 (47)	Beam
—	0.55 (48)	~1665°C (soft)	0.0033 (48)	—	Beam, suspension fibers
—	1.18 (49)	> 3500 (48)	—	—	Pivots
~0.01% at 20,000 psi psi 1000°C sintered (52)	8.7 (49)	2015 (48)	0.0723 sintered (48)	~10^{22} (14°C) (48)	(Beam suspension fibers)
~0.01% at 1200 psi 1100°C (53) sintered	9.7–11.4 (49)	2800 (48)	0.0860 (48)	~10^{14} (850°C) (48)	(Beam suspension fibers)
—	—	~2530 (48)	0.525 (48)	~10^4 (600°C) (48)	(Beam)

Table 3.8. Possible Bonding Agents for Microbalance Construction

Bonding agent	Application	Vapor pressure, torr	Maximum useful temp. °C	Ref.
Araldite epoxy resin AY105 and hardener HT972	W to dural	$\leq 10^{-6}$ (at 100°C after outgas)	—	60,61
AgCl	W to quartz	est. $\sim 10^{-7}$ (300°C)	<455	e.g. 62,63, 83,84
Varian torr seal	Metals or nonmetals	$\sim 10^{-8}$ (25°C)	~200	90
GE silicone resin GE SR-82	Metals or nonmetals	$< 10^{-10}$ (25°C)	~400	64
Metal pastes	Metals or nonmetals	est. $< 10^{-8}$ (25°C)	~400 est.	—

A. Environmental Considerations

Some of the important environmental considerations in the use of a microbalance are the operational temperature, bakeout requirements and outgassing properties, utilization in ultrahigh vacuum systems, sensitivity of balance materials to residual atmospheres, and background vibrations present in the laboratory area. A few important factors pertaining to the outgassing and resulting ultrahigh vacuum characteristics of various materials and bonding agents may be inferred from Tables 3.7 and 3.8. Ultrahigh vacuum may be achieved by a variety of techniques and extended low temperature bakeout permits a somewhat greater latitude in the selection of materials and bonding agents. However, the introduction of any relatively high vapor pressure material into an ultrahigh vacuum system may result in specimen contamination which seriously interferes with any experiments that are performed. Considerable care must be exercised in the choice of materials if any ultrahigh vacuum experiments are contemplated.

Any reaction between microbalance materials and the residual atmosphere can lead to apparent sample mass changes, or changes in several of the fundamental parameters governing balance operation. Such reactions may not only result in new chemical phases but one should also consider the effects of gas adsorption, absorption or dissolution depending on the porosity and other characteristics of the microbalance materials.

Another important environmental condition which must be considered is that of vibrations present in the laboratory area. It has been the author's experience that many mechanical devices in laboratory areas produce vibrations in the 25–50 cps frequency range. If the acceleration level at the balance location is of the order or greater than 10^{-2} g, serious difficulties can be expected in operating almost any microbalance, regardless of its deflection sensitivity. In a brief survey conducted by the author, it was found that when accelerations were less than approximately 10^{-4} g in the 20–50 cps frequency range, a sensitive beam microbalance could be operated at nearly its maximum resolution. In agreement with Czanderna (Chapter 2) it was also found that use of laboratory space on upper floors of most buildings seriously degrades the resolution of the balance. One should carefully assess this problem before establishing elaborate facilities for experimental microbalance work. If a seismic accelerometer is not available for a proper survey of potential experimental sites, a crude indication of the vibration level can be obtained using a beaker of water or mercury supported on a stable base. If any ripples or standing waves can be seen on the liquid surface, the area is probably unsatisfactory for use as a microbalance laboratory. (See Chapter 4 for additional discussion of vibrational problems.)

B. Experimental Considerations

Experimentalists interested in direct microweighing techniques should select an appropriate microbalance only after a detailed consideration of the experiment. One should consider the sample mass and the relative mass change that might be expected for the entire experiment as well as the resolution with which the mass change must be monitored. If the expected mass change represents a significant fraction ($\sim 10\%$) of the sample mass, one of the relatively simple spring or cantilever microbalance systems may be appropriate, as previously mentioned. Frequently relatively small mass changes ($\sim 0.1\%$) are expected and one of the beam microbalances will be more appropriate. Use of balances with greater deflection sensitivities implies that the minimum time between measurements will be increased, since the period is proportional to the square root of the deflection sensitivity (equation 3.24). Microbalances with greater deflection sensitivities are generally more fragile and successful operation requires considerably more experience and skill on the part of the experimenter.

C. Some Advantages and Disadvantages of Various Microbalances

The advantages and disadvantages of cantilever and spring microbalances are more concisely stated than in the case of other microbalance systems. Each of these microbalances has the possible advantage of simplicity, ruggedness, and a capability to be used for mass changes that are large in comparison to the sample mass. The principal disadvantage with the helical and cantilever balances is that both have a relatively low deflection sensitivity and the resultant system sensibility is correspondingly low. Mass changes of a few parts in 10^5 or 10^6 can be resolved with helical balances, as with the modern adaptations of Madorsky (65) and Moreau (66). Long-term calibration stability and repeatability are frequently excellent.

In discussing some of the advantages and disadvantages of various kinds of beam microbalances, we will distinguish between three types, depending on whether the beam is supported by horizontal or vertical fibers, or by a pivotal support. The more important factors related to these three microbalance types have been noted below after consulting investigators with several years experience in the operation of each particular balance.

1. Horizontal Fiber Suspensions

(*a*) ***All-Quartz Construction.*** The calibration of these balances has been known to be stable over periods of years and measurements are generally repeatable to the sensitivity limit. If a trussed beam construction is used, the calibration is quite insensitive to load. All quartz construction readily lends itself to ultrahigh vacuum application. It should be mentioned that repair of these balances requires some measure of skill and experience. Also, the use of an insulating material such as quartz necessitates the consideration of electrostatic charging and appropriate shielding and/or metallizing to minimize these effects. Untrussed beam microbalances require less space but have deflection sensitivities that are markedly load dependent.

(*b*) ***Wire Suspensions.*** Horizontal fiber suspensions employing tungsten fibers, for example, have the distinct advantage of being rugged and capable of withstanding shocks that would fracture the relatively fragile quartz fiber suspensions of the same geometry. Construction and repair are also simplified. If wire fibers are used, cements or other relatively high vapor pressure materials may limit the vacuum conditions or the cleanliness of the system in which the balance is used. (There are, however, other alternatives to use of cements which are discussed in the section dealing with fabrication.) Untrussed beam balances using wire suspensions are also subject to load dependent deflection sensitivity.

2. Pivotal Suspensions

Pivotal microbalances are perhaps the most sturdily constructed instruments in use today. Their construction is suitable for ultrahigh vacuum operation. They are capable of supporting relatively large loads and operating with deflection sensitivities comparable to those of horizontally suspended beam microbalances. The tungsten pivots must be prepared for use by initially loading the balance so plastic flow occurs at the supporting tungsten tips (43). This properly shapes the tips for subsequent operation at reduced loads. Other investigators have suggested the use of diamond, sapphire (67) and silicon carbide (94) as pivots. It has been found necessary to limit the angle through which these balances deflect in order to preserve stable zero point readings. When a pivotal balance swings into contact with a stop, one or both of the supporting tips shifts to a slightly different position in its cup resulting in a displaced balance reading when the null is restored. This movement of the supporting tips can be minimized by locating the beam stops at the center of percussion. Proper limiting of the deflection can reduce zero point deviations to about one part in 10^7 of the suspended mass. Additional information concerning zero shifts in pivotal balances may be found in the text associated with Figure 12.3 of Chapter 11.

3. Vertical Suspensions

Although vertical suspension microbalances are infrequently used, instruments of this kind have been fabricated with extremely high deflection sensitivities ($\sim$2.7 deg/μg) (44). The load capabilities of these balances will also be greater for a given fiber than that for horizontally suspended beams. Accordingly, microbalances utilizing vertical suspensions are useful in providing information concerning relatively smaller mass changes of the sample. These microbalances are fragile and fractures of the all quartz instruments can only be repaired by experts in the fabrication of such systems.

D. Fabrication of Microbalances

The fabrication and assembly of microbalances can often be accomplished by those with modest experience and dexterity. Among the important factors in successful construction are adequate preparation, proper equipment, and an appreciation for some of the important operational principles. Important tools in quartz balance fabrication include a microtorch, a universally mounted binocular microscope with long focal length objective lenses and reticuled eyepieces, cobalt or other glasses suited for viewing incandescent molten quartz, a variety of small tweezers, cutters, etc., and a working area surfaced with refractory material. Use of jigs to aid in the construction and adjustment is essential and significantly simplifies the task. It has been the author's experience that few glassblowers, for example, have had extensive experience in sealing small quartz fibers. Consequently, it is reassuring to the experimenter to develop some facility for repairing and modifying microbalances to suit particular applications. A good deal has been written about various aspects of microbalance construction and, if the use of quartz is contemplated, the reader is referred to the accounts of Moreau (66), Kirk and Schaffer (68), dealing with quartz helix balances, Neher (59) on the uses of quartz, Edwards and Baldwin (69) on quartz balances and fibers, Gulbransen and Andrew (70) on simple beam balances of quartz, Czanderna (71) on quartz pivotal balances, and Chappell (72) and Pettersson (44) on quartz torsion balances. Gulbransen and Andrew (70) have also described beam balances of invar in some detail. Langer (73) and Poulis et al. (67, 94) have discussed the use of various materials in pivots of pivotal balances. These representative and practical accounts of microbalance fabrication provide considerable information. Some helpful supplementary information may also be found in the following sections. Since there have been no detailed accounts of the construction of quartz trussed beam microbalances with quartz fiber suspensions, the author's procedure has been included in Section IV.D.4.

1. Spring Microbalances (Quartz or Tungsten)

These balances may be obtained from several commercial sources* or they can be fabricated by an experienced glassblower. W. A. Robertson (74) of Sandia Laboratory has recently developed an advanced technique that results in very uniform quartz helices of high quality in a single step. A stainless steel mandrel of the desired helix diameter is mounted in a lathe and rotated at a peripheral speed of approximately 10 cm/sec. A quartz rod approximately 0.5 cm in diameter is mounted with two torches (see Figure 3.5) on the longitudinal feed. In addition, the quartz rod is also fed transversely. The spiral is initiated by attaching a small rod of quartz to the mandrel from the 0.5 cm source rod and starting the three feeds after adjusting the torches. With a little experimentation it is possible to produce helices with fiber diameters as small as 100 μ varying less than 10% along the entire length of the mandrel. This method produces helices of considerably greater strength than those in which fibers are first drawn and subsequently annealed on cylindrical forms.

Tungsten spring microbalances have been fabricated by winding techniques such as that of Madorsky (65) and have the advantage of ruggedness. Of course, the specific mass and the relative torsional moduli (see Table 3.7) of various materials are worthwhile considerations in the design of helical balances, since self-deflection is an important consideration.

* See Chapter 14.

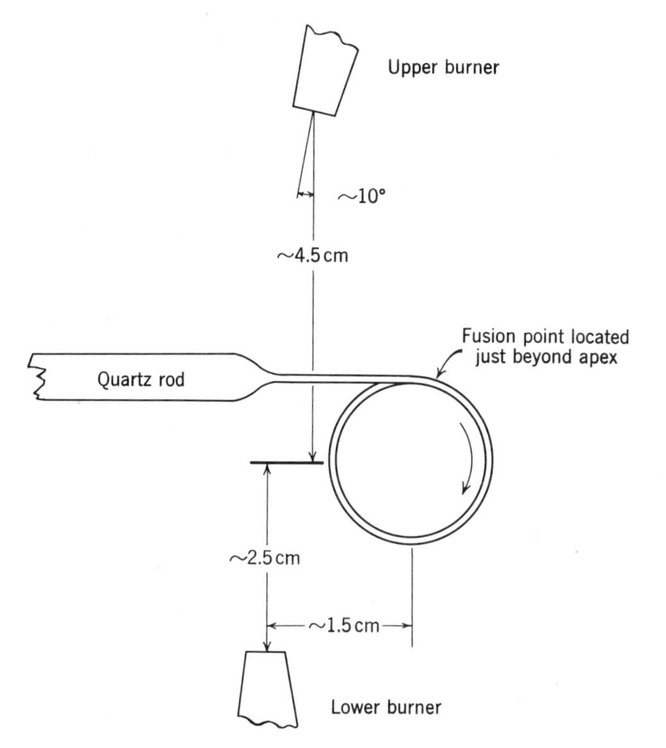

Figure 3.5. Very uniform quartz helices of extended length may be formed using a technique developed by W. A. Robertson. Fibers as small as 100 μ are drawn onto a rotating mandrel from the molten tip of a quartz rod translated in a lathe bed.

2. Cantilever Microbalances

Fibers useful in cantilever balances may usually be fabricated by simple drawing techniques unless the experimenter has some very special requirements. Neher (59) has described these techniques in considerable detail. With some practice, fibers of useful length and as small as 20 μ in diameter can in turn be "blown" from these fibers by holding them in a vertical flame, a technique familiar to glassblowers and again described in some detail by Neher.

3. Simple Beam Microbalances

The modern design of this type of microbalance is due primarily to Gulbransen (62). Construction is facilitated by use of an elaborate system of jigs through which the physical configuration of the beam and various fibers is precisely established. An excellent series of photographs showing the important jigs and detailing the constructional procedures has been published by Gulbransen and Andrew (70). Wolsky and Zdanuk (93) have also published a detailed description of the preparation of the precision grooves that are required in the beam. Those interested in building this type of balance will find these accounts to be most helpful. As can be inferred from the geometry of this balance, the important parameters determined by the jig system are the offset of the beam and the precise placement of each fiber with respect to one another. Typically, this balance utilizes tungsten fibers which are cemented to the quartz beam with AgCl. If the use of AgCl is not compatible with the intended use of the system, there are other suitable means of fusing quartz to tungsten. One possibility is use of low vapor pressure epoxy sealing materials commonly used to seal small leaks in ultrahigh vacuum systems. Of course, the maximum bakeout temperatures are also limited with most of these materials. A more satisfactory solution is to use GSC-4 grading glass and seal the tungsten fiber directly to the quartz. GSC-4 tubing is drawn into a capillary with an inside diameter such that it can be slipped over the tungsten fiber. The drawn and cut length of tubing should be adequate so that it will shield the fiber from the direct flame of the microtorch used for glassing the fiber. The glassed W fiber can then be butt sealed directly into the quartz beam. This technique permits use of somewhat more durable W fibers with a quartz beam and the bakeout temperature is now only limited to that for Pyrex.

4. *Trussed Beam Microbalances*

The fabrication of quartz trussed beam microbalances can be divided into two main steps. First, the beam framework and support are fabricated, and secondly, the load supporting hooks and fiber suspension are added to complete the microbalance assembly. Each step of the construction is facilitated by the use of appropriate jigs. These jigs have been fabricated from a variety of materials, e.g. graphite (Czanderna) or aluminum (Schwoebel) and the method of use in each case is quite similar. The fabrication of pivotal beam balances has been described elsewhere in some detail and a few of the important developments and associated publications will be noted in the succeeding paragraph. Fabrication of all quartz beam microbalances has not been described elsewhere and the author's technique, developed while at Cornell University, will be described in detail.

The modern form of pivotal microbalance, first described by Johnson and Nash (80), has subsequently undergone considerable development. Johnson and Nash incorporated tungsten tips supported in quartz cups for the pivotal support, as did the subsequent developments of Czanderna and Honig (39). Cochran (82) introduced the use of sapphire cups in the pivotal support. The development of pivotal supports was further extended by Poulis, Dekker and Meeusen (67) who have suggested the use of diamond tips supported on sapphire surfaces. They have determined the loading conditions under which such tips (18 μ radius) will not plastically deform the sapphire supports. These investigators have also examined some of the errors associated with use of pivotal supports as secondary fulcrums (at the beam ends) and concluded that concave supporting cups minimize errors associated with the relative positioning of the tips and tip supports. More recently, Poulis et al. (94), have studied the possibility of using supports of SiC and again have determined the elastic loading range over which such a support can be used. Czanderna (43) has published a very thorough and well-illustrated description of the construction of a pivotal microbalance using tungsten supports. Those interested in the constructional details are referred to this excellent account.

The technique used in our laboratory for the construction of all quartz microbalances with horizontal suspensions will now be described. Quartz cane is cut and mounted in grooves machined in the jig as illustrated in Figure 3.6. The ends are located at or near the center of drilled holes so that all are in contact with one another. When all rods are in place, a microtorch is used to heat and fuse each juncture. The surface tension of molten quartz is such that this operation usually proceeds in a fairly straightforward manner. The individual pieces of cane must be touching when heat is applied or melting will simply increase the separation. Small voids at joints can be filled in using the microtorch and a hand-held piece of cane. Then the entire joint is fused until the joint is smooth and well formed. The balance support assembly may be fabricated in an analogous manner. Usually a heavier quartz cane is used for this assembly to provide a comparatively massive structure on which to mount the beam. The design of the support assembly should include some provision for securely fastening the entire balance assembly in the experimental apparatus. The support should also include adjustable beam stops so the angular motion can be limited to a range consistent with the intended use. If the balance is to be operated as a null instrument, angular motion can be restricted to very small angles. In any case, the beam stops should be located near the center of percussion of the beam so that impulses to the suspension are minimized whenever the balance contacts the stops. This is especially important in reducing zero point shifts with pivotal suspension systems.

Assembly of the beam and support into a microbalance system requires use of a jig which will maintain the proper relative position of each component while the final and most difficult steps are completed. Minor adjustments in the completed balance are often necessary and it is convenient to design the assembly jig so that the balance may be tested without removing it from the jig. An example of such a jig used by the author is shown in Figure 3.7. In this jig, the upright supports may be moved away from the beam allowing observation of the unloaded and loaded periods. Period and sensitivity measurements are made by supporting a dummy mass on the left-hand side of the balance and suspending the compensator (a single

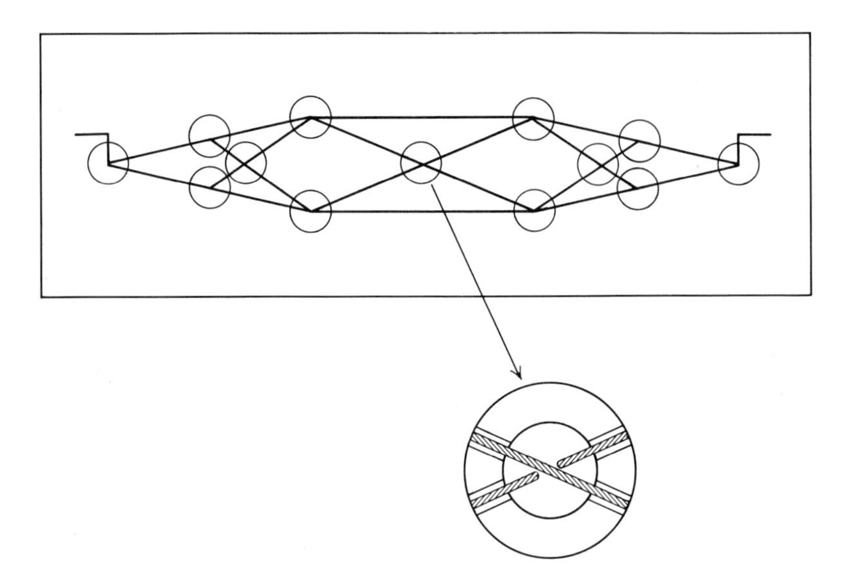

Figure 3.6. The fabrication of trussed quartz beams is readily accomplished using a template as illustrated in this figure. Quartz cane is mounted in grooves machined in an aluminium plate with all junctures located at holes that pass through the plate. Each juncture is then fused with a microtorch. This technique was developed at Cornell University in 1960–1962.

crystal of silicon-iron) on the right-hand side. The compensator hangs below the base plate in the field of a current solenoid. In this way one can rapidly check and adjust the balance for the desired period. This jig is used for the construction of horizontal quartz fiber suspension balances; however, the ideas involved would be essentially the same for other designs.

When the support and beam are mounted in the jig (Figure 3.7), the entire assembly is rotated 90° about axis AA' and a suspension fiber is drawn from the center of the lower side of the beam. The author draws these fibers by first fusing a length of quartz cane weighing about 0.5 g to the center of the beam. The cane is thinned near the juncture to a diameter of approximately 0.05 cm. The torch is then used to heat the zone immediately above the thin region such that the suspended cane abruptly falls approximately 4 in. In the course of this motion, the suspended quartz cane draws a straight, uniform fiber perpendicular to the plane of the beam. With a little practice the appropriate fiber size (approximately 25 μ in this case) can be obtained. When a fiber of the desired diameter has been drawn, the suspended quartz cane is separated from the fiber and the same process repeated on the other side of the beam after inverting the assembly jig.

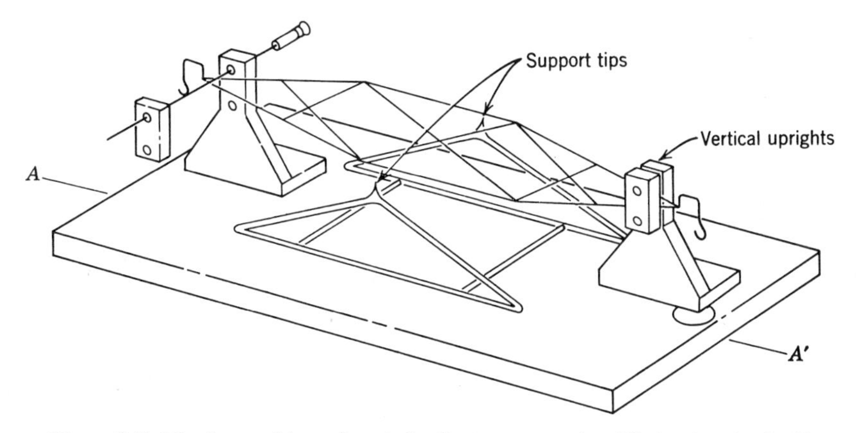

Figure 3.7. Final assembly and period adjustment are simplified using the jig illustrated here. The vertical uprights hold the beam in position until the quartz fibers have been drawn from the center of the beam and fused to the support tips. After the hooks have been fused to the beam ends, the uprights may be moved and the period measured and adjusted by suspending the compensating element through the hole in the base plate.

After the fibers have been successfully drawn from the beam, the assembly jig is returned to the normal orientation and the support elevated into position for fiber attachment. The support assembly (Figure 3.7) is located such that the fiber support tips are brought into contact with the fibers. After adjusting the binocular microscope so the fiber and fiber support tip are clearly visible, the microtorch is carefully brought near the end of the fiber. As the end of the fiber melts, it will immediately form a ball. This molten ball is advanced slowly toward the fiber support tip by proper manipulation of the torch. When the ball contacts the support tip it will cool since the tip acts as a large heat sink. The flame of the torch is very carefully applied to the support tip on the side away from the fiber. After a few seconds one can observe the ball fusing with the support tip and the junction between the fiber and the support tip moving to a position nearly opposite the beam. A common error is to apply too much heat during this step. The result is that a segment of the fiber nearest the support tip anneals and is considerably weaker than the remaining fiber. Heat should be applied only until the ball is fairly well incorporated into the support tip and then removed.

When the suspension fibers have been fused to the support, the balance can be quickly checked to determine the important operating characteristics. Perhaps the best test at this stage is to load the balance with masses nearly equivalent to those to be used in the experiment. If, even by the addition of individual microgram weights, the beam cannot be brought into some state of balance and the beam is either at one stop or another, the balance is unstable and one of the sample hooks must be lowered. On the other hand, if a balance can be established but the period is too short or too long, one of these hooks must be elevated or lowered, respectively. It is usually simplest to first adjust the vertical position of one hook until the balance is slightly unstable. The lowering of this hook can be performed using a reticuled microscope by heating the hook support until some slight sagging occurs (e.g. 20 μ). The instrument is then alternately tested and readjusted until it has the desired period. Raising or lowering the position of the support hook has the effect of decreasing or increasing the value of x in equation 3.23, the expression for the period of a gravity type beam microbalance.

V. MICROBALANCE OPERATION

In earlier sections of this chapter the primary concern has been with the important principles of microbalance operation and construction. Of equal importance is the procedure by which the deflection of the microbalance is related to a mass change. A variety of schemes have been devised for this purpose, from simple optical observations of deflection to sophisticated servo links which provide velocity damping and automatically restore the balance to a null position. Selection of a particular compensation technique depends in large measure on the nature of the experiments. For example, an extended series of routine experiments may clearly warrant the use of an automated system while less routine observations may best be handled by manual operation. The following discussion is divided into four major portions. These sections deal with deflection detectors, compensation and null operation, system sensibility or resolution, and balance calibration.

A. Deflection Detectors

Microbalance deflection has been detected by use of differential transformers (37,75,76) differential capacitors (77,30,33) angular motion of reflected light beams, intensity changes in partially shuttered light beams, optical microscopy, the observation of optical interference patterns, etc. The use of light beams represents a passive mode in which no force is introduced in the microbalance system by the observational technique. The observation of deflection using such an optical technique has been employed by a large number of investigators because of its simplicity and because it is readily compatible with direct deflection measurements as well as both manual and automatic null operation. In contrast to this technique, differential capacitors or transformers may exert some reaction force on the microbalance. These reaction forces must either be kept very small or they must be held constant to deviations less than the desired system resolution. The attainment of either of these two conditions must be carefully assessed when using position transducers which exert forces dependent on position.

B. Compensation and Null Techniques

The measurement of sample mass changes by direct observation of a deflection frequently represents a simple and appropriate scheme when cantilever and spring balances are used. These balances are generally used as deflection instruments; however, Griffiths (95), and Seanor and Amberg (96), have constructed magnetically compensated spring balances. Beam microbalances typically have much greater deflection sensitivities (in comparative terms of linear displacement of the beam end) and simple observation of deflection would severely limit the mass interval over which the instrument could be used. For example, if the microbalance used by the author were operated as a deflection instrument, mass changes could be observed only over a fifteen microgram interval, since the beam would be fully deflected for that mass change. The large deflection sensitivities obtainable with beam microbalances can best be utilized by operating the balance as a null deflection instrument and relating the compensating force to sample mass change. Null techniques also minimize errors due to nonlinearities in all balance systems. A variety of compensation schemes have been employed and three of these will be discussed here. Buoyancy and electrostatic compensation will be discussed briefly while more detail will be presented concerning magnetic compensation techniques.

Buoyancy compensation is effected by suspending or mounting a closed chamber of volume V at one side of the balance. Deflections resulting from mass changes of the sample may be compensated by varying the pressure of a gas of known density, ρ, in the balance chamber. If the gas pressure and temperature are known, the mass change corresponds to $\Delta\rho V$ at the null position. This compensation technique has limited application since, in many microbalance experiments, one is attempting to perform experiments either in an ultrahigh vacuum environment, or a gaseous atmosphere at a fixed pressure in which an alteration of the pressure may severely disturb or complicate the experiment. Buoyancy compensation is, however, an important means of calibrating microbalances as has been discussed by Honig (40).

Compensation may also be provided by purely mechanical means through the restoring torque of a fiber, for example. This method has been used for microbalance systems operated in the laboratory atmosphere. However, there are difficulties in adapting these techniques to balance operation in ultrahigh vacuum.

Electrostatic compensation has been employed by Mayer, Niedermayer (see Chapter 8 for experimental details) et al. (78) in which use is made of the repulsion or attraction between parallel charged plates. Neglecting edge effects, it can be shown that the attraction between two identical parallel plates charged by a voltage V will be

$$F = (\varepsilon/8\pi)(V^2A/d^2)$$

where ε is the permittivity of the intermediate media, A is the area of each plate, and d is the interplate spacing. If a differential capacitor arrangement is used in which plate A (attached to the balance) is positioned between two stationary plates B and C, the vertical force on the intermediate plate will be

$$F_p = \frac{\varepsilon A}{8\pi}\left(\frac{V_1^2}{x_1^2} - \frac{V_2^2}{x_2^2}\right)$$

where V_1, V_2 and x_1, x_2 are the voltages and spacings between plates AB and AC, respectively. Inspection of this equation shows that this method of compensation is statically unstable since a slight displacement of plate A from the equilibrium position toward B or C results in an increased force toward that plate. Consequently, a servo link may be required to maintain plate A in the balanced or null position. It is conceivable, however, that the stability of the balance can be adjusted so that this means of compensation may be employed using manual techniques. This condition will be attained by constructing the balance so the unbalancing torque due to the capacitance compensation is less than the restoring moments intrinsic to the balance itself.

Magnetic compensation schemes have been employed by several investigators (see Chapter 2) because it is simple and easily adaptable to either manual or automatic operation of a wide variety of balances. The usual arrangement is to suspend a quantity of some ferromagnetic material from one side of the balance in the field of a solenoid. Any balance deflection is compensated by varying the solenoid current and the current change is a measure of the sample mass change. If the balance calibration is to remain stable over long time intervals, it is important that the magnetic material have a low thermal and magnetic hysteresis. Addiss (26) examined a variety of materials for this purpose and found that both single crystals of aluminum kilned steel and silicon-iron ($3\frac{1}{4}\%$) were satisfactory. Magnetic counterweights of these materials showed no change in magnetic properties even after baking at 500°C. Similarly, magnetic hysteresis resulted in a calibration change of less than 2%. Addiss used rectangular rod shaped crystals ($0.014 \times 0.014 \times 1.0$ in.) of silicon-iron oriented with the [100] direction along the long axis. These small rods were mechanically and electrolytically polished, hydrogen annealed and finally sealed into glass tubing with a small hook at one end. The author has also used oriented single crystals of silicon-iron cut from plates of this material by an electron beam. No detectable changes ($< 1\%$) in microbalance calibration have been observed over an interval of 4 years, and after numerous bakeouts at 400°C. A variety of different magnetic materials have been used for this purpose by other investigators. However, the investigation of Addiss is the only reasonably complete study of the perturbing effects of thermal magnetic cycling on various magnetic compensators of which the author is aware. The important factor in utilizing any magnetic material may be that single crystals are used in preference to polycrystalline material. Then the effect of domain size changes induced by both temperature and field is minimized. In any case, the compensating material should be encased in glass or quartz tubing to insure that no chemical changes will occur due to other chemical species which may be present or introduced into the system.

The magnetic compensator typically is mounted rigidly within the beam or suspended from the balance. When the compensator is rigidly attached to the beam, one or more coils are positioned around the balance so that appropriate restoring moments may be applied to the beam. In the latter case, the compensator is often suspended from a fiber attached to one side of the balance and hangs in the center of a hangdown tube. A solenoid can then be slipped over the outside of the hang-down tube and properly positioned in relation to the compensator. If a simple solenoid arrangement is used in which the compensator moves along the solenoid axis, the relative position of the two can be adjusted to attain a stable equilibrium. That is, the restoring force will always be directed to an equilibrium position and the force increases with displacement from the equilibrium. Magnetic compensation has the inherent disadvantage that stray fields can perturb the balance and introduce significant errors in the results. For example, it is virtually impossible to use one of the magnetic sector type of residual gas analyzers in the close proximity of the compensator without damaging the balance or severely modifying the calibration.

C. Automatic Damping and Nulling Systems

Frequently investigators find that manual operation of a microbalance is satisfactory for a wide variety of experiments. In typical operation, a visual observation of a deflection is followed by manual adjustment of a compensating force to null the deflection and the mass change is determined by referring to the calibration curve. If manual operation is employed, the operator must cope with oscillation of the balance and the time required to reestablish the null will be at least of the order of the balance period. Manual operation is greatly simplified by utilizing automatic damping devices. Effective damping systems are relatively simple to design and incorporate into microbalance systems and their usefulness more than justifies the effort involved. A very simple damping system has been employed by Mayer et al. (78) in which a nonmagnetic conductor is suspended from the balance in the field of a permanent magnet. Motion of the conductor generates eddy currents opposing the motion and producing velocity damping. Those interested in employing this technique should consider, however, the mass of the conductor (preferably encased) compared to the balance capacity, the effect of field due to the permanent magnetic fields on any other magnetic components attached to the microbalance, and the problems associated with establishing the proper physical arrangement of the conductor and permanent magnet.

Another damping system which is somewhat more flexible and easily adaptable to microbalance systems operates by differentiating a signal from a displacement transducer such as an optical beam splitter. This is illustrated in the damping circuit portion of Figure 3.8. The differentiated signal, which is proportional to the angular velocity of the beam, is amplified and used to modify the compensating solenoid current and damp the motion. The amplifier gain determines the damping constant and can be adjusted for a wide variety of operating conditions. The dc amplifier should be of the null indicating type with low drift, stable zero, and high output impedance. A number of such amplifiers are available commercially. An analogous system has been described by Mauer (79).

Fully automatic microbalance systems may be of special value for extended experiments and two essentially equivalent systems have been used for this purpose. One of these utilizes a high gain feedback loop in which an error signal is amplified and the amplified signal used as the compensating force. The second of these automatic systems consists of an electromechanical servo loop which varies one of the compensating circuit elements until a null signal is obtained. Consider the general equation of motion,

$$\ddot{\theta} + A\dot{\theta} + B\theta = 0$$

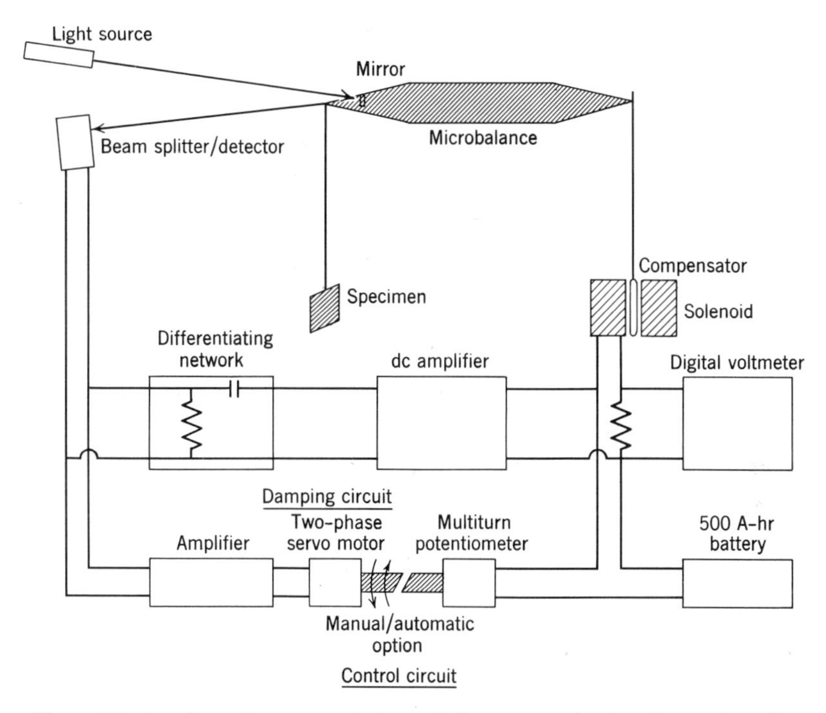

Figure 3.8. A schematic representation of the automatic damping and nulling system employed by the author. A beam splitter consisting of two photocells in a bridge circuit provides a signal proportional to the displacement of the microbalance from the null position. The displacement signal operates a control network to null the displacement, and the derivative of the displacement signal (angular velocity of the beam) is used to control the damping current.

in which θ is the deflection angle, A is a damping constant, and B is proportional to the restoring torque. A particular solution is that

$$\theta = e^{-At/2} \cos [(B - A^2/4)^{1/2}t]$$

when $B > A^2/4$. The period of the microbalance *system* is

$$T = 2\pi(B - A^2/4)^{1/2}$$

In the high gain feedback systems, it is frequently found that $B \gg A^2/4$ so that a displacement of the balance from the equilibrium position results in an oscillation about the equilibrium. In this case B is the sum of the restoring torques due to the balance plus a (large) term due to the compensating network. Consequently, the frequency of oscillation will be greater than that expected from a stable microbalance without compensation. When an unbalance is caused by sample mass change, the error signal of the displacement detector is amplified and used to return the balance near to, but not at, its original position. The new equilibrium position is displaced from the original position by an amount proportional to the inverse of the loop gain. Damping can be provided by proper phasing of the compensating force. Cahn and Schultz (85,86), Gruber and Shipley (88), Moret and Louwerix (36), Van den Bosch (37), and Langer (73), among others, have designed and discussed this kind of control circuit.

An electromechanical servo loop used by the author (Figure 3.8) has been designed so that $B \sim A^2/4$. That is, the system can be adjusted so that it is nearly critically damped and any displacement from the null position results in changes of the compensation force which restores the balance to the *same* null position. There are essentially no oscillations about the null and deviations from zero decrease exponentially with time. In this apparatus, the angular position of the balance is detected optically using a beam splitter. A difference signal proportional to any deviation from the null position is amplified and activates the control winding of a two-phase servo motor. The motor in turn drives a multiturn potentiometer which controls the current flowing through the compensating solenoid. Variable damping is provided by using the differentiated error signal and applying a damping voltage directly to the compensating solenoid. In this system, the damping is electronic while the set-point is reestablished by the electromechanical servo. Quite similar control systems have been previously described by Cochran (76), Moreau (66), and others.

If it is assumed that the resolution of the position transducer is infinite, both high-gain loop and the electromechanical systems are equivalent. Finite transducer resolution will introduce an error into each system but of different size. In the case of the high-gain loop, the error due to limited detector resolution will be proportional to the product of the loop gain and the resolution limit. In the case of the electromechanical system the error will be proportional to the resolution limit and will be a constant for all measurements. The primary advantage of the high-gain loop system is its inherent speed and capability of following significant weight changes occurring in a time interval less than the balance period. The primary advantage of the electromechanical system is that manual and automatic operation are identical and transitions between the two modes introduce no shears or oscillations into the system.

D. System Sensibility or Resolution

In Chapter 2, a distinction was made between sensitivity and sensibility. The term "sensitivity" has been reserved for describing the response of the microbalance to a given upsetting force or amount. "Sensibility" is that term which describes the *system* response (including the balance, position transducer, compensating circuitry, etc.) to a given upsetting moment of force. The sensibility of a microbalance system, therefore, is the error attributed to a mass change occurring at the specimen and depends on the amplifier deadband, transducer error and resolution, compensation force measurement, calibration stability and accuracy, and a number of other factors.

The schematic of a null operated microbalance system illustrated in Figure 3.9 is helpful in discussing possible sources of errors associated with each of the loop elements. A number of possible sources of error may be attributed to each segment of the measurement process. The kind

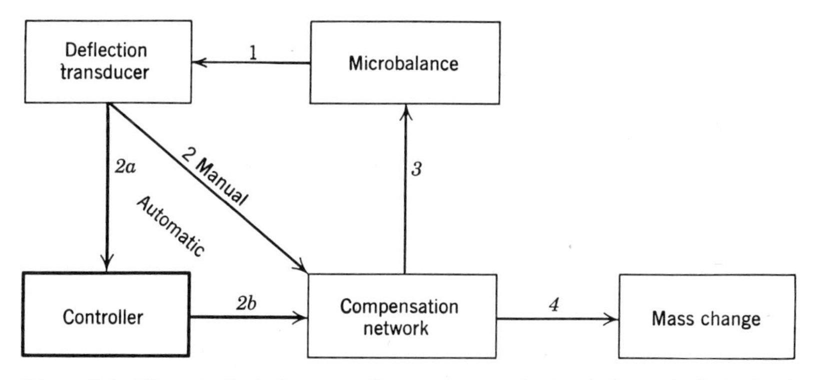

Figure 3.9. The principal elements of a compensated microbalance include the deflection transducer, controller and compensation network. Some of the principal sources of error introduced by each portion of the system are noted in the text.

of errors encountered at each step may be quite different in origin. For example, one could contrast the *random* error introduced by the drifting output of power supplies, the *systematic* error introduced by an erroneous calibration, and outright *blunders* occurring in manually recording the instrumental data. Similarly, it is important to recognize that some sources of error may be *dependent* and an elementary treatment of standard deviations assuming independence is not always appropriate.

The schematic (Figure 3.9) can be used to note the principal elements of a manual and/or automatic microbalance system. The microbalance response (*1*) to an upsetting moment is detected at the transducer. The deflection is returned to zero (if null operation is used) either by manual (*2*) or automatic (*2a, b*) adjustment of the compensation network. The compensation network (*3*) imposes countering moments on the microbalance so the deflection is nulled and a new equilibrium is established. The changes imposed on the compensation network are a measure (*4*) of the mass change occurring at the sample. Errors in the balance response to a given mass change may arise from (*a*) convection of gases in the microbalance enclosure, (*b*) temperature differentials, (*c*) mechanical vibrations in the laboratory and cross mode coupling, (*d*) electrostatic effects, (*e*) Brownian motion, (*f*) changes in the mechanical properties of the suspension fibers, etc. Errors introduced by the position transducer, regardless of the details of the device, are perhaps principally due to drift and limited resolution. If manual control is used one can consider approximate values for these errors and proceed to a consideration of the compensation network. If automatic control is used, errors arising from amplifier drift, stability, deadband, etc., must be included before proceeding to the compensation errors.

The compensation network may introduce uncertainties because of (*1*) variations in power supply voltages, (*2*) slight position changes in the compensation elements (solenoid, condenser plates, etc.), (*3*) hysteresis or temperature effects in the counterweight. Changes in the compensating circuit (measured in terms of, for example, current changes) are then associated with a mass change of the suspended sample using the calibration. The microbalance calibration is, of course, subject to time, temperature, and load effects which are distinct from the calibration errors associated with the use of improper standards. (See Appendix for discussion of calibration procedures.)

This listing of some possible sources of error emphasizes the fact that precise microbalance measurements require attention to detail and care in minimizing these and other errors. Furthermore, it is clear that statements regarding system resolution or sensibility must be based on more than a simple estimate of, for example, what deflection of the balance can be detected. A microbalance system may be able to resolve a very small mass change, but the accumulated drift of the system may be equivalent to a significantly greater mass change at some later time. If this is the case, and it frequently is, it is obvious that the system sensibility must be revised. In Chapter 4, errors associated with microgravimetry are discussed in detail.

VI. SUMMARY

The advanced design of microbalances is contingent on a host of considerations, some of which have been outlined in this chapter. While it would be convenient to have a format by which these instruments could be designed, it is perhaps quite clear that such a prescription could only be of the most general kind. It is important that the designer first establish the primary objectives relative to the experiment, and then proceed using the theoretical, material, and operational relations consistent with these objectives. The goal of this chapter has been to discuss and highlight some of the

UNIVERSITY OF VICTORIA
LIBRARY
Victoria, B. C.

important ideas pertinent to each of these areas. In the theoretical section, important relations have been derived for each of the three principal types of microbalances. This has included the derivation of relations that are particularly useful in the design and construction phases, as well as some of those relations describing the characteristics of completed instruments. In the following section, the properties of materials frequently used in microbalance construction have been compiled and compared to less frequently used materials as well as those which may hold some particular promise in future microbalance applications. A wide variety of material properties have been stated so that numerical computations concerning proposed balances can be made in conjunction with the section dealing with the theory of operation. The selection of a suitable type of microbalance is dependent not only on theoretical and material considerations, but also on important environmental and experimental considerations. Several of these conditions have been mentioned with their accompanying limitations in design, as well as some of the practical advantages and disadvantages of various microbalances. Some instructional examples of microbalance fabrication have been included and a number of excellent descriptions of microbalance fabrication have been cited. Important features of manual and automatic operation of microbalances have been discussed with an emphasis on deflection transducers, compensation techniques, and control systems. A variety of damping and control systems have been devised in recent years and the essential features of the two main categories of automatic systems have been described in general terms.

Each of these main topics could rather easily be expanded into an entire chapter and space limitations prohibit a broader and more extensive discussion. A choice and selection of topical material has been made in accord with these limitations, and with the author's intent to provide a fairly comprehensive introduction of these topics to those unfamiliar with microbalance techniques. Within this selection we have attempted to define important design areas and to briefly describe the critical factors and questions that will be of concern to those interested in microbalance design.

APPENDIX

A BRIEF NOTE ON THE CALIBRATION OF MICROBALANCES

The calibration of microbalances has been performed using a variety of techniques, but the use of calibrated microweights is perhaps the simplest and most frequently used method. This direct calibration method is relatively free of complicated sources of error and does not require auxiliary equipment as, for example, when a buoyance technique is employed. The authors of this book have been consulted concerning their individual calibration methods. These descriptions are generally quite similar and the essential elements of this technique are described in the following paragraphs.

1. A measured length of thoroughly degreased wire, usually tungsten, is weighed using a precision analytical balance and the average mass per unit length determined. Drawn tungsten wire is quite uniform in diameter, but the uniformity should be examined with a traveling microscope or other suitable means. In the case of 0.0025 cm diameter tungsten wire, a 2-m length will weigh approximately 20 mg, or the mass per unit length will be of the order of 100 μg/cm. The results of the mass, diameter, and length measurements should compare favorably with the theoretical density of tungsten.

2. Sections of desired length (e.g., 1 cm) are then cut from the initial wire sample and the length of each section accurately determined using a traveling microscope. These measurements can usually be made to about 10^{-3} cm without difficulty. At this point, it is convenient to separate individual microweights and keep them in thoroughly cleaned and labeled containers. The fractional error in the mass of each microweight is due to uncertainty in wire diameter, d, length, l, and computed density, ρ.

$$\frac{\Delta m}{m} \sim \sqrt{4\left[\left(\frac{\Delta d}{d}\right)^2 + \left(\frac{\Delta l}{l}\right)^2 + \left(\frac{\Delta \rho}{\rho}\right)^2\right]^{1/2}}$$

For example, the nonuniformity of the wire, $\Delta d/d$, may be the most significant term in this expression. If the nonuniformity is of the order of 1%, $\Delta m/m$ will be approximately equal to or greater than 0.02, or a 100 μg microweight will have a probable error of ± 2 μg.

3. It is possible to obtain nominal 100 μg weights from the mass section of the Bureau of Standards with the mass specified to within 1 μg. Each microweight can be compared to this standard by the following procedure. The microbalance is brought into midrange using a tare weight with uncompensated balances, and in the case of compensated balances, to a null position using a tare and compensation. Each microweight is then suspended and removed from the balance in turn, and the deflection or compensation for null noted in each case. Assuming for the moment that the microbalance is very nearly linear over the mass range of one of the microweights, the microbalance can be used to compare the relative mass of each designated microweight to the standard. These results should compare favorably with the previous calculations.

4. Following the calibration of each microweight, the tare weight (and compensation) should be adjusted so the balance is operating in the lowest portion of its range. Now each microweight is added in sequence and the corresponding deflection or compensation setting carefully noted. Similarly, the deflection or changes in compensation are noted as each microweight is removed in turn. During this procedure, it is convenient to use a multiple hook so one can easily distinguish between individual microweights. After a suitable number of points have been taken, a calibration curve can be plotted including the deviations associated with each point. It may be found that the calibration curve is not perfectly linear. In this case, the author finds it convenient to fit a second-order equation to the calibration points and to use this mathematical relationship in subsequent calculations. Actually the derivative of this curve is what is used in our computer program for processing experimental data. If the nonlinearity over the mass range equivalent to one microweight exceeds the uncertainty in the microweight mass, the calibration procedure should be repeated beginning with *3* (above) making appropriate comparisons between the standard and working microweights.

The calibration should be examined periodically and this can be done in a variety of ways. A qualitative calibration check that can be used during each experiment is to simply measure the period of the balance at some convenient point in the experiment. If the period compares favorably with previous observations, a detailed recalibration is probably unnecessary. A more precise check of the calibration can be made during the routine of sample changing between experiments. When a specimen is suspended from the balance, one can note the readings when a calibrated microweight is added to the specimen side of the balance. These two points can be compared to the calibration curve for consistency. If any appreciable deviations are noted, a complete recalibration should be performed. Most investigators perform a detailed calibration each year, regardless of the results of the less detailed calibration checks between and during experiments.

REFERENCES

1. A. Röever, "Beanspruchung Zylindrische Schraubenfedern mit Kreisquerschnitt," *Ver. Deut. Ing. Z*, **55**, 1907 (1913).
2. A. M. Wahl, *Trans. ASME*, 1929, and *Mechanical Springs*, Penton, Cleveland, 1944, Chapter 1.
3. O. Goehner, *Ver. Deut. Ing. Z.*, **74**, 272 (1932).
4. A. M. Wahl, Trans. ASME, 1929 and *Mechanical Springs*, Penton, Cleveland, 1944, Chapter 1.
5. F. Emich, *Montasch. Chem.*, **36**, 436 (1915).
6. J. McBain and A. Bakr, *J. Am. Chem. Soc.*, **48**, 690 (1926).
7. P. L. Kirk and F. L. Schaffer, *Rev. Sci. Instr.*, **19**, 785 (1948).
8. K. Hauffe and B. Ilschner, *Z. Elektrochem.*, **58**, 467 (1954).
9. P. W. Selwood, *Catalysis*, **1**, 353 (1954).

10. S. Madorsky, *Vacuum Microbalance Tech.*, **2**, 47 (1962).
11. C. Moreau, *Vacuum Microbalance Tech.*, **4**, 21 (1965).
12. A. Predwoditelew and A. Witt, *Z. Physik. Chem.*, **132**, 47 (1928).
13. O. Lowry, *J. Biol. Chem.*, **140**, 183 (1941).
14. E. Lord, *J. Textile Inst.*, **38**, T84 (1947).
15. T. N. Rhodin, *Advan. Catalysis*, **5**, 44 (1953).
16. E. Salvioni, Dissertation, Univ. of Messina (1901).
17. M. Bazzoni, *J. Franklin Inst.*, **180**, 461 (1915).
18. A. Friedrich, *Mikrochemie*, **15**, 35 (1934).
19. I. M. Korenman and Y. N. Fertelmeister, *Zavodsk. Lab.*, **15**, 785 (1949).
20. J. Donau, *Mikrochemie*, **9**, 1 (1931); **13**, 155 (1933).
21. H. Barrett, A. Birnie, and M. Cohen, *J. Am. Chem. Soc.*, **62**, 2839 (1940).
22. E. A. Gulbransen, *Trans. Am. Electrochem. Soc.*, **81**, 327 (1942); *Rev. Sci. Instr.*, **15**, 201 (1944); K. F. Andrew, *Vacuum Microbalance Tech.*, **1**, 1 (1961); *Advan. Catalysis*, **5**, 119 (1953).
23. P. J. Kirk, P. Craig, J. E. Gullberg, and R. Q. Boyer, *Ann. Chem.*, **19**, 427 (1947).
24. T. Rhodin, *J. Am. Chem. Soc.*, **72**, 4343 (1950); *Advan. Catalysis*, **5**, 39.
25. F. C. Edwards and R. R. Baldwin, *Ann. Chem.*, **23**, 357 (1951).
26. R. R. Addiss, "Oxidation of Magnesium Single Crystals and Evaporated Films," Thesis, Cornell Univ., Ithaca, N.Y., 1958.
27. R. L. Schwoebel, "Growth and Structure of Oxide on Single Crystals of Magnesium," Thesis, Cornell Univ., Ithaca, N.Y., 1962; *J. Appl. Phys.*, **34**, 2776 (1963).
28. T. J. Gray and T. J. Jennings (unpublished). See *The Defect Solid State*, by T. J. Gray, Ed., Interscience, New York, 1957, p. 486.
29. R. F. Walker, *Vacuum Microbalance Tech.*, **1**, 87 (1961).
30. N. J. Carrera, R. F. Walker, C. A. Steggerda, and W. M. Nalley, *Vacuum Microbalance Tech.*, **3**, 153 (1963).
31. O. M. Katz and E. A. Gulbransen, *Vacuum Microbalance Tech.*, **1**, 111 (1961).
32. L. Cahn and H. R. Schultz, *Vacuum Microbalance Tech.*, **3**, 29 (1963).
33. R. J. Kolenkow and P. W. Zitzewitz, *Vacuum Microbalance Tech.*, **4**, 195 (1965).
34. J. M. Thomas and B. R. Williams, *Vacuum Microbalance Tech.*, **4**, 209 (1965).
35. E. L. Evans, J. M. Thomas, and B. R. Williams, *J. Sci. Instr.*, **43**, 263 (1966).
36. H. Moret and E. Louwerix, *Vacuum Microbalance Tech.*, **5**, 59 (1966).
37. A. Van den Bosch, *Vacuum Microbalance Tech.*, **5**, 77 (1966).
38. V. A. Arslambekov, in V. I. Arkharov, and K. M. Gorbundua, Eds., *Surface Interactions Between Gases and Metals*, Consultants Bureau, New York, 1966.
39. A. W. Czanderna and J. M. Honig, *Ann. Chem.*, **29**, 1206 (1957).
40. J. M. Honig, *Vacuum Microbalance Tech.*, **1**, 55 (1961).
41. A. W. Czanderna, *Vacuum Microbalance Tech.*, **1**, 129 (1961).
42. A. W. Czanderna and H. Wieder, *Vacuum Microbalance Tech.*, **2**, 147 (1962).
43. A. W. Czanderna, *Vacuum Microbalance Tech.*, **4**, 175 (1965).
44. H. Pettersson, *Proc. Phys. Soc.*, **32**, 209 (1920).
45. A. G. Day, *J. Sci. Instr.*, **30**, 260 (1953).
46. K. H. Behrndt, *Vacuum Microbalance Tech.*, **1**, 69 (1961).
47. C. J. Smithells, *Metal Reference Book*, 3rd ed., Butterworths, Washington, D.C., 1962.
48. R. E. Weast, Ed., *Handbook of Chemistry and Physics*, 45th Edition, Chemical Rubber Company, 1964–65.
49. C. D. Hodgman, Ed., *Handbook of Chemistry and Physics*, 35th Edition, Chemical Rubber Company, 1953–54.

50. E. Ryshkewitch, *Ber. Deut. Keram. Ges.*, **23**, 243 (1942).
51. E. Ryshkewitch, *Ber. Deut. Keram. Ges.*, **22**, 54 (1941).
52. J. P. Roberts and W. Watt, *Ceramics and Glass*, Selected Government Research Reports, Ministry of Supply (London), **10** (1952); *Trans. Brit. Cer. Soc.*, **48**, 343 (1949); *Trans. Brit. Cer. Soc.*, **50**, 122 (1951).
53. *Refractory Ceramics for Aerospace*, Am. Ceram. Soc., 1964.
54. J. J. Gilman, "The Strength of Ceramic Crystals," in *Physics and Chemistry of Ceramics*, C. Klingsberg, Ed., Gordon and Breach, New York, 1963, p. 240.
55. F. O. Anderegg, *Ind. Eng. Chem.*, **31**, 290 (1939).
56. W. B. Hillig, *J. Appl. Phys.* **32**, 741 (1961).
57. S. S. Brenner, in *Growth and Perfection of Crystals*, R. H. Doremus, B. W. Roberts, and David Turnbull, Eds., Wiley, New York, 1958, p. 157.
58. E. Ryshkewitch, *Science and Technology*, Feb., 1962.
59. This table was published by H. V. Neher in Chapter 5, *Procedures in Experimental Physics*, J. Strong, Ed., Prentice-Hall, Englewood Cliffs, New Jersey, 1938.
60. E. L. Evans, J. M. Thomas, and B. R. Williams, *Rev. Sci. Instr.*, **43**, 263 (1966).
61. D. W. Aylmore, S. J. Gregg, and W. B. Jepson, *J. Inst. Metals*, **88**, 205 (1960).
62. E. A. Gulbransen, *Rev. Sci. Instr.*, **15**, 201 (1944); *Advan. Catalysis*, **5**, 120 (1953).
63. T. N. Rhodin, *J. Am. Chem. Soc.*, **72**, 4343 (1950).
64. J. R. Young, *Rev. Sci. Instr.*, **35**, 116 (1964).
65. S. L. Madorsky, *Vacuum Microbalance Tech.*, **2**, 47 (1962).
66. C. Moreau, *Vacuum Microbalance Tech.*, **4**, 21 (1965).
67. J. A. Poulis, W. Dekker, and P. J. Meeusen, *Vacuum Microbalance Tech.*, **5**, 49, (1966).
68. P. L. Kirk and F. L. Schaffer, *Rev. Sci. Instr.*, **19**, 785 (1948).
69. F. C. Edwards and R. R. Baldwin, *Anal. Chem.*, **23**, 359 (1951).
70. E. A. Gulbransen and K. F. Andrew, *Vacuum Microbalance Tech.*, **1**, 1 (1961).
71. A. W. Czanderna, *Vacuum Microbalance Tech.*, **4**, 175 (1965).
72. F. A. Chappell, *Vacuum Microbalance Tech.*, **2**, 19 (1962).
73. A. Langer, *Vacuum Microbalance Tech.*, **4**, 231 (1965).
74. W. A. Robertson, Scientific Glass Section, Sandia Laboratory, private communication.
75. A. H. Petersen, *Instr. Automation*, **28**, 1104 (1955).
76. C. N. Cochran, *Vacuum Microbalance Tech.*, **1**, 23 (1961).
77. J. H. Simons, C. L. Scheirer, and H. L. Ritter, *Rev. Sci. Instr.*, **24**, 36 (1953).
78. H. Mayer, R. Niedermayer, W. Schroen, D. Stünkel, and H. Göhre, *Vacuum Microbalance Tech.*, **3**, 75 (1963).
79. F. A. Mauer, *Rev. Sci. Instr.*, **25**, 598 (1954).
80. E. W. Johnson and L. K. Nash, *Rev. Sci. Instr.*, **22**, 240 (1951).
81. J. H. Simons, *Anal. Chem.*, **10**, 587 (1938).
82. C. N. Cochran, *Rev. Sci. Instr.*, **29**, 1135 (1958).
83. G. Martin, *Rev. Sci. Instr.*, **34**, 707 (1963).
84. W. Espe, *Werkstoffkunde der Hochvakuumtechnik*, Vol. III, Veb Deutscher Verlag der Wissenschaftsen, Berlin, 1961, pp. 335–338.
85. L. Cahn and H. R. Schultz, *Vacuum Microbalance Tech.*, **2**, 7 (1962).
86. L. Cahn and H. R. Schultz, *Vacuum Microbalance Tech.*, **3**, 29 (1963).
87. S. P. Wolsky, E. J. Zdanuk, C. H. Massen, and J. A. Poulis, *Vacuum Microbalance Tech.*, **6**, 37 (1967).
88. H. L. Gruber and C. S. Shipley, *Vacuum Microbalance Tech.*, **3**, 131 (1963).
89. T. Gast, *Feinwerktechnik*, **53**, 167 (1949).

90. Data Sheet 953-0001, Vacuum Products Division, Varian Associates, Palo Alto, California.
91. J. R. Thomas and B. R. Williams, *Vacuum Microbalance Tech.*, **4**, 195 (1965).
92. C. H. Massen, J. A. Poulis, and J. M. Thomas, *Vacuum Microbalance Tech.*, **4**, 35 (1965); *J. Sci. Instr.*, **41**, 302 (1964).
93. S. P. Wolsky and E. J. Zdanuk, *Vacuum Microbalance Tech.*, **1**, 35 (1961).
94. J. A. Poulis, P. J. Meeusen, W. Dekker, and J. P. de Mey, *Vacuum Microbalance Tech.*, **6**, 27 (1967).
95. D. Griffiths, *J. Sci. Instr.*, **38**, 463 (1961).
96. D. A. Seanor and C. H. Amberg, *Rev. Sci. Instr.*, **34**, 917 (1963).
97. J. Skelly, *Rev. Sci. Instr.*, **38**, 985 (1967).

Chapter 4

Instrumental and Environmental Factors Affecting the Sensitivity of Beam Type Balances

C. H. MASSEN AND J. A. POULIS

Physics Department, Eindhoven University of Technology,
Eindhoven, The Netherlands

Several centuries before the beginning of Christianity at that time when precious metal replaced cattle as a means of value exchange there was much concern about the accuracy of weighing in many an ancient marketplace. The unerring accuracy of weighing procedures was supported by theologians who roused an interest in the weight of abstractions. Their action culminated in the application of public weighing tests to those suspected of witchcraft.

Balance techniques have today become a separate and increasingly important scientific discipline. Abstractions or what are now more appropriately called spurious masses have become significant in the precise weighing of matter. Their importance is a logical result of the development of increasingly sensitive balances which allowed the detection of ever smaller mass variations. Some time ago the point was finally attained where the desired detectable mass variations became smaller than the spurious mass changes. An understanding of the nature of the spurious mass changes is necessary in order to utilize the microbalance in a meaningful manner.

In this chapter the more important disturbances resulting in spurious mass changes are discussed. Since there are many types of spurious disturbances associated with the large variety of experimental conditions under which balance measurements are made, it is impossible to provide a completely exhaustive discussion in a single chapter. The approach here has been to present simplified and idealized discussions wherever possible. The specific examples employed in this chapter have been chosen arbitrarily since a complete review of more representative situations would undoubtedly encompass a book rather than a chapter. These examples will, however, prove useful in illustrating the relative magnitude of such spurious effects under actual experimental conditions and the nature of the requirements for minimizing these effects.

I. BROWNIAN MOTION

The most fundamental cause of spurious mass changes is Brownian motion, i.e. the irregular movement of bodies due to thermal agitation. This effect is subject to analysis through the application of the law of

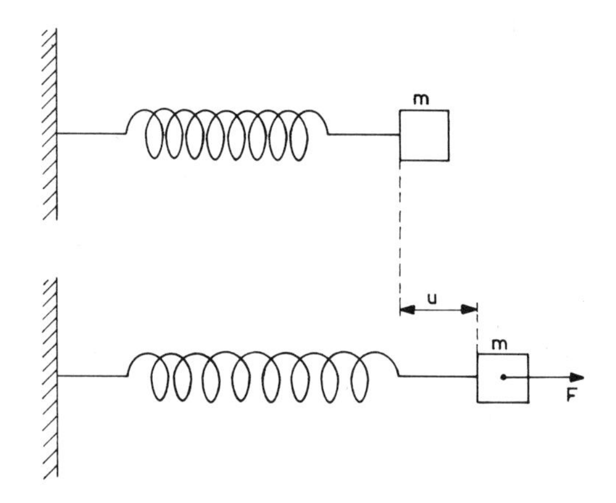

Figure. 4.1. Schematic of the spring balance for which the Brownian motion is calculated.

equipartition of energy which states that a body participates in the thermal motion in such a way that the mean energy in each degree of freedom is $\frac{1}{2}kT$. The following is an estimate of the magnitude of such irregular balance motions for the most simple form of a balance, i.e. the spring balance (Figure 4.1).

A body of mass m is connected to a massless spring with a known spring constant C. The force, F, acting on the body is related to the deflection or extension, u, of the spring as follows.

$$F = -Cu \tag{4.1}$$

The expression for the potential energy, ε, stored in the spring because of the deflection, u, is

$$\varepsilon = \tfrac{1}{2}Cu^2 \tag{4.2}$$

If u_B is the rms value of the irregular deflections caused by the Brownian motion, the law of equipartition reads:

$$\tfrac{1}{2}Cu_B{}^2 = \tfrac{1}{2}kT \tag{4.3}$$

When the deflection, u_B, is related to a spurious force, F_B, by equation 4.1,

$$(1/C)F_B{}^2 = kT \tag{4.4}$$

Equation 4.4 becomes more tractable by replacing C with an expression in terms of the period of oscillation, t_0, (equation 4.5) of the balance

$$t_0 = 2\pi(m/C)^{1/2} \tag{4.5}$$

This leads to:

$$F_B = (2\pi/t_0)(mkT)^{1/2} \tag{4.6}$$

Since there is little damping with the balance shown in Figure 4.1 and at least two reversal points have to be read in any one measurement, a time interval of the order of the period of oscillation is required. Using equation 4.6 and assuming $t_0 = 1$ sec, $m = 1$ g, $kT = 4 \times 10^{-14}$ erg (T approximately 300°K), the spurious force, F_B, is found to be of the order of magnitude of 10^{-6} dyne (approximately 10^{-9} g). Although this is only a rough estimate, a more detailed analysis dealing with practical types of balances shows that a 10^{-9} g mass variation is about the usual sensitivity limit (5,6).

Much attention is being directed to a type of balance which has a much higher senstivity, namely the quartz crystal oscillator (7,8). (See Chapter 5.) With the quartz crystal balance a sensitivity of 10^{-12} g seems to be obtainable. An explanation of this high sensitivity can be obtained by again considering Figure 4.1. Although both common balances and quartz resonators can be described by means of the same basic schematic representation, it must be remembered that the oscillation period of the two balance types differ by some six orders of magnitude. The mass of the sample on the crystal, included in the total mass m, is measured indirectly via the period of oscillation of the quartz crystal given by equation 4.5. Equation 4.5 is solved for m as follows.

$$m = \frac{1}{4\pi^2} Ct_0{}^2 \tag{4.7}$$

An analysis of the influence of the Brownian motion on the oscillation period will lead to the limit of the sensitivity. Figure 4.2 represents a simple crystal oscillator circuit. A high value of the electrical impedance of the crystal at resonance frequency insures selective amplification. The result is an oscillating voltage across the crystal, with a period, t_0, as shown in Figure 4.3. The mass determination is based on the measurement of the change of the frequency of this voltage. If the oscillation is perfectly harmonic, one can measure the time interval, t_p between oscillations and use equation 4.8 to determine t_0

$$t_0 = t_p/N \tag{4.8}$$

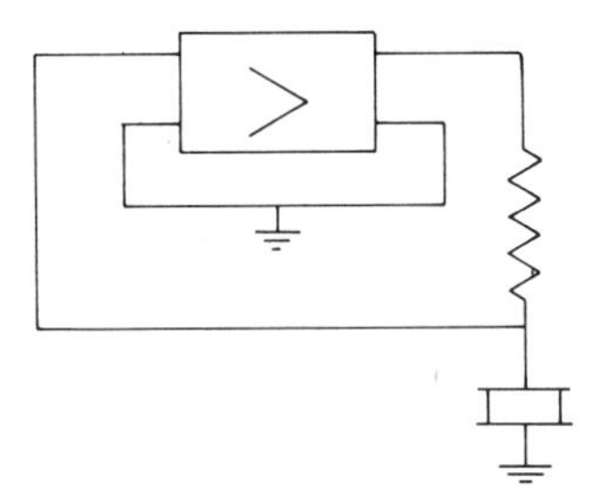

Figure 4.2. A simple oscillator using a resonating quartz crystal.

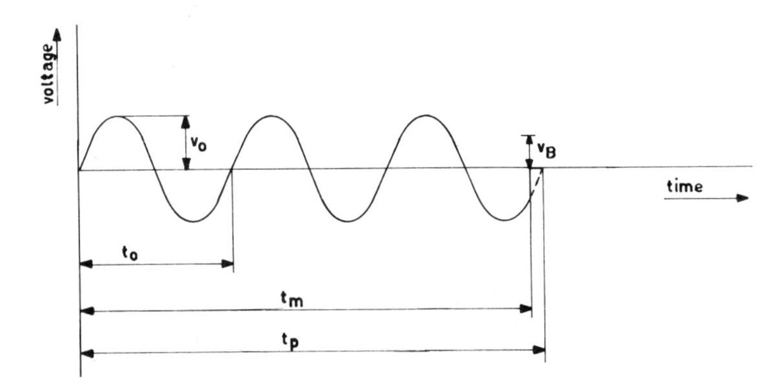

Figure 4.3. The voltage of a resonating crystal as a function of time and the influence of Brownian motion.

In reality, however, disturbances such as the Brownian motion of the crystal will cause the voltage to deviate from a perfect harmonic function. Remembering the schematic representation of the crystal oscillator (Figure 4.1), the irregular movements, u_B, of the mass as given by equation 4.3 can, with the use of equation 4.5, be expressed as follows:

$$u_B = \frac{t_0}{2\pi}\left(\frac{kT}{m}\right)^{1/2} \tag{4.9}$$

Since the quartz crystal is piezoelectric, the deflection u_B will give rise to a voltage, V_B, given by

$$V_B = \frac{1}{C_p} u_B = \frac{1}{C_p} \frac{t_0}{2\pi} \left(\frac{kT}{m}\right)^{1/2} \tag{4.10}$$

where C_p is the piezoelectric constant of quartz ($\sim 2 \times 10^{-10}$ cm/V). Figure 4.3 shows how the existence of the voltage resulting from Brownian motion can give rise to an error in the determination of t_0. Added to the perfectly harmonic voltage, V_0, V_B can cause the voltage to pass through zero at the measured time t_m instead of at t_p the time for a perfect harmonic oscillation. Using t_m for the calculation of t_0 the measured oscillation period t_{0m} is

$$t_{0m} = t_m/N \tag{4.11}$$

Comparing this with equation 4.8 the resulting relative error $\Delta t_0/t_0$ is

$$\frac{\Delta t_0}{t_0} = \frac{t_0 - t_{0m}}{t_0} = \frac{t_p - t_m}{N} \tag{4.12}$$

The numerator of this expression is easily calculated with the help of Figure 4.3. Assuming that V_B is much less than V_0, to a very good approximation

$$V_B = V_0 \sin\left\{2\pi \frac{t_p - t_m}{t_0}\right\} = 2\pi V_0 \frac{t_p - t_m}{t_0} \tag{4.13}$$

Equations 4.13 and 4.10 are used in equation 4.12 to obtain

$$\frac{\Delta t_0}{t_0} = \frac{1}{4\pi^2 C_p V_0} \frac{t_0}{N} \sqrt{\frac{kT}{m}} \tag{4.14}$$

Since $1/N = t_{0m}/t_m \cong t_0/t_m$, equation 4.14 reduces in a first-order approximation to

$$\frac{\Delta t_0}{t_0} = \frac{1}{4\pi^2 C_p V_0} \frac{t_0^2}{t_m} \sqrt{\frac{kT}{m}} \tag{4.14a}$$

It is seen from the second power in equation 4.7 which allows the calculation of m from t_0, that the relative error in the mass, m_B/m, will be twice the relative error in t_0 and therefore,

$$\frac{m_B}{m} = \frac{2}{4\pi^2 C_p V_0} \frac{t_0}{N} \sqrt{\frac{kT}{m}} \tag{4.15}$$

Finally using equation 4.14a, the error m_B resulting from the Brownian motion is found to be

$$m_B = \frac{t_0^2}{2\pi^2 C_p V_0 t_m} \sqrt{mkT} \tag{4.16}$$

Using as typical values $t_m = 1$ sec, $C_p = 2 \times 10^{-10}$, $V_0 = 1$ V, $kT = 4 \times 10^{-14}$ erg, $t_0 = 3 \times 10^{-7}$ sec, $m = 1$ g, it is found that $m_B = 5 \times 10^{-12}$ g, which value is in good agreement with the experimental results.

The high sensitivity of quartz oscillator balances is mainly dependent on having a high Q resonator but it also requires an amplifier in the oscillator circuit which can sense changes in the very short period of 10^{-7} sec. Electronic devices have, as has been discussed in Chapters 1 and 2, also been used with conventional balances to improve balance sensitivity (4).

II. KNUDSEN FORCES

This discussion is concerned with the effect of forces on the balance which originate from the surrounding gas. Buoyancy is the most obvious of these forces. Since the calculation of its magnitude is relatively simple if the gas pressure is known, buoyancy can hardly be classified as a problem of the type being considered in this chapter. A detailed discussion of buoyancy can be found in Chapter 2.

There are, however, specific situations where the effects of gas pressure require a more detailed analysis. We shall consider the typical situation in thermogravimetry (10–15) where there is a gas in a vessel the walls of which are not at a uniform temperature. The refinement of the conception of gas pressure necessary here has, as its most striking feature, the fact that the forces exerted by the gas upon the vessel are no longer only normal to the wall. The component parallel to the wall is usually called the longitudinal Knudsen force.

To illustrate the existence of longitudinal Knudsen forces consider a two-dimensional example of a single gas molecule moving inside a square vessel (see Figure 4.4). It is assumed that the molecule repeatedly follows the path *ABCDAB*. The schematic representation of a vertical temperature gradient ($T_C > T_A$) is embodied in the example by the different values of the velocity v_A, v_B, v_C and v_D the molecule has after striking the wall at *A*, *B*, *C*, and *D*, respectively.

The velocities can be expressed in term of a main part v_A and an additional part Δv such that

$$v_B = v_D = v_A + \Delta v \qquad \text{and} \qquad v_C = v_A + 2\Delta v \tag{4.17}$$

When Δv is small compared with v_A it can be inferred from equations 4.17 that the temperatures at *B* and *D* are half those at *A* and *C*. The calculation of the forces acting on the wall resulting from the motion of the gas molecule described above, will be restricted to the vertical component of these forces since the horizontal component is difficult to relate to the behavior of a real gas without including in the model a second molecule moving in the opposite direction. The contribution of each collision to the vertical force is shown in Figure 4.4. The vertical momentum transferred by the impact and the recoil of the molecule at B for instance is given by $\frac{1}{2}\sqrt{2}\, mv_A$ and $-\frac{1}{2}\sqrt{2}\, mv_B$, respectively. This results in a net vertical momentum transferred to the wall, which equals $-\frac{1}{2}\sqrt{2}\, m\, \Delta v$ and is the basis of the longitudinal Knudsen force at B, which acts parallel to the surface. An equal force acts at D, in the same direction.

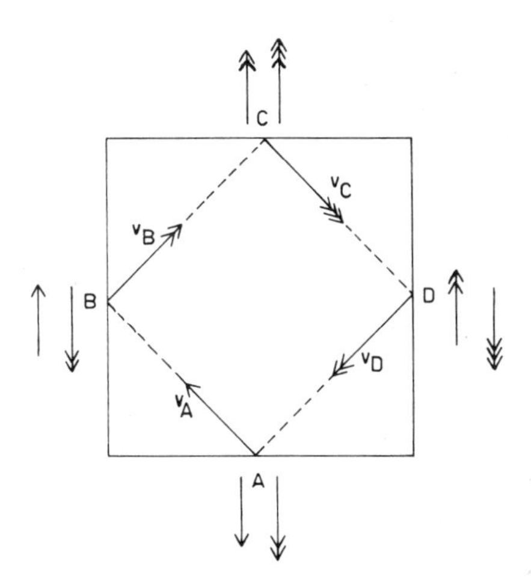

Figure 4.4. The one molecule model demonstrating the origin of longitudinal Knudsen forces at B and D and the Knudsen pressure difference between A and C. $T_C > T_B = T_D > T_a$.

It is also worth considering the vertical momenta which are related to the pressure of a real gas transferred to the top and bottom surfaces of the square. The momentum transferred at C is directed opposite to that at A. Their magnitudes, which are not equal, are given by $\frac{1}{2}\sqrt{2}\, m(2v_A + 3\,\Delta v)$ and $\frac{1}{2}\sqrt{2}\, m(2v_A + \Delta v)$, respectively. The algebraic sum of the momenta transferred at A and C equals $\sqrt{2}\, m\, \Delta v$. The fact that the two momentum transfers do not cancel is the basis of what is known as the Knudsen Pressure Difference, frequently referred to as Thermal Transpiration Effect or Transverse Knudsen Forces (16–20). The sum of all the vertical momenta at A, B, C, and D is zero, a result which shows that the total Longitudinal Knudsen Forces just cancel the total Transverse Knudsen Forces. From more general considerations it can be seen that this is an expected conclusion. When the stationary state conditions hold, the sum of all forces (transverse and longitudinal) on the wall of the vessel is zero.

This is a straightforward result of Newton's third law applied to gas and vessel, and of the fact that in the stationary state the total momentum of the gas is constant. In the example of Figure 4.4, the general statement does not exclude net forces on a part of the wall. From the above it can be concluded that in a weighing experiment, where a sample is suspended in a vessel filled with a gas, forces are exerted on both the sample and vessel. The above analysis deals with a vessel containing a single molecule. To apply it to a gas consisting of a large number of molecules, it is essential to employ the elementary kinetic theory of gases.

Of major importance in the kinetic theory of gases is the magnitude of the mean free path, λ, of the gas molecules as compared to the smallest dimension of the vessel W. The theory simplifies in (*1*) the low pressure region where $\lambda \gg W$ and (*2*) the slip or viscous flow pressure region where $\lambda \ll W$. In the pressure region where λ and W are of the same order of magnitude, the phenomenon is more complex. In practice extrapolation of the results from the two adjacent regions is not difficult.

In the low pressure region the number of molecules which strike the wall per unit area and unit time is the same all through the vessel even when the temperature is not uniform throughout (Figure 4.5). In the slip or viscous flow region the momentum transferred per unit area and unit time to the wall by the colliding molecules is the same all through the vessel, even with nonuniform temperatures (Figure 4.5). These two conditions hold in both regions when wall temperatures are homogeneous. These situations are visualized in a simplified manner in Figures 4.5 showing a vessel with a vertical temperature gradient. The influence of the vertical walls is eliminated by assuming that the molecules move only vertically.

From the law of conservation of mass it follows that in the low pressure range the number of gas molecules approaching the top per unit time and area equals the number of molecules leaving it. Obviously this must equal the number of molecules approaching the bottom per unit time and unit area. It is assumed that in the slip flow pressure range intermolecular collisions taking place nearest to the wall occur at a distance $\frac{1}{2}\sqrt{2}$ from it. Here the law of conservation of mass requires that the number of molecules at A traveling in the direction between the bottom of the vessel and the bulk of the gas and the number of those traveling in the opposite direction are equal. At B a similar equality is required for the number of gas molecules traveling between the top of the vessel and the bulk of the gas and vice versa. In the slip flow region it is the law of the conservation of momentum of the bulk of the gas that leads to the equality of the pressure along the wall. Assuming that the gas is in mechanical equilibrium, the momentum transferred to the bulk of the gas per unit time and unit area by the molecules at *A* must cancel that transferred by the molecules at *B*. Since momenta are equal to the gas pressures at the bottom and at the top, respectively, the gas pressure is homogeneous. It will be shown in the following analysis that the presence of the vertical walls in the vessel makes this result valid only as a first-order approximation.

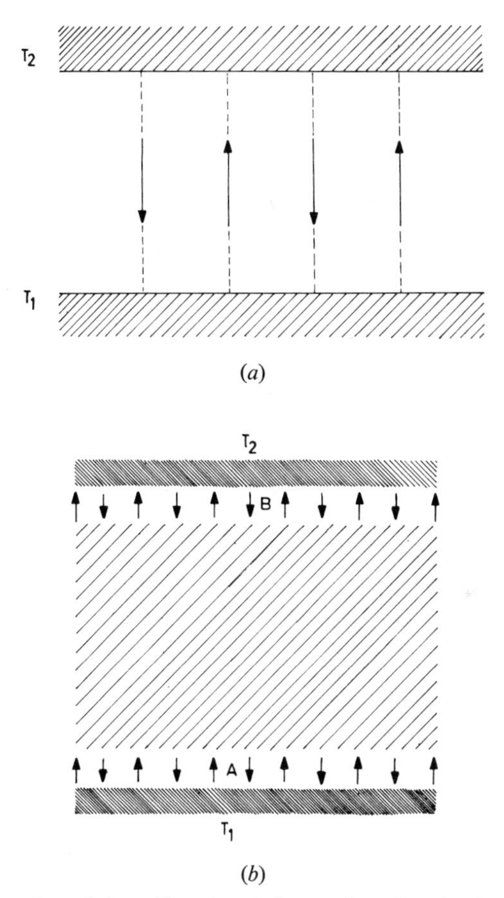

Figure 4.5. Illustration of the uniformity of the number of molecules to and from the container walls at low pressures (*a*) and of the quantity *p* at high pressures (*b*).

A more direct approach to weighing disturbances resulting from Knudsen forces is to calculate the forces acting on a sample inside the vessel (20–27). The presence of a body inside a vessel, however, inevitably complicates the calculations, without contributing to the understanding of the nature of the effects. Therefore only the forces on the walls of the vessel which contains nothing but the gas will be calculated. It will be kept in mind, however, that a sample would only contribute to the total surface affected and that a Knudsen force would act on it in very much the same manner that it acts on the walls.

The vessel chosen for the calculations is shown in Figure 4.6. It is rectangular with dimensions h, d, and w such that $h \gg w$ and $d \gg w$. The temperature gradient is assumed to be vertical implying that the isothermal planes are horizontal. For simplicity it is supposed that the gas molecules move only in planes parallel to the plane of the drawing, making angles of 45° with the vertical. Further their velocity, $v(z)$, is determined by the temperature at the height z where the last collision took place. Considering a molecule that hits the wall at the height z, it is seen that at low pressures (Figure 4.6) the previous collision was with the other wall at a height $z - w$ or $z + w$. At high pressures the previous collision took place with another gas molecule at a height $z - \frac{1}{2}\lambda\sqrt{2}$ or $z + \frac{1}{2}\lambda\sqrt{2}$ where λ is the mean free path. Let $M(z)$ be the number of molecules which leave the vertical wall at height z per unit time and area with equal distribution over the two directions allowed by the model of Figure 4.6. As explained previously, in the low pressure region M has a uniform value throughout the vessel. Consider the vertical force in the low pressure region caused by gas molecules on the part of the vertical wall at height z and of dimensions d and Δz where z is chosen not too near the top or bottom. The vertical momentum, dF_l, transferred to the wall resulting from the impact of the gas molecules equals

$$dF_l = \tfrac{1}{4}\sqrt{2}\, md\,\Delta z[Mv(z - w) - Mv(z + w)] \tag{4.18}$$

The total amount of momentum involved in the recoil of the same molecules is zero, since it is assumed that the velocity, $v(z)$, of all the molecules which leave the wall in each of the two directions allowed by our model have the same magnitude. Restricting the discussion to a linear relation between v and z:

$$v(z) = v(0) + z\frac{dv}{dz} \tag{4.19}$$

the total force F_l acting separately on each side wall is

$$F_l = -\tfrac{1}{2}\sqrt{2}\, mMwdh\frac{dv}{dz} = \tfrac{1}{2}\sqrt{2}\, mMwdv_{\text{bottom}}\left(1 - \frac{v_{\text{top}}}{v_{\text{bottom}}}\right) \tag{4.20}$$

Using a first-order approximation to calculate the pressure at the bottom, p_{bottom}, and, necessarily including the recoil momentum it is found that

$$p_{\text{bottom}} = \sqrt{2}\, mMv_{\text{bottom}} \tag{4.21}$$

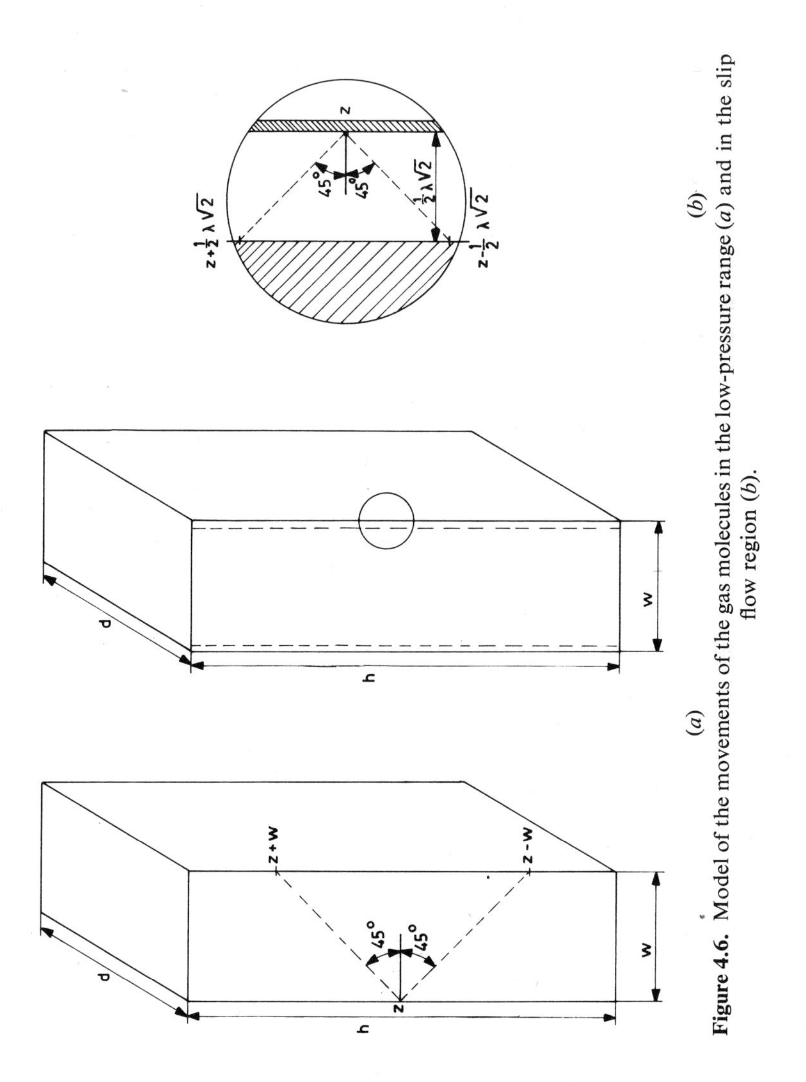

(*a*) (*b*)

Figure 4.6. Model of the movements of the gas molecules in the low-pressure range (*a*) and in the slip flow region (*b*).

from equations 4.20 and 4.21 the total longitudinal force on one of the side walls at low pressure ($\lambda > w$) is

$$F_l = \tfrac{1}{2}wdp_{\text{bottom}}\left(1 - \frac{v_{\text{top}}}{v_{\text{bottom}}}\right) = -\tfrac{1}{4}wd\frac{p_{\text{bottom}}}{T_{\text{bottom}}}(T_{\text{top}} - T_{\text{bottom}}) \quad (4.22)$$

This equation can be obtained from a simplified form of the equipartition theorem (equation 4.41e) and the assumption that $(T_{\text{top}} - T_{\text{bottom}}) \ll T_{\text{bottom}}$.

At higher gas pressures where λ is small compared with w, the vertical momentum, dF_l, transferred per unit time to the part of the wall of depth d and height Δz is given by

$$dF_l = \tfrac{1}{4}\sqrt{2}\, md\,\Delta z[M(z - \tfrac{1}{2}\lambda\sqrt{2})v(z - \tfrac{1}{2}\lambda\sqrt{2}) - M(z + \tfrac{1}{2}\lambda\sqrt{2})\, v(z + \tfrac{1}{2}\lambda\sqrt{2})] \quad (4.23)$$

where v is obtained from equation 4.19. To evaluate M it is necessary to develop a detailed model of the molecular motions. M not being uniform involves a gas flow along the wall. This is illustrated in Figure 4.6*b* which shows the transport through a horizontal rectangular plane of dimensions d and $\tfrac{1}{2}\lambda\sqrt{2}$. The number of molecules passing through this plane per unit time per unit area in the upward direction, $M_{h\uparrow}$, is from the following relationship:

$$\tfrac{1}{2}\sqrt{2}\,\lambda dM_{h\uparrow} = \tfrac{1}{2}\sqrt{2}\,\lambda dM(z - \tfrac{1}{2}\lambda\sqrt{2}) \quad (4.24)$$

In the downward direction the number passing through the plane per unit time, $M_{h\downarrow}$, is

$$\tfrac{1}{2}\sqrt{2}\,\lambda dM_{h\downarrow} = \tfrac{1}{2}\sqrt{2}\,\lambda dM(z + \tfrac{1}{2}\lambda\sqrt{2}) \quad (4.25)$$

The net transport of molecules through the plane equals:

$$\tfrac{1}{2}\sqrt{2}\,\lambda d(M_{h\uparrow} - M_{h\downarrow}) = \tfrac{1}{2}\sqrt{2}\,\lambda d[M(z - \tfrac{1}{2}\lambda\sqrt{2}) - M(z + \tfrac{1}{2}\lambda\sqrt{2})] \quad (4.26)$$

which in a first-order approximation becomes:

$$\tfrac{1}{2}\sqrt{2}\,\lambda d(M_{h\uparrow} - M_{h\downarrow}) = \frac{dM}{dz}\lambda^2 d \quad (4.27)$$

The fact that the result is not zero implies a net vertical gas flow along the wall which can be represented by the macroscopic velocity u_{wall} so that:

$$\tfrac{1}{2}\sqrt{2}\,\lambda dnu_{\text{wall}} = -\frac{dM}{dz}\lambda^2 d \quad (4.28)$$

where n is the number of gas molecules per unit volume.

Equation 4.28 can be modified to

$$u_{\text{wall}} = -\sqrt{2}\,\frac{1}{n}\,\lambda\,\frac{dM}{dz} \tag{4.29}$$

The macroscopic velocity of the gas is now equal at the two side walls of the vessel. Since the vessel is closed at the top and bottom, the gas flow along the walls must be compensated for by a gas flow in the center of the container in the opposite direction (see Figure 4.7). The driving force for

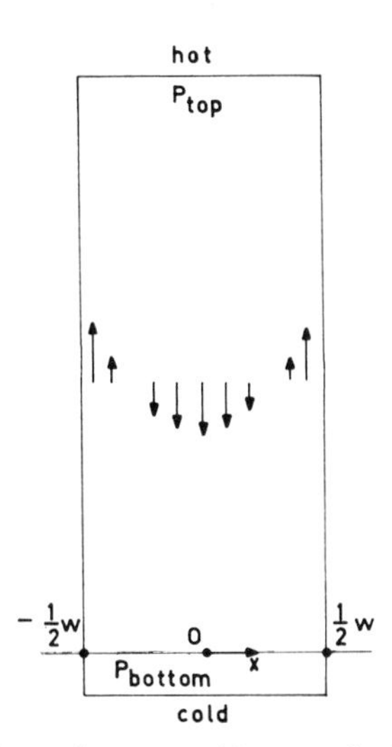

Figure 4.7. The macroscopic gas flow caused by a vertical temperature gradient in the slip-flow pressure region.

the central flow is a gas pressure difference ($p_{\text{top}} - p_{\text{bottom}}$) between the top and bottom of the vessel resulting from the flow along the wall. The vertical velocity $u(x)$ is therefore dependent upon the x coordinate, the zero of which is taken midway between the two walls (see Figure 4.7). For this situation in which the vessel is closed

$$\int_{-\frac{1}{2}w}^{+\frac{1}{2}w} u(x)\,dx = 0 \tag{4.30}$$

Assuming laminar flow conditions, it follows from Poiseuille's law:

$$u(x) = \frac{p_{\text{top}} - p_{\text{bottom}}}{2\eta h}\,(x^2 - \tfrac{1}{4}w^2) + u_{\text{wall}} \tag{4.31}$$

where η is the viscosity of the gas. Substituting equations 4.29 and 4.31

in equation 4.30 and assuming that dM/dz is independent of z, it follows that

$$p_{\text{top}} - p_{\text{bottom}} = -\frac{12\sqrt{2}\,\eta\lambda}{w^2 n}(M_{\text{top}} - M_{\text{bottom}}) \tag{4.32}$$

Comparing equations 4.32 and 4.23 it is seen that all the information about longitudinal and transverse forces, respectively, is given in terms of the unknown quantity M. However, one important condition has not yet been applied, namely, that the resultant force acting on the vessel must be zero. This condition follows from the stationary state of the gas and Newton's third law. To apply this the total longitudinal force F_l on each of the side walls is calculated. Using a first-order approximation for M and v, equation 4.23 yields

$$dF_l = -\tfrac{1}{2}m\lambda d\frac{d(Mv)}{dz}\Delta z \tag{4.33}$$

Integrating to obtain F_l

$$F_l = -\tfrac{1}{2}m\lambda d[(Mv)_{\text{top}} - (Mv)_{\text{bottom}}] \tag{4.34}$$

When differences in v and M are not too great

$$F_l = -\tfrac{1}{2}m\lambda d[M(v_{\text{top}} - v_{\text{bottom}}) + v(M_{\text{top}} - M_{\text{bottom}})] \tag{4.35}$$

As explained before, the total force on the wall must be zero, thus

$$2F_l + wd(p_{\text{top}} - p_{\text{bottom}}) = 0 \tag{4.36}$$

Substitution of equations 4.32 and 4.35 into equation 4.36 yields:

$$-m\lambda d[M(v_{\text{top}} - v_{\text{bottom}}) + v(M_{\text{top}} - M_{\text{bottom}})] = \frac{12\sqrt{2}\,\eta\lambda d}{wn}(M_{\text{top}} - M_{\text{bottom}}) \tag{4.37}$$

By rearranging equation 4.37

$$M_{\text{top}} - M_{\text{bottom}} = -\frac{M(v_{\text{top}} - v_{\text{bottom}})}{v + 12\sqrt{2}\,\eta/mwn} \tag{4.38}$$

which upon substitution into equation 4.35 results in

$$F_l = -\tfrac{1}{2}mM\lambda d(v_{\text{top}} - v_{\text{bottom}})\left(1 - \frac{1}{1 + 12\sqrt{2}\,\eta/mwnv}\right) \tag{4.39}$$

which according to equation 4.41 and for $\lambda \ll w$ leads in a first-order approximation to:

$$F_l = -\frac{6\sqrt{2}\,\eta M d\lambda}{wnv}(v_{\text{top}} - v_{\text{bottom}}) \tag{4.40}$$

From the elementary kinetic theory of gases the following relations are used

$$\eta = \tfrac{1}{2}\sqrt{2}\, mM\lambda \tag{4.41a}$$

$$\lambda = \Lambda \frac{T}{p} \tag{4.41b}$$

$$M = \tfrac{1}{4}\sqrt{2}\, nv \tag{4.41c}$$

$$p = \sqrt{2}\, mMv = \tfrac{1}{2} mnv^2 \tag{4.41d}$$

$$\frac{v_{\text{top}} - v_{\text{bottom}}}{v_{\text{bottom}}} = \tfrac{1}{2}\frac{T_{\text{top}} - T_{\text{bottom}}}{T_{\text{bottom}}} \tag{4.41e}$$

The unusual coefficients in these relations are the result of the model of molecules moving in two directions only. Equation 4.41e is only valid for small temperature differences. Substituting equations 4.41 in equation 4.40:

$$F_l = -\tfrac{3}{4}\frac{d}{w}\Lambda^2 \frac{T}{p}(T_{\text{top}} - T_{\text{bottom}}) \tag{4.42}$$

Equations 4.22 and 4.42 together completely describe the longitudinal Knudsen forces in the two pressure regions. Equation 4.22 shows the proportionality of the forces with pressure at low pressures. Equation 4.42 shows that at higher pressures, in the slip or viscous flow region, the longitudinal Knudsen forces fall off inversely with pressure. The maximum occurs in the region between the two pressure regions where the mean free path λ is of the order of magnitude of the distance w between the two walls. Comparison of equations 4.22 and 4.42 reveals a pronounced difference in the temperature dependences in the two pressure ranges. Further it is seen that only in the slip flow pressure range does the nature of the gas play a part through the specific constant Λ.

As noted previously, for simplicity, the longitudinal Knudsen forces acting on flat walls have been calculated rather than those acting on an actual sample or a hangdown wire which would have been more realistically related to a cylindrical model (21,22). This more complicated model leads to:

$$F_l = -R_i R_0 \frac{p}{T}(T_{\text{top}} - T_{\text{bottom}}) \qquad \text{low pressures} \tag{4.43}$$

and

$$F_l = -\frac{\pi}{2}\frac{\Lambda^2}{\ln(R_0/R_i - 1)}\frac{T}{p}(T_{\text{top}} - T_{\text{bottom}}) \qquad \text{slip flow region} \tag{4.44}$$

where R_0 is the radius of the vessel and R_i is the radius of the hangdown wire or of the sample.

The following two examples show the order of magnitude of longitudinal Knudsen forces. The first case considers the effect upon a sample with a 2 mm radius surrounded by a furnace, the inner radius of which equals 1 cm. The temperature difference causing the Knudsen forces originates here from nonuniformities of the furnace wall temperature, which are estimated to be 5°C at a furnace temperature of 500°K. If the

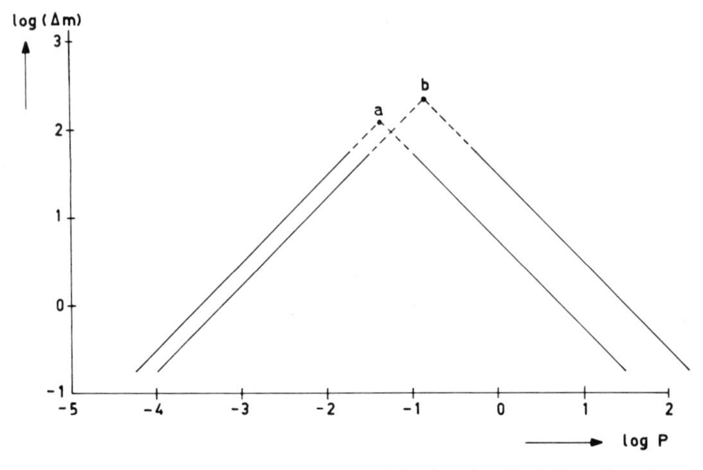

Figure 4.8. The spurious mass effect caused by longitudinal Knudsen forces. The numerical data are those from the examples dealt with in the text; *a* refers to the effect on the sample, *b* to the effect on the hangdown wire Δm in µg, *p* in torr.

gas in the furnace is oxygen, Λ is 3.35×10^{-2} dyne cm^{-1} °C^{-1}. Using these data in equations 4.43 and 4.44 in the low pressure region:

$$F_l = -p/500$$

and in the slip flow pressure region:

$$F_l = -7.2/p$$

These pressure dependences are illustrated in Figure 4.8. The intersection of the two lines occurs at $p_{max} = 60$ dynes/cm^2 $= 0.045$ torr corresponding to a Knudsen force of 0.12 dynes and a spurious mass effect of 120 µg. As a second example, the force acting on the hangdown wire (radius 20 µ) which passes through the orifice of the furnace ($T_{top} - T_{bottom} = -200$°C) is calculated. For a temperature of 400°K, the result in the low pressure region is:

$$F_l = p/1000$$

and in the slip flow pressure region:

$$F_l = 27/p$$

The intersection occurs at $p = 160$ dynes/cm^2 = 0.12 torr. The maximum force amount to 0.16 dynes which corresponds to 160 μg.

Apart from the longitudinal Knudsen force, a transverse Knudsen force F_t can be calculated from the equation

$$F_t = (p_{\text{top}} - p_{\text{bottom}})S \tag{4.45}$$

in which S equals the area of the horizontal cross section of the sample. In the low pressure region equation 4.45 gives:

$$F_t = p[1 - (T_{\text{top}}/T_{\text{bottom}})]S \tag{4.46}$$

This expression is a simple consequence of the fact that in the low pressure range the uniformity of M leads to the equation:

$$p_{\text{top}}/p_{\text{bottom}} = (T_{\text{top}}/T_{\text{bottom}})^{1/2} \tag{4.47}$$

Substituting in equation 4.46 for p the values of p_{max} calculated in the two previous examples, the maximum values of the transverse Knudsen force on the sample and on the hangdown wire are 0.038 dyne and 0.45×10^{-3} dyne respectively.

III. CAVITY FORCES

Cavity forces are by nature closely related to Knudsen forces. They occur when a temperature gradient exists perpendicular to the surface of a porous sample (28–30). Consider a vertical cylindrical pore with radius R (see Figure 4.9) in a sample with a vertical temperature gradient. For simplicity assume a uniform gas temperature T_g, above the sample. Stationary state conditions require that the number of gas molecules, A, entering the hole through the orifice equal the number of gas molecules, B, leaving it. The velocity of the A molecules is governed by the temperature of the gas $T_A = T_g$, while the velocity of the B molecules corresponds to the temperature somewhere on the wall of the hole, T_B. Again for simplicity suppose that the average place where the B molecules had their last collision with the wall is at a distance R below the orifice, the temperature T_B is then:

$$T_B = T_g - R\left(\frac{dT}{dz}\right) \tag{4.48}$$

Now consider the momentum transported by the A and B molecules through the orifice per unit time and area. Added together, these make up the pressure p_0 at the orifice. For simplicity, assume the case of a completely uniform temperature, i.e., $T_A = T_B$. Here each of the momentum transports correspond to half the bulk gas pressure p_g.

Returning to the situation of the pore with the temperature gradient, it is seen that the presence of the temperature gradient does not affect the numbers of molecules, A and B, passing through the orifice. The number of entering molecules, A, is only determined by the physical state of the gas above the sample and the number of leaving molecules, B, remains

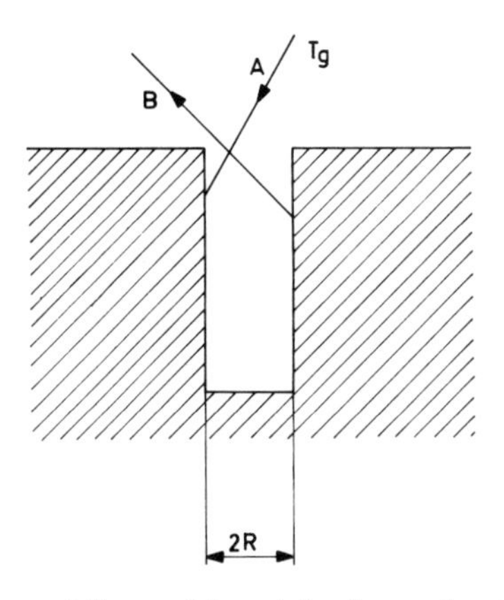

Figure 4.9. Illustration of the model used for the explanation of cavity forces.

equal to it because of the steady state conditions. The momentum transferred by the B molecules, however, has to be corrected with the factor $(T_B/T_A)^{1/2}$, because the velocities of the B molecules are related to the temperature T_B instead of T_A. This means that:

$$p_0 = \tfrac{1}{2}p_g + \tfrac{1}{2}(T_B/T_A)^{1/2}p_g \tag{4.49}$$

Using equation 4.48 and restricting the discussion to temperature gradients so small that first-order approximations are allowed, p_0 becomes

$$p_0 = p_g\left[1 - \tfrac{1}{4}\frac{R}{T_g}\left(\frac{dT}{dz}\right)\right] \tag{4.50}$$

It is seen that the pressure p_0 at the orifice differs from the bulk gas pressure p_g by the amount Δp as follows:

$$\Delta p = p_0 - p_g = -\tfrac{1}{4}p_g\frac{R}{T_g}\left(\frac{dT}{dz}\right) \tag{4.51}$$

Equation 4.51 shows that Δp is directly proportional to the gas pressure p_g. This proportionality only holds for the region where the derivation of equation 4.51 is valid, e.g., at pressures so low that the mean free path of the gas molecules is greater than the radius of the pore. At higher pressures the effect decreases rapidly with increasing pressure (30). The maximum effect will therefore occur not far from that pressure at which the mean free path equals the radius of the pore. Substituting equation 4.41b for p_g,

$$\Delta p = -\tfrac{1}{4}\Lambda \frac{R}{\lambda}\left(\frac{dT}{dz}\right) \tag{4.52}$$

and for the maximum value of Δp which occurs when $\lambda = R$,

$$\Delta p_{\max} = -\tfrac{1}{4}\Lambda\left(\frac{dT}{dz}\right) \tag{4.53}$$

It is apparent that although the maximum value of the effect is not directly dependent upon the pore radius, the radius does determine the pressure at which the maximum effect occurs. In order to obtain the resulting forces, it is necessary to multiply the expressions for Δp for single pores by the total surface area of the top of the sample that is covered by holes. For example assuming 0.1 cm^2 for this area, $\Lambda = 3.35 \times 10^{-2}$ (oxygen), and, $dT/dz = 5°C/cm$, the magnitude of the disturbing cavity force, $\Delta F_{c'}$, is -0.0042 dyne and that of the spurious mass effect, $\Delta M_{c'}$ is -4.2 μg.

For pores with a radius of 1 μ the maximum force will occur at about 75 torr. The cavity forces can be minimized by a furnace design which concentrates upon temperature homogeneity (31,32).

IV. ARMLENGTH EFFECT

Knudsen and cavity forces, because of their proportionality with pressure, disappear at low pressures. Other disturbances resulting from temperature inhomogeneities, therefore, become predominant in the very low pressures region.

This discussion will be concerned with the inequality of the arms of a balance beam arising from thermal expansion differences caused by temperature inhomogeneities. To understand the influence of the armlength variations upon force measurements, it must be remembered that a beam balance actually compares the moments of force. Apart from the obvious reduction of the armlength effect by reducing temperature gradients in the vicinity of the balance, much can be done through a suitable choice of the beam material (33,34). This is illustrated by the simplified example shown in Figure 4.10. Here one part of the beam is surrounded by a tube half of which is maintained at the temperature, T_t, while the rest of the tube is at room temperature, T_r. It is further supposed that the temperature of the beam at the fulcrum is room temperature. This last condition is especially difficult to realize but it can be approached if there is a similar but opposite temperature jump at the tube surrounding the other part of the beam. The heat flow shown by the dotted line in Figure 4.10 is of prime importance. Under low gas pressure conditions the heat is transported in the form of radiation, Q_r, between the hot part of the tube and the beam. The heat flow, through the beam, Q_c, to the fulcrum is by conduction. For simplicity, assume that the beam arm under consideration consists of equal

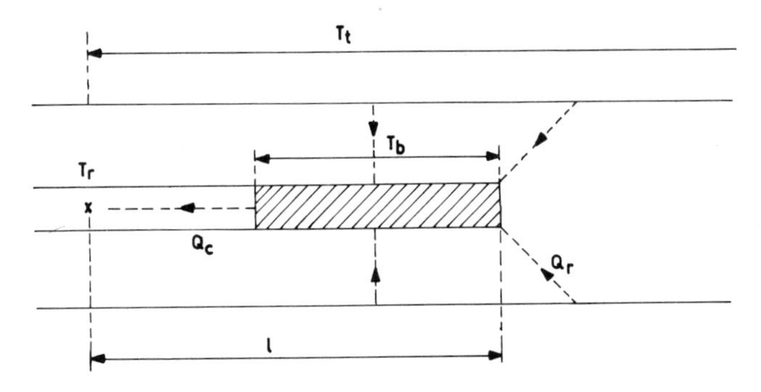

Figure 4.10. Schematic of the heat flow in the example for which the armlength effect is calculated.

parts having the following different physical properties: (*1*) The half at the end of the beam has the actual specific heat C of the beam material, but an infinite heat conductivity. Its temperature T_b is therefore uniform. (*2*) The other half, near the fulcrum, has the actual heat conductivity λ of the beam material but its specific heat is zero. $T_b - T_r$, therefore, represents the temperature difference across the latter part. The heat flow, Q_r, is, according to the Stefan–Boltzmann law, governed by the resistance R_r, which, when the temperature differences are small, equals:

$$R_r = \frac{T_t - T_b}{Q_r} = \frac{(D_t - D_b)^2}{\pi\varepsilon_t\varepsilon_b\sigma[(T_t + T_b)/2]^3 \, D_t D_b(\tfrac{1}{2}l)^2} \tag{4.54}$$

in which D_t and D_b are the diameters of tube and beam, respectively, l is the length of the arm, ε_b and ε_t are the emissivities of the beam and the tube, respectively, and σ, the Stefan–Boltzmann constant, is 5.6×10^{-5}. The heat flow, Q_c, governed by the resistance R_c is given by:

$$R_c = (T_b - T_r)/Q_c = 2l/\pi D_b^2\lambda \tag{4.55}$$

The heat capacity, C_b, of the end part of the arm is given by:

$$C_b = (\pi/8)D_b^2 l\rho C \tag{4.56}$$

where ρ is the density and C the specific heat of the beam material. With the help of these newly defined quantities, a simple equation for the beam end temperature, T_b is:

$$C\frac{\partial T_b}{\partial t} = \frac{T_t - T_b}{R_r} - \frac{T_b - T_z}{R_c} \tag{4.57}$$

From equation 4.57 it follows for stationary state conditions:

$$T_b - T_r = \frac{R_c}{R_r + R_c}(T_t - T_r) \tag{4.58}$$

The relaxation time interval, τ, characteristic of the establishment of the steady state situation is:

$$\tau = \frac{1/R_c + 1/R_r}{C_b} \tag{4.59}$$

In order to calculate the spurious mass effect, it is first necessary to determine the expansion of the arm under the conditions shown in Figure 4.10 in the stationary state. The expansion Δl_1 of the end part of the arm at the uniform temperature T_b is given by

$$\Delta l_1 = {}^1\!/_2\alpha l(T_b - T_r) \tag{4.60}$$

where α is the linear expansion coefficient of the beam material. The expansion Δl_2 of the other part of the arm, where the temperature drops (linearly) from T_b to $T_{r'}$ is given by:

$$\Delta l_2 = {}^1\!/_4\alpha l(T_b - T_r) \tag{4.61}$$

Neglecting the mass of the beam and incorporating equation 4.58, the corresponding spurious mass Δm, is:

$$\Delta m = m\frac{\Delta l_1 + \Delta l_2}{l} = {}^3\!/_4\alpha m\frac{R_c}{R_r + R_c}(T_t - T_r) \tag{4.62a}$$

where m is the mass of the sample.

The required information for specific beam materials and the results of the calculations are presented in Table 4.1. To illustrate the dependency of the effect on the beam material three different beams, viz. an aluminium beam (Al), a fused quartz beam (SiO_2) and a fused quartz beam covered with a thin gold layer (SiO_2–Au) have been considered. In the latter case the effective heat conductivity, λ_{eff}, is given by:

$$\lambda_{eff} = \frac{D^2_{SiO_2}\lambda_{SiO_2} + (D^2_{Au} - D^2_{SiO_2})\lambda_{Au}}{D^2_{Au}} \tag{4.62b}$$

where D_{Au} is the total diameter of the beam (0.2 cm), D_{SiO_2} is the diameter of the quartz body of the beam, taken to be 0.195 cm, and D_{Au} is 50 μ, i.e.,

Table 4.1.

Beam material	λ	ε_b	α	ρ	C
Al	2.4×10^7	0.5	2.3×10^{-5}	2.7	0.92×10^7
SiO_2	1.3×10^5	0.9	5.0×10^{-7}	2.2	0.80×10^7
SiO_2–Au	1.6×10^6	0.02	5.0×10^{-7}	3.1	0.59×10^7

a gold layer with a thickness of 25 μ. Inserting $\lambda_{SiO_2} = 0.013 \times 10^7$ and $\lambda_{Au} = 3 \times 10^7$ erg cm^{-1} sec^{-1} °C^{-1}, it follows that $\lambda_{eff} = 0.16 \times 10^7$ erg cm^{-1} sec^{-1} °C^{-1}. Effective values for the density, ρ, and the specific heat, C, have been calculated in a similar manner. The following values were used in the calculation to obtain τ and Δm: $D_b = 0.2$ cm, $D_t = 3$ cm, $l = 5$ cm, $(T_r + T_b)/2 = 300$°K, $\varepsilon_t = 0.9$ (fused quartz), $m = 1$ g and $T_t - T_r = 1$°C. The required data for specific beam materials and the results of the calculations are presented in Table 4.1.

From Table 4.1 it can be concluded that the combination of the fused quartz beam covered with a gold layer is most advantageous. This result is not surprising since such a beam combines the advantages of the high heat conductivity and low emissivity of a metal with the low expansion coefficient of fused quartz. This result has been verified experimentally (35).

V. RADIATION PRESSURE

This discussion will consider the influence of radiation pressure on mass determination with a microbalance. Let a light source be located at a distance d above one of the arms of a microbalance and the other arm be shielded from the radiation. The fraction f of the totally emitted radiation energy which is intercepted by the balance arm, is given by:

$$f = A/4\pi d^2 \tag{4.63}$$

where A is the effective area of the balance arm measured perpendicular to the radiation. The radiation is accompanied by a transfer of momentum per second, p, given by:

$$p = Q/c \tag{4.64}$$

where Q is the radiation power emitted by the bulb in both the visible and infrared part of the spectrum, and c is the velocity of light (3×10^{10} cm/sec). If the balance arm is a perfect reflector, the transferred momentum will be twice the momentum inherent in this radiation. So the resulting force F_r acting on the balance arm is:

$$F_r = 2fp = (A/2\pi d^2)(Q/c) \tag{4.65}$$

R_r	R_c	C_b	τ	Δm
9.6×10^{-4}	3.3×10^{-6}	1.9×10^{6}	6.3	6.0×10^{-8}
5.3×10^{-4}	6.1×10^{-4}	1.4×10^{6}	400.0	2.0×10^{-7}
2.4×10^{-2}	4.9×10^{-5}	1.4×10^{6}	69.0	7.5×10^{-10}

Taking the example of a bulb which emits $Q = 100$ W $= 10^9$ ergs of radiation while the distance $d = 10$ cm and $A = 2$ cm^2, equation 4.65 leads to $F_r = 10^{-4}$ dyne which corresponds to a spurious mass effect of 0.1 μg. This example shows that only in very special cases will radiation pressures be a source of appreciable disturbances.

VI. GRAVITATIONAL FORCES

The presence of a heavy body in the vicinity of a balance can cause gravitational forces, F_g, acting on parts of it as given by the general law

$$F_g = Gm_1m_2/d^2 \tag{4.66}$$

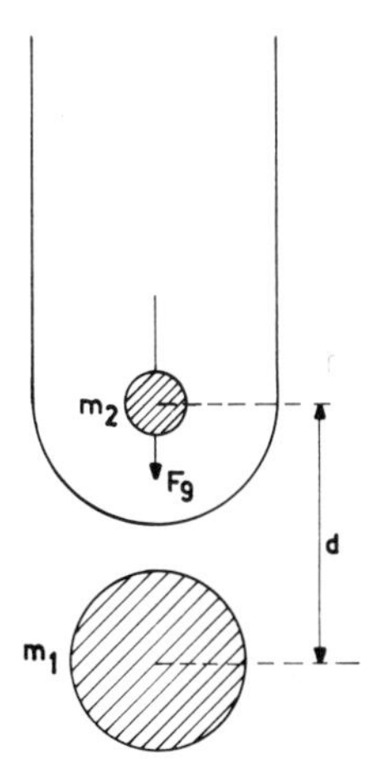

Figure 4.11. Illustration of gravitation force acting on the sample m_2 because of the presence of the body m_1.

in which m_1 is the mass of the heavy body, m_2 is the mass of the part of the balance involved, d is the distance between the two centers of mass, and G is the gravitation constant, 6.7×10^{-8} cm^3/g-sec^2.

The magnitude of the effect of this force is calculated using Figure 4.11.

Assuming that $m_1 = 10^5$ g, $m_2 = 0.3$ g, and $d = 20$ cm, it follows that $F_g = 5 \times 10^{-6}$ dynes, corresponding to a spurious mass effect of 5×10^{-9} g.

The fact that his value is so small even for the rather pronounced situation shown in the figure indicates that, in general weighing, disturbances of this kind are not significant.

VII. ELECTROSTATIC FORCES

When parts of the balance and its surroundings are electrically charged, forces described by Coulomb's law will be present. Such charges occur frequently with their origin usually related to the frictional electricity generated at the solid–gas interface.

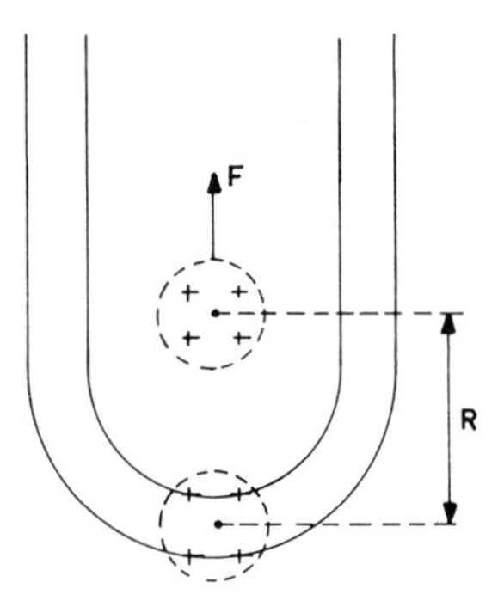

Figure 4.12. Model used in calculating the electrostatic force between charges on sample and balance case.

In order to give an estimate of these forces, consider the following simple hypothetical example in which the sample as well as a part of the bottom of the balance case below the sample are electrically charged. Assume that these charged parts can be represented by two 1 cm radius spheres 5 cm apart, as illustrated in Figure 4.12, with both spheres having a potential of 1000 V. The total charge of a sphere can be calculated from the equation.

$$V = 9 \times 10^{11} \frac{Q}{R_s} \tag{4.67}$$

where V is the electrical potential of the sphere in volts, Q is its charge in coulombs and R_s is its radius in cm.

In this example, $Q = 1000/(9 \times 10^{11}) = 10^{-8}/9$ coulomb. The force acting between the two spheres from Coulomb's law is

$$F = 9 \times 10^{18}\, Q_1 Q_2 / R^2 \tag{4.68}$$

where R is the distance between the spheres in cm and F is the mutual force in dynes. For the above example $F = 0.4$ dyne, equivalent to 400 μg. Although this has been a very simple example, it illustrates that electrostatic forces are among the more serious weighing disturbances and can result in substantial weighing errors.

The view that grounding of either the balance or its case alone would eliminate the electrostatic forces, ignores the fact that forces may also originate from the remaining induced electric charge. This may be seen by considering a capacitor with one of the plates grounded and the other electrically charged.

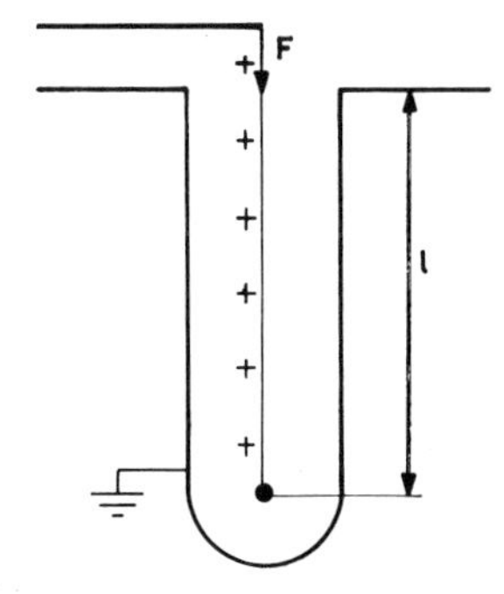

Figure 4.13. Illustration of the capacitor formed by a charged sample and hangdown wire and a grounded balance case which gives rise to an electrostatic force.

As an example the force will be calculated for the case in which the electrical charge is present only on the balance while the surrounding balance case is grounded. This is illustrated schematically in Figure 4.13. A constant charge is homogeneously distributed along the hangdown wire. The hangdown wire is coaxially surrounded by a grounded tube. The hangdown wire and the tube form together a cylindrical capacitor with the capacity C being given by:

$$C = 10^{-11} l/[18 \ln (R_t/R_h)] \tag{4.69}$$

where C is expressed in farads, l is the length of that part of the hangdown wire which is surrounded by the tube and R_t and R_h are the radii of tube and hangdown wire, respectively. The energy ε associated with this capacitor is given by:

$$\varepsilon = (10^7/2)(q^2 l^2/C) \tag{4.70}$$

where ε is expressed in ergs and q in Coulomb/cm. Using equation 4.69 in 4.70 yields:

$$\varepsilon = 9 \times 10^{18} q^2 l \ln (R_t/R_h) \tag{4.71}$$

The electrostatic force acting on the hangdown wire satisfies

$$F = \frac{\partial \varepsilon}{\partial l} = -9 \times 10^{18} q^2 \ln (R_t/R_h) \tag{4.72}$$

Using

$$q = \frac{CV}{l} = \frac{10^{-11} V}{18 \ln (R_t/R_h)} \tag{4.73}$$

this force can be expressed as a function of the potential V by

$$F = -10^{-4} V^2/[36 \ln (R_t/R_h)] \tag{4.74}$$

Substitution of the typical values:

$$V = 10^3 \text{ v}$$
$$R_t = 2 \text{ cm}$$
$$R_h = 5 \times 10^{-3} \text{ cm}$$

leads to an estimated value of $F = 0.5$ dyne, equivalent to 500 μg. This example demonstrates that grounding of the balance or case alone is not sufficient to eliminate electrostatic forces and that it is necessary to ground both the balance and its surroundings. When insulating materials are used, the ground connection is made with the aid of a conductive coating.

VIII. MAGNETOSTATIC FORCES

Before calculating spurious mass changes arising from magnetostatic effects, it is of value to discuss the origins of the disturbing magnetic fields. First, there is the earth's magnetic field which is of the order of a few tenths of an oersted and is constant within 1%. Inhomogeneities in the earth field may arise from the presence of ferromagnetic materials in the vicinity of the balance. Second, there are stray fields of electric currents which, though they are usually an order of magnitude smaller than the earth magnetic field, can be of importance because of their time dependency. For example, an electric furnace used to heat the sample can produce a sizeable disturbance. If it does not have a bifilar or noninductive winding, the magnetic fields may well be greater than 10 Oe. The magnitude of the field is then dependent upon the temperature of the furnace. The inhomogeneity of the field depends strongly upon the furnace design. Application of a bifilar winding can reduce the field by two orders of magnitude.

This discussion will cover two different effects of magnetic fields. First, a magnetic couple will arise when permanently magnetized ferromagnetic impurities are present in the beam. Such a couple corresponds to a spurious mass change, Δm, satisfying

$$981 l \, \Delta m = VIH \tag{4.75}$$

where l is the length of the arm of the balance in cm, V is the volume of the impurities, per cm^3, I is the permanent magnetization of the impurities in gauss, and H is the external field in oersted. Using the following typical values: $V = 10^{-4}$ cm^3 for a highly impure beam material, $I = 1000$ G, $H = 0.1$ Oe and $1 = 10$ cm, it follows that $\Delta m = 10$ μg. This clearly demonstrates the need for avoiding ferromagnetic impurities. Secondly, a force will be developed whenever paramagnetic and ferromagnetic materials are placed in a inhomogeneous magnetic field. Taking the magnetization of those materials to be proportional to the field, the resulting spurious mass variation is given by:

$$981\,\Delta m = V\chi H(\partial H/\partial z) \qquad (4.76)$$

where χ is the susceptibility of the material and $\partial H/\partial z$ is the vertical component of the magnetic field gradient.

As one example, consider the forces acting on a soft ferromagnetic sample when the earth magnetic field is slightly inhomogeneous. Using the typical data: $V = 1$ cm^3, $\chi = 10$, $H = 0.2$ Oe, $\partial H/\partial z = 0.01$ Oe/cm, it is found that $\Delta m = 2$ μg. As a second example, consider a paramagnetic sample in the field of an inductive wound furnace. Substituting in equation 4.76 the values: $V = 1$ cm^3, $\chi = 10^{-5}$, $H = 10$ Oe, $\partial H/\partial z = 1$ Oe/cm, results in $\Delta m = 0.1$ μg. An appropriate furnace design can limit the magnetostatic disturbances, so that only the ferromagnetic materials in the balance need cause concern.

IX. BUILDING VIBRATIONS

As noted in Chapters 2 and 3, the elimination of building vibrations is of extreme importance to a successful weighing procedure. There has been, however, little detailed consideration of vibration problems.

The words "building vibrations" cover vibrations of three different origins. First, there are the vibrations originating from the ground on which the laboratory is built. Heavy traffic, for instance, is an important source of vibration of the soil. Second, some vibrations have their origin in the building itself, e.g., machinery, the slamming of doors, or the variation of wind pressure against the walls. Finally, the furniture or framing associated with the balance can generate vibrations or modify those from the other sources.

Vibrations can be reduced by an appropriate choice of the location of the balance laboratory and of the construction of the experimental setup. Since unfortunately these choices usually involve different people with widely diverging responsibilities, optimum vibration-free conditions are not always within reach of an experimenter. In general he must concentrate therefore on reducing the localized effects of the building vibrations.

Many of the various ways for achieving this reduction are based upon the "heavy table—weak spring" method. This method is illustrated with the example depicted in Figure 4.14. The table on which the balance is

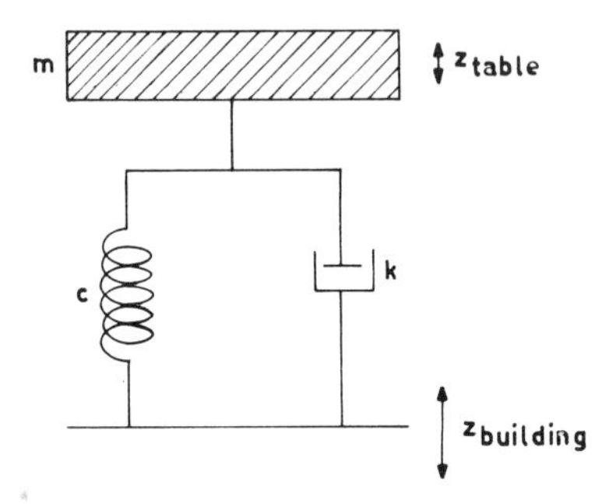

Figure. 4.14. Schematic picture of the "heavy table–weak spring" method for the reduction of building vibrations.

built is represented by the body of mass m. This body is supported by a spring with a constant C. The other side of the spring is connected to the building which is supposed to take part in a vertical vibration given by the equation:

$$z_{\text{building}} = z_{\text{building}} e^{j\omega t} \tag{4.77}$$

A damping device with a damping constant k is mounted parallel to the spring. We get the following equation of motion of the table:

$$m\ddot{z}_{\text{table}} + k\dot{z}_{\text{table}} + Cz_{\text{table}} = Cz_{\text{building}} + k\dot{z}_{\text{building}} \tag{4.78}$$

Using equation 4.77 the particular solution of equation 4.78 reads

$$z_{\text{table}} = \hat{z}_{\text{table}} e^{j(wt + Q)} \tag{4.79}$$

where

$$\hat{z}_{\text{table}} = \hat{z}_{\text{building}} \frac{C + j\omega k}{C + j\omega k - \omega^2 m} \tag{4.80}$$

The reduction of building vibrations is successful when the amplitude of the table is smaller than that of the building so when $\hat{z}_{\text{table}} < \hat{z}_{\text{building}}$ which with the use of equation 4.80 leads to

$$\omega > (2c/m)^{1/2} = \sqrt{2}\omega_0$$

This shows that the device is effective as long as the frequency of the oscillation w is greater than 1.4 times the free oscillation frequency ω_0 of the tablespring system. A low value of ω_0, which requires a large value of m and a small value of C, is therefore desirable.

REFERENCES

1. T. N. Rhodin, *Advan. Catalysis*, **5**, 39 (1953).
2. J. M. Thomas and B. R. Williams, *Chem. Soc. Quart. Rev.*, **19**, 231 (1965).
3. H. Pettersson, *Proc. Phys. Soc. (London)*, **32**, 209 (1920).
4. C. W. McCombie, Rept. Progr. Phys., **16**, 266 (1953).
5. J. A. Poulis and J. M. Thomas, *Vacuum Microbalance Tech.*, **3**, 111 (1963).
6. J. A. Poulis, *Proc. Phys. Soc.*, **80**, 918 (1962).
7. G. Sauerbrey, *Z. Physik*, **155**, 206 (1959).
8. A. W. Warner and C. D. Stockbridge, *Vacuum Microbalance Tech.*, **2**, 71; 93 (1962); **3**, 55 (1963).
9. I. Haller and P. White, *Rev. Sci. Instr.*, **34**, 677 (1963).
10. O. M. Katz and E. A. Gulbransen, *Vacuum Microbalance Tech.*, **1**, 111 (1961).
11. A. W. Czanderna, p. 129.
12. S. P. Wolsky, p. 143.
13. W. E. Boggs, p. 145.
14. A. W. Czanderna and J. M. Honig, *Anal. Chem.*, **29**, 1206 (1957).
15. A. W. Czanderna and J. M. Honig, *J. Phys. Chem.*, **63**, 620 (1959).
16. M. H. C. Knudsen, *The Kinetic Theory of Gases*, Methuen, London, 1934.
17. S. C. Liang, *J. Appl. Phys.*, **22**, 148 (1951).
18. S. C. Liang, *J. Phys. Chem.*, **57**, 910 (1953).
19. M. J. Bennett and F. C. Tompkins, *Trans. Faraday Soc.*, **53**, 185 (1957).
20. J. A. Poulis and J. M. Thomas, *J. Sci. Instr.*, **40**, 95 (1963).
21. J. A. Poulis, B. Pelupessy, C. H. Massen, and J. M. Thomas, *J. Sci. Instr.*, **41**, 295 (1964).
22. J. A. Poulis, C. H. Massen, and J. M. Thomas, *J. Sci. Instr.*, **43**, 234 (1966).
23. J. M. Thomas and J. A. Poulis, *Vacuum Microbalance Tech.*, **3**, 15 (1963).
24. C. H. Massen, B. Pelupessy, J. M. Thomas, and J. A. Poulis, *Vacuum Microbalance Tech.*, **5**, 1 (1966).
25. K. H. Behrndt, C. H. Massen, J. A. Poulis, and T. Steensland, *Vacuum Microbalance Tech.*, **5**, 33 (1966).
26. T. Steensland and K. S. Førland, *Vacuum Microbalance Tech.*, **5**, 17 (1966).
27. J. A. Poulis and C. H. Massen, *Proc. 1965 Trans. 3rd Intern. Vacuum Congr.*, Vol. 2, Pergamon Press, New York, 1966, p. 347.
28. J. M. Thomas and B. R. Williams, *Vacuum Microbalance Tech.*, **4**, 209 (1964).
29. C. H. Massen and J. A. Poulis, *Vacuum Microbalance Tech.*, **6**, 17 (1967).
30. J. A. Poulis and C. H. Massen, *J. Sci. Instr.*, **44**, 275 (1967).
31. J. A. Poulis, C. H. Massen, and B. Pelupessy, *Vacuum Microbalance Tech.*, **4**, 41 (1964).
32. J. A. Poulis, *Appl. Sci. Res. A*, **14**, 98 (1965).
33. C. H. Massen, J. A. Poulis, and J. M. Thomas, *J. Sci. Instr.*, **41**, 302 (1964).
34. C. H. Massen, J. A. Poulis, and J. M. Thomas, *Vacuum Microbalance Tech.*, **4**, 35 (1964).
35. S. P. Wolsky, E. J. Zdanuk, C. H. Massen, and J. A. Poulis, *Vacuum Microbalance Tech.*, **6**, 37 (1967).

Chapter 5

Micro Weighing with the Quartz Crystal Oscillator—Theory and Design

A. W. WARNER

Bell Telephone Laboratories, Inc.,
Murray Hill, New Jersey

I. THE QUARTZ RESONATOR

A. Introduction

The fact that an oscillating quartz plate may be used to detect a change in mass has been known almost as long as quartz crystal oscillators have been in use. For example, in the early days of radio it was common practice to lower the transmitter frequency by marking the surface of the controlling quartz plate with a pencil, thus adding an adhering mass of graphite. In this chapter a review of the underlying design parameters affecting the sensitivity and accuracy of such a frequency–mass relationship is presented, as well as information useful in the practical application of quartz plates to both simple and highly sophisticated micro weighing.

The quartz resonator used for micro weighing is basically a simple device. It consists of a quartz plate having a mechanical resonance frequency inversely proportional to its thickness, i.e., $f = K/t$. For the commonly used flat AT cut plate,* $K = 1670$ m Hz, and because of the very high Q (1) (low internal friction) of quartz the resonant frequency may be easily measured electrically through the piezoelectric effect. A simple view is that an added mass in the form of a thin film has an effect on frequency very nearly that of an equivalent mass of quartz. The added mass may be roughly determined by translating frequency change into an equivalent thickness of quartz and then into mass by means of the known density of quartz.

Among our basic physical quantities, frequency is probably the one most precisely measurable. The frequency of the 133 resonance in cesium is known and usable to 12 significant figures (2). One of the factors contributing to the extremely high sensitivity for the quartz resonator method of mass measurement is that the frequency of resonating quartz plates may be compared to a reference frequency to at least one part in 10^{10}. Another is the high degree of sophistication and development already achieved in the design of quartz crystal units for frequency control (3), making possible crystal plates having a frequency stability of a few parts in 10^{10}, over intervals as short as 1 sec or as long as 1 month, unaffected by the normal conditions of shock, vibration, and temperature encountered in the laboratory.

For purposes of quantitative micro weighing, only AT or BT cut (4) quartz plates are useful (AT and BT refer to the orientation of the plate with respect to the crystal structure). These are the two high-frequency mode plates which vibrate in shear about an axis parallel to the major surface, have low or zero temperature coefficients at the temperature of use†, and have surfaces that are antinodal in displacement (i.e., zero strain at the surface of the plate). The AT cut is superior in temperature coefficient and in mass sensitivity to the BT. Data on these cuts are given in Figure 5.1. Each curve represents a very slight change in orientation with respect to the crystallographic axes of quartz.

* See Section III.A of this chapter and reference 4.

† At high temperatures, e.g. > 300°C, it may be necessary to use a specially designed plate. A.E.Ü., **17**, 75–84 (1963).

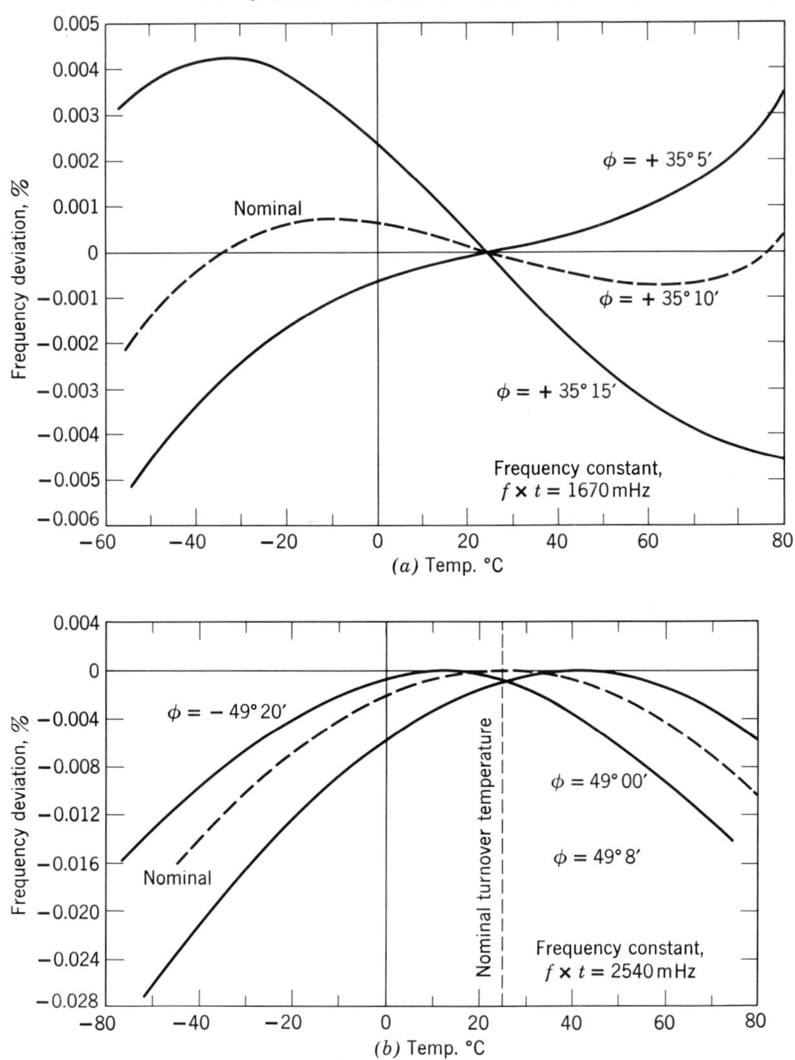

Figure 5.1. Frequency vs temperature data for (*a*) AT cut and (*b*) BT cut fundamental mode quartz plates. Each curve represents a slightly different angle of rotation about the X axis.

A theoretical calibration of frequency vs. mass could, of course, be based on the elastic and piezoelectric constants (5) of quartz. However this is a rather involved procedure, for although the constants are accurately known, the calculations for a particular quartz plate would involve a knowledge of all of the physical dimensions of the plate and the crystallographic orientation of the major surfaces.

A preferable method is the use of a small-mass perturbation analysis (discussed in Section II.C) which can be verified for masses that can be experimentally determined and which is increasingly valid as the added mass decreases. The application of this calibration involves only a knowledge of frequency and mass. It is usually sufficient to show the linearity of the added mass versus change in frequency relationship over some measurable range in order to establish a frequency constant that may be used for any lesser mass. In applying this calibration, the only significant limitation on accuracy will be due to the nature of the experiment being performed, i.e., the nature and distribution of the added mass, since frequency can be very accurately determined and the physical dimensions of the quartz resonator need not be involved.

The quartz oscillator, under the best conditions that the state of the art permits (6), is capable of resolving mass changes as small as 1 pg (10^{-12} g), with an accuracy of perhaps 0.1%. In other words, a change in mass of about 1 pg can be detected, but the absolute value of the total mass is not likely to be known to more than three significant figures.

II. PRINCIPLES OF MICRO WEIGHING WITH QUARTZ RESONATORS

A. General

The mass that is determined by the use of a quartz resonator is inertial mass. It is therefore independent of gravity, and may differ in some theoretical sense from gravitational mass. It is the mass in the Newtonian equation stating force equals mass times acceleration. By its nature, this method of mass determination in its strictest sense is limited to tightly adhering, rigid materials. Further, by the nature of the quartz resonator and its calibration, the method is limited to uniformly distributed thin films if quantitative results are required.

On the other hand, since any adhering added mass will cause some change in the resonant frequency, the sensitivity of the quartz crystal resonator may be used to great advantage in a number of closely related measurements, where the accuracy of mass measurement is not of prime importance. For example, the resistive and reactive loading resulting from the pressure of an inert gas can be readily observed (7),* and for an active gas where adsorption takes place at the crystal electrode, the mass change due to adsorption can be separated from the reactive loading by careful analysis. Meaningful measurements have also been made by coating a resonator with a liquid or viscous substrate and subsequently measuring the change in mass upon interaction with various gases (8).*

The quartz resonator also responds in a characteristic manner (Figure

* Also see the discussion of applications in Chapter 7.

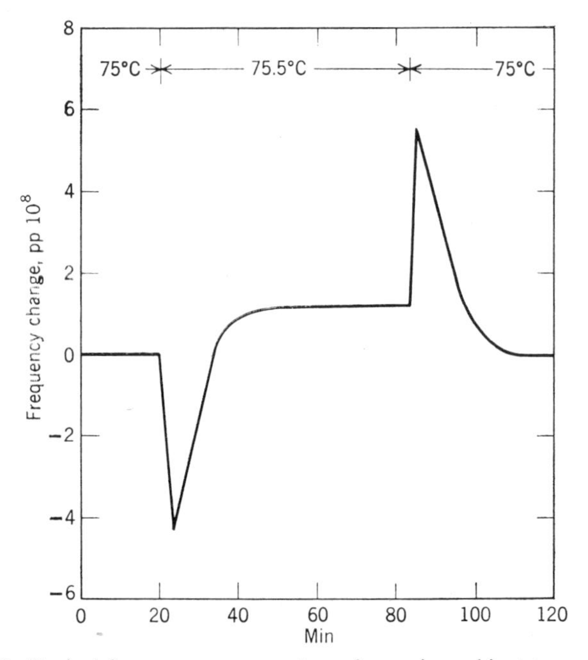

Figure 5.2. Typical frequency response to a change in ambient temperature. Any thermal disturbance which produces a thermal gradient in the quartz will produce a similar effect.

5.2) to very small temperature gradients in the quartz. This sensitivity to thermal shock can be used to indicate when a particular reaction on its surface is endothermic or exothermic. Sudden changes in temperature as small as 0.001°C can be detected. This temperature gradient effect is a function of relative plate thickness and has not been calibrated. The effect is undoubtedly related to a transient temperature gradient within the quartz which alters the elastic constants. Stockbridge (9) carried out an analysis assuming a small discontinuity in temperature at the boundary of the plate in order to relate the peak frequency excursion to the temperature change at the surface. He obtained the relationship

$$\alpha\,\Delta T = (1/\beta)(\Delta f/f),$$

where α is the depth of the surface layer having a temperature difference ΔT from the center of the plate, and β is the coefficient of linear expansion divided by twice the thickness.

B. Choice of Design Parameters for Optimum Accuracy

The usefulness of a quartz plate resonator in detecting and measuring small mass changes will in large measure depend on the ability to keep all other sources of frequency change low in comparison to that caused by mass loading. Therefore the highest accuracy will not be at the highest sensitivity, but at the sensitivity which produces the most significant numbers. For example, the use of extremely thin high-frequency plates may defeat their purpose by introducing more errors due to larger variations in frequency with amplitude of vibration and lowered Q than could be compensated for by increased sensitivity to mass.

The choice of quartz as the piezoelectric material for microweighing stems from the requirement that all other sources of frequency change be small. Quartz is chemically and physically stable to a high degree, has an inherently high Q (3×10^6 at room temperature for the AT cut at 5 MHz) and can be made to have a low temperature coefficient of frequency at the temperature of interest (e.g. 1 part in $10^8/°C$ over the range of 38–42°C).

By choosing a frequency such that no one cause of frequency instability predominates, the effective sensitivity will be high. For example, should it be technically necessary to work with a temperature uncertainty of $\pm 2°C$, which would limit frequency measurement to 10^{-8}, one would not choose the high Q 5 MHz plate which is capable of frequency stability of 10^{-10} with 0.01°C control. Instead one would choose a thinner plate operating at a higher frequency and thus obtain more sensitivity to mass. The limitation on raising the frequency would of course be the increasing sensitivity to amplitude control and lowered Q mentioned above. This subject is discussed in more detail in Section III.B.

C. Quantitative Relationship between Frequency and Added Mass

As pointed out at the beginning of this chapter, there is a rough equivalence between a simple thickness-shear mode quartz plate characterized by frequency proportional to thickness, and a *loaded* thickness–shear mode quartz plate in which the added thickness does not store strain energy or participate in the piezoelectric effect. However, the effect of an added thin film cannot be considered the *exact* equivalent of the same mass of quartz. In the all quartz case, the proportional change in frequency with mass is a function of the total thickness of the plate, and therefore the sensitivity to mass varies as the mass is increased. On the other hand, an analysis, given below, of a composite resonator shows that frequency is affected to a first degree only by the ratio of the added mass to the fixed mass of a given quartz resonator.* Consequently the thickness–shear mode quartz re-

* A recent work by Eer Nisse (10) applies the approximation techniques of variational calculus to this problem and will be of interest to those wishing to consider the formulation of a theoretical calibration of frequency versus mass beyond the analysis given here.

sonator when used under the proper conditions approaches a true mass sensing device, i.e., the density of the added uniform film is not involved.

Onoe (11) has considered the problem of the quartz plate loaded by a surface film of material for the general case, which includes not only the high-frequency thickness modes but also the contour modes. (For the contour modes the surface is not nodal in strain but is distorted, thus the resonant frequency depends upon more than just the mass of the added material.)

From a consideration of the energy relation in a vibrating body (12), the change of the resonant frequency, f, due to mass deposited uniformly on the surface can be expressed in the form:

$$(\Delta f/f) \approx \tfrac{1}{2}[(P'/P) - (K'/K)] \tag{5.1}$$

in which K and P are, respectively, the time averages of kinetic and potential energy in the quartz plate itself. The primes indicate the corresponding quantities in the added mass. For low-frequency contour modes, the change in frequency upon adding mass is difficult to predict and may even reverse direction because of a change in stiffness.

However, for a thickness–mode high Q resonator, where a nondissipating mass is added at the surface, P' is zero and only the ratio of the kinetic energies need be considered. The ratio of kinetic energies will be:

$$\frac{K'}{K} = \frac{(2\pi f)^2 \rho' \iiint u^2\, dv}{(2\pi f)^2 \rho \iiint u^2\, dv} \tag{5.2}$$

in which u is displacement, v is volume, and ρ is density. Without measurable error this becomes:

$$K'/K = \rho' u_{\max}^2 A' t' / \tfrac{1}{2} \rho u_{\max}^2 At \tag{5.3}$$

in which A and t are area and thickness, respectively. Therefore combining equations 5.1 and 5.3

$$\Delta f/f \doteqdot -(\rho' A' t' / \rho A t) = -m/M \tag{5.4}$$

in which M is the *applicable* mass of the quartz resonator and m the uniformly distributed added mass giving rise to the change in frequency, Δf, and f is the resonant frequency of the unloaded quartz plate. By using a uniformly distributed mass, A' may be assumed to be equal to A and

$$(m/A') = -(\Delta f/f)\rho t \tag{5.5}$$

In other words, mass per unit area of an unknown material may be determined without knowing the area A. If the area exposed and used for M and m is *much* less than the active area of the plate, a second-order correction may be in order.

Where contoured or shaped quartz plates, desirable for high Q (1) and uniform temperature characteristics, are used, the average value for the thickness of the quartz plate may be derived from the known frequency constant for flat plates, e.g., 1670 kHz mm for AT cut quartz. In any case, the errors are of the order of 1–2%, and are principally due to nonuniformity of the added mass. Sauerbrey (13), using a vacuum torsional microbalance, so placed as to receive an identical film to that placed on the quartz plate, and accurate to $\pm 2\%$, found no systematic deviations from theoretical values. Stockbridge (14) using gold on a high Q, contoured, 2.5 MHz overtone quartz plate also found agreement within 2%, using a mass range of 700–7000 μg and an external microbalance. Other investigators (15) have used copper, silver, and other metals and have found good agreement with theory.

Having established this validity over the range of 700–7000 μg for the contoured 2.5 MHz unit (0.15–1.5 mg/cm^2), one may safely extrapolate the calibration to smaller masses, where the assumptions used in deriving equation 5.4 become increasingly valid.

At some upper limit where the added mass is an appreciable fraction of the mass of the vibrating system, the calibration curve deviates from linearity for one or more possible reasons: (*1*) potential energy may be stored in the added material, (*2*) the bond between the quartz and the film may be severed, or (*3*) second-order terms may become significant. One can differentiate somewhat between these factors by observing whether such nonlinearity is a function of mass or a function of the particular material, and whether or not a change in Q (observable as a change in electrical resistance) takes place.

D. Measurement Techniques

The quartz plate has a mechanical resonance, sensitive to mass loading, whose properties may be described in electrical terms by making use of the piezoelectric effect. The properties of interest are only those of the loaded piezoelectrically driven resonator, and the electronic measuring equipment need have no effect on the value of the resonance frequency.

The equivalent electrical circuit of a crystal plate with electrodes deposited directly on its major surfaces is shown on Figure 5.3. Such a circuit will have a series resonance (f_s) and an antiresonance (f_a) separated by (16)

$$(f_a - f_s)/f_s = C_1/2C_0 \tag{5.6}$$

The series resonance frequency is defined as the frequency of zero reactance in the series arm. If the added mass is nondissipative, and if it is added at a displacement antinode, the only change will be in the inductance, L_1, and the proportional change in f_a will be the same as in f_s. Under these conditions, any oscillator circuit may be used, regardless of where the operating point may lie in relation to series resonance and antiresonance, providing some means is available to monitor the stability of the circuit.

On the other hand, if there are changes in Q (such as would be caused by gas loading) which would not only detune an oscillator circuit but which would also alter the amplitude of vibration, a bridge circuit may

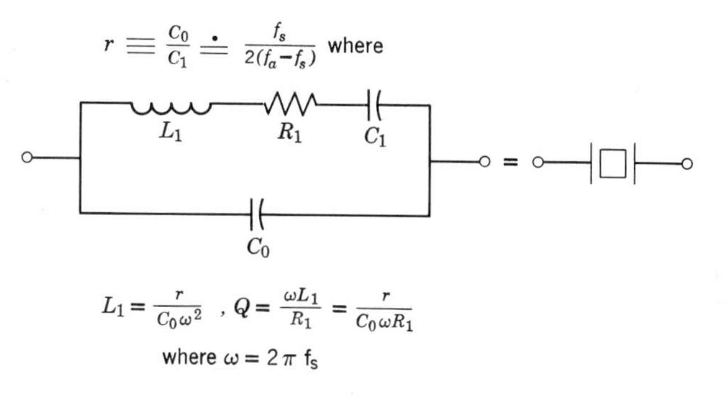

$$L_1 = \frac{r}{C_0\omega^2}, \quad Q = \frac{\omega L_1}{R_1} = \frac{r}{C_0 \omega R_1}$$

where $\omega = 2\pi f_s$

Figure 5.3. The equivalent electrical circuit of a lightly damped, piezoelectrically excited resonant quartz plate.

become necessary in order to exactly define the operating conditions. Again it does not matter what operating condition is chosen, as long as the ratio of C_1 to C_0 remains fixed, although the series resonance frequency, f_s, is probably the easiest to use.

E. Summary of Advantages and Disadvantages of Quartz Crystal Micro Weighing

The principal limitation is that the mass to be measured *quantitatively* must be in the form of an adherent uniformly distributed thin film. The maximum mass is probably limited by adherence problems to 2 mg/cm^2. Further, corrections must be made for both static and dynamic effects of gas pressure, if there is any appreciable change in pressure or composition of an ambient gas.* Within these limitations, however, there are several outstanding advantages.

The advantages of the quartz microbalance are the following: (*1*) The sensitivity is very high (under ideal conditions, which include temperature control to 0.01°C, 10^{-12} g/cm^2 change in mass can be measured. This is, perhaps, four orders of magnitude beyond the sensitivity of microbalances). (*2*) The cost in dollars per magnitude of sensitivity is quite reasonable. (*3*) The apparatus is unaffected by moderate shock and vibration or by position. The quartz sensing element may be made very

* See Chapter 7, equation 7.10.

small, and may be positioned to best advantage in an experiment. (*4*) The apparatus may be used for the dynamic measurement of a changing mass. (*5*) In general, the contamination introduced into an experiment by the sensing element is negligible—only such materials as quartz, gold, and stainless steel are used and all exposed parts are smooth and easily vacuum baked.

While it becomes obvious that quartz resonator micro weighing is best suited to measurements of thin films, there are other uses that may be of particular importance in specialized instances. For example, if an independent calibration or criterion exists, the high sensitivity may be used for detecting changes in a wide variety of gases and liquid films, as well as in materials not expected to meet the requirement of being nondissipative. Almost any material, gas, liquid, or solid, when in adhering contact with the crystal will change its frequency. Unknown inert gases may be identified by density and viscosity by observing frequency versus pressure. Known reactive gases may be studied by observing dynamic mass changes due to sorption, and studies involving complex molecules may be carried out by observing complex viscosity. These and other applications are discussed further in Chapter 7.

III. SPECIFIC DESIGNS

A. Design of a Simple Quartz Resonator for Micro Weighing

For moderate weighing sensitivity, capable of detecting a change of $\pm 1 \times 10^{-7}$ g/cm^2, quite simple procedures can be followed. The quartz plate may be flat. An AT cut (3) is preferable because of its low temperature coefficient over a wide range of temperature and its ability to respond linearly to a change in the mass of a uniform film. An AT cut in the IRE system (17) is $YXl + 35°$, the plate having an X axis in the plane of the plate and an orthogonal Y' axis also in the plane of the plate $+35°$ from the crystallographic Y axis. The exact angle will depend on the desired operating temperature as shown on Figure 5.1 or on Figure 5.4. A semi-polished surface is desirable, but not essential, to facilitate cleaning and stabilizing the quartz unit. A frequency of 5–10 MHz is best, with a ratio of width to thickness of about 50 in order to maintain reasonable Q. This results in a quartz plate between 8 and 15 mm in width. The shape of the plate is unimportant, except that a wider frequency range free from interference by other modes may sometimes be obtained by varying the dimensions or shaping or bevelling the edges.*

A thin metallic electrode for electrical contact is first deposited by vacuum evaporation directly on each major surface of the quartz plate,

* See Chapter 7, Section II.

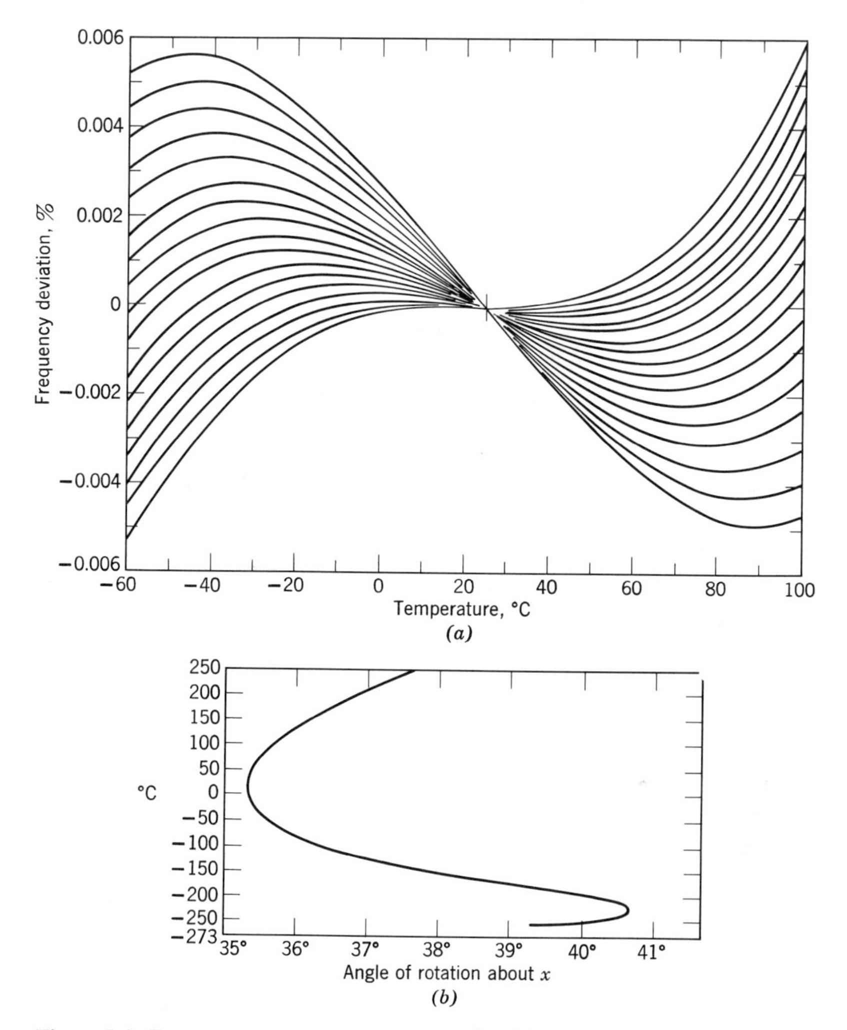

Figure 5.4. Frequency vs temperature curves for AT cut quartz plates. Each curve in (*a*) represents a slightly different orientation about the *X* crystal axis. The angle of rotation to give a zero temperature coefficient of frequency at a particular temperature is given in (*b*).

and then some masking device is usually provided to limit added mass to an area on one face somewhat less than this electrode area. The mask also prevents shorting of the electrodes when the added mass is a conducting film. The equations of interest are:

$$\text{frequency} \times \text{thickness} = 1670 \text{ MHz} \tag{5.7}$$

$$(\Delta f/f) = -(m/A)\,(1/\rho_q t_q) \tag{5.8}$$

where m/A is the mass per unit area added to one side giving rise to the proportional change in frequency, $\Delta f/f$, and ρ_q and t_q are the density and thickness of the quartz plate, respectively. A frequency of 6.65 MHz will result in a ratio of frequency change to mass change of 100 Hz/μg per cm^2 or 1 Hz for each 10^{-8} g/cm^2 change in mass.

Adequate temperature control can usually be obtained by holding the quartz plate against a water stabilized stainless steel block which serves as one electrical connection. A spring clip serves as the other connection and holds the plate in place.

A very simple oscillator circuit will suffice, since the frequency need be measured only to a few parts per million. Frequency measurement is accomplished by comparison to another oscillator by means of the usual frequency meter or counter.

A simple oscillator circuit is often built into the water-cooled block that holds the crystal plate. A single shielded wire is used to conduct both power in and signal out. The mounted crystal and circuit are usually placed in a vacuum chamber, and any changes in mass on the crystal surface may be continuously monitored. Apparatus of this sort for thin film monitoring is commercially available from a number of manufacturers.*

In 1948 the author constructed a small self-contained thin-film calibrator for use in standardizing the electrodes used in various designs of crystals units. The apparatus consisted simply of a "crystal duplicator" (two oscillator circuits, one mixer, frequency meter, and power supply) equipped with a special holder to accept a 12 mm square quartz blank with 6 mm diam round electrodes. The unit of measurement, borrowed from electroplating usage, was the milligram per square inch (msi). An array of eight such quartz blanks had an area of 1 sq. in., and could be weighed for a direct calibration. BT cut quartz plates, 6.2 MHz were used and 3 msi, the standard coating, equalled 50 kHz change in frequency. The 50 kHz scale on the duplicator was remarked to read in msi. By use of an array of blanks, all coated at the same time using a single multi-crystal mask, not only the average weight of coating but also the distribution could be checked in any given vacuum coating apparatus.

B. Design of Quartz Plates and Circuitry for Ultrahigh Sensitivity to Mass

1. General

As pointed out at the beginning of this chapter, the inherent stability and low internal friction of quartz resonators is such that an evenly distributed mass as small as 1×10^{-12} g/cm^2 can be detected and measured. To achieve this "state of the art" sensitivity, however, is by no

* See Chapter 14.

means simple either in design or in subsequent use. As the required sensitivity goes from 10^{-8} to 10^{-12} g/cm^2, the practice of quartz micro weighing necessarily becomes more specialized and limited in its application. The vacuum baking and subsequent stabilization, for example, may take as long as 1 month. On the other hand, for the study of surface phenomena involving small mass changes, thermal disturbances, and/or effects of various gases, the method has no equal and in the opinion of the author is deserving of more widespread use.

One must now consider in greater detail every factor that can change or mask the resonant frequency of the quartz plate. The objective is to arrive at a quartz plate design that will have the highest *significant* mass sensitivity for the particular experimental environment involved, i.e., a plate which will have the least frequency change due to unknown or uncontrollable factors in comparison to its frequency sensitivity to mass.

Often the use of more than one crystal plate, each of a different design, is necessary in order to differentiate among the many possible sources of frequency change, as well as to aid in identifying the process responsible for a particular small change in mass.

2. *Temperature Coefficient Considerations*

The limiting factor most difficult to deal with is temperature control. If the study involves the production of a large amount of heat, there is no way to avoid a period of time for temperature equilibrium. Of course, predetermined temperature information may be used (18) to apply correcting factors for the effect of temperature and temperature gradients on the resonant frequency, but in general this is not very practical. For small amounts of heat, a successful system has been the use of three dissimilar quartz plates to (*1*) detect the thermal disturbance, (*2*) detect when static conditions have returned, and (*3*) indicate the shift in temperature, if any. However, before describing this three-crystal system, other factors must be considered.

The temperature coefficient of an AT cut quartz plate is a function of orientation with respect to its crystallographic axes. This "angle of cut" not only controls the temperature at which the temperature coefficient is zero, but at the same time controls the slope of the temperature versus frequency curve away from zero. The result is a family of curves about as shown on Figure 5.4*a*. The locus of the zero temperature coefficient follows the curve in Figure 5.4*b*. The exact shift in zero temperature coefficient point with angle, and for any given angle the temperature range over which a tolerance of ± 1 part is 10^8 can be maintained, is shown on Figure 5.5. The best conditions are obtained at room temperature, but to produce such plates is nearly impossible because only a few seconds change in orientation angle will shift the zero coefficient by 15°C. A good compromise has been found to be 40°C, where the temperature coefficient of frequency is better than one part in 10^8/°C over a 4°C temperature range. If the experiment requires it, the temperature of zero temperature coefficient can be put at any temperature up to 500°C, although as the operating temperature departs from room temperature, the slope away from

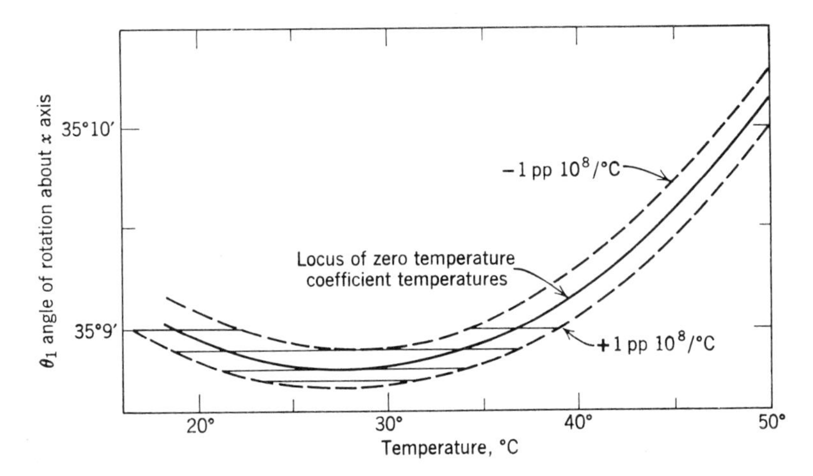

Figure 5.5. Detailed information on AT cut quartz plates near room temperature. The solid curve is the locus of the zero temperature coefficient orientation angle plotted as a function of temperature. The dotted curves permit an estimate of angular tolerances and operating range.

the zero point becomes much steeper. An exception to this occurs at temperatures near absolute zero. At temperatures below about 12°K the temperature coefficient is negligible regardless of "angle of cut," although the temperature gradient effect remains.

3. *Balanced Design*

It is important to note that the temperature coefficient of frequency is independent of the operating frequency. Since the mass sensitivity in terms of proportional frequency change per unit of mass varies directly with the frequency of the unloaded quartz plate, the use of thinner, higher frequency plates would reduce the adverse effects of temperature fluctuations.

Unfortunately, the Q of quartz is an inverse function of frequency as can be seen on Figure 5.6 for AT cut quartz plates. As the frequency increases, one rapidly loses the ability to define or delineate the exact resonance frequency, not only because the Q decreases, but also because requirements on circuit elements for phase stability increase. In other words, where a higher Q is needed to hold a given percentage frequency stability, the Q actually becomes lower. Also, unfortunately, it is an inherent property of quartz that as the frequency increases by virtue of

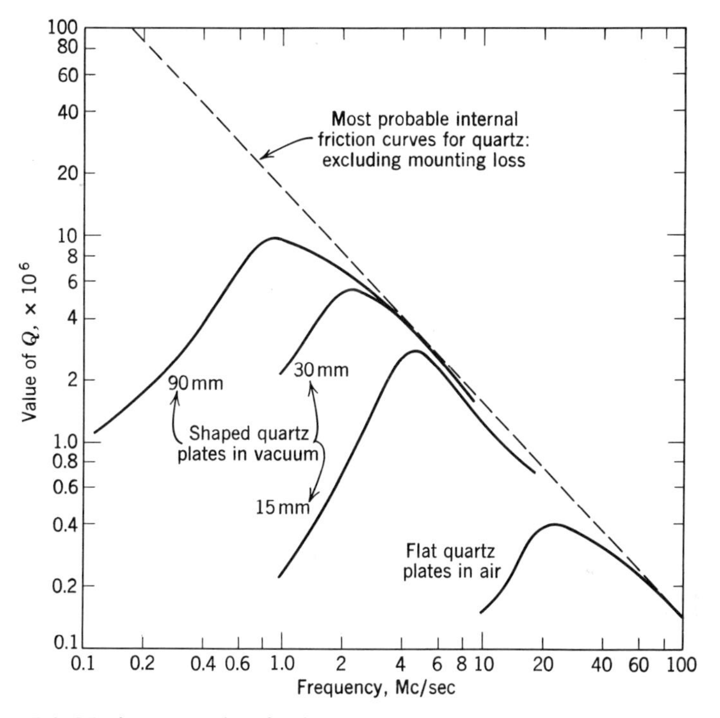

Figure 5.6. Maximum Q values for AT cut quartz plates 15, 30, and 90 mm in diameter as a function of frequency.

making the plate thinner, the frequency becomes increasingly sensitive to amplitude of vibration. This means that the oscillator circuitry will be required to maintain a more constant, low level, noise-free crystal current.

Figure 5.7, which gives data for a fundamental mode AT plate, is an attempt to ullustrate (*1*) the conflicting factors in selecting a frequency for a given experiment, and (*2*) that selection is mostly a matter of the available temperature control. Operating in the shaded area shown in Figure 5.7 offers no advantage, unless by more sophisticated circuit design, the limits, which are an estimate of present state of the art, are improved. For this reason, 2.5 and 5 MHz crystals are usually used, both for micro weighing and for frequency standards.

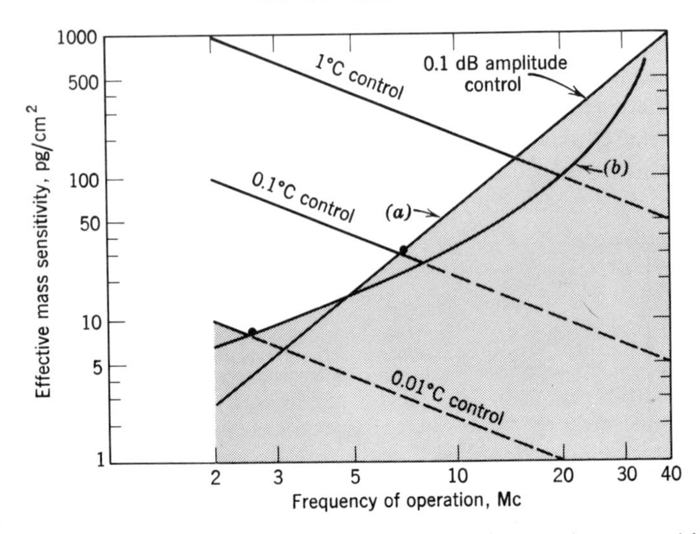

Figure 5.7. Engineering estimate of the limitation to increased mass sensitivity by an increase in the frequency of a fundamental mode, thickness–shear quartz plate, for various degrees of temperature control. Curve (*a*) is the limit due to frequency sensitivity to amplitude of vibration and curve (*b*) is the limit due to crystal Q and lowered impedance levels.

4. Energy Trapping

In the above discussion, the limiting Q was assumed to be that of the quartz itself. To achieve this Q, which is 5×10^6 at 2.5 MHz for an AT cut, rather severe contouring or energy trapping is used to keep from losing acoustic energy at the edge or mounting points of the quartz plate. This purposeful deviation from a flat plate, fortunately, has very little effect on the calibration or sensitivity of the quartz crystal plate, and is very necessary to keep the size within practical bounds. Figure 5.6 gives some data relating size, contouring, frequency, and Q.

5. Overtone Mode of Operation

Often, reference is made to the use of an overtone mode of vibration both for mass sensing (19) and for frequency standards (2). A proper understanding of the use of overtone modes is rather involved, but essentially it is the operation of a given quartz plate at some odd multiple of its fundamental resonance frequency. The advantages are rather subtle, being, at a given frequency, an increase in electrical impedance and a stiffening, or improvement, in the shift of frequency with a shift in circuit phase. The increased impedance for a given Q, however, is accompanied by an increase in resistance, requiring more circuit gain so that the overall result is an improvement from better impedance matching to connecting cables and circuits rather than from any inherent resonator characteristic.

An overtone resonator plate, operating at 5 MHz on its 5th overtone is five times as thick and 1/5 as sensitive to mass as a 5 MHz fundamental plate, and has been found to be five to ten times more stable as indicated by the oscillator output frequency. This is a large gain for frequency standard use, but it is an even choice as far as mass measurement is concerned. The thicker plate is somewhat more sensitive to thermal shock, which may or may not be desirable, depending on the experiment. Also, detuning of the circuit by gas loading on the quartz plate has less effect when the overtone unit is used. The principal use of overtone operation is to provide a quartz plate whose temperature coefficient, mass sensitivity, and thermal shock characteristic is radically different from that of a fundamental unit.

6. Use of Parallel Field

In order to excite an AT cut quartz plate, it is not necessary to use an electric field perpendicular to the major faces such as is provided by placing electrodes on each face. Essentially the same mechanical resonance can be excited by a field parallel to the major faces as shown on Figure 5.8. Here

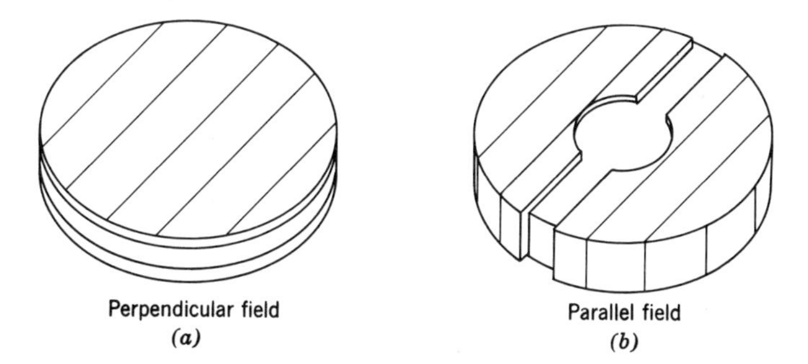

Figure 5.8. Arrangement of electrodes for perpendicular and parallel field excitation of the quartz plate.

for practical reasons, the electrode is brought around on the major faces so that a reasonably low impedance, about that of the 5th overtone perpendicular-field plates, is obtained. Although the usual perpendicular-field is the easiest to apply and use, the parallel-field excited plate has certain advantages. The principal one is that it removes the electrode material from the mass sensing portion of the crystal plate, thus permitting experiments where mass change at a quartz surface is to be measured. Another advantage, as shown on Figure 5.9 is that the sensitivity to thermal shock is drastically reduced. The frequency versus temperature characteristics are about as shown on Figure 5.5 for fundamental perpendicular-field plates, except that an angle of cut 7 min of arc higher is

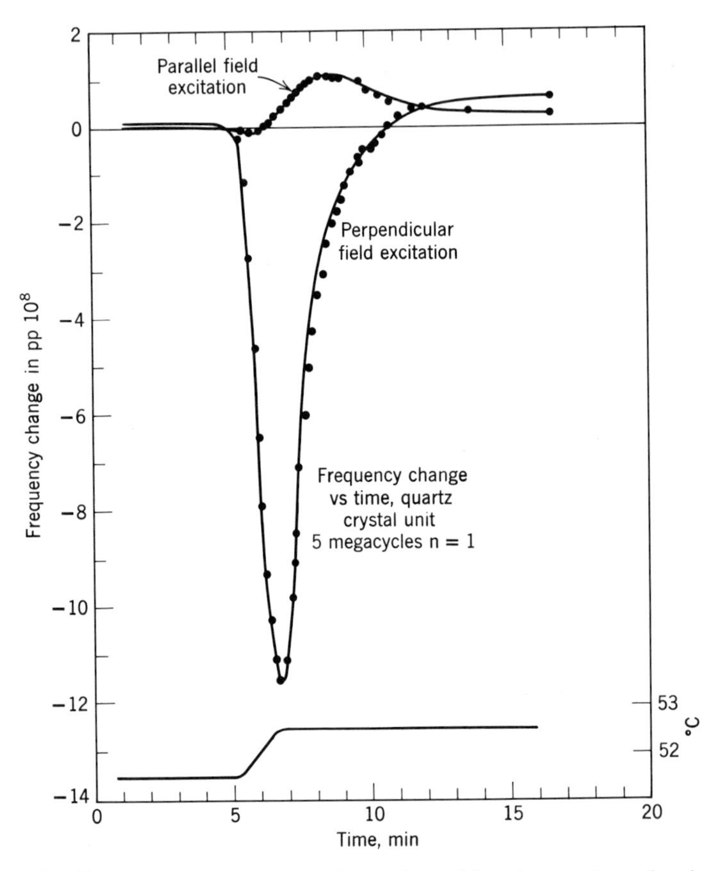

Figure 5.9. Frequency response to a change in ambient temperature showing one advantage of the parallel-field excited plate.

required. Again, as in the case of the overtone unit, the principal use for a parallel-field excited plate for micro weighing is to provide a quartz plate of contrasting characteristics for use in a multiple-crystal, ultrasensitive, mass-sensing head.

7. Multiple Crystal, Mass Sensing Unit

The design of the crystal plate for ultrasensitive micro weighing is, as indicated on Figure 5.7 for the fundamental mode, a compromise, taking into consideration the inherent properties of quartz as well as the many possible variations in construction and assembly. The use of several crystal plates of different design in one sensing head aids greatly (*1*) in differentiating among thermally induced, stress induced, and mass induced frequency change; (*2*) in providing a confirmation of true mass change; (3) in differentiating between various kinds of sorptions; and (4) in general providing for a wide variety of experiments.

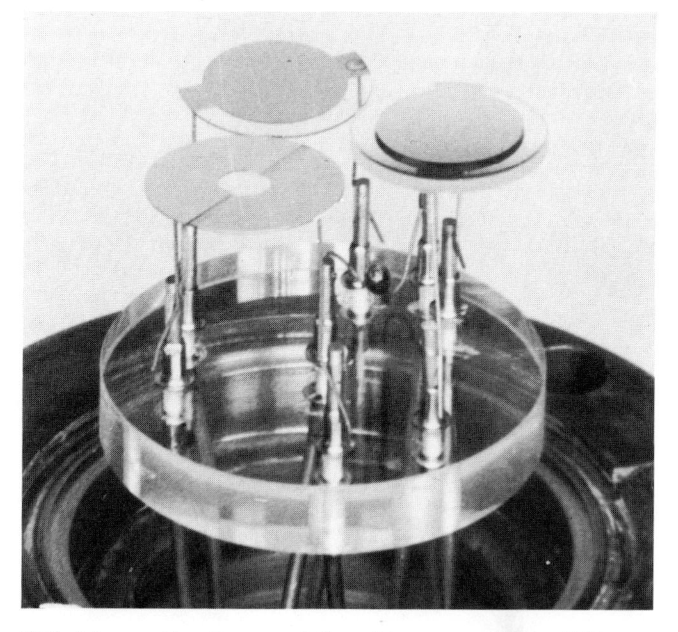

Figure 5.10. Mass sensing array consisting of two perpendicular-field quartz plates, (fundamental and overtone mode) and one parallel-field plate, and a temperature sensing thermistor.

The sensing head used by the author (20) is shown on Figure 5.10. This head consists of three 5 MHz AT cut crystal plates whose characteristics are shown in Table 5.1. The thick plate is the 5th overtone mode. The plate with the electrode absent from the center is the parallel field. The remaining plate, a fundamental mode, perpendicular field unit is the principal mass sensing element.

The crystal plates are mounted using a headed wire, thermocompression mounting technique (21), and a polished glass baffle plate. This in turn is mounted on a high vacuum flange so that the whole assembly may be vacuum baked.

It is essential that the electrical connections be thermally locked to the

Table 5.1. Frequency Characteristics

Crystal parameter	Overtone mode 5 Mc/sec plate	Fundamental mode 5 Mc/sec plate	Parallel field excited 5 Mc/sec plate
Thermal shock effect	Slow recovery	Fast recovery	Fast recovery, reduced effect
Temperature coefficient	Selected for $+10^{-8}$/°C	Selected for -10^{-8}/°C	Near zero
Added mass to change frequency by -10^{-10}	20 pg/cm^2	4 pg/cm^2	4 pg/cm^2
Active area surface	Gold	Gold	Quartz

Figure 5.11. Mass sensing head, showing coaxial cable connectors, thermal transfer wiring box, and vacuum sealing flange (Varian).

temperature-controlled housing, for in vacuum a large part of the heat transfer to and from the quartz plates is by way of the mounting wires. In the header pictured in Figure 5.11, thermal lock was accomplished by filling the wiring box with silver-loaded Dow Corning compound. Other methods include (*1*) the coating of a portion of the outside of the oven with a thin coat of alumina and bringing the connections from the top terminals to the outside by means of a metallic film deposited on the alumina, and (*2*) the use of beryllia (beryllium oxide) which is a good thermal conductor. A circuit attached to the header was also tried, but it was found that even the small power involved in the transistor circuits was enough to upset drastically the desired ($\pm 0.01°C$) temperature control.

8. Vacuum System for the Multiple Crystal Sensing Unit

The rest of the assembly, which was initially used to study the various causes of aging in quartz crystal units (20), is shown in Figure 5.12. Some of its features are:

1. A vacuum of 10^{-10} torr can be readily achieved.
2. The entire apparatus can be baked to 400°C.
3. The crystal plates can be isolated from the pumps and gauges by bakeable 1 in. valves.
4. The crystal plates can be temperature controlled to $\pm 0.01°C$ at any temperature from 30 to 100°C.
5. The crystal plates can be continuously monitored for frequency and

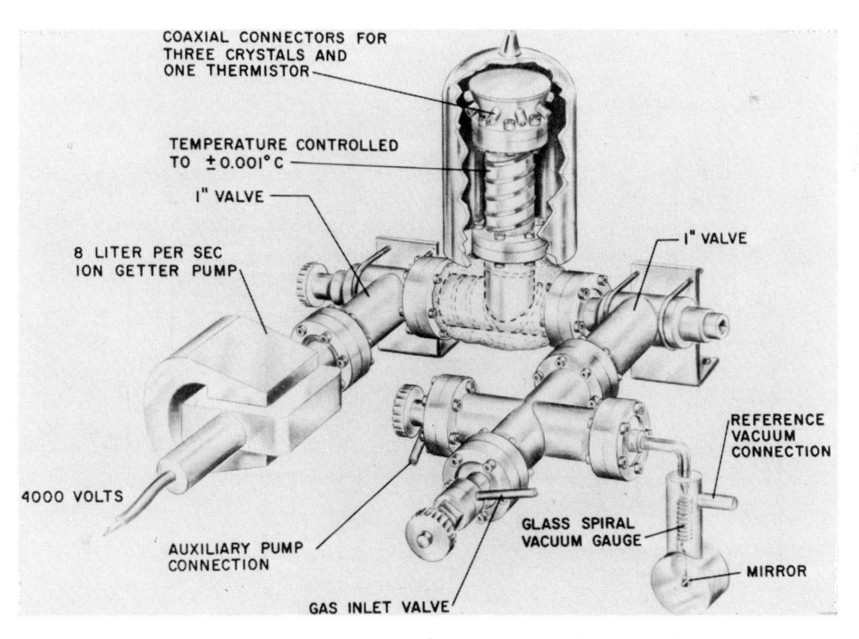

Figure 5.12. Vacuum system used for extreme environmental control during micro weighing.

Q at any temperature to 200°C, with a frequency accuracy of 1 part in 10^{10}.

6. The crystal environment can be analyzed by means of a mass spectrometer.
7. The crystal environment can be altered by the introduction of gases of known composition and pressure.

A known pressure of a gas or contaminant can be prepared and analyzed in the first section, introduced through heat baffles into the main temperature controlled chamber containing the sensing head, and finally pumped out. Frequency readings from all three crystals can be continuously recorded—before, during, and after the gas has been introduced.

9. *Circuitry for Multiple Crystal Sensing Unit*

Measurement of the resonant frequency of the quartz plates to one part in 10^{10} (equivalent to about 10^{-12} g/cm^2), required some rather sophisticated, although well-known, circuit techniques. An oscillator shown in block form in Figure 5.13, operating in class A, rigidly controlled in

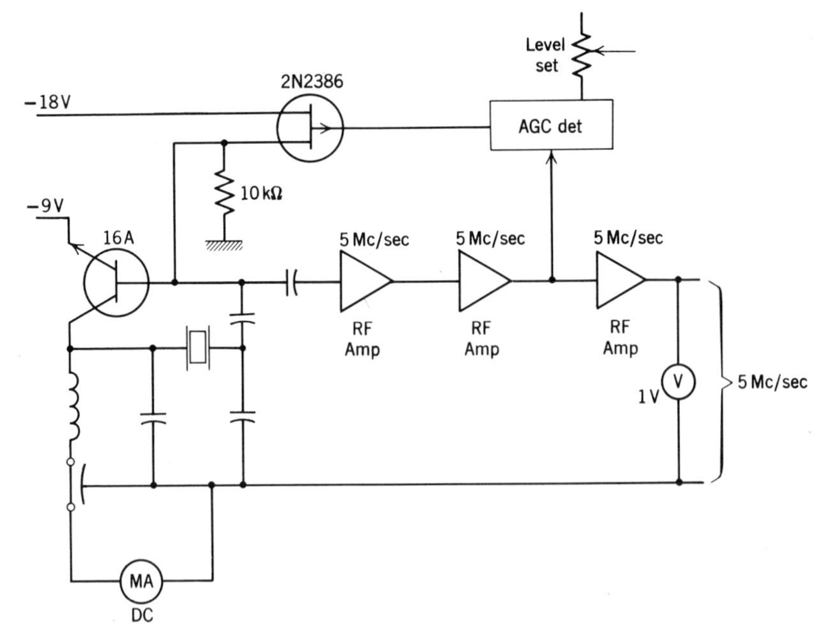

Figure 5.13. Typical oscillator circuit for quartz plates with amplitude control and isolation amplifier.

amplitude and isolated from the measuring circuit was entirely adequate as long as the Q of the quartz plates remained constant, i.e. as long as there was a vacuum or a constant pressure. Under these conditions the oscillator method is unexcelled in producing fast, dynamic information on a changing mass.

Although the oscillators used were compensated for resistance changes should they occur, doubt existed as to whether the compensation was adequate under all conditions, i.e., a factor of 2 or 3 change in resistance. An alternative method which eliminates the phase versus resistance problem is the use of an admittance bridge, using the very precise output of a frequency synthesizer (± 0.0001 Hz), and a high gain tuned voltmeter as detector. With this equipment and the high Q, 5 MHz plates, frequency could easily be determined to one part in 10^9 (± 0.005 Hz).

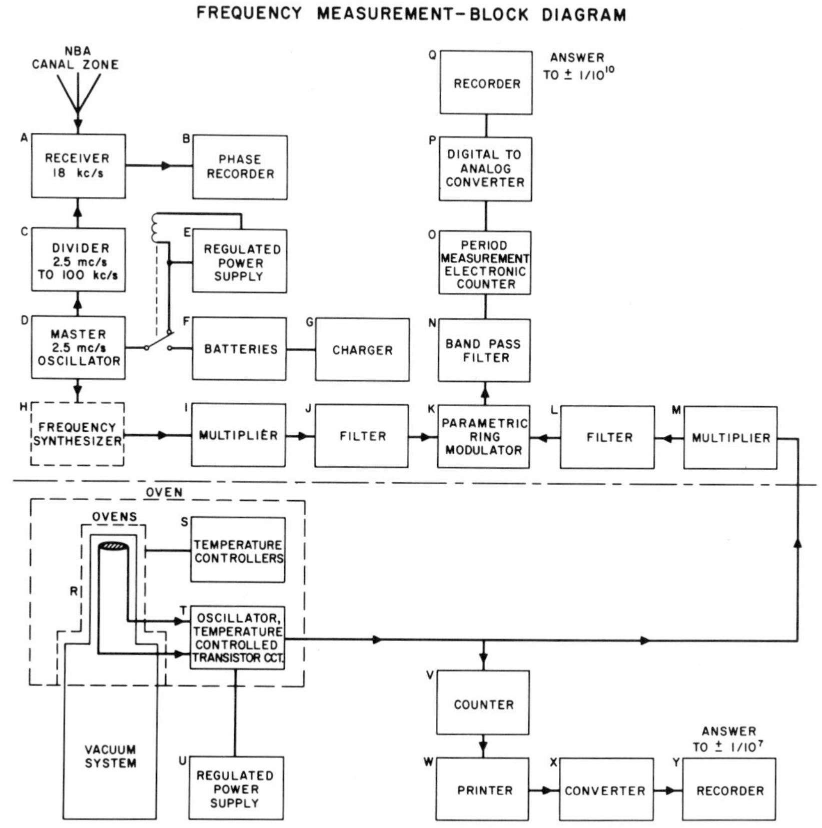

Figure 5.14. Complete block diagram of equipment used for absolute frequency measurement.

The bridge method requires continual balancing of the bridge and recording of data, but is necessary when small frequency changes and large pressure changes are involved or where the accuracy of the compensated oscillators is in doubt and requires checking.

Figure 5.14 is a complete block diagram showing the auxiliary apparatus

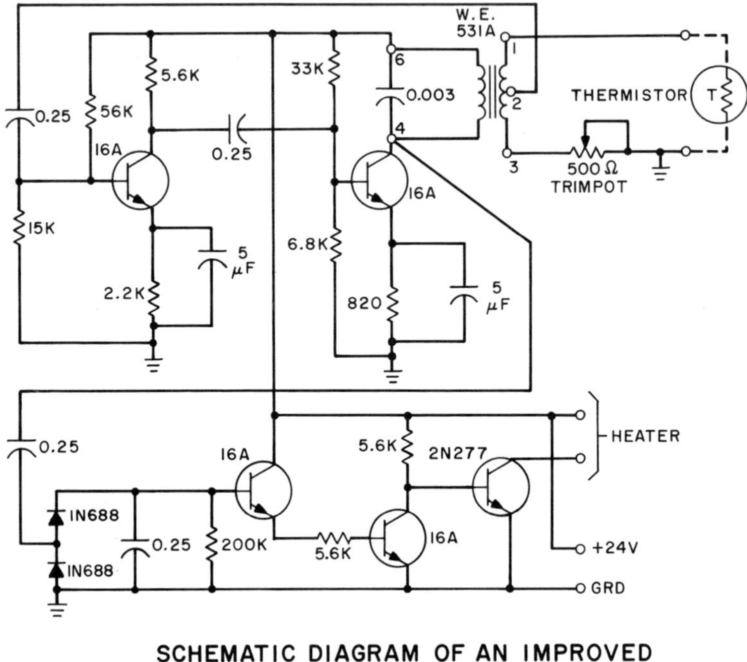

Figure 5.15. Schematic diagram of an improved transistorized oven control.

needed both to measure the frequency simply to ± 1 part in 10^7 and to measure the frequency very accurately to ± 1 part in 10^{10}. When frequency changes as small as 10^{-10} are involved, it is well to use a reference frequency traceable to the United States Frequency Standard. A low-frequency radio receiver, sold for this purpose, is entirely adequate to correct for any small drift in the frequency of the master oscillator.

Figure 5.15 shows the type of temperature control circuit used to supply continuous dc power to the oven windings. The sensing element is a tiny thermistor bead, which can be located to best advantage. Separate circuits were used for control of the upper and lower sections of the oven, and another used to monitor temperature at the crystal plate location. The monitoring thermistor can be seen in Figure 5.10, fastened to the center terminal.

10. Summary

Information necessary for an introductory understanding of the design and use of quartz crystal plates for micro weighing has been presented. The material has been made purposely concise, to (*1*) simplify the approach by considering only quartz plates that are practical to use, and (*2*) provide an easily understood basis on which particular experiments may be designed. The description of apparatus has been divided into two categories, first, apparatus which may be purchased or easily assembled, the use of which does not require extensive preparation or analysis, and, second, apparatus that will fully exploit the excellent physical and chemical stability of quartz and the sophisticated design knowledge of present-day quartz frequency standards.

REFERENCES

1. H. E. Bömmel, W. P. Mason, and A. W. Warner, *Phys. Rev.*, **102**, 64 (1956).
2. R. Beehler et al., *Proc. IEEE* (*Letter*), **54**, 301 (1966). R. J. Harrach, *20th Ann. Symp. Frequency Control*, 1966, p. 424.
3. A. W. Warner, *Bell System Tech. J.*, **39**, 1193 (1960). A. W. Warner, *IEEE Trans. Sonics Ultrasonics*, **SU-12**, No. 2 (1965).
4. R. A. Heising, *Quartz Crystals for Electrical Circuits*, Van Nostrand, New York, 1946, p. 24.
5. IRE Standards on Piezoelectric Crystals, 58 IRE 14.S1, *Proc. IRE*, **46**, 765 (1958).
6. A. W. Warner and C. D. Stockbridge, *Vacuum Microbalance Tech.*, **3**, 67 (1963).
7. C. D. Stockbridge, *Vacuum Microbalance Tech.*, **5**, 147 (1966).
8. W. H. King, Jr., *Anal. Chem.*, **36**, 1735 (1964).
9. C. D. Stockbridge and A. W. Warner, *Vacuum Microbalance Tech.*, **2**, 107–113 (1962).
10. E. P. Eer Nisse, *IEEE Trans.*, **SU-14**, 59 (1967).
11. M. Onoe, *Proc. IRE.*, **45**, 694 (1957).
12. J. W. Strutt (Lord Rayleigh), *The Theory of Sound*, rev. ed., Dover, New York, 1945, sec. 90, p. 113.
13. G. Sauerbrey, *Z. Physik*, **155** (2), 206 (1959).
14. C. D. Stockbridge, *Vacuum Microbalance Tech.*, **5**, 202 (1966).
15. H. L. Eschbach and E. W. Kruidhof, also R. Niedermayer, N. Gladkich, and D. Hillecke, *Vacuum Microbalance Tech.*, **5**, 207–229 (1966).
16. IRE Standards on Piezoelectric Crystals, 57 IRE 14.S1, *Proc. IRE*, **45**, 356 (1957).
17. IRE Standards on Piezoelectric Crystals, 49 IRE 14.S1, *Proc. IRE*, **37**, 1379 (1949).
18. C. D. Stockbridge, *Vacuum Microbalance Tech.*, **5**, 169 (1966).
19. A. W. Warner and C. D. Stockbridge, *Vacuum Microbalance Tech.*, **2**, 81 (1962).
20. A. W. Warner, D. B. Fraser, and C. D. Stockbridge, *IEEE Trans. Sonics Ultrasonics*, **SU-12**, 52 (1966).
21. J. P. Griffin, 13th and 17th Annual Frequency Control Symposium sponsored by Solid State and Frequency Control Division, U.S. Army Electronics R & D Laboratory, Fort Monmouth, New Jersey.

Chapter 6

Vacuum and Controlled Environment Systems

W. KREISMAN

GCA Corporation
Bedford, Massachusetts

I. INTRODUCTION

In order to provide a controlled environment system with a minimum impurity gas level, it is necessary first to remove the air or other undesired gas from the system (before creating the gas environment that is required). The manner in which this is done will be governed primarily by the type of system, e.g. the system size, materials of construction, maximum impurity levels that can be tolerated, etc. For those cases in which the final environment is to be a gas mixture at a pressure below atmospheric pressure, the air in the system is removed by one or more of the pumping processes that fall within the domain of vacuum technology. The creation of controlled gas environments at pressures above atmospheric pressure do not necessarily require initial evacuation of the system, since gas purging techniques may be adequate to remove the air. This chapter, however, will be concerned with those controlled environments that require the use of high vacuum and ultrahigh vacuum technology.

The purpose of this chapter is to acquaint the reader with the fundamentals of vacuum technique as required to provide controlled environments. Due to the limitations of space, many topics are treated in an abbreviated fashion. In these cases, references to general textbooks on vacuum are provided so that the subject matter may be studied in greater depth. Most of the high vacuum textbooks provide references to the periodical literature in which specialized subjects are discussed in great detail. In brief, the aim has been to provide the reader with the immediate, important information and facts needed to help him select and design his controlled environment system. The various phenomena of high vacuum are discussed to acquaint the experimenter with potential problems and their solution rather than to make him an "expert" on these matters. In the last analysis, experience with vacuum components and systems is the best teacher.

Section II of this chapter contains a discussion of the fundamental elements of vacuum science such as the kinetic theory of gases and the manner in which gases are evolved, and transported within, and removed from, the gas phase. Sections III and IV contain discussions of the methods and equipment currently in use to pump gases at low pressures, and to measure the degree of evacuation and the composition of the residual gases of a system. Included in Section IV is a description of gas handling and leak detection techniques. Section V contains descriptions of the elements of vacuum system design and construction, including a discussion of the most commonly used materials and their fabrication, the availability of various pumps, valves, flanges, gauges and other vacuum components, and some examples of typical system designs.

The degree of vacuum required to establish a controlled environment system is often the single most important parameter in determining the design of the system. The American Vacuum Society has defined various vacuum regions, as shown in Table 6.1. Pressures below atmospheric

Table 6.1. Classification of Vacuum Regions

Region titles	Pressure range, torr
Rough or low vacuum (torr range)	760 to 1
Fine or medium vacuum (millitorr range)	1 to 10^{-3}
High vacuum (microtorr range)	10^{-3} to 10^{-6}
Very high vacuum (nanotorr range)	10^{-6} to 10^{-9}
Ultrahigh vacuum	below 10^{-9}

pressure are currently measured in torr units (named in honor of Torricelli), with the torr being defined as 1/760 of standard atmospheric pressure. (Current standard atmospheric pressure is defined as 1,013,250 dynes/cm^2.) Within 1 part in 7 million, the torr is equal to the pressure exerted by a column of mercury 1 mm high at 0°C under standard acceleration of gravity of 980.665 cm/sec^2. The difference in the two units is due to different definitions of the standard atmosphere. For all practical purposes, the units torr and mm Hg are used interchangeably today. Table 6.2 shows the "vacuum spectrum," so to speak, and associates with each pressure level of nitrogen gas some pertinent data concerning the amount of gas present, the rate at which molecules strike the surface (the so-called impingement rate), time constants for the formation of a monolayer of gas

Table 6.2. Important Parameters of Nitrogen Gas at Room Temperature

Pressure, torr	No. density n, molecules/cm^3	Mean free path, cm	Impingement rate, molecules per cm^2 sec	Time constant for monolayer formation time, sec[a]	Ratio of mono-layer to gas phase molecules in 1 liter flask[b]
760	2.5×10^{19}	6.3×10^{-6}	2.9×10^{23}	1.7×10^{-9}	9.7×10^{-6}
1	3.3×10^{16}	4.8×10^{-3}	3.9×10^{20}	1.3×10^{-6}	7.3×10^{-3}
10^{-3}	3.3×10^{13}	4.8	3.9×10^{17}	1.3×10^{-3}	7.3
10^{-6}	3.3×10^{10}	4.8×10^{3}	3.9×10^{14}	1.3	7.3×10^{3}
10^{-9}	3.3×10^{7}	4.8×10^{6}	3.9×10^{11}	1.3×10^{3}	7.3×10^{6}
10^{-12}	3.3×10^{4}	4.8×10^{9}	3.9×10^{8}	1.3×10^{6}	7.3×10^{9}
10^{-15}	3.3×10^{1}	4.8×10^{12}	3.9×10^{5}	1.3×10^{9}	7.3×10^{12}

[a] Calculated on the assumption that all impinging molecules that strike vacant surface sites will stick, i.e. a sticking coefficient of 1.

[b] A monolayer of nitrogen gas is taken to be 5×10^{14} molecules/cm^2.

atoms on an initially clean surface, and the ratio of the amounts of gas in a surface monolayer to that in the volume of a spherical 1-liter flask.

II. FUNDAMENTAL CONSIDERATIONS

A. Kinetic Theory of Gases

The kinetic theory of gases applies the laws of mechanics to the atoms and molecules of a gas system. From the application of these laws, together with a number of basic assumptions, expressions for the pressure, internal energy, and specific heat capacity of a gas may be derived.

The relationship between the parameters of the system, the pressure P, the volume V, and the temperature T of a gas, is known as the equation of state of the gas. The equation of state for an ideal gas has been obtained by extrapolation of experimental results and has also been derived theoretically using the kinetic theory. The theoretical derivation requires the following assumptions:

1. Any finite volume contains a large number of molecules. For example, Avogadro's number N_0, the number of molecules in 1 g-mole of any substance is 6.023×10^{23} molecules. Since it is found that 1 g-mole of any gas occupies 22,414 cm^3 under standard conditions, there are approximately 2.7×10^{19} molecules per cm^3 at atmospheric pressure and 0°C (273°K). In Table 6.2, the number of molecules per cm^3 at atmospheric pressure (760 torr) and 20°C (293°K) is listed as 2.5×10^{19}, since the number density varies inversely with the absolute temperature.

2. Molecules exert no forces on one another except when they collide. Between collisions with other molecules or with the walls of a container, and in the absence of external forces, the molecules move in straight lines.

3. Collisions of molecules with one another and with the walls of a container are perfectly elastic. The walls of a container are considered to be perfectly smooth so that there is no change of tangential velocity in a wall collision.

4. In the absence of external forces, the molecules are distributed uniformly throughout a container.

5. All directions of molecular velocities are assumed to be equally probable.

On the basis of the above assumptions, one can derive the equation of state of an ideal or perfect gas in several forms:

$$PV = NkT \tag{6.1a}$$

$$P = nkT \tag{6.1b}$$

or

$$PV = n_M RT \tag{6.2}$$

where the units for the pressure, volume, and temperature determine the values of the gas constants k and R. Here N is the total number of molecules in the volume V, so that $n = N/V$ in equation 6.1b is the molecular number density. Note that for any given container of volume V, fixed values of P and T imply a fixed number of gas molecules independent of the gas. The constant k is the Boltzmann constant, 1.3803×10^{-16} erg/°K or 1.03×10^{-22} torr-liters/°K. The quantity n_M in equation 6.2 is the number of moles, and R is the gas constant per mole. As for k, the numerical value of R depends on the units and is equal to 8.31×10^7 erg/°K-mole or 2.0 g-cal/°K-mole. From equation 6.2, it can be seen that the quantity PV is equivalent to the translation kinetic energy of a gas and can be expressed in units of ergs or gram calories using the appropriate value of the gas constant R. From equations 6.1a and 6.2, the relationship between k and R is $k = R/N_0$, that is, the Boltzmann constant is just the gas constant per molecule. One other important quantity that is frequently used in vacuum practice is the number of molecules in 1 torr liter of gas at 20°C. Using the value of $k = 1.03 \times 10^{-22}$ torr liters/°K in equation 6.1a, it can be verified that there are 3.3×10^{19} molecules/torr liter at 20°C. This quantity will vary inversely with the absolute temperature.

The equations of state 6.1 and 6.2 hold only for ideal gases. There is non-ideal behavior in the gas phase and in the interaction of the gas with environmental surfaces. At lower pressure equilibrium conditions gases generally behave more like ideal gases. If they attain the saturation vapor pressure of their solid or liquid phases, a two-phase behavior occurs. For example, if large amounts of water are present in a vacuum system, the equilibrium pressure of the water vapor at 20°C will be 17.5 torr, at Dry Ice temperature (−78°C) 6×10^{-4} torr and at liquid nitrogen temperature (−195°C or 77°K) about 10^{-18} torr. The pressure of a gas must be below its saturated vapor pressure for the total amount present to behave like a gas. At the other extreme, when the pressure of a gas is very low, the gas phase is nearly ideal, but the surface effects of adsorption and desorption make the total amount of the gaseous material deviate appreciably from the behavior of an ideal gas.

Making use of the kinetic theory of gases, including the five assumptions listed earlier, it can be shown that the pressure exerted on a container wall by the change in momenta of the molecules striking that wall and rebounding elastically, is given by the expression:

$$P = \tfrac{1}{3} n m \overline{v^2} = N m \overline{v^2} / 3V \tag{6.3}$$

Here m is the mass of the molecules and $\overline{v^2}$ is the mean squared velocity of the molecules. Using equations 6.1a and 6.3 and multiplying by the factor 3/2 yields the basic relationship between the translational kinetic energy of a molecule and the absolute temperature:

$$\mathrm{KE} = m\overline{v^2}/2 = 3kT/2 \tag{6.4}$$

Since the mean squared velocity is by definition the square of the root-mean-square (rms) velocity,

$$v_{\mathrm{rms}} = (3kT/m)^{1/2} = 15{,}800(T/M)^{1/2}\ \mathrm{cm/sec} \tag{6.5}$$

where $M = mN_0$ is the molecular weight. The kinetic theory leads to two other important velocities that characterize the speed distribution of a gas in equilibrium. These are the most probable speed v_m and the average speed v_a:

$$v_m = (2kT/m)^{1/2} \tag{6.6}$$

$$v_a = (8kT/\pi m)^{1/2} = 14{,}551(T/M)^{1/2}\ \mathrm{cm/sec} \tag{6.7}$$

The kinetic theory also leads to the following relatively simple expression for the total number of molecules of all velocities that collide with the wall per unit area, per unit time, per unit solid angle:

$$\nu_1 = (1/4\pi)nv_a \cos\theta \tag{6.8}$$

where θ is the direction between the normal to the area and the cone from which the molecules arrive. Considering a number of cones of equal solid angle, the greatest number of molecules arrive from the cone centered about the normal ($\theta = 0$), while the number of molecules decreases to zero for the cone tangent to the surface ($\theta = 90°$). If the surface of the wall contains a hole that is small enough so that the outflow of gas does not affect the equilibrium within the container, then every molecule arriving at the hole escapes through it and equation 6.8 describes the distribution in the direction of molecules leaving the hole. Assuming that the container wall is infinitely thin, it can be seen that the number of molecules leaving the hole are a maximum in the direction of the normal and fall to zero in the tangential direction.

The total number of molecules striking the wall per unit area per unit time from all directions and with all velocities is obtained by integrating equation 6.8 over the total solid angle. The result is:

$$\begin{aligned}\nu &= nv_a/4 = (n/4)(8kT/\pi m)^{1/2} = P(2\pi mkT)^{-1/2} \\ &= 14{,}551\ \mathrm{n}\ (T/M)^{1/2}/4 = 3.5 \times 10^{22} P_{\mathrm{torr}}(MT)^{-1/2}\mathrm{cm}^{-2}\mathrm{sec}^{-1}\end{aligned} \tag{6.9}$$

making use of equations 6.1b and 6.7. The quantity ν is the impingement rate that has been mentioned earlier. This quantity is a key parameter in vacuum technology as will be shown after the concept of mean free path has been defined and described.

The mean free path of a molecule between collisions with other molecules is defined as the average length of the many straight line movements the molecules make between collisions. If the molecules are assumed to be elastic spheres, the following development lends some insight into collision phenomena.

Molecules have a finite size and travel with relatively high velocities of the order of 1 km/sec at room temperature. Assume that at a certain instant all of the molecules except one are "frozen" in position, while the remaining molecule moves among the stationary ones with a speed equal to the average velocity v_a. For any collision, the center-to-center distance of the colliding molecules is $2r = D$, where r and D are the radius and diameter, respectively, of the molecules. The collision phenomena would be unchanged if the moving molecule had a radius of $2r$ and all stationary molecules were considered to be geometrical points. The *effective* cross-sectional area of the moving molecule, the collision cross section σ, would be $\sigma = 4\pi r^2 = \pi D^2$. In a time interval t, the moving molecule moves a total distance $v_a t$ along a zig-zag path and sweeps out a volume $\sigma v_a t$. If there are n molecules per unit volume, the number of collisions is $\sigma n v_a t$. The number of collisions per unit time, the so-called collision frequency, is equal to $\sigma n v_a$. One should notice the similarity between collision frequency and impingement rate. If the collision cross section σ becomes unity, the two quantities become comparable. The average distance between collisions, the mean free path, equals the total distance covered in time t divided by the number of collisions, so that

$$\lambda = v_a t / \sigma n v_a t = 1/\sigma n \tag{6.10}$$

If the Maxwellian distribution of velocities is taken into account, a careful derivation yields the expression

$$\lambda = 1/\sqrt{2}\,\sigma n = kT/\sqrt{2}\,\pi D^2 P \tag{6.10a}$$

Equation 6.10a shows that the mean free path varies directly with the absolute temperature and inversely with the molecular diameter and the gas pressure. Table 6.2 lists the mean free paths for nitrogen gas at various pressures. It is convenient to remember that for nitrogen the mean free path is $4.8 \times 10^{-3}/P$ cm where the pressure P is in torr units. At a pressure of 10^{-6} torr, the mean free path for nitrogen is 48 m, much larger than any ordinary vacuum equipment. At pressures of this order or lower, most collisions occur between the molecules and the walls of the system rather than between molecules. Gas flowing under these low pressure conditions is defined to be "molecular flow" and is quite different from the so-called "viscous" flow which is governed by collisions among the molecules at higher pressures. Microbalance behavior in these two regions is discussed in detail in Chapter 4.

Another useful expression is that for the mean free path of one type of molecule colliding with molecules of another type in a mixture of two gases of molecular weights M_1 and M_2 and of partial pressures P_1 and P_2:

$$\lambda_{1,2} = 4kT/\pi P_2(D_1 + D_2)^2(1 + M_1/M_2)^{1/2} \tag{6.11}$$

The above equation can be used to show that the mean free path of a small percentage of helium in nitrogen is approximately twice the mean free path for the nitrogen. In addition, the equation verifies that the mean free path of an electron in a gas is $4\sqrt{2}$ times the mean free path of the gas molecules.

The state of a gas at higher pressures is normally described by the gas pressure and temperature through equation 6.1. In a static closed system, the pressure is independent of the location within the system. At high vacuum, on the other hand, it is not the pressure which is constant throughout the system but rather the impingement rate. Equilibrium is achieved when molecules pass through any cross-sectional area equally in both directions.

Different wall temperatures of a high vacuum chamber lead to different gas pressures and densities. If two high vacuum chambers at different temperatures T_1 and T_2 are connected together, the molecules within the chambers will redistribute themselves until the impingement rates in either direction are equal. From equation 6.9, it can easily be shown that this equilibrium condition will lead to different pressures as follows:

$$P_1/P_2 = (T_1/T_2)^{1/2} \tag{6.12}$$

By combining equations 6.12 and 6.1 the relationship between number density and temperature under conditions of high vacuum (mean free path much greater than the dimensions of the containing vessel) will be seen to be

$$n_1/n_2 = (T_2/T_1)^{1/2} \tag{6.13}$$

The variation of pressure and density with the wall temperature under high vacuum conditions is known as thermal transpiration.

Other gaseous transport phenomena that are of importance in vacuum work are viscosity, thermal conductivity, and diffusion. These properties of a gas can be explained in terms of the transport of molecular momentum, energy, and mass across an imaginary surface within the gas volume.

When the mean free path in a gas is very small compared with the dimensions of the containing walls, the flow of gas is dominated by collisions between the molecules. If the mass movement of the gas does not proceed at too high a speed, a smooth laminar flow will develop in which the various layers of gas exert an influence on one another through the phenomenon of viscosity. The coefficient of viscosity is defined as the tangential force (parallel to the direction of flow) per unit area per unit velocity gradient normal to the area (and thus normal to the direction of flow). The viscosity of a gas does not derive from any "frictional" forces between its molecules but rather from the momentum carried across a surface as a result of random thermal motions. It would then be expected that the viscosity will vary with the gas temperature. A calculation of the net rate of transfer of momentum per unit area per unit time by the kinetic theory yields the following coefficient of viscosity:

$$\eta = nmv_a\lambda/2 = mv_a/2\sqrt{2}\,\sigma = (mkT)^{1/2}/\pi^{3/2}D^2 \tag{6.14}$$

In the cgs system, the unit of viscosity is the poise (1 P = 1 dyne-sec/cm^2). The expression for the viscosity holds for pressures from about 10 atm down to the region where the mean free path is comparable to the size of the container. The viscosity (and the heat conductivity, as will be shown) is proportional to $\sqrt{T}$ and independent of pressure over the entire region where intermolecular collisions are the primary phenomenon.

The manner in which heat is transferred in a vacuum is often of great importance. At pressures near atmospheric pressure, hot objects are cooled efficiently by gaseous convection. The gravitational forces arising from density changes due to temperature gradients set up circulating currents. As the pressure is lowered, convection effects become smaller and finally disappear. Heat is then transferred by the intermolecular collisions that can be treated via the kinetic theory. The theory leads to the expression

$$K = \varepsilon\eta c_v, \qquad \varepsilon = (9\gamma - 5)/4 \tag{6.15}$$

where γ is the ratio of the specific heat at constant pressure to the specific heat at constant volume. For monatomic gases $\gamma = 5/3$ and for polyatomic gases γ approaches unity so that $1 \leq \varepsilon \leq 2.5$. The specific heat per unit mass at constant volume c_v increases with temperature. The thermal conductivity K is the heat flux per unit area per unit time for unit temperature gradient. It is most often given in the units of watts per square centimeter for a temperature gradient of 1°C per cm. Since the viscosity η varies directly with $\sqrt{m}$ and the specific heat *per unit mass*, c_v, varies inversely with m, the smaller, lighter molecules such as hydrogen and helium have the largest values of K.

If gas is admitted into a vacuum system for the purpose of rapidly cooling a hot object, there is no point in raising the pressure more than about 1 torr. Hydrogen or helium will produce the fastest heat transfer.

Helium is customarily used in cryogenic systems when it is desired to cool elements that are normally thermally isolated by a vacuum.

In the process just described, the gas in the vicinity of the hot object takes on the temperature of the object and sets up a temperature gradient in the gas which serves to transport the heat. When the pressure is reduced so that the mean free path approaches the size of the container, the individual molecules that strike the hot object can travel directly to the cooler surfaces. Heat transported in this way is less efficient than transfer by intermolecular collisions and so the thermal conductivity decreases.

When molecules that are originally at a temperature T_i strike a wall that is at a temperature T_w, there is only a partial interchange of energy. The extent to which reflected or reemitted molecules have their energy adjusted or accommodated to the wall surface is given by the accommodation coefficient

$$\alpha = (T - T_i)/(T_w - T_i) \tag{6.16}$$

where T is the temperature of the molecules as they leave the wall, and it is assumed that the velocity distribution of the molecules leaving the wall is Maxwellian. From equations 6.4, 6.9, and 6.16 one finds the cooling rate per unit area per unit time to be

$$W = 3\nu k\alpha(T_w - T_i)/2 = \text{constant}\cdot\alpha\cdot P(T_w - T_i)/(T_i M)^{1/2} \tag{6.17}$$

It is shown in treatises on kinetic theory that the energy transferred from a surface at temperature T is given by $E = 2kT$ instead of by $3/2 kT$, which is the average energy of molecules in a volume, and so the constant of equation 6.17 must be adjusted accordingly. The fact that the energy transferred is proportional to the pressure is utilized in thermal conductivity-type vacuum gauges. The coefficient of accommodation is generally of the order of a few tenths for helium and hydrogen gas, but approaches unity for the heavier gases. It varies with the nature of the surface as well as with the gas.

When gas pressures become very low, say about 10^{-5} torr, or below, in vacuum systems of laboratory size, heat is transferred primarily by radiation according to the Stefan–Boltzmann law:

$$W = \sigma\varepsilon(T_1^4 - T_2^4) \tag{6.18}$$

$\sigma = 5.67 \times 10^{-12}\text{W/cm}^2\,{}^\circ\text{K}^4$ is the Stefan–Boltzmann constant, ε is the ratio of the emissivity of the surface to that of a blackbody, and T_1 and T_2 are the temperatures of the hot and cold portions of the system. When the temperatures T_1 and T_2 are sufficiently dissimilar (as when cold traps or hot filaments are employed in the system), heat transfer by radiation cannot be neglected even at higher pressures.

Another basic process occurs when two gases of molecular weights M_1 and M_2 are mixed together. For example, if nitrogen and helium are introduced at opposite ends of a long tube of small diameter, it is found that concentration gradients of each gas will be established along the length of the tube even though the total pressure is constant along the tube. As a result of the random motion of the molecules and their collisions with one another, there will be a slow diffusion of each gas into the other until eventually a uniform concentration will be established. Kinetic theory shows that the rate at which the partial pressure of one of the gases changes at a given point is given by the expression:

$$\frac{\partial P_1}{\partial t} = \text{constant} \cdot \frac{T^{3/2}}{P} \frac{(1/M_1 + 1/M_2)^{1/2}}{(D_1 + D_2)} \frac{\partial P_1}{\partial x} \qquad (6.19)$$

where $P = P_1 + P_2$ is the total pressure, D_1 and D_2 are the molecule diameters, x is the distance along the tube, and t is the time. Generally, $\partial P_1/\partial x$ is not known (although it could be determined by various techniques) and the above equation cannot be used quantitatively. Expression 6.19 does show the role of the various factors in gaseous diffusion, however. It can be said, as a word of caution, that it often takes hours to mix thoroughly two or more gases at atmospheric pressure, whereas it may only take minutes or seconds to mix gases at lower pressures where the mean free path approaches the size of the containing vessel. At very low pressures, mixing is practically instantaneous. Hydrogen, helium, and neon diffuse faster than heavier gases because their average molecular velocities are higher and their mean free paths are longer.

B. Gas Flow Equations

There is a close analogy between the flow of electrical current in a dc circuit and the gas flow in a vacuum system. In the electrical case, the electrical current consists of electrons in motion around a closed circuit that generally contains one or more sources of emf and one or more resistances. The current is neither created nor destroyed, and the potential changes along the circuit in a continuous, single-valued fashion. Kirchoff's laws and the rules for combining resistances in series and parallel are shown to apply to dc circuits.

In the case of gas flow in a vacuum circuit at a constant temperature, the same situation exists as for electrical current flow. Consider, for example, a system that is used to calibrate vacuum gauges for air. Air from the atmosphere is allowed to enter a vacuum chamber through a "variable leak" type of valve. A diffusion pump backed by a rotary mechanical pump may be used to remove the incoming air from the chamber and discharge it back into the atmosphere. The pressure of the air is analogous to the electrical potential while the flow of the air molecules is analogous to the electron current flow. The drops in the pressure as the air flows through constricting orifices and long narrow tubes of the vacuum system are analogous to the drops in potential as the electrical current flows through resistances. The vacuum pumps play the same role as sources of emf. They create differences of pressure (potential) in the circuit.

The characteristic quantities of resistance, conductance, impedance, and current of electrical dc circuits have their counterparts in vacuum theory. The resistance of a vacuum element such as a small orifice or a long narrow tube is defined as the pressure drop ΔP across the element per unit of gas current flow Q through the element. The term impedance is used synonymously with resistance here. The conductance of a vacuum element is just the reciprocal of the resistance. It turns out to be more convenient to work with vacuum conductances than with vacuum resistances. Gas current flow or gas "throughput," as it is commonly termed in a vacuum circuit, can be defined in several ways. The most often used measure of gas throughput, Q, is the product of the volumetric flow rate past a given cross section of the circuit, dV/dt, and the gas pressure, P, at that plane.

$$Q = P(dV/dt) \tag{6.20}$$

Most commonly, the pressure P is expressed in torr units while the volumetric flow dV/dt is expressed in liters per second. As shown earlier, 1 torr liter of gas at 20°C consists of 3.3×10^{19} molecules, and so it is easy to convert torr liters/sec of flow into molecules per second at a temperature of 20°C. The actual mass flow equivalent can be found from the fact that 1.70×10^4 torr liter = 1 g mol wt. The current flow in molecules/sec at any absolute temperature T is obtained by combining equations 6.1b and 6.20 to get

$$n(dV/dt) = Q/kT \tag{6.21}$$

The conductance C of a vacuum element, as described earlier, is the ratio of the gas current flow through the element to the pressure drop ΔP across the element

$$C = Q/\Delta P = Q/(P_1 - P_2) \tag{6.22}$$

This is the "Ohm's law" of vacuum circuits. As is true for electrical circuits, conductances in parallel add together while the reciprocal of conductances in series add together:

$$C = C_1 + C_2 + \cdots \quad \text{for parallel conductances} \tag{6.23a}$$

$$1/C = 1/C_1 + 1/C_2 + \cdots \quad \text{for series conductances} \tag{6.23b}$$

A vacuum pump is a device which creates a flow of gas in a vacuum circuit (and concomitantly creates a pressure drop in the circuit). The effectiveness of a vacuum pump is expressed in terms of its pumping speed S at a given pressure P. The speed is simply the volume of gas removed by the pump per unit time at the pressure P. That is,

$$S = (dV/dt)_P = Q/P \tag{6.24}$$

The gas flow Q can be measured at any plane in the system, but it is understood that the pressure P is to be measured at the inlet to the pump.

Vacuum conductance and pumping speed are both measured in the same units of liters per second, but they are not equivalent quantities. Conductance is a geometrical property of a vacuum element and is associated with a pressure drop developed across the element as a result of gas flow through the element. Pumping speed refers only to the flow of gas at a particular position (plane) for a given value of pressure. For many vacuum pumps, the pumping speed changes as the pressure changes. On the other hand, in the molecular flow regime, the vacuum conductance of an element is independent of pressure (but does vary with the molecular species and the temperature).

When a vacuum pump is connected to a vacuum chamber via a length of tubing, a cold trap, a valve, or any other vacuum element that has a specific vacuum conductance C for the gas being pumped, the *effective* pumping speed of the pump S at the chamber where the pressure is P is always less than the intrinsic pumping speed S_p of the pump at its inlet where the pressure is P_p. From the relationship

$$Q = SP = S_pP_p = C(P - P_p)$$

one can derive the basic equation used for the determination of the *effective* speed of the pump at the chamber:

$$1/S = 1/S_p + 1/C \tag{6.25}$$

Expressions for the conductance C of a vacuum element depend on the shape of the element and the flow regime. In this chapter, it will only be possible to list the conductance of a few of the more common flow restrictions. The reader is referred to texts such as references 5 and 15 for more detailed information.

The gas flow regime which will exist in the flow through an aperture or a constricting tube depends primarily on the diameter of the aperture or tube, D, and the gas pressure (which determines the mean free path). The temperature and the composition of the gas can also influence the flow. There will be molecular flow if $D < 3\lambda$. If the pressure in a system is sufficiently high so that $D > 200\lambda$, the flow will be continuum or viscous flow. As noted earlier, in this flow regime, molecules move along under the action of a pressure gradient by colliding frequently with one another. At very high velocities and pressures, viscous flow can become turbulent. For air at 25°C, the flow is turbulent when $Q > 200D$, where Q is the throughput in torr liters/sec and D is a tube diameter in cm. The maximum viscous flow that an orifice can pass is called critical or choked flow, and is of importance in vacuum systems.

In the molecular flow regime, the flow of gas through an infinitely thin orifice is simply the flow due to the net impingement of molecules from either side. In terms of the kinetic theory, the conductance of an orifice for molecular flow is

$$C = (kT/2\pi m)^{1/2} A = 3.64(T/M)^{1/2} A \text{ liters/sec} \tag{6.26}$$

which, for nitrogen gas at 293°K (20°C) becomes: $C_{N_2} = 11.8A$ liters/sec. Note that the conductance is independent of the pressure. The conductance of an orifice for continuum (viscous) flow is somewhat greater than that calculated above by a factor of 1.3–1.7. In this case, there is an organized motion of the molecules along stream lines, and the flow is a complicated function of the ratio of specific heats of the gas and the ratio of the upstream and downstream pressures. Critical or choked flow occurs when the gas velocity reaches the sonic velocity in passing through the orifice.

When the infinitely thin walls of an ideal orifice take on a finite thickness, the conductance of the orifice decreases, as would be expected. As the orifice thickness increases, a tube is formed, and the conductance decreases more and more. One can derive an expression for the molecular conductance C_m of a tube of any length based on the assumption that molecules are diffusely reflected from the walls.

$$C_m = 11.4Kr^2(T/M)^{1/2} = 3.64K(T/M)^{1/2} A \tag{6.27}$$

where K is Clausing's factor, a quantity that depends on the ratio of the length l to the radius r of the tube. Values of Clausing's factor are listed in Table 6.3. For rough calculations, one may use the approximation

$$K = 1/(1 + 3l/8r) \tag{6.28}$$

Another convenient expression for the molecular flow conductance of long tubes is

$$C_m = 3.8(T/M)^{1/2} D^3/L \text{ liters/sec} \tag{6.29}$$

which becomes $C_{mN_2} = 12.3\ D^3/L$ liters/sec for nitrogen gas at 293°K. Here D is the tube diameter and L is the tube length, both expressed in cm.

The viscous flow conductance of a long tube is shown to be

$$C_v = \pi D^4 \bar{P}/128\eta L \tag{6.30}$$

Table 6.3. Values of Clausing's Factor K as a Function of the l/a Ratio for Long Tubes[a]

Ratio, l/a	K	Ratio, l/a	K
0.0	1.0	3.2	0.4062
0.1	0.9524	3.4	0.3931
0.2	0.9092	3.6	0.3809
0.3	0.8699	3.8	0.3695
0.4	0.8341	4.0	0.3589
0.5	0.8013	5.0	0.3146
0.6	0.7711	6.0	0.2807
0.7	0.7434	7.0	0.2537
0.8	0.7177	8.0	0.2316
0.9	0.6940	9.0	0.2131
1.0	0.6720	10.0	0.1973
1.1	0.6514	12.0	0.1719
1.2	0.6320	14.0	0.1523
1.3	0.6139	16.0	0.1367
1.4	0.5970	18.0	0.1240
1.5	0.5810	20.0	0.1135
1.6	0.5659	30.0	0.0797
1.7	0.5518	40.0	0.0613
1.8	0.5384	50.0	0.0499
1.9	0.5256	60.0	0.0420
2.0	0.5136	70.0	0.0363
2.2	0.4914	80.0	0.0319
2.4	0.4711	90.0	0.0285
2.6	0.4527	100.0	0.0258
2.8	0.4359	1000.0	0.002658
3.0	0.4205		$(8a/3l)$

[a] From D. J. Santeler, D. H. Holkeboer, D. W. Jones, and F. Pagano, *Vacuum Technology and Space Simulation*, NASA SP-105, U.S. Govt. Printing Office, Washington, D.C., 1966, p. 88.

in fluid flow theory (Poiseuille flow). The quantity η is the gas viscosity and $\bar{P}$ is the average pressure in the tube. For nitrogen at 20°C, and with D and L expressed in cm and $\bar{P}$ in torr,

$$C_{vN_2} = 188 D^4 \bar{P}/L \tag{6.31}$$

Except for gases such as hydrogen and water vapor, the viscosities of the usual atmospheric gases are not too different from that of nitrogen

($\eta = 175\ \mu$P at 20°C). Hydrogen and water vapor have viscosities which are about one-half that of nitrogen.

In the region between pure molecular flow and pure continuum flow, there is a transition flow region which occurs over a range of two or three decades. Empirical equations have been developed to describe this region where molecules make significant collisions with both the walls and with each other. The net result is that the effective conductance increases from the molecular flow value to the viscous flow value as the ratio of D/λ increases (increasing pressure for a given tube).

C. Adsorption–Desorption, Diffusion, Permeation, and Evaporation

Surfaces of solids and liquids exert forces of attraction that cause impinging molecules to stick and remain on the surface for various lengths of time (residence times). The capture of molecules by a surface is generically termed sorption while the release of such molecules is called desorption. Sorption or adsorption may be either physical adsorption or chemical adsorption (also called chemisorption). Physical adsorption involves a weak bonding via van der Waal forces between the adsorbed molecules and the surface. The heats of adsorption for physical adsorption are under 10 kcal/mole. Adsorption occurs spontaneously, the free energy and the entropy decrease (the molecules lose one degree of freedom), and thus the adsorption is exothermic. Inert gases generally adsorb physically on surfaces.

Under certain conditions, molecules can react with a surface to form strong chemical bonds. Such strong bonding is termed chemisorption and is quite specific for certain gases and solid surfaces. The heats of chemisorption are much greater than those of physical adsorption by roughly one order of magnitude. In addition, an activation energy is often required to permit the gas molecules to become chemisorbed. As shown in Figure 6.1, a molecule may be initially adsorbed physically, then chemisorbed with the supply of additional energy of activation to overcome the potential barrier. Molecules can also be dissociated at a surface and chemisorbed as atoms. Similarly, atoms can recombine and be desorbed as a molecule or they can be desorbed as atoms depending on the temperature and pressure.

Thermal desorption is the process whereby adsorbed gas is released from a surface by the application of heat. Other methods of removing adsorbed gases are by bombardment with electrons or other charged particles and by irradiation with high energy (ultraviolet) photons. The energy of desorption E_D required to remove a molecule from a surface is related to the residence time t_r by the expression

$$t_r = \tau_0 \exp(E_D/RT) \tag{6.32}$$

Here τ_0 is the period of oscillation of the molecule normal to the surface and has a value of about 1×10^{-13} sec. The rate at which molecules leave a surface of fractional monolayer coverage θ is directly proportional to the total number of molecules on the surface $N\theta$, where N is the number of molecules in a monolayer (approx 5×10^{14} molecules/cm^2 equivalent to

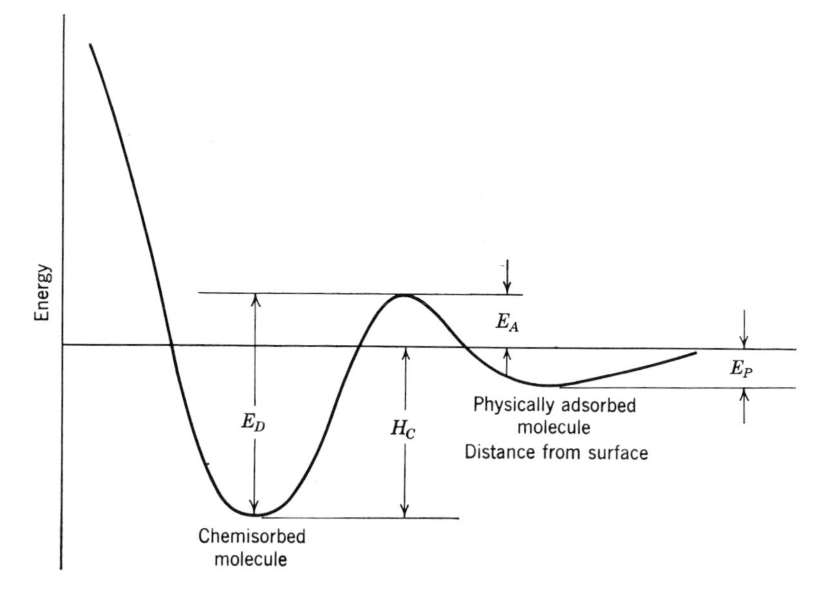

Figure 6.1. Potential energy for activated chemisorption for molecular adsorption. (From G. Lewin, *Fundamentals of Vacuum Science and Technology*, McGraw-Hill, New York, 1965, p. 37.)

1.5×10^{-5} torr liter/cm^2 at 293°K), and inversely proportional to the residence time:

$$dn_D/dt = N\theta/t_r \tag{6.33}$$

A surface having a constant residence time t_r will have an exponential desorption decay rate, and gas will be removed from the surface in much the same way that it would be removed from a small volume through a capillary leak. Since the residence time depends sensitively on the absolute temperature, one would expect that the desorption of gas (surface outgassing) from a surface would depend sensitively on temperature. It has been found experimentally that for each 10% increase in absolute temperature, there is approximately a tenfold increase in desorption or surface outgassing. A study by Hobson showed that if the desorption energy E_D of adsorbed gas is less than 15 kcal/mole, the gas is easily removed in an ordinary pumpdown at room temperature (295°K) with a nominal pumping speed. This includes all physically adsorbed gases. If E_D is greater than 25 kcal/mole, the desorption is negligible since the molecules are strongly bonded to the surface. In the desorption energy region from 15 to 25 kcal/mole, gas is only slowly removed from a surface by a pumpdown at room temperature. According to equations 6.32 and 6.33, raising the absolute temperature T of the wall surfaces is equivalent to reducing the desorption energy E_D and hence the process of baking a system during pumpdown at temperatures of 300–450°C permits gases with $E_D = 30$ kcal/mole to be removed as easily as gases with $E_D = 15$ kcal/mole at room temperature.

The amount of gas that remains on a surface in a vacuum system depends on the equilibrium between adsorption and desorption. The amount of gas N that remains is found by equating the following adsorption rate to the desorption rate of equation 6.33.

$$dn_A/dt = s\nu = 3.5 \times 10^{22} sP(MT)^{-1/2} \text{ sec}^{-1} \text{ cm}^{-2} \tag{6.34}$$

where ν is the impingement rate previously defined (cf. equation 6.9) and s is the so-called "sticking factor," the fraction of molecules striking the surface that adsorbs. The result of equating the two rate expressions is

$$\theta = 7 \times 10^{7} sP(MT)^{-1/2} \tau_0 \exp(E_D/RT) \tag{6.35}$$

The above equation will show that for the inert gases with $E_D < 10$ kcal/mole, the residence time at 300°K is $< 10^{-6}$ sec and for xenon ($M = 131$, $s = 1$) the surface coverage is $\theta \leq 0.35P$. At a pressure of 10^{-6} torr, the coverage of physically adsorbed gases will be less than 10^{-6} monolayer. On the other hand, at very low temperatures, the surface coverage becomes large even for small values of E_D, and this is the basis of cryogenic sorption pumping and surface area measurements.

The sticking factor or sticking probability s is a function of the species of gas and the type of surface and may vary with the surface coverage. Typical values of s lie between 0.1 and 1 for clean (low coverage) surfaces and decrease when monolayer coverage is approached.

At very high temperatures (for example, flashing a tungsten filament at 2300°K), the residence time, even for strongly adsorbed species, becomes small. This technique of high temperature flashing is used to prepare clean surfaces for various experiments. If the flashing is carried out in a low pressure environment (say, below 10^{-10} torr), the freshly cleaned surface will remain clean for many hours or days.

All solids and liquids generally contain gases in solution. The total gas content in metals, for example, may be as high as 0.1 atom % with the exception of hydrogen in certain metals such as titanium and columbium.

Plastics and elastomers may contain about 1% by weight of various gases. The solubility of a gas in a solid is defined as the quantity of gas in cm^3 at standard temperature and pressure (STP, 273°K and 1 atm) that is dissolved in 1 cm^3 of the solid at a pressure of 1 atm. When solids are placed in a vacuum, a diffusion gradient of the dissolved gases is set up within the material resulting in eventual release or outgassing of these gases. The outgassing rate for a semiinfinite wall will decay according to the relation

$$Q = C_0 D^{1/2} (\pi t)^{-1/2} \tag{6.36}$$

where C_0 is the initial concentration of gas in torr liter/cm^3 and D is the diffusion coefficient in cm^2/sec. The diffusion coefficient varies exponentially with temperature according to the relationship

$$D = D_0 \exp(-E/RT) \tag{6.37}$$

The quantity D_0 varies between 0.01 and 10 cm^2/sec for gases in solids.

It can be seen that the diffusion coefficient varies with temperature in the same manner as the residence time of an adsorbed molecule (equation 6.32). Again, a 10% change in absolute temperature can cause a 10-fold change in the diffusion coefficient, but the outgassing rate Q due to diffusion will only change by a factor of $\sqrt{10}$. The solubility of gases in solids is only slightly affected by temperature changes.

One important example of the diffusion of gases from within a solid into a vacuum is the diffusion of water vapor and other gases from elastomer O-rings. It has been found experimentally that the outgassing rate of O-rings per linear inch is of the order of 1×10^{-6} torr liters/sec at room temperature. If the O-ring is cooled, the outgassing will be decreased. However, an O-ring should not be cooled below the temperature at which it loses its elasticity.

Having described the phenomena of adsorption, desorption, solubility and diffusion, it can now be pointed out that it is possible for gas to be transferred from the outside of a vacuum enclosure through the wall and into the vacuum. The total process of gas transmission is called permeation. Permeation is possible whenever the gas in question will dissolve (has a finite solubility) in the wall material. Several very general statements can be made concerning permeation. First, it has been found that all gases permeate through all polymers. Second, none of the rare gases appreciably permeate any metal. Third, certain gases have relatively high permeation rates through specific materials. Some of the important gas–solid combinations that exhibit especially high permeation rates are H_2–Pd, O_2–Ag, and He–quartz. In fact, tubes of palladium, silver, and quartz are used as so-called "diffusion leaks" to transmit hydrogen, oxygen, and helium, respectively, into a vacuum system.

The permeability of various materials can be found in the vacuum literature and are generally expressed as the quantity of gas in cm^3 (STP) passing per second through a wall of 1 cm^2 area and a thickness of either 1 cm or 1 mm when a pressure difference of 1 atm exists across the wall.

The vapor pressures of substances, both solids and liquids, are important quantitites in vacuum work. The vapor pressure may be expressed conveniently as follows:

$$P_v = C \exp(-H/RT) \tag{6.38a}$$

or

$$\log P_v = A - (B/T) \tag{6.38b}$$

Tables and charts of vapor pressures of the elements, both metals, nonmetals and condensed gases, are available in most of the books on vacuum technique. At room temperature, solid and liquid chemical elements, with the exception of liquid mercury, cesium, sulphur, rubidium, and potassium, have vapor pressures below 10^{-10} torr. Some common metals, such as cadmium and zinc, have relatively high vapor pressures of 10^{-7} and 10^{-8} torr, respectively, at temperatures of about 100°C. These metals, or alloys containing them, are to be avoided in vacuum systems. The refractory metals such as molybdenum, tantalum, and tungsten have the lowest vapor pressures. For example, tungsten at a temperature of 2000°K has a vapor pressure of less than 1×10^{-11} torr.

The vapor pressures of the common gases as functions of temperature are quantities of great concern in vacuum work since many of these gases can be "frozen out" with the use of cold traps and cryogenic pumps. At a temperature of 20°K, for example (the boiling point of liquid hydrogen), all gases, with the exceptions of helium, hydrogen, and neon, have vapor pressures below 10^{-10} torr. At a temperature of 4.2°K, only helium and hydrogen can have vapor pressures above 10^{-10} torr.

Since vapor pressures vary exponentially with temperature, changes of the order of 10% in the absolute temperature cause vapor pressure changes of a decade. One common technique for testing for the presence of vapors within a vacuum system is to cool a portion of the system with Dry Ice or some similar refrigerant. A sharp decrease in the pressure indicates the presence of condensables.

Evaporation is the process whereby a solid or liquid loses molecules to the gas phase. This process is similar to desorption and has the same exponential dependence on temperature. There will be a net evaporation from a solid or liquid substance as long as the pressure of its gaseous phase is less than the equilibrium vapor pressure of the substance at the system temperature. This situation occurs when the vapors are being continually removed, or when the temperature of the substance is suddenly raised. At equilibrium, there are as many molecules being condensed from the gas phase as there are evaporating so that the pressure of the gas phase remains constant at the equilibrium vapor pressure. The evaporation rate under equilibrium, being equal to the condensation rate, can then be easily expressed as the impingement rate associated with the equilibrium vapor pressure multiplied by a sticking coefficient s_v (which is unity for metals in equilibrium with their own vapor).

$$Q_e = 3.638 s_v (T/M)^{1/2} P_v \text{ torr liters/cm}^2 \text{ sec} \tag{6.39a}$$

or

$$Q_e = 0.058 s_v (T/M)^{1/2} P_v \text{ g/cm}^2 \text{ sec} \tag{6.39b}$$

where P_v is the equilibrium vapor pressure in torr, T is the absolute temperature in °K, and M is the gram molecular weight of the substance.

D. Evacuation of a System, Outgassing Procedures

In the previous section that dealt with the topics of gas flow and pumping speed, it was shown that the inherent pumping speed of a pump was generally reduced by the gaseous conductances (resistances) of elements that might be interposed between the pump and the chamber being evacuated. Although the usual tortuous geometry of cold traps and sorption traps reduces the pumping speed for the noncondensable and nonsorbable gases, respectively, it should be remembered that *additional* pumping speed is provided by these traps for the condensable and sorbable gases.

Suppose that a vacuum chamber is to be evacuated by a "vacuum system" which could consist of several pumps in series or parallel together with various vacuum traps, valves, etc. The effective pumping speed S of the system *at the chamber* is first calculated by adding the total of all the final pump speeds that appear in parallel, as modified by the combined conductance of all traps, valves, connections, etc., that lead from the pumps to the chamber, using equation 6.25.

Having determined the effective pumping speed, the next step is to consider the sources of gas that must be pumped. First of all, the fixed amount of atmospheric gas within the chamber of volume V must be removed. Next, it may be desired to admit gas into the chamber via a fixed leak (or there may be an undesirable atmospheric leak) at a rate of Q_L torr liters/sec. Experience has shown that there is always some wall outgassing or desorption from surfaces within a vacuum chamber. This outgassing rate may be labeled Q_0. Finally, most vacuum pumps are imperfect in the sense that they become sources of gas at low pressures and actually emit new gases, or reemit old gases that have been previously pumped, into the vacuum chamber at a rate that will be designated as Q_P. If, now, one considers the balance of gas flow out of and into the vacuum chamber, it will be seen that the following relationship must hold:

$$-\frac{d}{dt}(PV) = -V(dP/dt) = SP - (Q_L + Q_0 + Q_P) \qquad (6.40)$$

This is a first-order linear differential equation. If the pumping speed S is taken to be constant (it may very well not be constant for certain types of pumps to be discussed later) and the Q's are constants, the solution is

$$P = (P_0 - P_u)\exp(-S/V)t + P_u \qquad (6.41a)$$

where

$$P_u = (Q_L + Q_0 + Q_P)/S \qquad (6.41b)$$

In these equations, P is the chamber pressure at any time t, the initial chamber pressure being P_0 at time $t = 0$. The pressure P_u is the ultimate or lowest pressure that can be achieved by balancing the pumping speed S against the residual sources of gas $Q_L + Q_0 + Q_p$.

It can be seen from equation 6.41 that the pressure P approaches its ultimate value P_u in an exponential fashion. Considering the pressure difference $P - P_u$ as the principal variable, the time constant for this variable is $\tau = V/S$, e.g. $P - P_u$ will be reduced to $1/e$ (about 36.8%) of its initial value in V/S sec (V in liters and S in liters/sec). If a vacuum system is clean, dry, and tight, it should pump down from atmosphere to the forevacuum pressure region below 100 μ in about 10τ sec. The average vacuum system which has been exposed to atmospheric air will generally pick up enough water vapor so that a considerably longer pump down time will be required. When the pressure of a room temperature vacuum chamber reaches a value of about 17 torr (the vapor pressure of water at 20°C), it will remain at that value until the water is pumped out, and only then will it resume its exponential decrease. If a chamber is being emptied through a constricting orifice or tube of conductance C by a pump or by connection to a second chamber at a much lower pressure, the time constant for the evacuation process is $\tau = V/C$.

The ultimate pressure of a leaktight vacuum system is most generally limited by outgassing from the walls of the system, including the pump walls, or by outgassing from materials placed within the system, including pressure measurement devices. An idea of the magnitude of the outgassing can be obtained from a study of Table 6.4, which lists outgassing rates of typical vacuum materials. It can be seen at once that glass, ceramics, and metals offer low outgassing rates, especially after they have been baked out under vacuum. Plastics and elastomers can provide high outgassing rates.

Table 6.4. Representative Outgassing Rates of Common Vacuum Materials in torr liters/cm^2 sec at Room Temperature.

Material	Rate after a few hours of pumping	Rate after bakeout	Typical bakeout conditions	Gas evolution or permeation
Metals, including stainless steel	10^{-9} to 10^{-8} (10^{-10} after 50–100 hr)	10^{-15} to 10^{-14}	400°C for 24 hr	CO, H_2O, H_2, N_2 (O_2 in iron)
Glass (Pyrex)	10^{-9} to 10^{-8}	10^{-15} to 10^{-14}	400°C for 24 hr	H_2O, primarily, He, H_2, CO
Ceramic	10^{-9} to 10^{-8}	10^{-15} to 10^{-14}	400°C for 24 hr	
Teflon (polytetrafluoroethylene)	10^{-8} to 10^{-7}	$< 10^{-8}$	250°C for 24 hr	
Nylon (polyamide)	10^{-8} to 10^{-6}	4×10^{-11}	120°C for 24 hr	CO_2, N_2, H_2O
Viton	10^{-8} to 10^{-7}	10^{-10} to 10^{-9}	150°C for 24 hr	H_2O, CO_2 plasticizers
Neoprene	10^{-7} to 10^{-6}			CO_2, N_2, H_2, He, O_2, A plasticizers
Lucite (methyl methacrylate)	10^{-6}			H_2O, CO_2 plasticizers
Epoxy (Araldite)	10^{-8} to 10^{-6}	10^{-10}	100°C for 24 hr	Plasticizers
Kel-F	10^{-8}			O_2, N_2, H_2O
Silicone rubber	10^{-6}			H_2O, CO_2 plasticizers

For attainment of the lowest pressures, glass or metal systems are generally baked at approximately 450°C for about 24 hr. Lower temperatures and/or shorter bakeout periods provide less outgassing and result in higher ultimate system pressures. Very often one or more components present in a system will limit the maximum bakeout temperature that can be used. At the termination of the bakeout period but while the system is still hot, it is standard procedure to degas all hot filament gauges and mass spectrometers. One should be careful, however, to degas a gauge or bake out a system only when the system pressure is low enough ($\sim 1 \times 10^{-4}$ torr) to avoid oxidation of internal surfaces.

Fundamentals of kinetic theory, gas flow, permeation, diffusion, adsorption–desorption, evaporation, and other phenomena basic to vacuum technology are described in detail in Dushman and Lafferty (5), which is the definitive treatise on vacuum technology. Other books which have especially good treatments of these fundamental topics are Lewin (12), Santeler et al. (18), Guthrie and Wakerling (8), and Turnbull et al. (20).

As a conclusion to this section on fundamental considerations, the reader may find it useful to have the most important vacuum constants and conversion factors grouped together in a single table (see Table 6.5).

III. METHODS OF PRODUCING VACUUM

The pumping equipment available today to produce and maintain vacuum conditions takes many different forms. In any given system, it is common procedure to use more than one type of pump if pressures in the high or ultrahigh vacuum region are required. A listing of pump types would include mechanical positive displacement pumps, aspirator pumps, sorption pumps, vapor stream pumps, chemical getter pumps, ion pumps, cryogenic pumps, and turbomolecular pumps. Some of the more common pumps are described below.

A. Mechanical Positive Displacement Pumps

Positive displacement type pumps operate by isolating a fixed volume of gas from the system being evacuated, compressing the gas to a smaller volume and higher pressure and then exhausting the gas through a check valve arrangement to atmospheric pressure (or the input of a backing pump in some cases). The most common of these pumps is the rotary oil-sealed variety in which a low vapor pressure pumping oil is used to seal the surfaces vanes, rotors, pistons, etc. that compress the gas. A second type of positive displacement pump is actually a blower in which there are close clearances between the moving and fixed parts. These pumps have no oil seal. They are used almost exclusively on very large chambers.

Positive displacement pumps are used for the initial evacuation of a system and to back other low pressure pumps that cannot exhaust directly to atmospheric pressure. Modern oil-sealed rotary pumps of the two-stage design can reach blankoff pressures in the 10^{-4} torr region. Mechanical pumps are available with capacities that range from a few tenths of a liter per second to over 1000 cfm.

It might be expected that the pumping speed of a mechanical pump

Table 6.5. Useful Constants, Formulas, and Conversion Factors For Vacuum Calculations

(In the expressions below, the pressure P is given in torr units; the volume V is in liters; M is the molecular weight in grams/mole; A is area in cm^2; D is diameter in cm; and L is length in cm)

Avogadro's number, N_o (molecules per mole)	6.023×10^{23}
Boltzmann constant, k	1.38×10^{-16} erg/°K, molecule
	1.03×10^{-22} torr liter/°K, molecule
Molar volume at STP (760 torr, 273°K)	22.414 liters
Gas constant R	62.36 torr liters/°K, mole
	8.314×10^{7} erg/°K, mole
	1.987 cal/°K, mole
Atomic weight unit	1.660×10^{-24} g
Energy of gas at 293°K	0.6 kcal/mole
	0.026 eV/molecule
Electron volt equivalent	23 kcal/mole/eV
	1.60×10^{-12} erg/eV
Approximate no. of molecules in monolayer	5×10^{14} molecules/cm^2
	1.5×10^{-5} torr liters/cm^2 at 293°K
Mass of gas at 20°C	5.44×10^{-5} MPV g
	1.56×10^{-3} g/torr liter for air
Mean free path in air at 20°C	$4.86 \times 10^{-3}/P$ cm
Molecular flow conductance of orifice at 20°C for air	$11.6A$ liters/sec
Molecular flow conductance of long tube at 20°C for air	$12.1\ D^3/L$ liters/sec
Viscous flow conductance of orifice at 20°C for air	$20A$ liters/sec

Units are: 1 torr = 1/760 std atm = 133 N/m^2 = 1 mm Hg
= 0.03937 in Hg = 0.01934 lb/m^2 = 10^3 μ
= 1,333 dynes/cm^2 = 1,333 microbar
1 torr liter/sec = 1.316 atm cm^3/sec = 1.27×10^5 μ ft^3/hr
1 atm cm^3/sec (STP) = 760 μ liters/sec = 0.760 torr liters/sec
1 liter = 61.0 in^3 = 0.0353 ft^3
1 liter/sec = 2.12 ft^3/min = 3.6 m^3/hr

(defined as the volume of gas at the inlet pressure that flows through the pump per unit time) would be constant and independent of pressure since the pump rotor sweeps out a fixed volume per unit time. Realistically, the effective pumping speed decreases as the pressure decreases due to a number of effects including leakage paths within the pump at the various sealing surfaces, the release of gas that has been dissolved in the oil, and the decomposition of the oil by local high temperatures. Many pumps have an optional feature known as gas ballasting, which allows them to pump water vapor without contaminating the oil.

In order to prevent mechanical pump oil liquid and vapor from entering the vacuum chamber being pumped, it is standard practice to interpose a cold trap or adsorption trap between the pump and the system. Such a trap serves the additional function of preventing condensables such as water vapor and mercury vapor, for example, from entering the pump and contaminating the oil. Dry Ice or activated zeolite sorption traps are most commonly used.

B. Aspirator and Sorption Pumps

As an alternative to the use of mechanical pumps to "rough out" or initially evacuate a chamber from atmospheric pressure, it is becoming increasingly popular to use so-called "clean or dry" forepumping techniques that avoid the use of pumping oils. One such technique is an adaptation of the old water aspirator principles to a pump which uses dry air or other dry gas to entrain the vacuum chamber gases. By means of the venturi effect, a low pressure is created at some point in the flow of the dry air through the pump. Such aspirator pumps, operating with compressed air or gas of 70—100 psi, can reduce the pressure of a system to about 125 torr.

Sorption forepumps that are available at present use a granulated sorbent or molecular sieve material such as artificial zeolites (porous aluminum silicate compounds) or activated charcoal, which present an extremely large surface area to the gases being pumped. Initially, gases are sorbed on the highly porous sieve material in an amount that is in equilibrium at room temperature. To activate the sieve material, it is baked at a temperature of 200–300°C while exposed to atmospheric air, and then sealed off while hot. The sorption pump is placed in operation by cooling the sorbent material to liquid nitrogen temperature and valving it into the system. Gases such as nitrogen, oxygen, argon, krypton, carbon dioxide, and water vapor are readily sorbed; in fact, 1 g of zeolite can adsorb about 100 torr liters of nitrogen at -195°C (liquid nitrogen temperature) over the pressure range from 10^{-2} to 10^{3} torr. At pressures below 10^{-2} torr, the amount adsorbed decreases according to the adsorption isotherm. Gases such as neon, helium, and hydrogen, which are present in normal atmospheric air in small amounts (see Table 6.6) are sorbed to only a

Table 6.6. Composition of Dry Air at Sea Level[a]

Gas	Percent by volume	Partial pressure, torr
Nitrogen	78.1	593.0
Oxygen	20.9	159.0
Carbon dioxide	0.03	0.23
Argon	0.93	7.1
Krypton	1.0×10^{-4}	7.6×10^{-4}
Neon	18.0×10^{-4}	137.0×10^{-4}
Helium	5.24×10^{-4}	40.0×10^{-4}
Hydrogen	0.5×10^{-4}	3.8×10^{-4}

[a] From *Handbook of Chemistry and Physics*, 42nd ed., Chemical Rubber Publishing Co., Cleveland, Ohio, 1960, p. 3391.

small degree, so that the composition of the residual ambient contains a large percentage of helium. A single stage of sorption pumping with a pump of the proper size will reduce the system pressure into the 10^{-2} torr range. A second stage and additional stages of sorption pumping may be used to lower the system pressure to 10^{-4} or 10^{-5} torr. Sorption pumps are used most often in combination with ion pumps to provide the low pressure level (10^{-3} to 10^{-2} torr) at which the ion pumps can start and operate. Once the ion pump is started, the sorption pump is isolated from the system.

C. Vapor Stream Pumps

Diffusion pumps and ejector pumps are classified as vapor stream pumps. In these pumps the dense vapor of a hot working fluid such as oil or mercury emerges at supersonic velocities from one or more jet nozzles. The heavy molecules of the pumping fluid vapor are directed in such a way that they collide with molecules of the gas being pumped and impart to them a large component of momentum in the direction of the forevacuum or backing pump. The pumping fluid vapor is condensed and returned to the pump boiler where it is again evaporated and the cycle is repeated.

Diffusion pumps cannot exhaust to atmospheric pressure, and it is necessary to provide a forepressure or backing pump, usually a mechanical positive displacement type pump, to maintain the system forepressure at some value of the order of 0.1 torr or less. As long as the foreline pressure is maintained below a critical value called the limiting or tolerable fore-pressure, the diffusion pump will operate independent of the forepressure, except for very low pressures in the ultrahigh vacuum region where back-diffusion of light gases such as hydrogen and helium can occur. If the forepressure is too high, the jet vapor stream is disrupted, the pumping action stops, and there will be unusually high backstreaming of the vapor from the jet assembly toward the mouth of the pump. The size of forepump to be used with a given diffusion pump is a function of the maximum throughput ($Q = PS$) that the diffusion pump can handle. The same gas load that flows through the diffusion pump must be carried away at a much higher pressure by the forepump. Generally, one simply follows the manufacturer's recommendation, but it is possible to substitute a larger forepump if faster system pumpdown is desired or even use a smaller forepump to back the diffusion pump once it has been pumped down to low pressures where the throughput is low. At very low pressures, it is even possible to isolate the forepump from the diffusion pump and let the diffusion pump exhaust into an evacuated reservoir. The mechanical pump can then be turned off if its vibration interferes with delicate measurements.

Not all of the gas molecules that enter the mouth of a diffusion pump are pumped away. The ratio of the actual pumping speed to the theoretical maximum is called the speed factor or the Ho coefficient. In modern oil diffusion pumps, the speed factor is about 0.5 for unbaffled, untrapped pumps. The inherent pumping speed of a diffusion pump is constant at low pressures but decreases to zero as the limiting forepressure is approached. Diffusion pump speeds are generally given for air, but rated speeds are approximately correct for gases such as nitrogen, oxygen, argon, carbon dioxide, and carbon monoxide. Lighter gases, such as helium and hydrogen, are pumped somewhat faster, since the vacuum conductance of the pump mouth and annular regions surrounding the jets varies inversely with the square root of the molecular weight. However, the efficiency with which pumped molecules are entrained (the capture efficiency) is less for smaller molecules and depends on the pump geometry. The final result is that each pump design has a unique set of pumping speed ratios for various gases.

Mercury diffusion pumps are not as commonly employed as oil diffusion pumps chiefly because of the danger of contaminating the system being pumped with mercury. Mercury amalgamates easily with copper, gold, and aluminum, three materials that are commonly found in high vacuum systems. Despite the fact that mercury pumps have a lower pumping speed than oil diffusion pumps of comparable size (the mercury molecule is smaller), there are many applications where mercury pumps have decided advantages. Mercury does not decompose and does not contribute to the noncondensable gas load. Mercury does not react with most other gases or with hot filaments. It is possible to keep mercury vapor completely out of the system being pumped with the use of suitable cold traps. A system pumped with a mercury diffusion pump has no complicated hydrocarbon impurities that can be troublesome in mass spectrometer type analyzers or wherever hydrocarbons will affect the reactions or analyses being performed. Mercury diffusion pumps can create pressures as low as 10^{-13} torr in properly designed UHV systems, on the one hand, and can operate with limiting forepressures as high as 30 torr, on the other hand.

Diffusion pumps are frequently used in combination with matching watercooled baffles and liquid nitrogen cooled traps. The use of such baffles and traps reduces the effective pumping speed, perhaps by a factor of 2 in some cases, but prevents the pump fluid from entering the system in any large quantity. Experience has shown that it is difficult to keep all oil vapors or oil decomposition products out of an oil diffusion pumped system especially during startup or shutting-down procedures. Liquid nitrogen cooled traps serve the additional function of pumping water vapor or other condensables that may be present in the system and would tend to contaminate the diffusion pump oil.

Many new diffusion pump oils have been developed in the past decade. Some of the fluids, especially the silicones, have great resistance to heat decomposition and oxidation. In addition, these fluids have low vapor pressures of the order of 10^{-10} torr at room temperature. Such low vapor pressure oils have been used as lubricants in vacuum systems.

Some typical commercial oil diffusion pumps have the following diameters and associated pumping speeds: 2 in., 150 liters/sec; 4 in., 750 liters/sec; 6 in., 1500 liters/sec; and 10 in., 4200 liters/sec. Diffusion pumps are useful primarily where large amounts of gas must be pumped continuously, where constant pumping speeds are required, and where "memory effects" such as reemission of previously pumped molecules are to be avoided.

D. Getter-Ion and Titanium Sublimation-Type Pumps

Although the earliest ion pumps that were used to attain ultrahigh vacuum in the early 1950's were Bayard Alpert-type hot filament ionization gauges, the commercial ion pumps of today take several other forms. A true ionization type pump is one which converts a fraction of the molecules being pumped into positive ions and then accelerates these ions in an electric field with sufficient energy to penetrate a target or pumping surface and become trapped in the target lattice. True ion pumping is a relatively inefficient process and generates low pumping speeds.

Modern ion pumps combine true ion pumping with the very efficient process of chemical gettering. Titanium and tantalum metals, especially, when freshly evaporated or sputtered to form a newly deposited surface, combine with active gas molecules to form stable chemical compounds. Some other materials which have been used historically for gettering are barium, zirconium, aluminum, magnesium, and mischmetal.

One class of getter-ion or sputter-ion pumps, as they are commonly called, uses a magnetically supported cold cathode discharge process to ionize and excite the gas being pumped. The pumping action of a sputter-ion pump comes about when positive ions produced in the discharge bombard the titanium cathode plates. Some of these ions (e.g. the rare gas atoms of helium, neon, argon, etc.) penetrate into the cathode and are lost from the volume. The impinging ions also cause atoms of titanium to be sputtered from selected areas of the cathode onto the anode and also onto unsputtered portions of the cathode. The freshly sputtered titanium atoms combine with gas atoms and molecules that have been excited and dissociated by the discharge.

The pumping speed of an ion pump is chiefly a function of its size. Pumps are available presently with speeds ranging from a few tenths of a liter/sec to thousands of liters/sec. The wide range of speeds available is a distinct advantage in matching a pump to a given system. One specific example of this is the use of a very small 1 liter/sec pump as an appendage to an evacuated vacuum tube such as a traveling wave tube or klystron.

One undesirable characteristic of ion pumps is the occurrence of re-emission of the inert gases such as argon, neon, and helium. In some cases, when large amounts of argon have been pumped, instabilities occur in the pumping operation. Special ion pumps have been developed that have increased pumping speed and capacity for pumping argon. Standard diode ion pumps have an argon pumping speed that is about 1% of their nitrogen pumping speed. Special slotted-cathode diode pumps have an argon pumping speed that is about 6% of their nitrogen pumping speed while triode ion pumps have an argon pumping speed that is about 30% of their nitrogen pumping speed. Relative pumping speeds of diode and triode ion pumps are given in Table 6.7. The pumping speed of sputter ion pumps is generally low at high pressures of the order of 10^{-4} torr, increases to a maximum at pressures of 10^{-6} to 10^{-7} torr, and then decreases again at lower pressures. This variation in pumping speed with pressure may be a disadvantage in some applications. The life of a sputter ion pump is eventually limited by the erosion of the cathode material and the formation of flakes of deposited metal which may short out the discharge. The rated life of commercial diode pumps is about 50,000 hr at 10^{-6} torr and will vary inversely with the pressure. One other undesirable characteristic

Table 6.7. Relative Pumping Speeds of Diode and Triode Sputter-Ion Pumps for Various Gases

Gas	Diode pump	Triode pump
N_2	100	100
Air	100	105
O_2	57	115
H_2	270	210
D_2	190	
CO_2	100	
H_2O	100	
He	10	21
Ne	4	
A	1	31

of sputter ion pumps is their generation of small amounts of hydrocarbons, especially methane.

Quite often, sputter-ion pumps are combined with titanium sublimation pumps in order to increase the pumping speed for the active gases. Very high speeds may be achieved with equipment of a relatively small size since each square inch of coated surface pump pumps gases like nitrogen at the rate of about 15 liters/sec.

A second class of ion pump combines titanium evaporation, rather than sputtering, with true ion pumping. The most recent pumps in this category use a hot filament to supply electrons for gas ionization purposes (instead of creating a crossed electric and magnetic field cold cathode discharge). The new pumps place the electrons into long spiral paths without the use of a magnetic field and also use the kinetic energy of the electrons to sublimate pure titanium within the pump. Hallmarks of this latest ion pump are a high pumping speed (a 6-in. diam pump about 24 in. high has a speed of 800 liters/sec for air) in a small light package, a relatively constant pumping speed for pressures below 10^{-6} torr, and the absence of a magnetic field.

Summing up, one can say that ion pumps are especially suitable in those applications where the absence of pumping fluids such as oil and mercury is essential, and where large amounts of gas at high pressures need not be pumped. Refrigerants or special traps are not required with ion pumped systems. The ion pump current in sputter-ion pumps can serve as a measure of the system pressure. In the event of power failure, the system will not be vented to atmosphere nor will the pump normally be damaged.

Sputter-ion pumps have no vibration problems nor do they introduce heat gradients. When used in combination with valved sorption pumps to produce the required starting pressure of less than 10^{-2} torr, a completely oil-free system results. Scientific instrumentation, such as vacuum microbalances or equipment used to study surfaces or thin films, might advantageously be pumped with ion pumps.

E. Cryopumps

Cryogenic pumping is the general process of removing atoms and molecules from the gas phase by condensing them onto a cold surface. Liquid nitrogen cooled surfaces at a temperature of 77°K are often used to pump water vapor and CO_2. (The vapor pressure of CO_2 at 77°K is 10^{-8} torr.) Liquid hydrogen cooled surfaces at 20.4°K will condense all gases except helium, neon, and hydrogen. Liquid helium cooled cryopumps (4.2°K surfaces) will pump all gases except helium. As with sorption and getter-type surfaces, the pumping speed for a given surface area depends on the sticking coefficient or condensation coefficient s. For nitrogen and hydrogen, the pumping speeds are $11.8s$ and $44s$ liters/sec, cm^2, respectively, for a pure gettering or sorption surface at the temperature of the vacuum chamber. For a cryosurface at a low temperature, the ultimate pressure P_u of a gas that is being pumped is somewhat higher than its vapor pressure at the surface temperature due to the thermal transpiration effect. The pumping speed of a cryosurface is given by the expression

$$S = 3.64s(T/M)^{1/2}(1 - P_u/P) \qquad (6.42)$$

Cryopumping is an ideal method for pumping relatively large chambers such as low density wind tunnels and space simulators due to its inherently high pumping speed. Cryopumping is also the cleanest method of pumping since the reemission and hydrocarbon generation characteristics of sputter-ion pumps are not present. Cryogenic surfaces can handle relatively heavy gas loads that would saturate ion pumps or sorption pumps. In order to attain ultrahigh vacuum with a cryopump, however, it is necessary to use an auxiliary pump to remove gases such as helium or hydrogen. (The vapor pressure of hydrogen at 4.2K is 3.5×10^{-7} torr.) Small ion pumps or panels of bonded molecular sieve material can be used to remove these noncondensable gases. The latter method of pumping is termed cryosorption. One final low temperature pumping phenomenon that should be mentioned is cryotrapping, the trapping of noncondensable gases by a condensing vapor. For example, nitrogen may be trapped and thus pumped by condensing water vapor at 77°K, and hydrogen and helium may be trapped by condensing argon at 4.2°K.

F. Turbomolecular Pumps

In the turbomolecular pump, a high speed rotor imparts a directed momentum to impinging gas molecules. A forevacuum pump, generally a mechanical positive displacement pump, is used to lower the gas pressure at the outlet of the high speed rotor and thus reduce heating due to gas friction. Characteristics of the turbomolecular pump (or molecular drag pump, as it is sometimes called) are a constant pumping speed over a wide pressure range, the ability to pump heavier gas molecules more effectively than light molecules and atoms, resulting in an oil-free background gas spectrum, and the limitation of pumping speed due to the size and weight of the pump. Turbomolecular pumps are especially useful where it is necessary to obtain low pressures of the order of 10^{-9} torr rapidly and with a minimum of skill being exercised in the operation of the pump. Such pumps are practically "push-button" type devices requiring a minimum of vacuum technology. On the other hand, these pumps are relatively expensive for the pumping speed provided, they are heavy and space consuming, and they often generate high frequency vibrations that may affect delicate measurement apparatus.

Summing up the various pumping methods, it can be said categorically that no single pumping system is the best for all occasions. The most suitable method of pumping will depend on the application. The various textbooks on vacuum theory that are listed in the bibliography contain much more detailed material on specific pumps than that listed herein. The reader will find especially complete discussions of mechanical displacement pumps in references 5, 7, 15, 19, 20, 22, and 24. Diffusion pumps are well described in references 5 and 19. References 2, 12, 16, and 18 contain good descriptions of ion pumps, sublimation pumps, and sorption pumps while references 2, 12, and 18 discuss cryopumping (and cold traps and baffles) in great detail. Commercially available pumps are listed in reference 21.

IV. METHODS OF MEASURING HIGH VACUUM AND CONTROLLED ENVIRONMENTS

Vacuum gauges can be classified according to a number of different characteristics, but perhaps the best way to group them is according to the general physical process involved in the measurement. There are four principal classes of gauges.

One group can be categorized as mechanical manometers. In this class one would find liquid manometers, the Dubrovin gauge, the McLeod gauge, Bourdon gauges, aneroid capsule gauges, bellows gauges, and membrane and diaphragm gauges.

A second group of gauges can be classified as thermal conductivity gauges. Pirani, thermocouple, and thermistor gauges belong here.

A third group of gauges can be called momentum transfer gauges. The Knudsen gauge, molecular gauge, and decrement or viscosity gauges are in this class.

Finally, there are a number of gauges which can best be classified as gas ionization gauges. Members of this class are hot filament and cold cathode ionization gauges, mass spectrometers, radioactive source (alphatron type) gauges, and discharge tubes (Tesla coil).

Low gas pressures can also be measured by other techniques such as surface reaction techniques, but here there are no well-developed instruments.

A. Mechanical Manometers

Mechanical manometers measure gas pressures by actually balancing the force exerted by the gas with a mechanical or hydrostatic force. These gauges are true pressure gauges in that they actually measure the pressure or force exerted by a gas. The mercury manometer, the oil manometer, and the McLeod gauge are hydrostatic devices that use a head of liquid to balance a gas pressure. More will be said about these instruments later because they are useful as primary pressure measurement devices. The Dubrovin gauge balances a liquid buoyant force against the gas pressure on an inverted, evacuated cylinder that is submerged in mercury. The pressure is linearly related to the cylinder position:

$$P \simeq 4W\rho gh/D \text{ dynes/cm}^2 \quad \text{if} \quad W \ll D \tag{6.43}$$

Here W is the wall thickness of the cylinder of diameter D, h is the depth of submersion, ρ is the liquid density, and g is the local acceleration of gravity. Units in the cgs system are used in this equation.

The principle of operation of aneroid capsule gauges, bellows, membrane, and diaphragm gauges are rather obvious. Usually mechanical forces are used to oppose the gas pressure force. In the past 10 years or so, beginning with Alpert's work in ultrahigh vacuum, the diaphragm manometer has become popular as a pressure measuring device. Many of these gauges use the null technique, where the thin metal diaphragm is restored to its central position by an electrical or gas pressure restoring force. Capacity measuring techniques are usually used to determine the deflection of the diaphragm. Because some types of this gauge can be made to give reproducible readings over a relatively wide pressure range, and the gauges are portable and robust, they should make good secondary standard gauges within their range of operation. Their pressure range extends from about 10^{-3} to 760 torr with 1% accuracy or better.

B. Thermal Conductivity Gauges

Thermal conductivity gauges are of two principal types, (*1*) the Pirani gauge and (*2*) the thermocouple and thermistor gauges. In these gauges a fine wire filament mounted within the gauge is heated by a current from an external source until it reaches some operating temperature above ambient. Within a certain range of pressures, depending on the gauge construction, the heat loss of the wire, due to conduction through the gas is sufficient to change the wire temperature. Since the thermal conductivity of a gas under molecular flow conditions is proportional to the gas pressure, a measure of the operating temperature of the wire provides a measure of the gas pressure. In the Pirani gauge, the temperature of the wire is measured via its change in resistance, a Wheatstone bridge circuit being used for this purpose. In the thermocouple or thermistor gauge, a thermocouple or thermistor mounted on the heated wire measures its temperature.

Hot-wire, thermal conductivity gauges are very useful in vacuum work because they are inexpensive, have a rapid response, indicate the pressures of vapors as well as permanent gases, and can be used to operate relays, etc. However, these gauges are not suitable for making accurate pressure measurements. Their sensitivity depends on the mass, specific heat, and accommodation coefficient of the gas being measured, and the gauge calibration is changed by changes in the surface of the hot wire (oxidation, sorption of gases) that affect its accommodation coefficient and the emissivity.

C. Momentum Transfer Gauges

Momentum transfer gauges such as the Knudsen gauge and the molecular gauge are not in widespread use today. In the Knudsen gauge, a radiometric force is created between a fixed hot plate or heated coil and a cooler moveable plate. Molecules which come from the hot plate have greater kinetic energy and velocity than those that come from the cool walls of the gauge. The net molecular force exerted on the moveable plate is such as to push it away from the hot surface and toward the cooler surface. The Knudsen gauge is slightly dependent on gas composition because the accommodation coefficient is different for different gases. The pressure P_m on the moveable plate of a Knudsen gauge is given by the expression

$$P_m = P[(T_1/T)^{1/2} - 1]/2 \qquad (6.44)$$

where P is the gas pressure and T_1 and T are the hot plate and wall temperatures, respectively.

The molecular gauge is based on the transfer of momentum from a rotating disk to a moveable vane. This gauge cannot be considered an absolute gauge since the reading is proportional to $P(M)^{1/2}$. Knudsen and molecular gauges are very delicate and must be shielded from mechanical vibrations. For the molecular gauge, the torque α is given by the equation

$$\alpha = K\tau^2 r^4 . P . \omega . (M/T)^{1/2}/I \tag{6.45}$$

where K is a constant, τ is the period of oscillation, r is the rotating disk radius, I is the disk moment of inertia, and ω is the angular velocity of rotation of the disk.

It is important to note that thermal conductivity gauges and momentum transfer gauges do not measure actual gas pressures, but rather some property of the gas that is related to the pressure via kinetic theory.

D. Total Pressure Ionization Gauges

This class of ionization gauges includes perhaps the most important instruments in use today, especially for very low pressure measurement. In an ionization instrument, a constant fraction of the gas molecules present are ionized, and the positive ion current so created is measured. The number of ions formed depends on the number density of gas molecules present and thus indirectly on the gas pressure. The sensitivity of these instruments depends on the particular gas being measured. Ionization gauges are completely electrical, and thus they have certain advantages over mechanical and optical instruments. In addition, ionization instruments are generally linear within a suitable range of pressures. Ionization instruments have been used to measure total number densities that correspond to pressures of 10^{-13} torr, and partial pressures of individual molecular species less than 10^{-13} torr have been measured with mass spectrometers. Pressures as high as about 760 torr have been measured with radioactive source types of ionization gauge.

Among the more common types of ionization gauges are the thermionic or hot cathode gauges, the cold cathode or glow discharge gauges, and the radioactive ionization gauges. Thermionic gauges ionize the gas within the gauge with electrons that are emitted from a directly heated filament and accelerated through a potential of about 100–200 V. In the popular Bayard Alpert-type gauge (abbreviated to B-A gauge), the filament is located outside of a cylindrical grid which is maintained at a potential of about 150 V. A fine wire suspended axially along the centerline of the grid structure acts as the collector of the positive ions that are formed within the grid ionization region. The filament is maintained at a potential of about 30 V above ground while the collector is kept at ground potential. The Bayard-Alpert gauge was originally designed to reduce the x-ray created residual current to the ion collector, which limited the low pressure capability of earlier gauge designs. The B-A gauge can measure pressures as low as 10^{-11} torr and as high as 10^{-3} torr at a reduced electron emission. The positive ion collector current i_+ of this gauge is linearly related to the pressure P by means of the expression

$$i_+ = S i_- P \tag{6.46}$$

where i_- is the electron current from the filament to the grid and the proportionality constant S is called the sensitivity of the gauge and has the dimensions of torr^{-1}. The sensitivity S varies with gauge geometry (it will change if the gauge filament moves or the grid structure sags after repeated degassing), the grid voltage, and the ionization cross section of the gas. The ionization efficiencies of electrons for all gases follow a fairly general pattern. The ionization maximum occurs in the vicinity of 100 V for most gases. A typical B-A gauge has an electron emission current of 10 mA and a sensitivity $S = 10$ for nitrogen gas so that it would produce an ionization output current of 0.1 A per torr pressure. Another important type of hot filament gauge is the Schulz-Phelps high pressure type of ionization gauge which covers the range from 10^{-6} to 1 torr. If pressure readings are to be made at the extremes of the ranges covered by these gauges, it would be wise to become more familiar with the details of gauge operation including such effects as high and low pressure nonlinearities, gauge pumping, change in wall potential (two mode operation), dissociation of gases, desorption of surface gases by electron bombardment, release of hydrogen and carbon monoxide from the filament, oscillations in the ionization current, poisoning of the filament, the effect of wall deposits, and the effect of nearby magnetic fields.

The hot filament of an ionization gauge has a limited life and can be damaged easily by exposure to air while in operation. The cold cathode or glow discharge gauge avoids this difficulty by providing electrons via the mechanisms of cold field and secondary emission. In the Penning-type gauge, a magnetic field of approximately 1000 Oe in a direction normal to the cathode plates, is obtained with a permanent magnet. A potential difference of about 2000 V between the anode and cathodes furnishes the electric field required to accelerate the electrons and start the gas discharge. The ion current output of a typical cold cathode gauge is generally greater than the output current of a typical hot filament gauge operating at the same pressure. Besides the Penning gauge, there are several other cold cathode gauge designs such as the Redhead magnetron and inverted magnetron gauges that have been developed for special purposes. A magnetron gauge consists of a cylindrical anode surrounding a smaller diameter coaxial cylindrical or spool-shaped cathode immersed in a magnetic field whose direction coincides with the axis of the two electrodes. The magnetron gauges can measure pressures as low as 10^{-13} torr, but they are nonlinear for pressure below about 1×10^{-9} torr. A typical magnetron gauge might have an output current of 5 A per torr pressure. Cold cathode gauges are especially suitable where the heating effect of hot filament gauges cannot be tolerated, where gas reactions that occur at a hot filament must be avoided, and where power is limited, such as in satellite applications.

Many experimental types of total pressure ionization gauges have been developed in recent years. These include Redhead's modulated BA gauge, Lafferty's hot filament magnetron gauge, Schumann's suppressor gauge, Helmer's bent beam gauge, and Herb's orbitron gauge. All of these new gauges extend the low pressure measurement limit. These gauges are described in references 12, 16, and 18.

Radioactive type ionization gauges make use of radioactive substances such as radium or tritium to ionize the gas, and thus they present a slight radiation hazard. Such gauges operate over the pressure region from 10^{-4} to 10^3 torr. As with other ionization gauges, their sensitivity depends on the gas species.

Table 6.8 lists the characteristics of important total pressure measurement instruments.

E. Partial Pressure Ionization Instruments

Devices which can electrically or electromagnetically separate the various gases in a mixture on the basis of their mass for the purpose of yielding the composition of the mixture are known as mass spectrometers, partial pressure analyzers, or residual gas analyzers. Since almost all mass spectrometers that are used for gas identification and analysis work use an electron beam to ionize the atoms and molecules, these instruments have many of the characteristics of hot filament ionization gauges. Mass spectrometers, as well as ionization gauges, have differing sensitivities for the various gas species. Due to ion analyzer characteristics, the sensitivities of a mass spectrometer and a typical ion gauge for various gas species usually differ considerably. In the ion gauge, the sensitivity generally increases with the mass number (and ionization cross section) of the gas, while the opposite is generally true for a mass spectrometer.

All partial pressure analyzers have three basic components: an ion source where the neutral gas is ionized, an analyzer section in which ions

Table 6.8. Characteristics of Several Pressure Measuring Devices

Name	Pressure in torr		Principle of operation
	Minimum	Maximum	
Manometer	10^{-1}	10^{3}	Height of Hg column
Spark discharge	10^{-3}	10^{2}	Color of gas discharge
Diaphragm gauge	10^{-5}	10^{3}	Moving membrane
McLeod gauge	10^{-6}	10	Height of Hg column after compression
Pirani gauge	10^{-6}	10^{-2}	Resistance measured } Heating element cooled by gas
Thermocouple gauge	10^{-4}	1	Temperature measured } Heating element cooled by gas
Alphatron gauge	10^{-4}	10^{3}	Alpha ray ionization
Bayard Alpert	10^{-11}	10^{-3}	Ionization by electrons from hot filament
Penning or PIG gauge	10^{-12}	10^{-2}	Cold cathode discharge in crossed fields
Magnetron and inverted magnetron ionization gauges	10^{-13}	10^{-3}	Cold cathode discharge in crossed fields
Helmer ionization gauge	10^{-13}	10^{-5}	Ionization by electrons from hot filament; deflected ion beam
High pressure (Schulz) ionization gauge	10^{-6}	1	Ionization by electrons from hot filament

of differing mass-to-charge ratio are separated, and a collector or ion current amplifier. Many instruments have almost identical ion sources and collectors. It is the analyzer section which is unique for each type of mass spectrometer.

Two important quantities which characterize all mass spectrometers are their sensitivity and resolution. Sensitivity is defined as the amount of output current produced for a given partial pressure of a known gas species such as nitrogen. Sensitivities of 10^{-4}–10^{-5} A per torr are typical for most mass spectrometers that have a simple collector. This number is to be compared with the much higher typical sensitivity of 10^{-1} A per torr for a hot filament Bayard Alpert gauge. When an electron multiplier is used as a collector, the ion current may be amplified by a factor of 10^{5}–10^{6}, thus giving an overall sensitivity of up to 100 A/torr. When no multiplier is used, the lowest pressure that an instrument can detect is generally limited by the low current limit of the electrometer or amplifier used to amplify the ion current. Typical vacuum tube electrometers can read currents of about 1×10^{-13} A, and so partial pressures as low as 10^{-9} torr can be measured with such instruments. With the use of vibrating reed electrometers that can measure currents as small as 10^{-17} A, it becomes possible to lower the low pressure limit by another 4 decades. The use of electron multipliers, however, permits even a few ions per second or minute to be counted, with suitable precautions, and so even lower partial pressures (values of 10^{-17} torr have been mentioned in the literature) can be measured. Multiplier equipped mass spectrometers generally have a low partial pressure limit of 10^{-13}–10^{-14} torr, providing that the total pressure in the instrument is not too high. The detection of these low pressures, however, depends on the cleanliness of the spectrometer.

Resolution is the ability of an instrument to separate or resolve closely adjacent masses. Resolution can be defined in terms of the largest mass at which a given criterion is met. For example, one can specify that there be no more than a given percentage contribution from one mass to the next or that there be no more than a given percentage valley between adjacent peaks. Since mass peaks have an approximate triangular shape, it is also common to speak of a peak width at half height and specify the spacing between the centers of the two mass peaks in terms of the peak width. For careful analytical work, it is best to have adjacent mass peaks fully resolved, that is, completely separated, so that there is no contribution from one peak to the next. In general, resolution is a function of the size of an instrument, so that large heavy mass spectrometers may be expected to have high resolutions of hundreds (or thousands) of mass units. Small appendage-type instruments may only have resolutions of 10 or 20 mass units. A low resolution instrument may be adequate for identifying the various gas species present even though it cannot be used to measure the partial pressures of closely adjacent and overlapping masses.

Mass spectrometer analyzers can be classified in many ways, but one of the simplest methods is to group them into magnetic and nonmagnetic categories. Magnetic instruments require the use of magnets with field strengths ranging from 1000 to 5000 g. The nonmagnetic instruments generally require the use of high frequency or short duration pulsed electric fields.

For a beam of positive ions that has been accelerated by a potential V (in an electron bombardment-type ion source) and then injected, usually through collimating slits, into a magnetic field B, ions having different values of mass-to-charge ratio m/e will follow circular trajectories, having different radii of curvature given by the expression:

$$r = 143.6(Vm/e)^{1/2}/B \tag{6.47}$$

with the radius given in centimeters, B expressed in gauss, V in volts, and m/e expressed in atomic mass units ($m/e = 1$ for H^+, $m/e = 4$ for He^+, etc.). Magnetic instruments may deflect the ions through total angles of 60°, 90°, 180°, etc. depending on the extent of the magnetic field sector.

Aside from the purely "magnetic analyzer"-type mass spectrometer, some instruments use a combination of magnetic and electric fields. The omegatron, for example, is a relatively low cost partial pressure analyzer. A beam of electrons is injected into a boxlike structure parallel to a magnetic field and perpendicular to an rf electric field. Ions formed near the center of the box are acted on by crossed electric and magnetic fields, and thus they begin to move in curvilinear fashion normal to the magnetic field. The frequency of the rf field is adjusted so that each ion of charge-to-mass ratio e/m is subjected to its cyclotron frequency $\omega_c = eB/m$ during a mass scan. When the rf field is in resonance with an ion's cyclotron frequency, the ion is caused to spiral outward since it moves in phase with the electric field and gains energy during each revolution. Such resonance ions are collected by a small collector electrode located near one of the rf plates. The resolution of the omegatron is inversely proportional to the ion mass and is of the order of 40 or 50. The low pressure limit of the ordinary omegatron is about 10^{-9} or 10^{-10} torr, and one cannot use a multiplier to decrease this limit. There is an upper pressure limit of about 10^{-5} torr due to the long spiral path length of resonant ions.

A second type of mass spectrometer that uses both a magnetic and an electric field is the coincident crossed-field or trochoidal spectrometer. In this instrument, ions that are formed are deflected in combined magnetic and electric fields, so that the overall size of the spectrometer is small, though not as small as the omegatron. A special feature of this instrument is the presence of velocity focussing, which increases both the sensitivity and resolution. Resolutions of 80–100 are typical. The instrument sensitivity is about 1 or 2×10^{-4} A/torr and its low pressure limit is in the 10^{-12} torr region. The trochoidal mass spectrometer can be fitted with an electron multiplier.

Nonmagnetic-type mass spectrometers are becoming increasingly popular for a variety of reasons. Since no magnet is present, they are easily attached to a vacuum system and easily baked at high temperature.

The Bennett-type linear rf mass spectrometer uses a multiplicity of grids (up to 20 grids in some designs) in series. The grids are connected to an rf source with alternating polarity. As an ion beam passes through the series of grids, only those ions with the correct velocity will gain enough energy from the changing electric fields to overcome a final retarding grid and reach the collector. Mass resolutions of about 100 can be attained over the pressure range from 10^{-4} to 10^{-8} torr.

A second type of nonmagnetic mass spectrometer is the time-of-flight instrument. This spectrometer is unique in that it creates pulses of ions that last less than 1 μ sec at a repetition rate of about 10^4 pulses/sec. The ions are all accelerated to the same energy and are then permitted to drift in an electric field-free region until they arrive at the collector, which is usually an electron multiplier. Lighter ions attain a higher velocity during the acceleration phase and they arrive at the collector before heavier ions. If the drift tube is long enough, ions of masses up to 100, or so, are separated into bunches that do not overlap. An oscilloscope with a fast sweep is used to display the current pulses that result when each bunch of ions reaches the multiplier. The time-of-flight instrument is especially useful for studying high speed gas reactions and kinetics. Partial pressures down to 10^{-12} torr can be detected, and completely resolved masses up to mass 200 are attainable with a 100 cm time of flight instrument.

The most recently developed mass spectrometers of the nonmagnetic type are the quadrupole and monopole instruments. In the quadrupole instrument, the ion beam passes along the axis of four symmetrically spaced rods. Opposite rods are connected electrically, and each pair of rods is supplied with a dc voltage upon which is superimposed a high frequency rf voltage. Ions that pass along the axis of this alternating electric field are forced into making lateral or transverse oscillations. Except for ions that have a specific mass-to-charge ratio, all other ions experience an increasing amplitude of oscillation until they strike one of the four rods and are collected. In this way, the quadrupole electric field tends to filter out all ions except one select group. By varying either the dc and rf voltages or the rf frequency, the full mass scale can be scanned. This instrument accepts ions that have a wide range of entrance angles and velocities. The resolution of this instrument can be made quite high, of the order of 200, and with the use of a multiplier, it can detect partial pressures as low as 10^{-14} torr. The monopole mass spectrometer is a simplified version of the quadrupole instrument in which only a single rod and a V-shaped channel simulate the electric fields present in one quadrant of the full 360° quadrupole instrument. The operational characteristics of the monopole are generally similar to those of the quadrupole. Both of these mass spectrometers are susceptible to contamination of the rods, and they are really most suitable for use in clean, low pressure systems.

All gases, when analyzed with a mass spectrometer, produce a characteristic mass spectrum or "cracking pattern," as it is called, for hydrocarbons. When molecules are bombarded in the ionization region, some of them lose electrons and become positively charged while others break up into two or more ionized fragments. Under the same conditions of ionization, a given gas will always ionize and break up according to a fixed pattern, its characteristic mass spectrum for that instrument. In order to carry out quantitative analysis, it is necessary to find the sensitivity of the particular mass spectrometer for each pure gas that is expected to be present in the mixture being analyzed. Knowing the sensitivity and pattern for each gas, it becomes possible to set up a series of simultaneous equations which, when solved, gives the percentage of each constituent of the mixture. A few gases, such as nitrogen and carbon monoxide, have approximately the same mass, and their parent mass peaks at $m/e = 28$ overlap and are not resolved. However, the nitrogen spectra contains some ionized atomic nitrogen N^+ of mass 14, while the carbon monoxide spectrum contains ionized carbon C^+ of mass 12 and ionized atomic oxygen O^+ of mass 16. Again, if the individual characteristic spectra of nitrogen and carbon monoxide are known, the amounts of each gas present can be found.

Mass spectrometers are extremely useful to have as part of a vacuum system. They yield a measure of the total pressure in the system since the total pressure is just the sum of the partial pressures as measured by a calibrated instrument. A word of caution is in order here since occasionally a mass spectrometer of limited mass range may not see some very light gases such as hydrogen or helium nor some heavy hydrocarbons, mercury vapor, or other heavy molecules that may be present. Some instruments have a special collector that collects a fraction of all of the ions formed, and thus acts as an ionization gauge to measure the total pressure.

Mass spectrometers can be used as leak detectors when they are connected to a system. Small leaks show up on the instrument as a characteristic air spectrum with nitrogen being the largest peak. If the exterior of the system is probed with helium or some other suitable gas that is not present in the air spectrum or the background spectrum, the leak can be found very quickly.

In general, a mass spectrometer is a wonderful diagnostic tool for vacuum work. With it, one can determine the residual gases present in a system after it has been pumped to a low pressure. Heating various portions of the system while watching the mass spectrum often locates the sources of many of these gases. Closing valves that isolate specific portions of the system or turning off various gauges or other apparatus will often locate additional sources of unwanted gas. Very often, cold traps will evolve unwanted gases as the level of the refrigerants changes, and some pumps will contribute gases that can be detected with a mass spectrometer. At the lowest pressures of 10^{-11} and 10^{-12} torr, the mass spectrometer itself may be the largest source of unwanted gas in the system.

F. Calibration of Pressure Gauges

In addition to obtaining high or ultrahigh vacuum in a particular system and having total pressure and partial pressure gauges to measure the degree of vacuum and the composition of the gas, it is necessary to have some means of calibrating the gauges and mass spectrometers in an absolute fashion. Calibrated gauges and mass spectrometers must be used when performing quantitative experiments that involve the pressure of the gas(es) present. Most of the gauges that have been discussed earlier do not give absolute readings of pressure or number density.

The American Vacuum Society is currently in the process of establishing primary pressure standards for the various vacuum regions and specifying standardized procedures for calibrating hot filament ionization gauges and other gauges. Mercury manometers and McLeod gauges are under consideration as primary pressure standards.

Mercury manometers can be read accurately to approximately 0.1 torr. Oil manometers have a higher sensitivity since the density of the fluid is lower by about an order of magnitude, but great care must be taken to degas the oil and to keep the gases being measured from dissolving in the oil. Cold traps must be used to keep mercury or oil vapors from affecting the vacuum system or the gauges being calibrated.

The McLeod gauge is really the only well developed gauge which can measure pressures directly in an absolute fashion in the range from 10^{-3} to 10^{-7} torr. By an absolute measurement, we mean one that is based on the measurement of fundamental quantities such as mass, length, and time. In this sense, the Knudsen gauge, which also covers the range from 10^{-3} to 10^{-7} torr, is not an absolute gauge because its reading depends on the accommodation coefficient of the particular calibration gas on the moveable vane.

The McLeod gauge is essentially a liquid column manometer having a built-in pressure amplification arrangement. The pressure amplification portion of a typical low pressure gauge consists of a large volume (1–2 liter) compression bulb and a small volume ($1\text{–}2 \times 10^{-5}$ liter) closed capillary tube that has a diameter of the order of 0.5 mm.

The gas (at an unknown pressure p) that is to be measured is introduced into the compression bulb and closed capillary (having a total volume V). The gas is then compressed by raising the liquid level until it occupies a much smaller volume v_f in the closed capillary and has attained a much higher pressure p_f. The pressure p_f is measured by comparing the height of the liquid in the closed capillary with the height of the liquid in an adjacent identical (same diameter) open capillary tube that is in communication with a low pressure. This is a simple U-tube manometer type of measurement. Boyle's law is assumed to hold, so that the unknown pressure is computed from the proportion $p/p_f = v_f/V$.

Although the McLeod gauge is a well-tested and widely accepted calibration standard gauge, it does have serious shortcomings. Thus, in spite of the fact that the gauge compression bulb and capillaries can be calibrated with accuracies of the order of 0.1%, unknown surface forces (sticking, etc.), static electrical charges on the glass capillaries, varying meniscus shapes, etc. degrade this accuracy by at least an order of magnitude.

Other, less important, shortcomings of the McLeod gauge are its inability to measure the pressure of condensable gases such as low pressure oil vapors, or to allow continuous readings, its slow speed of response, and a certain dependence on the skill of the operator in taking consistent, reliable readings. However, when used correctly, the McLeod gauge allows direct pressure calibrations and can be employed indirectly with other schemes of calibration.

The McLeod gauge and U-tube manometers are used to calibrate gauges directly by a comparison of their readings. In the case of high pressure ionization gauges of the Schulz type, there is a good overlap between the ranges of the McLeod gauge and the ionization gauge. For Bayard Alpert-type ionization gauges and most mass spectrometers, there is only about 1 decade of pressure overlap (from 1×10^{-5} to 1×10^{-4} torr) in which both the ionization gauge and the McLeod gauge have reasonable accuracy.

In order to extend the range of a pressure standard such as a manometer or McLeod gauge toward lower pressures, one creates a lower pressure that has a known relationship to a much higher pressure that can be measured with the pressure standards. In the case of a static system, a known low pressure can be established by expanding a small measured volume of gas at a measured high pressure into a larger known volume of a test chamber on which the gauges to be calibrated are mounted. This is the well-known Knudsen method that is based on Boyle's law. Adsorption of gas on the walls of the test chamber and outgassing of the walls makes this method questionable for accurate work at lower pressures.

The best method of creating two or more related low pressure levels for calibration work is to use a continuous flow system. The known pressure in the test chamber(s) is established by controlling the rate at which gas flows into and out of the chamber(s) (generally by using orifices of various sizes), and is measured by measuring the pressure in one of the higher pressure chambers preceding the test chambers. A series of differentially pumped chambers can be used in this way to obtain related pressures in each chamber of the series. In lieu of measuring the pressure in one of the higher pressure chambers with a pressure standard gauge, one can equally well measure the actual gas throughput into the test chamber. If the amount of gas removed from the test chamber per unit time is fixed by a "limiting conductance" of known value placed in series with the chamber pump, the known test chamber pressure can be easily determined.

There is a simple and classical method that can be used to verify the linearity of gauge pressure readings. The test chamber containing the gauge is connected to a gas-filled volume through a low conductance path. If the pressure difference between the two chambers is large and the flow is molecular, the pressure in the test chamber increases essentially linearly with time, and a linear increase of the gauge reading with time verifies the gauge linearity. Care must be taken to minimize gauge pumping during a test of this kind by reducing the electron emission current to a value no larger than 0.1 mA and operating the gauge intermittently. If two gas inlet chambers are used instead of a single chamber, the pressure in the test chamber will increase quadratically with time. A larger range of pressures can be covered with the quadratic arrangement.

G. Sources of Pure Gas and Gas Handling Techniques

The creation of controlled atmospheres requires that known, pure gases be introduced into a vacuum system after it has been pumped down to a suitable vacuum level. The vacuum level required, in the first approximation, is such that the background gas pressure constitutes the maximum impurity allowable at the final operating pressure of the system. For ex-example, if the system is to operate in a pure gas environment at a pressure of 10^{-2} torr and the maximum impurity content is to be 100 ppm, then the system background pressure can be no higher than 10^{-6} torr. Pure gases for introduction into vacuum systems are most conveniently obtained from manufacturers of industrial and medical gases.

When small quantities of very high purity gas are to be used in an experiment or process, gases can be obtained in sealed glass flasks with analyses of the contents at extra cost. Metal valves are widely used to control the flow of gas from such a gas bottle.

It has been found that even the best sealed glass flasks contain gas with impurities of up to 100 ppm. Higher purity for certain selected gases can be obtained through the use of commercially available "diffusion" leaks for H_2, He, and O_2. Gases that are filtered through diffusion leaks may have only a few ppm of impurities.

In addition to determining the basic source of gas, some thought must be given to the manner in which the gas is to be admitted. Will cold traps or drying agents be required to remove water vapor? Must the gas pressure be kept constant (via regulating valves, for example)? Will it be necessary to bake out the tubing that joins the gas tank or bottle to the gas flow valve? If glass gas bottles are used and have been installed by a glass blower, a great deal of moisture can be deposited in the adjoining tubing.

Leak valves that are used to admit gas into a vacuum system are many and varied. Small metal–plastic needle valves have been used historically for this purpose, but these valves are best suited for higher pressure and higher leak rate applications. For low pressure or UHV applications, all metal bakeable valves are necessary. Some very excellent adjustable leak valves are available that have a stainless steel to sapphire seat. Within the past few years a servo-driven leak valve has been marketed. The flow rate through this valve can be controlled automatically with the use of a sensor (pressure gauge) that monitors the pressure in a chamber into which the gas is flowing. Use of the feedback principle permits a constant gas pressure to be maintained with this device.

Gas flows can also be limited with small diameter capillary tubing, fine pinhole orifices in gold foil, or porous ceramic plugs. The leak rate of a porous ceramic plug can be varied by varying its immersion depth in a liquid metal such as mercury, indium, or gallium. Such liquid metals also have been used for valving action.

In some experiments, it is necessary to isolate thermally a source of gas or some apparatus from the bakeable portion of the system. Hydrocarbon gases, for example, could be decomposed if they were in contact with system elements that were being baked at 450°C. In cases of this sort, two valves or glass breakoff seals are used in series with an evacuated length of tubing between them. The valve or breakoff nearest the system can be baked.

Known quantities of gas can be admitted to a system by using a calibrated volume adjacent to the main source of gas. This known volume is separated from the gas source and the system by valves, and it may contain a nonpumping pressure measurement device such as a liquid manometer, a metal diaphragm gauge, or a thermocouple gauge.

Gas admitted to a system can often be further purified, when required, by using the selective gas sorption of certain getters or sorption material such as zeolite. When working with certain highly reactive gases, such as atomic oxygen, for example, it is necessary to provide nonmetal surfaces to prevent recombination of the atoms. Glass, ceramic, and Teflon or other plastic components are used in these situations.

H. Leak Detection

The location and removal of leaks in a vacuum system is often an unwelcome but necessary activity.

The first step in dealing with leaks is to attempt to avoid them by giving each major vacuum system component a thorough leak check at its expected operating temperature before system assembly. When a large number of individually leaktight components have been assembled, the odds are very good that any leaks present will be found only in the joints between the components, and the leak hunting process is simplified considerably.

There are many techniques of finding leaks, and the most popular of these will be listed below. However, from a practical point of view, the fastest and best method for finding ordinary leaks is to use a helium mass spectrometer-type leak detector. This instrument is simply a mass spectrometer that has been tuned to detect helium and is used together with associated pumping equipment. The mass spectrometer leak detector is an indispensable piece of equipment if any large amount of vacuum work is to be done. With the average mass spectrometer type leak detector one can measure leaks as small as 10^{-10} torr liters/sec. Even smaller leaks can be found using "accumulation" techniques in which the leak detector pumping speed is reduced. When small leaks are suspected, one should permit sufficient time for the helium gas to penetrate into regions around metal gaskets and other constricted openings. Entire elements of the system can be enclosed in plastic bags filled with helium gas. Porous metals or ceramics are often discovered in this way. Care should be exercised when probing near rubber or plastic hose connections and O-rings. Only fast response signals should be interpreted as air leaks. The one limitation of the mass spectrometer leak detector is that the leak must be small enough so that the system under test can be pumped down to a pressure below at least 10^{-4} torr.

Large leaks are most easily found by pressurizing system components to about 20 to 90 psi, depending on the construction, and then painting the suspected area with a soap solution or, better yet, with one of the many proprietary leak detecting solutions that are available commercially. Leaks are evidenced by the formation of bubbles. Leaks as small as 10^{-4} torr liters/sec can be detected. Occasionally such pressurized components are placed under water to test large surface areas quickly. Very large leaks in a system can often be detected directly by ear or with the use of a stethoscope.

A spark coil or Tesla coil is generally used to find pinhole leaks in glass. The leak is evidenced by a concentration of the normally diffuse discharge at the leak site. Care should be taken not to use too high a coil voltage or the discharge may actually puncture the glass where the wall is thin.

The isolation and pressure rise technique can be used to help determine the presence and location of air leaks. In this method various portions of the system are isolated from the pump(s) by closing system valves, and the rate of pressure rise in the closed-off portions is monitored. If the rate of pressure rise is constant for a long period of time, there is probably a true leak present. If the pressure rise appears to be asymptotic to a limiting value, there is either high vapor pressure material in the system or there may be an internal virtual leak due to unvented spaces such as blind threaded holes. A knowledge of the system and its normal behavior is the best aid one can have in using this technique.

Finally, when operating at very low pressures of 10^{-10} torr and below, extremely small leaks are most easily found by painting or spraying suspected areas with pure acetone. When acetone covers a small leak, there is usually a dramatic decrease in the system pressure as the acetone temporarily plugs the leak.

Good descriptions of pressure gauges and partial pressure analyzers may be found in almost all of the general texts on vacuum technology. Leck (11) deals exclusively with pressure gauges while Duckworth (6) is one of many modern texts that discusses mass spectrometry in detail. Reference 16 has good descriptions of gas handling techniques and arrangements for UHV experiments. The subject of leak detection is covered very well in references 2, 8, and 18. Commercially available equipment for vacuum measurement, gas handling, and leak detection is listed in reference 21.

V. VACUUM SYSTEM DESIGN AND CONSTRUCTION

A. General Considerations

The pros and cons of the different kinds of vacuum systems will be discussed in the following pages to give the reader a broad view of the possibilities of system size, materials, pumps, etc.

1. System Size

Ideally, the size of a vacuum system, or rather let us say the working chamber size, depends on the operations that are to be performed in vacuum. Since the cost of a system obviously increases as the size increases, the working chamber size should be adequate for the job at hand, with perhaps an extra margin of volume for unexpected requirements.

Very often, system size is dictated by the availability of certain vacuum pumps and components. In this case, the important consideration is to be sure that the pumps on hand are large enough to do the job. If they are much too large, then their use is simply a trade-off between equipment costs and operating costs.

Once the size of the working chamber has been determined, the amount of pumping speed required at the chamber can be determined using the following rule of thumb. One should provide at least 1 or 2 liters/sec of pumping speed for each 100 cm^2 of internal surface area of the system (including all pump, valve, and manifold areas). Rarely does one require as much as 10 liters/sec per 100 cm^2 of internal surface area. Of course, the higher the pumping speed, the faster the pump down and the lower the ultimate pressure. The highest pumping speeds will be obtained with the pump connected directly to the working chamber. Diffusion pumps, mechanical pumps, and sorption pumps are generally mounted vertically below the chamber, but ion pumps, sublimation pumps, and cryopumps can usually be mounted in any direction.

One of the next steps to be taken is to determine the materials and thickness of the chamber walls. For small glass systems, where the chamber volume may vary from about 1 liter to 30 liters or so, the standard spherical glass flasks have sufficient thickness to withstand atmospheric pressure when evacuated. Glass containers of other shapes, such as ehrlenmeyer flasks and cylindrical bottles, may have flat portions which are not strong enough to withstand evacuation. Whenever possible, cylindrically shaped chambers should have bell-shaped ends for added strength.

In the case of metal systems, there are many stress and strain formulas available in mechanical engineering handbooks such as L. S. Marks' *Mechanical Engineers' Handbook* (McGraw-Hill). The book by Steinherz (reference 19) also has a section on vacuum chamber design including information on stiffening rings, head design, and ports.

2. Metal vs Glass Vacuum Systems

In the early days of vacuum work, most vacuum systems were small and made of glass. These systems were used primarily in chemical research. With the advent of the electron tube industry, the use of glass in vacuum work took on added importance. However, as the requirements for larger and larger vacuum chambers in industry increased, the use of metal chambers and components became a necessity. Today, there are available components for both glass and metal systems, with the latter dominating in the market place.

All large vacuum systems with working chamber sizes of approximately 10 cubic feet, or more, are made of metal. Systems with working chamber sizes between 1 and 10 ft^3 are usually made of metal, with the exception of glass bell jar systems and systems using either glass pipe or large glass reaction flasks. Glass bell jar systems could be labeled "hybrid" systems since most of the system components, except for the bell jar itself, are made of metal.

Small vacuum systems, arbitrarily defined as those with working chamber sizes of less than 1 ft^3, may be constructed of either metal or glass. For some applications, glass systems have definite advantages over metal systems, although the advantages appear to be slowly disappearing as more and more metal system accessories are developed.

Among the advantages of glass over metal in vacuum system construction are the following. Glass systems are (*1*) less expensive to build, primarily because of the availability of low cost, standardized parts, (*2*) comparatively simple to assemble in unusual shapes, (*3*) transparent thereby simplifying complex manipulations, (*4*) easily "sealed off" under vacuum, (*5*) excellent insulators thereby minimizing leakage between voltage feedthroughs, and (*6*) easily leak detectable with a Tesla coil.

Glass, especially pyrex and high silica content glass, is chemically inert, low in vapor pressure, free from corrosion, and is thermally stable. Pyrex systems are regularly baked at 450°C when the systems contain graded seals that have strain points below that of Pyrex. Pyrex itself can be baked under vacuum up to about 550°C, somewhat above its strain point temperature of 519°C. Aluminosilicate glasses are regularly baked at 600°C. Furnace tubes of Vycor glass or fused silica may be attached to a Pyrex vacuum system via graded seals, and temperatures up to 1000°C can be maintained within these tubes.

There are several disadvantages to the use of glass in vacuum systems. The most important of these are the limitations in size, strength, and dimensional accuracy. The limited strength of glass (tensile strength of ~1000 psi) and its brittle nature make it very susceptible to mechanical shock and vibration. Extended elements of a glass vacuum system should be protected with metal shields or enclosures.

The limited dimensional accuracy possible with glass precludes its use in many applications. Parts cannot be fabricated with the close tolerances that are more or less routine in metal working. In addition, the processes of softening the glass during joining and shaping operations and of annealing it afterwards tend to move elements out of position and alignment.

3. *Demountable and Bakeable Systems*

Vacuum systems may be constructed in either demountable or nondemountable fashion. In the latter case, metal system components are soldered, brazed, or welded together in more or less one piece. Early metal systems were invariably constructed this way, just as glass systems are constructed even today.

During the past 10 or 12 years, the emphasis in metal system construction has been placed on the use of flanged modular components. Flanged demountable vacuum systems have the advantage of flexibility in changing components (thus offering competition to glass systems in this respect). Disadvantages of flanged systems include the increased cost and weight (Large metal-gasketed UHV flanges, for example, are very expensive and very heavy) and the increased probability of leaks at the flanges.

B. Available Vacuum Components

1. Ready-Made Systems

The reader should be aware of the existence of many completely developed, commercially available vacuum systems. Most of the vacuum equipment manufacturers who manufacture pumps and valves (reference 21) have ready-made systems of different sizes which incorporate their products. Special purpose vacuum systems are available for inert gas, brazing, welding, and heat treatment of metals, evaporation coating, sputter coating, electron beam welding, vacuum gauge calibration, and many other specialized applications. These systems can usually be modified by the manufacturer to suit the customer's needs.

2. Pumps, Valves, and Vacuum Accessories

Many vacuum equipment manufacturers produce matching sets of pumps, cold traps, valves, and fittings that can be easily assembled to produce complete pumping stations. The customer need design only the working chamber.

Vacuum valves of various types are available in great profusion. Valves are often classified as being bakeable or nonbakeable. All-metal UHV valves (which may contain some ceramic and glass components) are generally bakeable to 450°C. Some of these valves may be baked in either open or closed positions while others can be baked at high temperatures only when open. Welded bellows or diaphragms are used in these valves to afford the required motion. Copper, silver, and gold gaskets are generally used to make the vacuum tight seal at the valve seat. The price of these valves increases rapidly with increasing port diameter.

Among the family of nonbakeable vacuum valves one finds both large gate valves and butterfly-type valves that provide a high conductance when open as well as quite small needle valves used for gas flow control. Perhaps the most common valve in this category is the low cost block-type, bellows-sealed valve that is made of brass. This is a large aperture valve in which the seal at the seat is made with an O-ring and in which the stem retracts a considerable distance.

Flanged metal vacuum systems are becoming more and more useful as new accessories and instrumentation are developed that permit a wider class of vacuum operations to be performed. New pressure gauges and partial pressure analyzers have extended the range and sensitivity of gas analysis. Pyrex, quartz, and sapphire windows permit radiation experiments to be performed. Flanges with all sorts of electrical feedthroughs allow complex electrical and electronic operation in vacuum. Liquid feedthroughs permit liquid helium, liquid nitrogen, and cooling water to be passed through the vacuum chamber walls. Motion feedthroughs are now available which permit accurately measured and controlled translational and rotational motion within the vacuum environment. Servo-driven valves are available which work in conjunction with a pressure transducer to regulate the pressure of gas that is introduced into a system. All of these accessories and instruments are mounted on standard metal gasketed flanges which mate with those installed on the system.

C. Materials of Construction

The four general categories of materials used in vacuum systems are metals, ceramics, glass, and plastics. Ultrahigh vacuum systems do not contain plastics.

Three important characteristics of good vacuum materials are nonporosity, low vapor pressure, and elevated temperature strength. The vapor pressures of various materials are listed in most books on vacuum technology. Roberts and Vanderslice (16), have a good presentation of vapor pressure data in addition to many other tables and graphs dealing with vacuum system materials of construction.

1. Metals

The material most commonly used in the construction of vacuum systems is 304 stainless steel. The austenitic stainless steels (300 series) contain about 16–20% Cr, 6–13% Ni, about 0.1% C, about 2% Mn, and the balance Fe. Extra-low carbon 304 stainless steel welds more easily than 304 and is more suitable for ultrahigh vacuum systems. The 300 series stainless steels are nonmagnetic (they will become slightly magnetic when work hardened); fairly inert chemically (unattacked by mercury at lower temperatures and thus suitable for mercury diffusion pumped systems); free of rust and corrosion; easily machined, bent, and spun; easily welded (except possibly 303 stainless); and sufficiently strong for most applications. Series 303 stainless steel should not be used for vacuum applications since it has a high sulfur content that will contaminate sensitive surfaces when it is baked at high temperatures. Stainless steels form a stable protective oxide of chromium on surfaces exposed to air at temperatures of about 300–450°C.

Iron and mild steel are not good materials for vacuum system construction because of rusting and hydrogen permeation. Iron slugs are often enclosed in glass (or stainless steel) when used within vacuum systems. Pure nickel and kovar are used within ultrahigh vacuum systems when magnetic materials are required. Nickel is an excellent vacuum system material in general. It can be obtained in pure form (99.999% purity) and is easily machined, formed, and welded. Alloys of nickel such as 'Monel' and 'Inconel' have specialized applications. Some of the grades of Monel are nonmagnetic while Inconel has exceptional high temperature strength.

Oxygen-free high-conductivity copper (OFHC copper) is frequently used in vacuum systems where its high electrical or thermal conductivity is advantageous. It is often used to handle high currents at low voltages and to provide low temperature shields in cryopumps. Copper is commonly used for the metal gaskets of UHV flanges and valves. Copper starts to oxidize at 200°C and oxidizes heavily at 450°C in air. After three or four long-term bakeouts of a flanged system at 450°C, it is not unusual for copper oxidation to have progressed to the point where leak paths are formed through a superficial oxide layer at the knife edge of the flange. In this respect, gold O-ring type seals are superior to copper seals. Copper also has a tendency to migrate within a vacuum system when baked in the presence of water vapor due to the formation of a volatile oxide.

Aluminum is seldom used in vacuum systems, chiefly because it cannot be easily welded or brazed vacuum tight or joined to other metals. In addition, many aluminum alloys have high outgassing rates. Recently, some new techniques have been developed to join aluminum to stainless steel so that the use of aluminum may increase. Aluminum gaskets are occasionally used in place of copper gaskets.

Silver, gold, platinum, iridium, and other precious metals are used for specialized applications in vacuum systems. Some UHV valves have silver or gold seals at the valve seat. Gold O-rings are commonly used with certain UHV flanges. Mass spectrometer ionization sources may be gold plated to reduce contact potentials, secondary electron emission and oxidation. Glass and ceramic insulating surfaces may be plated with platinum or gold to prevent charging up or to provide electrostatic shielding. Metal surfaces may be plated with gold to obtain high reflectivity (low emissivity) or to keep certain surfaces clean and unoxidized. Iridium is used in nonburnout filaments of ionization gauges and in those applications requiring resistance to oxidation at high temperatures.

The refractory metals are widely used in high vacuum components such as ionization gauges, mass spectrometers, high temperature furnace elements, and vacuum coating sources. Tungsten has a high melting point and the highest "hot strength" of any of the elements. It is used extensively for heated filaments. Molybdenum and tantalum, due in part to their machinability and ease of fabrication, are also used extensively in vacuum tube fabrication. Molybdenum has a thermal expansion coefficient which approaches that of aluminosilicate glass and certain ceramic materials, and thus is used in making glass-to-metal and ceramic-to-metal seals.

Low melting point metals and alloys are rarely used in vacuum work with the exception of mercury and indium. Indium is used as a metal gasket material and as a liquid metal valve sealant in alloys containing tin and gallium. Indium and the alloys just mentioned have low vapor pressures at ordinary temperatures. Brass is not recommended for high vacuum work because the zinc evaporates when it is heated. Brass and cadmium plated steel hardware should not be used inside high vacuum systems.

Table 6.9 presents a concise summary of the physical properties of many of the metals and alloys that are useful in vacuum work.

2. *Ceramics*

Ceramics are nonmetallic inorganic materials that have been heat treated to give them permanent shape and hardness. Ceramics include noncrystalline glass, glass-bonded crystalline aggregates such as porcelain, and single-phase compounds such as oxides, sulfides, and halides.

Ceramic materials are used in vacuum work chiefly as electrical insulators. High density alumina (Al_2O_3) is rapidly becoming the most important and useful ceramic material. It has high mechanical strength relative to that of other ceramics, is gastight, has a high-volume resistivity, low dielectric losses, and good thermal shock resistance. Polycrystalline high purity alumina (99.9% Al_2O_3) that is translucent is available from the General Electric Company under the trade name Lucalox. Transparent single crystal sapphire (99.99% pure Al_2O_3) is available from a number of manufacturers. Sapphire is sensitive to heat shock because its thermal expansion parallel to the axis is 9% greater than its expansion transverse to the axis. Sapphire is especially useful for windows that must pass UV radiation (down to about 1450Å) and/or IR radiation.

Another class of ceramics that is useful in vacuum work is the "machinable" ceramics. This class includes natural lava (soapstone or talc) and special formulations of the aluminum silicates that are produced by various manufacturers such as the American Lava Corporation. This company sells unfired lava which can easily be machined and fired afterwards to

Table 6.9 Properties of Useful Metals and Alloys

No.	Name	Atomic weight	Density, g/cm³ at 20°C	Melting point, °C	Latent heat of fusion, g-cal/g	Boiling point, °C	Latent heat of vap., g-cal/g	Specific heat		Coef. of linear expans.		Thermal conductivity	
								g-cal/g or Btu/lb	Temp., °C	$\times 10^4$/°C	Temp., °C	g-cal/sec cm² °C/cm	Temp., °C
1.	Aluminum	26.97	2.70	657.1	93.0	2056.0	2000.0	0.226	0–100	0.2545	20–300	0.52	50
2.	Beryllium	9.02	1.85	1285.0	318.0	2780.0		0.425	0–100	0.123	20		
3.	Copper	63.57	8.94	1083.0	50.6	2595.0	1756.0	0.0918	18–100	0.1642	20	0.923	18
4.	Gold	197.2	19.3	1063.0	16.11	2966.0	446.0	0.0308	18	0.144	17–100	0.707	17
5.	Iridium	193.1	22.42	2408.8	26.1	4900.0	340.0	0.0323	18–100	0.0641	20	0.141	17
6.	Iron	55.84	7.87	1535.0	65.0	2998.0	1110.0	0.1075	20	0.119	0–100	0.19	0
7.	Magnesium	24.32	1.74	651.0	64.8	1107.0	1500.0	0.249	0–100	0.257	20–300	0.376	0–100
8.	Mercury	200.61	13.546	−38.87	2.66	356.9	71.0	0.0332	17			0.020	0
9.	Molybdenum	95.95	10.2	2620.0		4803.0	176.8	0.0647	0	0.0549	25–100	0.346	17
10.	Nickel	58.69	8.85	1452.0	73.8	2900.0	1010.0	0.112	0	0.137	25–100	0.14	0–100
11.	Platinum	195.23	21.45	1773.5	27.1	4389.0	637.0	0.0319	20–100	0.088	20	0.1664	18
12.	Silver	107.88	10.5	960.5	24.3	2001.0	551.6	0.0558	0	0.189	0–100	0.974	18
13.	Tantalum	180.88	16.6	2850.0		6093.0		0.0356	58	0.0655	0–100	0.130	17
14.	Titanium	47.90	4.5	1800.0		5100.0	1320.0	0.142	0–100	0.0714		0.040	20
15.	Tungsten	183.92	19.3	3370.0	44.0	5927.0	1183.0	0.034	100	0.0444	27	0.476	17
16.	Kovar		8.3	1450.0				0.16		0.052	30–450	0.046	
17.	304 S.S.		7.9	1430.0				0.12		0.180	20–500	0.039	
18.	Monel		8.84	1350.0						0.145	20–300	0.062	
19.	Inconel		8.51	1425.0				0.109		0.137	20–300	0.036	
20.	Yellow brass		8.47	930.0				0.09		0.203	20–300	0.28	
21.	Graphite		1.75	3700.0				0.20				0.4	

No.	Electrical Resistivity		Temp. Coef. of Resistivity		Yield strength, annealed, psi 10^{-3}	Tensile strength, annealed, psi 10^{-3}	Modulus of elasticity, psi 10^{-6}	Temp. for vapor press. of 10^{-8} torr, °K	Temp. for vapor press. of 10^{-2} torr, °K	Spectral emissivity at 0.66 μ, 1000°C	Electron work function, eV
	μohm cm	Temp., °C	Temp., Coef.	Temp., °C							
1.	2.655	20	0.00445	0–100	~5	10–16	8.5–9.7	958	1490		4.2
2.	18.5	20						980	1500		3.9
3.	1.682	20	0.00382	20	10	33	17–18	995	1530	0.105	4.6
4.	2.42	20	0.0034	20	Very low	18–20	12.0	1080	1670	0.140	4.9
5.	6.08	0	0.00411	0–100			75.0	1850	2770	0.30	5.40
6.	9.8	20	0.0065	0–100	21	26–41	32.0	1165	1750	0.39	4.5
7.	4.461	20	0.0040	20		27	6.25	458	712		3.6
8.	95.783	20	0.00089	20				201	319		4.5
9.	4.77	0	0.0034	0–100	Up to 57	~120	46.0	1865	2800	0.378	4.2
10.	6.9	20	0.006	20	10–30	59–81	26–33	1200	1800	0.375	4.6
11.	9.83	0	0.003	20		24	15–26	1565	2370	0.29	5.32
12.	1.629	18	0.0038	20	Low	19–23	9–12	847	1300	0.055	4.5
13.	15.5	20	0.0031	20		51	28.0	2230	3330	0.45	4.1
14.	3.0	20			63	79	17.0	1335	2010		3.9
15.	5.48	20	0.0045	18		260	46.0	2390	3500	0.44	4.5
16.	49.0	20			55	85	22.0				
17.	72.0	20			30	80	29.0			0.35	
18.	48.0	20	0.0011		25–40	70–88	27.0				
19.	98.0				25–50	82–100	32.0				
20.	6.4				~17	~49	15.0				
21.						2.2–3.7		1930	2730		

ceramic hardness as well as machinable fired ceramics (Alsimag 222). Most of the machinable ceramics are porous and cannot be used as a portion of the vacuum wall.

Mullite, beryllium oxide, magnesium oxide, and pyroceram are other ceramics which have specialized applications in vacuum work. Boron nitride, glass-bonded mica and quartz are other vacuum-tight materials that can serve as high temperature insulators.

Ceramic materials (and glasses) have much higher compressive strengths than tensile strengths. Whenever possible, ceramics should be placed under compression rather than in tension.

Table 6.10 lists the more important properties of the ceramic materials that have been mentioned above.

3. Glasses

The most common glasses used in vacuum work are the low expansion borosilicate glasses (Pyrex, for example). High silica content glasses are used for high temperature and UV radiation transmission applications. Lower temperature glasses have been specially formulated to match the expansion coefficients of certain metals such as tungsten, Kovar, iron, etc.

The softness (or hardness) of glass as it changes from a solidlike material to a flowing liquid is represented by four values of viscosity. The strain point (viscosity of $10^{14.50}$ P) is the temperature at which internal stress is relieved in 4 hr and represents the upper limit of serviceability for annealed glass. At the annealing point (viscosity of $10^{13.00}$ P), internal stress is relieved in 15 min. At the softening point (viscosity of $10^{7.6}$ P), the glass deforms rapidly and begins to adhere to other bodies. At the working point (viscosity of 10^{4} P), the glass is readily fabricated.

Table 6.11 summarizes important properties of some of the more common types of laboratory glassware.

D. Fabrication Techniques

The metals used in vacuum system construction can be joined together in several ways. By and large, the most common way of joining stainless steel elements of a system is to weld them together. High temperature brazing in hydrogen or vacuum is still used to join small specialized pieces.

1. Welding

Welding is the joining of metals by fusion. In heliarc (or TIG) welding, which is a form of inert gas-shielded arc welding, the joint is heated by electrical resistance under a protective atmosphere of helium or argon. No

Table 6.10. Properties of Useful Ceramics

	Lava grade A	Alsimag 222 machinable $AlSiO_3$	Mullite ~40% Si ~60% Al_2O_3	Sapphire	Lucalox	99.5% BeO	Pyroceram 9606	99% Al_2O_3
Density g/cm³	2.3	2.0	3.2	4.0	4.0	2.9	2.6	3.8
Porosity	Not gastight	Highly porous	Gastight	Gastight	Gastight	Gastight	Gastight	Gastight
Tensile strength, psi 10^{-3}	2.5	2.5	18	65	40–55	10	20	35
Compressive strength, psi 10^{-3}	40	10	110	300	325	225		300
Flexural strength, psi 10^{-3}	9	5	20		23	30	20	52
Modulus of elasticity, psi 10^{-6}				50–56	57	51	17.3	50
Specific heat, cal/g°C				0.18	0.32	0.31	0.185	0.21
Coef. of thermal expansion, $°C^{-1} \times 10^6$, 25–700°C	3.6	10	~4	~8	~7	9	5.7	~8
Thermal conductivity, g cal/cm sec°C	0.003	0.005	~0.014	~0.015	~0.05	0.60	0.0087	~0.07
Maximum working temp., °C	1200	1300	1700	1900	1900	1850	1260	1725
Volume resistivity at 100°C, ohm-cm	6×10^{11}	$>10^{14}$			$>10^{14}$	$>10^{17}$		$>10^{14}$
Volume resistivity at 300°C, ohm-cm	2×10^9	6.0×10^{11}		$\sim 10^{14}$	2×10^{13}	$>10^{15}$	$\sim 3 \times 10^9$	1.0×10^{13}
Volume resistivity at 500°C, ohm-cm	5×10^6	4.6×10^9	$\sim 10^7$	10^{11}	1.5×10^{12}	5.0×10^{13}		6.3×10^{10}
Dielectric strength, V/mil				1200	1700	610		500
Dielectric constant, 1 Mc	5.3	5.5	6.6	10.3	9.9	6.7	5.58	9.5
Dissipation factor, 1 Mc	0.01	0.0002		0.00004	0.000025	0.0003	0.0015	0.0002
Loss factor, 1 Mc	0.053	0.001				0.002	0.009	0.002

Table 6.11. Glasses of Corning Glass Works

Glass code	Type	Principal use	Thermal expansion coef.-/°C from 0 to 300°C	Upper working temperatures (mechanical considerations only) Annealed: Normal service, °C	Annealed: Extreme limit, °C	Viscosity: Strain point, °C	Viscosity: Annealing point, °C
0010	Potash soda lead	Soft glass; general purpose	93×10^{-7}	110	380	395	435
0080	Soda lime	Soft glass; general purpose	92×10^{-7}	110	460	470	510
0120	Potash soda lead	Soft glass; general purpose	89×10^{-7}	110	380	395	435
1720	Alumino-silicate	Low helium permeation	42×10^{-7}	200	650	670	715
1990	Potash soda lead	Iron sealing	124×10^{-7}	100	310	330	360
3320	Boro-silicate	Tungsten sealing	40×10^{-7}	200	480	500	540
7050	Boro-silicate	Sealing	46×10^{-7}	200	440	460	500
7052	Boro-silicate	Kovar sealing	46×10^{-7}	200	420	435	480
7070	Boro-silicate	Low loss electrical	32×10^{-7}	230	430	455	495
7720	Boro-silicate	Tungsten sealing	36×10^{-7}	230	460	485	525
7740	Boro-silicate	General purpose	33×10^{-7}	230	490	515	565
7750	Boro-silicate	Kovar sealing	40×10^{-7}	230	400	431	467
7910	96% silica	UV transmission	8×10^{-7}	800	1090	820	910
7940	Fused silica	UV transmission	5.5×10^{-7}	900	1100	990	1050

data							Dielectric properties at 1 MHz and 20°C			
Softening point, °C	Working point, °C	Modulus of elasticity, psi	Log of volume resistivity			Power factor, %	Dielectric constant	Loss factor, %	Glass code
			25°C	250°C	350°C				
625	985	8.9×10^6	17.+	8.9	7.0	0.16	6.7	1.0	0010
695	1005	10.0×10^6	12.4	6.4	5.1	0.9	7.2	6.5	0080
630	980	8.6×10^6	17.+	10.1	8.0	0.12	6.7	0.8	0120
915	1190	12.7×10^6	—	11.4	9.5	0.38	7.2	2.7	1720
500	755	8.4×10^6	—	10.1	7.7	0.04	8.3	0.33	1990
780	1155	9.4×10^6	—	8.6	7.1	0.30	4.9	1.5	3320
705	1025	8.7×10^6	16.0	8.8	7.2	0.33	4.9	1.6	7050
710	1115	8.2×10^6	17.0	9.2	7.4	0.26	4.9	1.3	7052
—	1070	7.4×10^6	17.+	11.2	9.1	0.06	4.1	0.25	7070
755	1140	9.1×10^6	16.0	8.8	7.2	0.27	4.7	1.3	7720
820	1245	9.1×10^6	15.0	8.1	6.6	0.50	5.1	2.6	7740
704	—	—	17.0	9.5	7.7	0.24	4.3	—	7750
1500	—	9.6×10^6	17.+	11.2	9.2	0.024	3.8	0.091	7910
1580	—	10.5×10^6	—	11.8	10.2	0.001	3.8	0.0038	7940

flux is required and usually no filler rod is required if the joint is properly designed. The parent metal at the edges of the joint are melted and allowed to flow together. The 300 series stainless steels weld nicely to themselves, to Kovar, to OFHC copper, to K.R. Monel, to platinum, and to palladium. Even the refractory metals can be welded in weld boxes in which air is excluded. Fusion welding provides the strongest, cleanest joints between metals provided that the proper techniques are used. Trapped volumes and dirt-catching crevices are to be avoided in designing the welded joints. Wherever possible, welds should be made inside the vacuum system unless, of course, it is planned to remove a welded piece at a later date and the inside of the system will not be accessible. Very specialized welding of dissimilar metals and metals having widely divergent thicknesses can be performed by electron beam welding in a vacuum chamber. Welds of high penetration and narrow width are obtained with this method, and there is a minimum of metal distortion. One of the disadvantages of the welding process, as compared with brazing techniques, is the distortion of the metal at the joint by the highly localized heat of the torch.

Spot welding is commonly used to mechanically join together wires and small metal parts. Whereas fusion welding is used primarily to provide vacuum tight seams, spot welding is used to make mechanical and electrical connections. In spot welding, the two metals are joined together by overlapping them and applying pressure and a large electrical current to a relatively small "spot" area. Table 6.12 lists the various metals which can be spot welded together. It is common practice to use thin nickel or platinum foil to facilitate the welding of the refractory metals W, Mo, and Ta to each other. The foil acts as a brazing filler material.

2. *Brazing*

Brazing is the process whereby two metals are joined together by a third nonferrous filler metal that has a melting point below that of the base metals. The filler metal distributes itself by capillary attraction between the fitted surfaces of the joint. Spacing between the fitted base pieces should not be more than about 0.003 in. Brazing is done in a vacuum chamber or hydrogen chamber, and the parts may be heated either by induction with an rf coil or by radiation from a tungsten coil.

The brazing filler metal has to wet the base metals to be joined so that at least one constituent of the brazing metal must alloy with the base metal. Brazed joints should be designed so that there are large areas of contact. If several brazes are to be made on the same equipment, brazing fillers of successively lower flow points are used. When a brazed joint is remelted, a higher temperature is generally required since there has usually been some

Table 6.12. Spot Welding of Various Metal Combinations[a]

1 \ 2	Al	Ag	Cu	Cu Be	Cu Ni	Ni Cu	Fe Cr Ni	Ni Cr	Fe Ni	Fe Ni Co	Ni	Fe Cr	Fe	Pt	Mo	Ta	W
W	4	4	4	3	3		2, 5			3, 4	2		3	2	3, 5, 6	3, 5	3, 5
Ta	4	3, 4	3, 5	1	2		1			2, 3	2		3	2	3, 5, 6	3, 5	
Mo	4	4	4	3	3	2	2		2	3	2		3	2	3, 5, 6		
Pt	3, 4	3, 4	2	1	2	1	1	1	1	2	1	1	1	1			
Fe	3	3	2, 3	3	2	2	1	2		3	1	2	1				
Fe Cr (70/30)	3, 4		4						3		2	2					
Ni	3, 4	3	3	1	2	2	1	2	2	2	1						
Kovar Fe Ni Co	3, 4	4	3, 4	3	3			2		3							
Invar Fe Ni	4					2		2	2								
Ni Cr (80/20)	4			1	2	2	1	2									
V 2 A Fe Cr Ni	4	3	3, 4	1	2		1										
Monel Ni Cu	4					2											
Constantan Cu Ni	2	3, 4		2	2												
Cu Be (97.5/2.5)	3	3	2, 3	1													
Cu	4	4	3														
Ag	4	3															
Al	3																

Explanation:
1. Very good weld
2. Good weld
3. Adequate weld
4. Poor weld or no weld
5. Weld under a protective fluid
6. Weld with an intermediate layer of tantalum foil 0.002 in. to 0.004 in. thick

[a] (Adapted from M. Von Ardenne, *Tabellen der Elektronenphysik* Vol. 2, VEB Deutscher Verlag der Wissenschaften, Berlin, 1956)

solution of the base metal and some depletion of the more volatile constituents. There are literally hundreds of brazing alloys prepared by a dozen or so companies, each alloy being suitable for certain base metals. Two popular brazing alloys are the copper–silver eutectic alloy (28 Cu, 72 Ag), which can be used on stainless steel in dry H_2 or vacuum and on nickel and copper in H_2 or vacuum, and the nickel–gold eutectic alloy (18 Ni, 82 Au) which brazes Mo, W, Cu, Fe, Kovar, Monel, Ni, Pd, Ta, stainless steel, and other metals. In eutectic brazing alloys, the melting points and the flow points coincide, the same as they do for pure metals. For other alloys, only a partial liquid phase is formed at the melting point, and the braze material becomes pastelike with solid metal in suspension. If the flowpoint of the alloy is much higher than the melting point, it may take considerable time to reach this higher temperature, and during this long period a large amount of the base or joint metal can be dissolved. Successful high-temperature brazing, as well as heliarc welding, requires a fair amount of knowledge and experience, and one should consult with specialists in the field before designing brazed joints.

3. Ceramic-to-Metal Seals

There are three important techniques in use for joining ceramic to metal. The first is a diffusion seal as exemplified by the RCA Corporation proprietary "ram seal." A silver-plated Inconel X sleeve is forced over the tapered end of a ceramic tube and makes a high-pressure line contact with the corner of the ceramic where the taper meets the cylinder. The process is carried out at a high temperature so that there is a diffusion of the metal into the surface of the ceramic. Small seals of this general type have been made at the GCA Corporation using copper and aluminium pressed into intimate contact with the ceramic.

A second type of ceramic-to-metal seal is the active metal seal which makes use of the reactivity of titanium and zirconium toward ceramics. In one of these processes, the ceramic is first coated with titanium hydride. Next, a brazing alloy or pure metal foil is placed on the titanium hydride, and the metal to be joined to the ceramic is placed in position on the brazing alloy. The entire assembly is placed in a vacuum furnace and heated. The titanium hydride decomposes between 370 and 450°C. The temperature is raised to about 850–900°C depending on the brazing alloy to perform the braze. Suitable metals for joining to ceramic in this process are platinum, 430 stainless steel, tantalum, Kovar, and other nickel–iron alloys.

The most popular ceramic–metal seal used today is the moly-manganese sintered metal powder seal. In this process, the ceramic is first metallized by coating it with a mixture of 80% Mo and 20% Mn powder suspended in a volatile organic binder. The coating is generally sprayed onto the ceramic to a thickness of 0.001–0.002 in. The coated ceramic is then sintered in H_2 of 25°C dewpoint at about 1250–1500°C, depending on the softening point of the ceramic used. Finally, the metallized ceramic is plated with copper or nickel and brazed to a metal having a matching coefficient of expansion, using conventional brazing technique. Most commercially available ceramic–metal feedthroughs are made by the moly-manganese method.

REFERENCES

1. American Vacuum Society, Inc., Committee on Standards: Glossary of Terms Used in Vacuum Technology, Pergamon Press, 1958.
2. A. E. Barrington, *High Vacuum Enginering*, Prentice-Hall, Englewood Cliffs, N.J., 1963.
3. W. G. Brombacher, D. P. Johnson, and J. L. Cross, *Mercury Barometers* and *Manometers*, NBS Monograph No. 8, U.S. Dept. Com., May 20, 1960.
4. Cleaning of Electronic Device Components and Materials, *ASTM Spec. Tech. Publ.*, **246** (1959).
5. S. Dushman and J. M. Lafferty, *Scientific Foundations of Vacuum Technique*, 2nd ed., Wiley, New York, 1962.
6. H. E. Duckworth, *Mass Spectroscopy*, Cambridge Univ. Press, New York, 1958.
7. A. Guthrie, *Vacuum Technology*, Wiley, New York, 1963.
8. A. Guthrie and R. K. Wakerling, *Vacuum Equipment and Techniques*, McGraw-Hill, New York, 1949.
9. M. Knoll, *Materials and Processes of Electron Devices*, Springer, Berlin, 1959.
10. W. H. Kohl, *Materials and Techniques for Electron Tubes*, Reinhold, New York, 1960.
11. J. H. Leck, *Pressure Measurements in Vacuum Systems*, 2nd ed., Chapman and Hall, London, 1964.
12. G. Lewin, *Fundamentals of Vacuum Science and Technology*, McGraw-Hill, New York, 1965.
13. L. B. Loeb, *Kinetic Theory of Gases*, 2nd ed., McGraw-Hill, New York, 1934.
14. L. H. Martin and R. D. Hill, *A Manual of Vacuum Practice*, Cambridge Univ. Press, New York, 1949.
15. M. Pirani and J. Yarwood, *Principles of Vacuum Engineering*, Reinhold, New York, 1961.
16. R. W. Roberts and T. A. Vanderslice, *Ultrahigh Vacuum and Its Applications*, Prentice-Hall, Englewood Cliffs, N.J., 1963.
17. F. Rosebury, *Handbook of Electron Tube and Vacuum Techniques*, Addison-Wesley, Reading, Mass., 1965.
18. D. J. Santeler, D. H. Holkebaer, D. H. Jones and F. Pagano, *Vacuum Technology and Space Simulation*, NASA SP-105, U.S. Govt. Printing Office, Washington, D.C., 1966.
19. H. A. Steinherz, *Handbook of High Vacuum Engineering*, Reinhold, New York, 1963.
20. A. H. Turnbull, R. S. Barton, and J. C. Riviere, *An Introduction to Vacuum Technique*, Wiley, New York, 1962.

21. Vacuum Technology 1966 Directory and Specification Catalog, *Research/Development*, F. D. Thompson Publications, Chicago, Ill.
22. C. M. Van Atta, *The Design of High Vacuum Systems*, Kinney Vacuum Division, The New York Air Brake Co., Boston, Mass., 1960.
23. E. L. Wheeler, *Scientific Glass Blowing*, Interscience, New York, 1958.
24. J. Yarwood, *High Vacuum Technique*, 3rd ed., Wiley, New York, 1956.

PART II

Applications

Chapter 7

Micro Weighing with the Quartz Crystal Oscillator—Applications

A. WARNER

Bell Telephone Laboratories
Murray Hill, New Jersey

I. INTRODUCTION

In Chapter 5 the design parameters affecting the accuracy and sensitivity of oscillating quartz plates when used for micro weighing were reviewed. In this Chapter specific applications are discussed and working equations are given in each case as well as a more complete description of problems likely to be encountered. Equations 7.1 and 7.11 are the principal working equations for AT quartz plates.

II. THIN-FILM MONITOR

The piezoelectric quartz crystal microbalance is probably best known as a device for monitoring and controlling the thickness of sputtered or evaporated films (1). Since apparatus with instructions for use is commercially available, only a few useful points will be made here. First, the best quartz plate is the AT cut, because of its favorable temperature coefficient and its ability to respond linearly to a change in the mass of a uniform film. The proportional change in frequency, $\Delta f/f$ (see equation 5.5,

Chapter 5), upon adding to one side a uniform film of mass, m, in g/cm^2 is

$$\Delta f/f = -(f/443)m \tag{7.1}$$

where f is the frequency of the crystal plate in kHz. (This is 2.26 Hz/μg per cm^2 for a 1 MHz plate and increases with the square of the plate frequency.) If both sides are used, the effect on frequency is additive (see Section IV.A). The thickness of the added film is $t = m/\rho$, where ρ is the density of the added material. Second, the temperature coefficient of frequency is likely

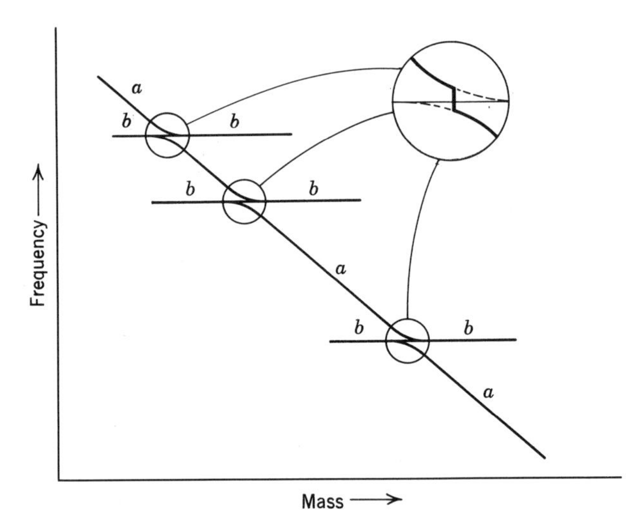

Figure 7.1. Effect of interfering low frequency resonances on the frequency versus mass curve. Curve a is the mass sensitive high frequency resonance and curves b represent various longitudinal or flexual resonances sensitive only to edge dimensions of the plate.

to be one part in 10^7/°C. The effect of a sudden heat shock may be considerably greater but recovery will be rapid. Third, one should be on the alert for signs of an interfering mode of vibration, particularly if the experiment covers a wide range of frequency. There are other modes of vibration, particularly low frequency flexure and face shear modes, that may be coupled electrically or elastically to the thickness–shear mode used for micro weighing. Normally these other modes are both very low in Q and also free of resonances that coincide with the frequency being used. If and when this is not the case, a sharp drop in Q and a characteristic shift in frequency will occur as energy is transferred to the other mode. Figure 7.1 shows the general nature of an unwanted mode. The curve labeled a represents the behavior of the wanted mode, which changes uniformly with mass.

The curves labeled *b* represent the behavior of high overtones of a low-frequency mode whose frequency is a function of edge dimensions and changes very little with mass loading. At certain frequencies the oscillator will tend to follow a *b* curve, with a loss in activity or Q, and perhaps even cease to oscillate completely. As the mass is increased and the interfering frequency is left behind, normal conditions will again prevail. The location of interfering modes of vibration may be changed by beveling or grinding the edge of the quartz plate, or by changing the operating temperature. Fourth, in multilayer work it is of interest to note that the change in frequency will be the same regardless of the position of the layer, since Δf responds only to mass. In other words, $\Delta f/m$ is a constant for a given quartz plate. Masses as great as 2 mg/cm^2 can be applied to a 5 MHz AT plate and measured with less than 2% error. Much of this error may be accounted for by second-order terms which have been omitted, for practical reasons, from the equations used.

Finally, if the thin film is not uniform, or if the film is of such nature that it dissipates energy (lowers the Q of the resonator) then it may be necessary to empirically determine the frequency variation with mass for each experimental investigation.

III. SORPTION PHENOMENA STUDIES

A. General Types of Sorption

When a surface is exposed to an ambient different from that with which it has reached equilibrium, mass changes may occur due to sorption or desorption. This may be absorption, chemisorption, or physisorption, with each type having its own characteristics. Quartz micro weighing may be used to great advantage in differentiating these types of sorption by observing and recording the mass changes with time and temperature.

There is a whole class of measurements that may be made using essentially the following procedure:

1. The substrate on which the reaction is to take place is put down, usually by an evaporation technique, on one or both faces of the quartz plate. If the substrate is metallic, it may also serve as the electrode. If not, a thin electrode, usually of gold, is first deposited by vacuum evaporation before the substrate of interest is applied.

2. The assembly is cleaned, vacuum baked if applicable, and measured for frequency and resistance in vacuum. Stability with time and temperature are noted.

3. The assembly is exposed to some active gas or other reagent according to the nature of the experiment and then returned to vacuum.

4. The change in mass on the crystal surface and the permanence of this change with time (or temperature) are measured.

Where exposure to a gas at a fixed pressure and temperature is involved, the relative mass change during exposure can also be measured, although there will be a certain shift in frequency due to the pressure of the gas itself.

If the process is chemisorption, the mass change will be irreversible with a subsequent decrease in pressure, and is unlikely to proceed beyond one molecular layer. For physisorption, the process will be reversable with pressure at constant temperatures. If the substrate is porous the mass change will be much larger due to the considerable internal surface area.

B. The Sorption Detector

Another application, that of the piezoelectric sorption hygrometer (2), involves a constant pressure and temperature with only the composition of the gas being changed. King (2) has used quartz plates coated with various liquids, powders, and oxides, in some cases applied with glue or cement, as selective gas detectors, gas chromatographic detectors, and sorption

Table 7.1. Sorption Detector Coating Materials

Hydrogen detection, nonselective to compound type	Squalane Silicone oil Apiezon grease
Selective detection of polar molecules such as aromatic, oxygenated, and unsaturated compounds	Polyethylene glycol Sulfolane Dinonyl phthalate Aldol-40 Tide (alkyl sulfonate)
Water vapor	Silica gel Molecular Sieve Alumina Hygroscopic polymers[a]
Hydrogen sulfide	Lead acetate Metallic silver Metallic copper Anthraquinone–disulfonic acid

[a] Natural resins, glues, cellulose derivatives, and synthetic polymers.

hygrometers. This class of micro weighing devices usually operates in a region of mass sensitivity, 10^{-7} g/cm^2 or greater, where it is not necessary to correct for temperature and pressure changes. Table 7.1 shows a partial listing of substances that have been used as coatings. To relate the frequency change quantitatively to the sorption isotherm of the coating material, King uses the following equation:

$$\Delta f = \Delta f_0 (\Delta W / \Delta W_0) \tag{7.2}$$

Figure 7.2. Sensing element of a piezoelectric sorption detector.

where Δf_0 = frequency change due to application of coating, ΔW_0 = weight of coating, Δf = frequency change due to sorption of solute vapor, and ΔW = weight increase due to sorbate.

Δf_0 is likely to be 1–100 kHz. Of particular interest are the gas chromatography substrates which show a linear relationship between the concentration of solute vapor and the fractional weight pickup $\Delta W/\Delta W_0$. Advantages are said to be (*1*) higher senstivity at higher molecular weight, (*2*) linear response, (*3*) selectivity, and (*4*) speed.

Figure 7.2 shows one version (2) of the sensing element in a piezoelectric sorption detector. Typical absorption isotherms measured with coated quartz crystals are shown in Figure 7.3.*

* The Du Pont 510 Moisture Analyzer is based on piezoelectric sorption detection.

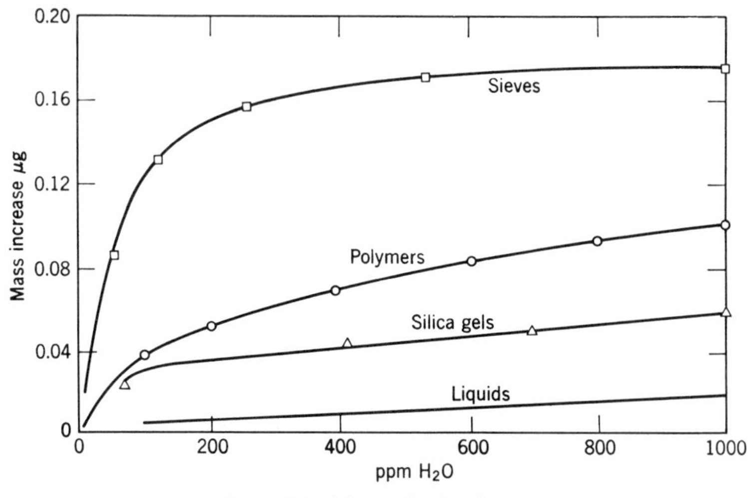

Figure 7.3. Absorption isotherms.

C. Chemisorption of CO

The study of aging in quartz crystal units for frequency control is largely one of sorption phenomena. In this connection and as an example of the type of study possible with the three-crystal apparatus described in Chapter 5 and shown in Figure 5.12, the measurement of the chemisorption of carbon monoxide on gold electrodes is described. Carbon monoxide was chosen as the contaminant because of its suspected chemisorption on gold. Although the chemisorption is small, very little is known about it and its magnitude is of the order of the aging found in quartz crystal frequency standards.

In preparation for the experiment, the entire appartus including the crystals were first baked at 250°C for three days, using external pumps to attain a pressure of about 10^{-7} torr. The system was then allowed to stabilize at 40°C and 10^{-10} torr and the various cables and temperature controls were attached. Temperature coefficients of the crystals were measured, unwanted resonances noted, and the aging curves established. The lower and upper oven voltages, the temperature near the crystals, the temperature at the oven shell, the system pressure, and the resistance and frequency of the three crystals were monitored.

A stabilization period of 7 weeks followed, by which time the frequency aging was a few parts in 10^{10} per day. Analytical reagent grade carbon monoxide was admitted to a pressure of about 5×10^{-2} torr. This resulted in an immediate lowering of the frequency, which differed from crystal to crystal. Increasing the amount of CO in steps to 300 torr over a period of several days and then pumping it out, resulted in a further decrease in frequency. After a period of stabilization of 8 days, an attempt was made to remove some of the CO by raising the oven temperature from the 40°C operating temperature to 90°C for 1 hr. The total effect on each crystal of

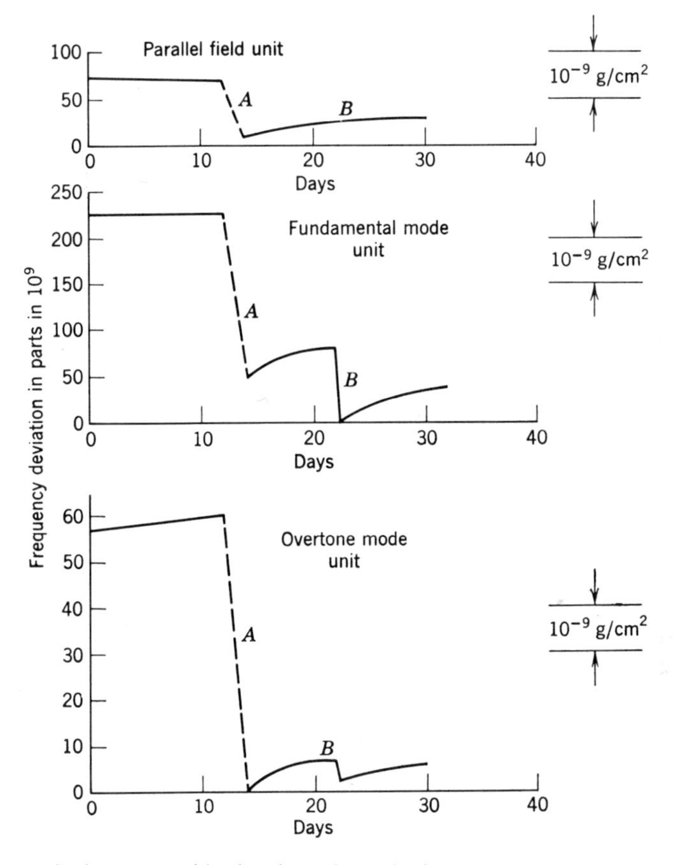

Figure 7.4. Carbon monoxide chemisorption. The frequency scales have been chosen to show uniform mass sensitivity.

exposure to 300 torr CO is given in Figure 7.4 at *A* and the effect of the mild bake at *B*. The variation of the frequency with time following each operation is also shown. The following tentative conclusions have been drawn from this one experiment.

1. Carbon monoxide was heterogeneously chemisorbed on gold surfaces at 40°C. On the two crystals with gold electrodes over the active area, the total amounts adsorbed were 36 and 60 × 10^{-10} g/cm^2. This is approximately one to two tenths of a monolayer. The effect on the crystal having gold electrodes only at the periphery of its active area was much less.

2. Even in the 1 to 9 $\times$ 10^{-2} torr pressure range, the effect of carbon monoxide on frequency was significant and measurable, and contributed to the aging of the quartz resonator.

3. Mild heating in vacuum for up to 24 hr had no measurable effect on mass, that is, the CO was not desorbed. The lowering of the frequency and subsequent recovery for the two gold-surfaced resonators is attributed to the relief of thermally induced strain since the parallel-field unit showed no effect of the 90°C bake.

IV. STUDY OF GASES, AND THE EFFECT OF PRESSURE ON FREQUENCY

A. The Effect of Pressure on Frequency

When a quartz microbalance plate goes from vacuum to a controlled constant pressure of some gas, its resonant frequency shift will consist of three distinct components, (3,4). The frequency will (*1*) increase linearly with pressure due to the effects of hydrostatic pressure on the elastic moduli of quartz, (*2*) decrease linearly with the half power of the pressure due to the shear impedance of the gas, and (*3*) decrease with any increase in mass due to sorption. This may be expressed as

$$\frac{\Delta f}{f} = \left(\frac{\Delta f}{f}\right)_p + \left(\frac{\Delta f}{f}\right)_x + \left(\frac{\Delta f}{f}\right)_m \tag{7.3}$$

where p, x, and m refer to the hydrostatic pressure, the shear impedance, and the mass, respectively. The frequency change with hydrostatic pressure has recently been shown by Stockbridge (3) to be

$$\left(\frac{\Delta f}{f}\right)_p = +13.5 \times 10^{-10} \text{ per torr} \tag{7.4}$$

for an AT cut quartz plate at 50°C, with a nearly linear variation with temperature of -0.015×10^{-10}/°C.

The frequency change due to the shear impedance of the gas, apart from any sorption effects, is a function of the density and viscosity of the gas as well as the frequency of operation. For a purely viscous gas, the classical expression (4) is

$$R_m = X_m = (\pi f \rho_g \eta_g)^{1/2} \tag{7.5}$$

where ρ_g and η_g are the density and viscosity of the gas, respectively, and R_m and X_m are the mechanical resistance and reactance of the gas. Since $\rho = PM/RT$, the resistance and reactance increase with the half power of the pressure P and molecular weight M of the gas, and decrease with a rise in temperature T.

In order to arrive at a numerical value for the frequency change due to the shear impedance of the gas, it is convenient to consider this loading effect as though it were an increase in the inertia of the surface electrodes. Following this procedure the reactance, X_m, will equal $2\pi f(m/A)_g$, where $(m/A)_g$ is the effective mass per unit area of the gas. This is, of course, apart from any sorption effect which would result in a real change in mass on the crystal surface. Therefore, combining with equation 7.5.

$$(m/A)_g = (\pi f \rho_g \eta_g)^{1/2}/2\pi f \tag{7.6}$$

Before proceeding further, it should be pointed out that equation 5.8 Chapter 5, i.e., $(\Delta f/f) = -(m/A)(1/\rho_q t_q)$, and equation 7.1 in this chapter which was derived from equation 5.8, are for a uniform mass per unit area applied to one side only of the quartz plate. When gas loading and various sorption phenomena are involved, both sides of the quartz plates are affected, and since the quantity measured is mass per unit area, the change in frequency will be double that of one side exposure. This results from the area A being used to calculate the mass of the resonating quartz, i.e. $M = \rho_q A_q t_q$ (see equation 5.4 of Chapter 5). Therefore when both sides are covered, there is twice as much mass for the defined area A. The equation for both sides exposure is

$$\frac{\Delta f}{f} = \left(\frac{m_1}{A} + \frac{m_2}{A}\right)\frac{1}{\rho_q t_q} = 2\left(\frac{m}{A}\right)\frac{1}{\rho_q t_q} \tag{7.7}$$

where m/A is the quantity measured in units of mass per unit area.

Combining equations 7.6 and 7.7, then

$$\Delta f/f = (\pi \rho_g f \eta_g)^{1/2}/\rho_q f t_q \tag{7.8}$$

For a check on this procedure, the same equation was derived by Tiersten (5) from more fundamental relationships without using the analogy of mass. For fundamental AT cut quartz, $ft_q = 167{,}000$ Hz cm, $\rho = 2.67$ g/cc and equation 7.8 becomes

$$\left(\frac{\Delta f}{f}\right)_x = 0.720 \times 10^{-6}\,(\pi f \rho_g \eta_g)^{1/2} \tag{7.9}$$

where f is in Hz, ρ is in g/cc, and η is in poise.

The equation for frequency change at 50°C for a fundamental AT cut quartz plate due to introduction of a gas, determined from equations 7.4, 7.9, and 7.7. is

$$\Delta f/f = 13.5 \times 10^{-10} P_{(\mathrm{torr})} - 0.72 \times 10^{-6}\,(\pi f \rho_g \eta_g)^{1/2} - 4.52 \times 10^{-6}(m/A)f \tag{7.10}$$

with the units in Hz, g, cm, poise, and torr. The density, ρ, of the gas, may be calculated from

$$\rho = PM/RT \tag{7.11}$$

where P is the pressure in torr, M is the molecular weight, T the absolute temperature in °K and $R = 6.23633 \times 10^4$ cc-torr/mol-°C.

Mass changes may be measured under differing gas pressures using equation 7.10, provided the molecular weight and viscosity of the gas are known. For air, the fractional change in frequency from atmosphere to vacuum is about -2×10^{-7} for a 5 MHz AT cut plate ($+8 \times 10^{-7}$ for hydrostatic pressure, and -10×10^{-7} for mechanical impedance).

B. Determination of the Characteristics of Gases

Using equation 7.10 many special studies are now possible. For an inert gas, for example, the viscosity and density may be determined by measurements at several pressures. Another interesting study is that of complex viscosity. At low pressure, the viscosity becomes a function of frequency, according to

$$\eta_* = \eta(1/1 + j\omega\tau) \tag{7.12}$$

where τ is a relaxation time. When the space between gas molecules is the same order of magnitude as the wavelength of the radiated shear wave in the gas, the gas is no longer a simple viscous fluid but must be regarded as a viscoelastic fluid. By plotting the resonant frequency of the crystal versus the square root of pressure, the pressure at which the viscosity must be treated as a complex quantity may be found. Recently Stockbridge (3) found upon measuring the crystal frequency versus gas pressure of a wide variety of relatively inert gases, that the lighter gases began to exhibit this complex viscosity at pressures below about 300 torr. The actual frequency versus pressure for these gases, H_2, D_2, He, Ne, Ar, Kr, SF_6, Xe, and Freon C318, is shown in Figure 7.5. The elastic modulus effect is shown by a dotted line. Since there is negligible sorption of these gases, that portion of the frequency change due to the shear impedance of the gas will be represented by the departure from the dotted line.

From the known molecular weight and viscosity of the gases, the actual mass of gas sorbed at 1000 torr over that *in vacuo* was calculated and is listed in Table 7.2. The anomalous behavior of H_2, D_2, and He may be due to some sort of "mechanical sorption" in the gold electrodes. The change in mass is very small and corrections for this would not in most cases exceed the width of the line in Figure 7.5.

Freon C318 (C_4F_8) exhibited unambiguous physical adsorption, and a

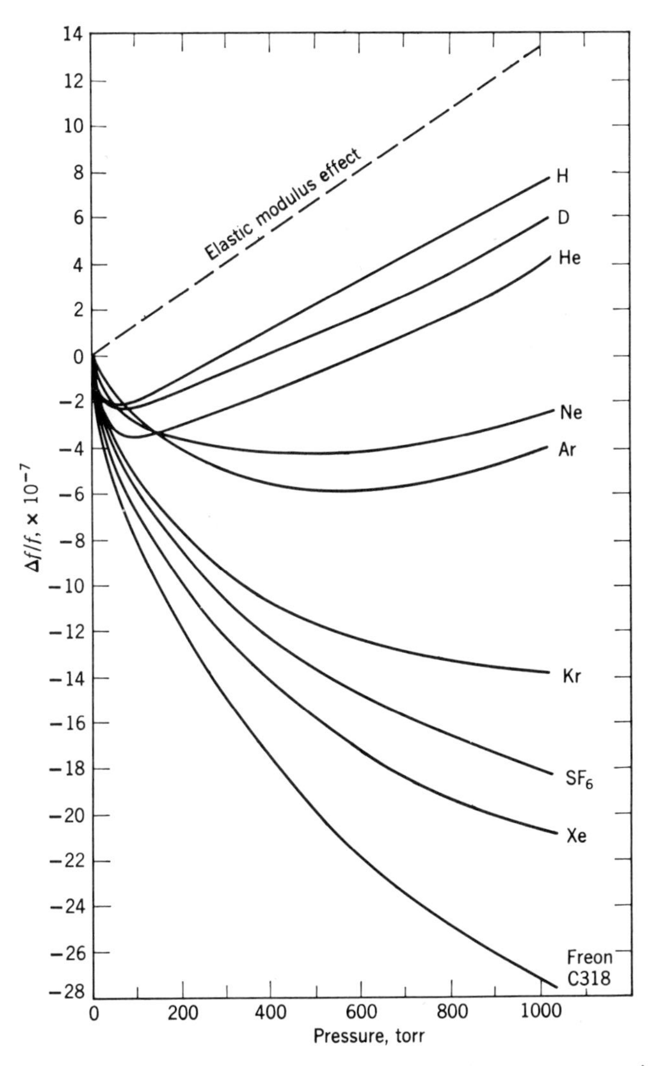

Figure 7.5. Experimental frequency change observed versus pressure in different gases with a 5 MHz fundamental AT cut quartz crystal. If there were no viscous loading, all the curves would lie close to the dotted line.

plot of mass versus pressure is shown in Figure 7.6. The data obtained for Freon C318 show that this gas with a boiling point of -6°C is reversibly physisorbed on gold at 55°C. The number of molecules adsorbed at 1000 torr is approximately 7×10^{13}.

In studies of the gas CO, it was expected that considerable chemisorption by the gold electrodes would take place. However, at 1000 torr CO the

Table 7.2. Increase in Mass at 1000 torr over Mass in Vacuum

Gas	ng/cm²	θ[a]
H_2	10	2.0
D_2	11	1.1
He	13	0.83
Ne	6	0.11
CO	4	0.11
N_2	4	0.11
Ar	5	0.088
Kr	2	0.021
Xe	2	0.019
SF_6	6	0.12
C_4F_8	23	0.49

[a] $\theta \equiv$ fraction of a close-packed monolayer of gas.

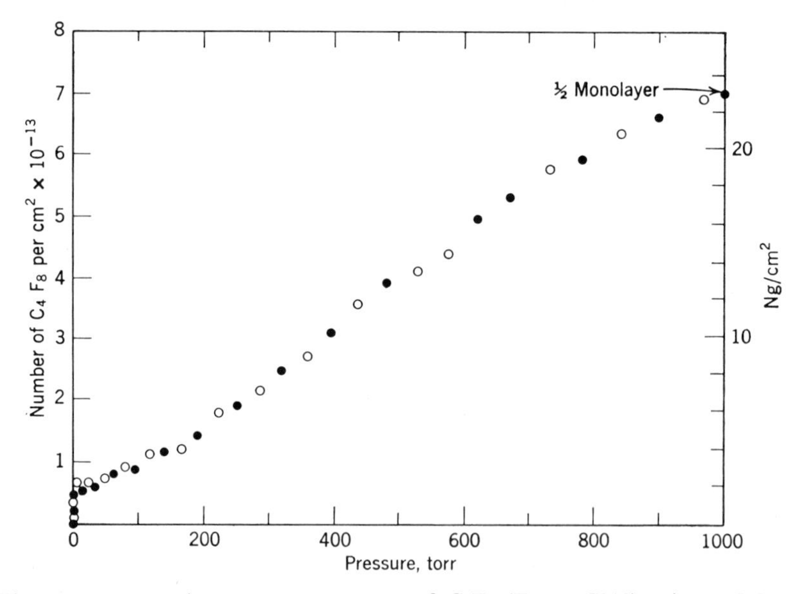

Figure 7.6. Mass change versus pressure of C_4F_8 (Freon C318) using a 5 MHz fundamental AT cut quartz crystal.

mean mass of CO adsorbed was only 4 ng/cm². On pumping out the CO to high vacuum the crystals were found to be permanently heavier by about 3 ng/cm². That is, exposure to CO of the evaporated-gold electrodes resulted in irreversible chemisorption of about 0.08 monolayers of CO. This result is consistent with that obtained from the aging studies discussed above.

It may also be noted that equation 7.10 contains frequency terms that make it possible to emphasize or deemphasize the three basic contributions, hydrostatic pressure, mechanical impedance, and mass, to frequency change. Within the limitations discussed in Section II.B of Chapter 5, higher frequency will better indicate a true mass change, while a lower frequency will show more prominently the hydrostatic effect.

V. CONCLUSIONS

The quartz crystal microbalance may be used for the quantitative measurement of (*1*) thin films deposited on its surface, (*2*) sorption and desorption at a surface, and (*3*) the characteristics of various gases. The physical and chemical stability of quartz, along with its low temperature coefficient, make possible experiments with a high degree of sophistication. In addition, the very high effective sensitivity of the quartz crystal plate to any disturbance on its surface makes it useful even in situations where the nature of the disturbance is not completely understood and cannot be directly calibrated.

REFERENCES

1. K. H. Behrndt and R. W. Love, *Vacuum*, **12**, 1 (1962).
2. W. H. King, Jr., *Anal. Chem.*, **36**, 1735 (1964).
3. C. D. Stockbridge, *Vacuum Microbalance Tech.*, **5**, 147 (1966).
4. W. P. Mason, *Trans. Am. Soc. Mech. Engr.*, **69**, 359 (1947).
5. H. F. Tiersten, private communication.

Chapter 8

Mass Thickness Determination in Ultrahigh Vacuum

R. NIEDERMAYER AND D. HILLECKE

Technische Hochschule
Clausthal, Germany

I. INTRODUCTION

Conventional microbalances have only rarely been used in thin film investigations. However, the quartz crystal oscillator (see Chapters 5 and 7) microbalance has recently found broad utility application in thin film work. As noted in earlier chapters popularity of the quartz crystal microbalance arises primarily from the fact that, in contrast to the conventional microbalance, it is mechanically very simple, small enough to fit in any vacuum system, and can be baked at high temperatures in order to obtain ultrahigh vacuum. Most conventional microbalances unless specially constructed cannot be baked out at sufficiently high temperature to obtain ultrahigh vacuum conditions.

A number of typical applications of microbalances and quartz oscillators to thin film studies are discussed in this chapter. As will be seen, there are certain instances where a conventional microbalance cannot be replaced by the quartz oscillator.

II. DESCRIPTION OF VARIOUS MICROBALANCES

The following types of microbalances have been used for determination of the mass thickness of thin films.

1. Beam balances with horizontal torsion suspension.
2. Rotors with vertical suspension.
3. Quartz oscillator microbalances.

A brief description of these balances as applied in thin film studies will be given in this section.

A. Beam Balances with Horizontal Torsion Suspension

This type of balance has been developed by Gast (1), Cahn (2), and Mayer et al. (3). The Gast and Cahn balances have a galvanometer-type compensation. The current through the galvanometer coil required to keep the balance in the null position is proportional to the torque exerted on the balance by the weight. This type of balance has been discussed in Chapters 2 and 3 of this book. Since the permanent magnets and the coil required for this balance cannot be baked out at high temperatures, their application in ultrahigh vacuum is limited. Recently Cahn has developed a commercially available balance that can be baked out to 100°C and therefore allows some limited use in high and very high vacuum. Mayer et al. (3) avoided the bakeout problem in his magnetic compensation type of balance by placing the coil outside the vacuum system. In cases where the magnetic compensation system interfered with other measurements electrostatic compensation was used.

B. Quartz Fiber Suspended Magnetically Compensated Balances

Thin quartz fibers are the most favored microbalance suspension material. The advantages of quartz have been discussed in Chapter 3. One of the most important factors is the reproducible elastic behavior of quartz. The quartz suspensions, however, are very fragile and therefore require very careful handling.

The mechanical strength is highly dependent upon the fiber tension (4).

The following is a quantitative treatment of the relationship between mechanical strength and fiber tension.

Figure 8.1 shows the deformation of a torsion fiber under the influence of a weight B and a horizontal stress K. The fiber thickness is always much smaller than its length. The tensions, therefore, within the fiber near the point of attachment can be described by the Bernoulli formulas. The

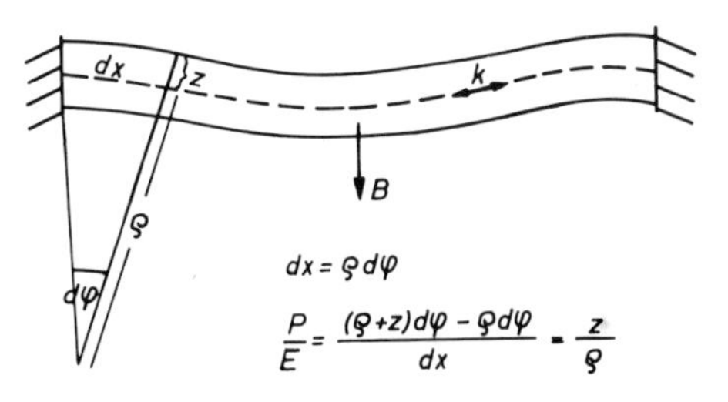

Figure 8.1. Deformation of a torsion fiber under the influence of weight B and horizontal stress k; dashed line is the neutral fiber.

bending stress P that is developed in the distance z from the neutral fiber is given by:

$$P = Ez/\rho \tag{8.1}$$

where E is the modulus of elasticity and ρ is the radius of curvature. The horizontal stress, K, must be added to the bending stress. The total stress is then:

$$P = Ez/\rho + K \tag{8.2}$$

The mechanical strength of the torsion fiber must be greater than the total stress, that is:

$$P = Ez/\rho + K \leq Er/\rho + K < P_{ts} \tag{8.3}$$

with r and P_{ts} being the radius and tensile strength of the quartz fiber, respectively. The radius of curvature, ρ, can be calculated and depends on the horizontal stress K and the applied weight B. Normally the torsion fiber is almost horizontal, that is the force exerted by the horizontal stress ($r^2\pi K$) is much greater than the weight B. In this case the radius of curvature, ρ, has its minimum value, given by equation 8.4, at the point of attachment of the fiber:

$$1/\rho_{\min} \simeq 2B/r^3\pi(EK)^{1/2} \tag{8.4}$$

The requirement that the tensile strength must be greater than the total stress leads to the following condition:

$$B \leq \{[(\pi^2/4)K(P_{ts} - K)^2 r^4]/E\}^{1/2} \tag{8.5}$$

where B is the maximum weight that can be suspended by a quartz fiber of tensile strength P_{ts} and horizontal stress K. It can be seen that this maximum has its highest value if

$$K = P_{ts}/3 \tag{8.6}$$

High mechanical strength can be achieved if springs having small force constants maintain a constant horizontal stress especially in the presence of seismic or other vibrations.

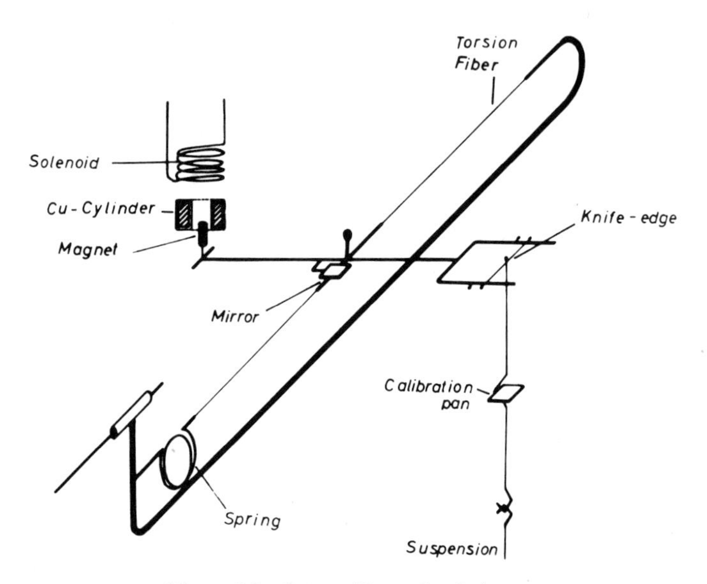

Figure 8.2. Quartz fiber microbalance.

Figure 8.2 shows a microbalance in which the principles derived above were employed to obtain the proper fiber tension. The torsion fiber was fixed between the frame and a quartz spring by melting. During construction the quartz spring was loaded so that the fiber after unloading had the correct stress. The position of the beam was followed by the reflection of a light beam from a mirror attached to the balance. A solenoid rigidly connected to the glass balance case outside the vacuum system exerted a compensating force on a small permanent magnet to maintain the balance in the null position (5). A copper cylinder attached to the frame produced eddy current damping.

The balance was calibrated with small pieces of aluminium wire, the length of which were determined with a microscope. One millimeter of this wire weighed approximately 10^{-5} g. The very small calibration weights were handled with the aid of a vacuum nozzle. The compensating current was proportional to the weight on the pan in a load range of 0.05–500 μg. The reproducibility of the measurements with this balance is shown in Figure 8.3 for an experiment where mercury was evaporated from a Knudsen cell at constant temperature onto a cooled substrate (5). The increase of the condensed mass with time was linear and identical in three independent runs.

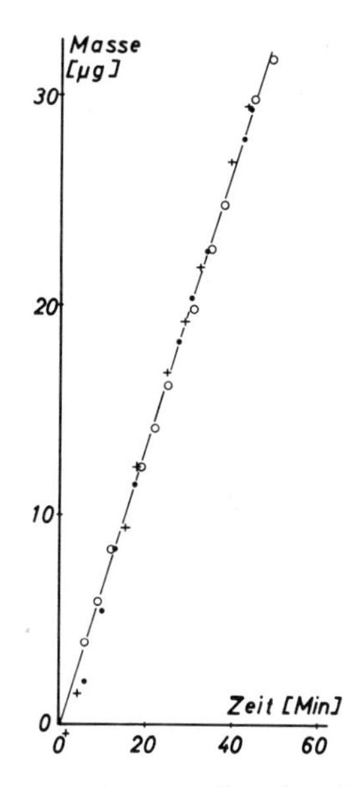

Figure 8.3. Dependence of mass on evaporation time for three different evaporations (source temperature kept constant).

As noted above a magnetically compensated microbalance can be used in ultrahigh vacuum experiments only if the permanent magnet does not change its magnetic properties as a result of the required bakeout treatments. In the balance shown in Figure 8.2, Göhre successfully used Koerzit 400 of Krupp Widia as the magnet material. Schwoebel has previously discussed in detail in Chapter 3 the work of Addiss in relation to this same problem. A circulating cooling water system effectively prevented oxidation of the solenoid during bakeout. Microbalances built in this manner have been baked out repeatedly up to 400°C without any changes in their properties.

C. Tungsten Wire Suspended Electrostatically Compensated Balances

For some applications where stray magnetic fields may be undesirable, electrostatic rather than magnetic compensation is preferred. Stünkel (6) and Niedermayer (7) have developed two types of electrostatic compensated microbalances based on the principle of a simple electrometer, that is, the beam of the balance moved between two electrodes rigidly connected to the frame of the balance. Voltages were applied to the beam and the electrodes so that the field of the two rigid electrodes exerted a force on the beam. Since an electrically conducting suspension is required for this kind of compensation, tungsten wires of 8–20 μ diam were used. Two different methods of null detection were employed.

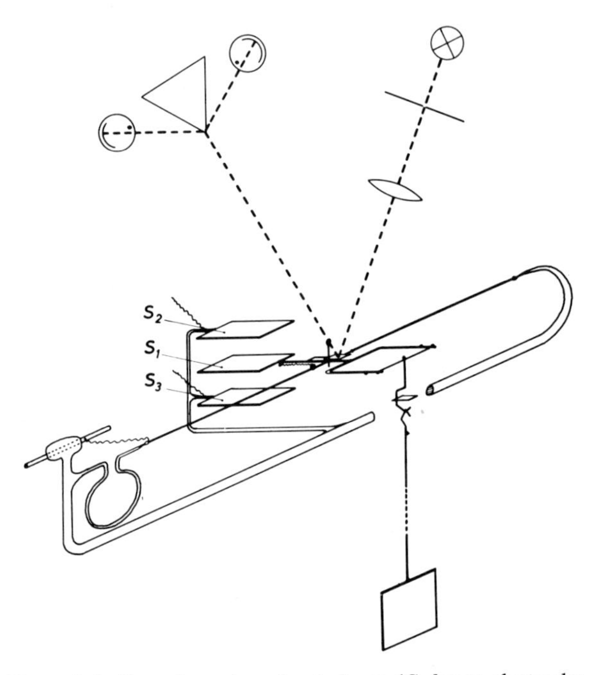

Figure 8.4. Tungsten wire microbalance (S_1 beam electrode; S_2, S_3 frame electrodes).

The first model, Figure 8.4, had a mirror fixed to the beam. A light beam reflected by the mirror was divided by a prism onto two photocells which served as two arms of a bridge circuit. The bridge voltage was amplified and fed back to the beam electrode to compensate the deflection caused by a weight change. The voltage necessary to keep the beam midway between the two electrodes was directly proportional to the weight change (6).

A compensating system consisting of the fixed electrodes and the middle electrode was used in a second model (Figure 8.5) as a capacity bridge. The bridge signal was rectified in a phase discriminator and the amplified dc voltage was used for compensation in the same manner as in the first

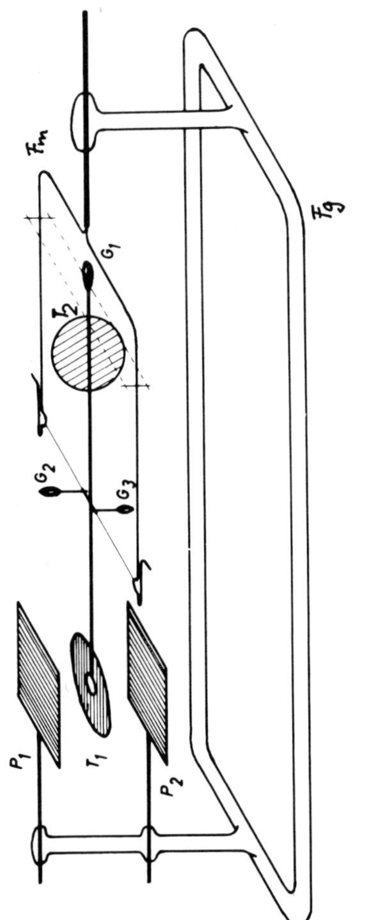

Figure 8.5. Tungsten wire microbalance (T_1 beam electrode; P_1, P_2 frame electrodes; G_1, G_2, G_3 glass weights for adjusting the mechanical zero position; T_2 readjustment pan; Fm molybdenum frame; *Fg* glass frame).

model (Figure 8.6). The compensation voltage was proportional to the weight change. Czanderna and Schwoebel have discussed electrostatically compensated balances in Chapters 2 and 3. Additional detailed information can also be obtained from Niedermayer's work (7).

The electrostatic force K on the beam electrode midway between the rigid electrodes is given by:

$$|K| = 2\varepsilon_0 FVv/l^2 \tag{8.7}$$

where ε_0 is the dielectric constant of vacuum, F is the area of the beam electrode, $2V$ is the voltage between the frame electrodes, v is the voltage between the beam electrode and mean voltage of the frame electrodes, and

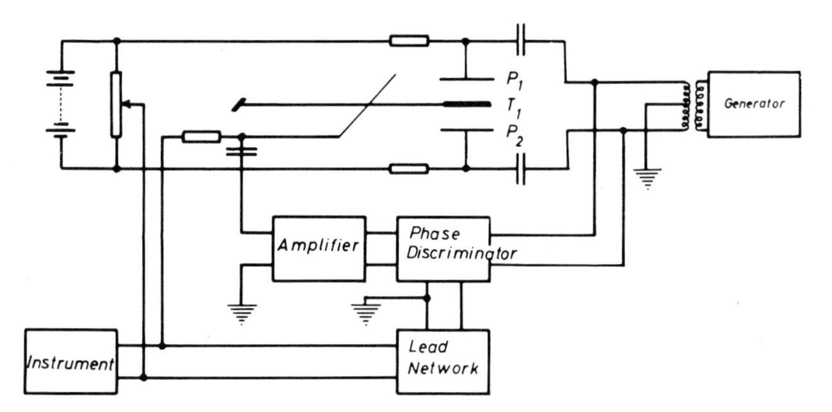

Figure 8.6. Feedback compensation circuit.

$2l$ is the distance between the frame electrodes. From this expression the sensitivity, γ, and its dependence on the voltage of the frame electrodes can be expressed by:

$$|\gamma| = l^2/2\varepsilon_0 FV \tag{2.8}$$

At $V = 210$ V, an area, F, of 1.2 cm^2, and a distance l of about 1 cm, the sensitivity was 1.22 V/μg. The results of a typical experiment at constant deposition rate are shown in Figure 8.7. 3×10^{-6} g were deposited in 10 min. The mean square deviation of 5×10^{-8} g resulted mainly from the seismic vibration of the building. The effect of a bigger shock, e.g., closing of a door, is shown in the middle of the curve.

The tungsten torsion wires were wound around the frame and then attached permanently by means of "Einbrennsilber." Since these wires can be plastically deformed, the balance is stronger mechanically than a quartz fiber balance. On the other hand, the method of attachment and the plasticity increase the possibility of an irreversible change in the zero point. This drawback can be minimized by using very thin wires so that the elastic torque of the wire is much smaller than the torque exerted on the balance by the center of mass. Balances constructed as described can be baked out to 400°C without any effects on balance characteristics. Both balances were calibrated in the manner described previously in this chapter.

As noted in Chapter 2 of this book, surface charges accumulated on the glass balance envelope exert force on the balance. In order to minimize

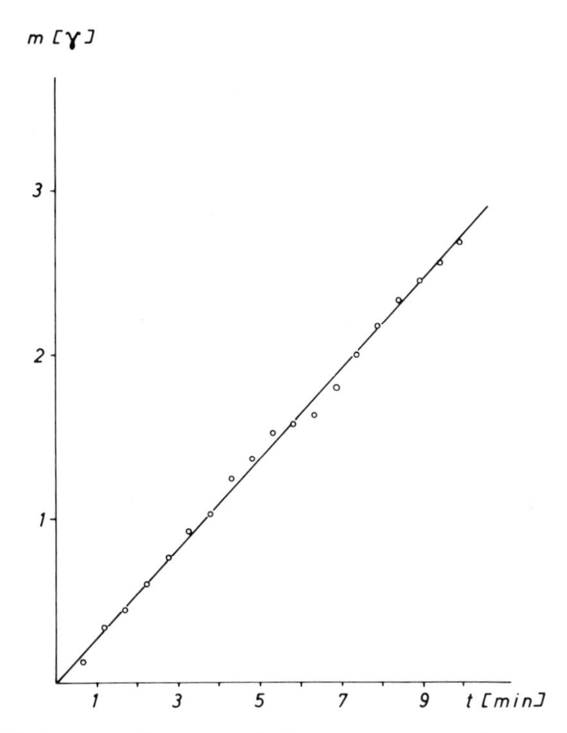

Figure 8.7. Increase of mass m with time t in an evaporation experiment.

such undesirable effects the glass walls are coated with a conducting material, such as tin oxide, or a metallic housing is used for the microbalance. The electrical conductive surroundings of the balance should be kept at a known constant potential.

Readjustment of the beam after a certain amount of weight has been deposited onto the pan is a difficulty common to all of the microbalances discussed above. This limitation, especially severe with electrostatically compensated balances, was overcome partially by evaporation of some material (e.g., silver) onto a readjustment pan, as shown in Figure 8.5. This method, however, is irreversible. A second common difficulty is the existence of parasitic pendulum oscillations, which occur if the pan is suspended on a very long quartz fiber as shown in Figures 8.2 and 8.4. Such oscillations may cause a much lower reproducibility than that shown in Figure 8.7. A final limitation arises from the difficulty in attaching leads to the pan in order to make electrical measurements.

To overcome these disadvantages Schmider (29) has constructed an improved type of electrostatic compensated microbalance. His design objective was to eliminate parasitic oscillations. To this end he combined two microbalances at right angles, as shown in Figure 8.8. System I is an electrostatic compensated balance with a tungsten wire suspension analogous to that in Figure 8.5. Compensation is achieved by the capacity bridge detector–electrometer method. System II is mounted at right angles about 15 cm below system I. This arrangement stabilizes the sideward movements of the pan suspension. Both systems are joined by the quartz frame, see no. 6 in Figure 8.8, in such a way that each balance supports half of the total weight of the pan and the pan suspension.

The torsion wires of a microbalance can be used as leads for electrical measurements of material on the sample pan. The special construction shown here provides two leads from each balance, so that either a Thompson-bridge measurement of film resistance or a current voltage characteristic in addition to a resistance–temperature measurement can be made. It is important to note, however, that the resistance of the tungsten torsion wires is about 300 ohms, so that small resistances can be determined only with a Thompson-bridge. When all four leads were used in this manner, the compensation voltage for the beam electrode was fed by an additional lead, for which a 7 μ gold wire (no. 3 in Figure 8.8) was used. To minimize the torque exerted by this wire it was wound around the fulcrum axis several times.

This balance design offers an additional possibility for readjustment and calibration. In system II a narrow quartz tube (see no. *7* in Figure 8.8) in which a thin quartz fiber (length = 80 mm, diameter = 82 μ, mass, m_s = 1.06 mg) could be moved, was fused to the beam. A shift, Δx of the quartz fiber corresponded to a weight change ΔM at the pan of

$$\Delta M = (m_s/l)\Delta x \tag{8.9}$$

where l is the length of the lever arm (for system II, l = 34.5 mm). For Δx = 0.1 mm, ΔM is 3.07 μg. The movement of the quartz fiber was achieved by means of bimetals and measured with a cathetometer. This method allowed a more sensitive calibration of the balance and a zero change adjustment of 0.5 mg. Calibration as well as readjustment could be performed under working conditions. Although this arrangement seems complicated, the system has high mechanical strength and can be baked out

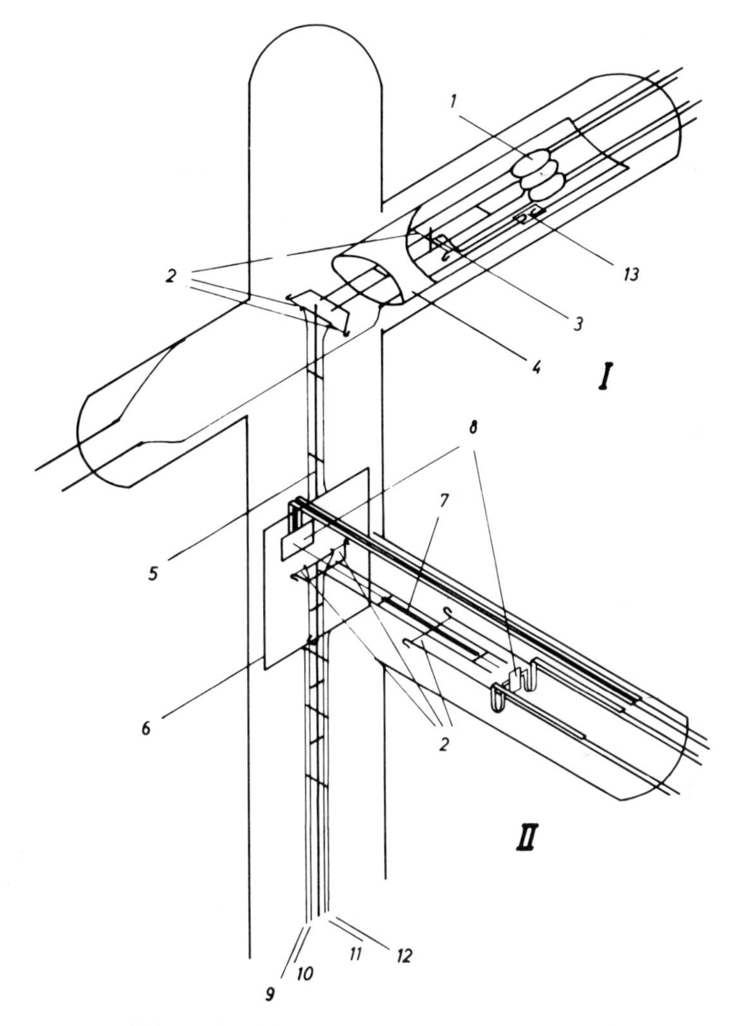

Figure 8.8. Stabilized microbalance system: I, compensating system, II, stabilizing system, (*1*) capacity bridge detector and electrometer arrangement, (*2*) tungsten torsion wire suspension, serve at the same time as leads (100 μA maximum current), (*3*) gold wire connection to beam electrode, (*4*) electrostatic shield, (*5*) plate suspension, (*6*) quartz frame, (*7*) narrow quartz tube with moveable quartz fiber, (*8*) bimetal pushers, (*9*), (*10*), (*11*), (*12*) leads to pan, (*13*) attachment for gold wire lead.

at high temperatures. Its performance is shown in Figure 8.9, where the compensation voltage is plotted against the mass, ΔM, and the quartz fiber shift, Δx. The mean square deviation of this calibration curve is 0.2 μg. The sensitivity of the balance was 0.25 V/μg.

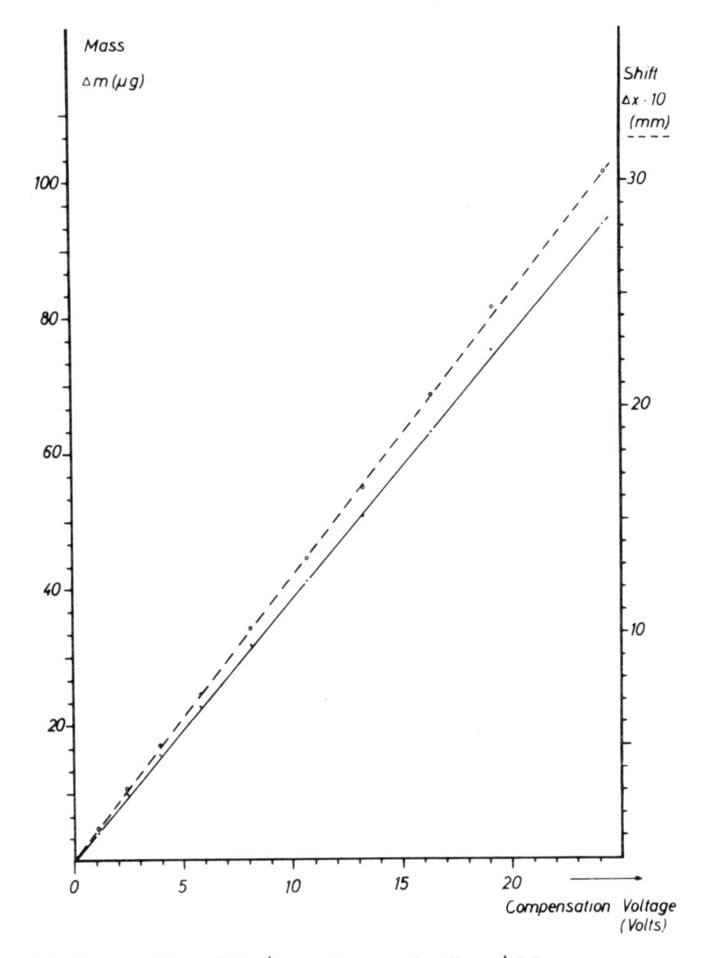

Figure 8.9. Quartz fiber shift Δx and mass loading ΔM, versus compensation voltage.

D. Rotors with Vertical Suspension

Since it is often not possible in thin film studies to connect a substrate directly to a conventional microbalance, measurement of the vapor stream density of the evaporant is more convenient than direct mass thickness determination. From the time integral of the vapor stream density the mass thickness can be concluded if the condensation coefficient is known.

The vapor stream density can be determined from the mass changes on the pan of any conventional microbalance. On the other hand, it is possible to determine the vapor stream density directly by using a sensitive microbalance measuring the momentum transferred from a vapor stream. This principle has been developed to a high degree by Campbell and Blackburn (8) and Neugebauer (9).

The force K which is exerted by a vapor stream of N atoms/cm^2-sec on an area of 1 cm^2 is given by:

$$K = N(8kTM/\pi)^{1/2} \tag{8.10}$$

This formula assumes that all atoms condense, i.e., that the condensation coefficient equals 1. The formula is modified by a constant factor if a vapor beam is formed inside the evaporation source. The force, K, which is directly proportional to the vapor stream density N, is also explicitly proportional to the square root of the temperature, T, and the mass, M of the evaporated atoms. The relationship between N and T is

$$N = N_0 \exp(-\lambda/kT) \tag{8.11}$$

where N_0 is weakly dependent on source temperature and λ is the heat of evaporation.

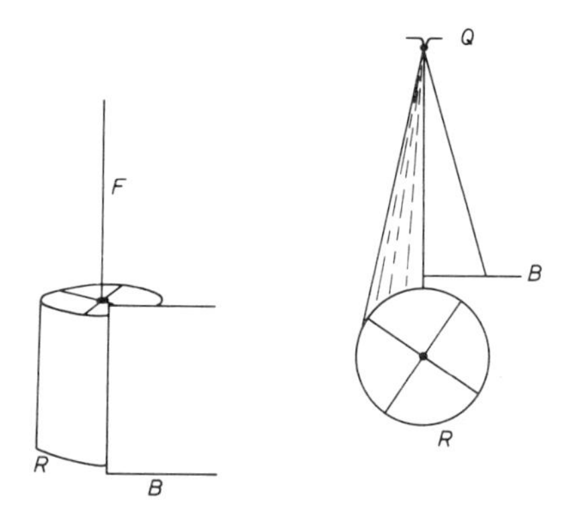

Figure 8.10. Principle of a suspended cylindrical rotor. F, tungsten torsion wire; R rotor; B mask; Q atom beam oven.

As the dependence of the vapor stream density N on the temperature T is much stronger than the dependence of the mean momentum $(8kTM/\pi)^{1/2}$ on T, this momentum may be considered constant for a given material. For example, for gold N increases by a factor of 100 between 1700 and 2000°K while the momentum changes by a factor of 1.08. Therefore the mean momentum $(8kTM/\pi)^{1/2}$ may be incorporated into the calibration factor for a given material.

Figure 8.10 demonstrates the principle of the method. A cylindrical rotor is suspended on a thin tungsten wire. One-half of the cylinder is shielded by a mask. The other half is struck by the atoms of the vapor stream. The resulting force exerts a torque on the suspension. Because of the symmetry of the arrangement, this torque does not change if some of the impinging atoms reevaporate symmetrically with respect to the surface normal. Since Hurlbut (10) has shown that the reflection of thermal molecules from metal surfaces follows approximately a cosine law, the above mentioned symmetry condition is fulfilled (Figure 8.11).

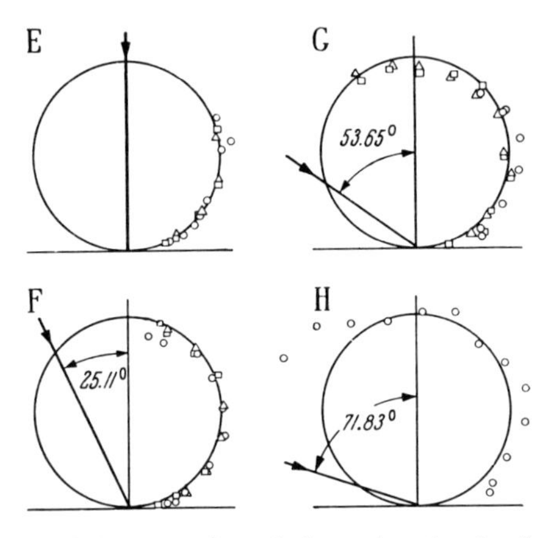

Figure 8.11. Polarplot of the scattering of thermal molecules from aluminum. The arrows indicate the incident angle.

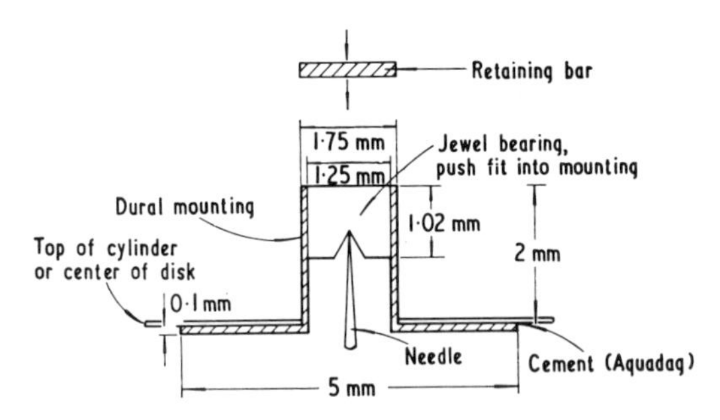

Figure 8.12. Rotor with pivot bearing.

The following apparatus designs have been used for these measurements:

1. Cylindrical rotor with torsion suspension (Figure 8.10)
2. Disk rotor with torsion suspension
3. Cylindrical rotor with pivot bearing (Figure 8.12)
4. Disk rotor

All rotors are eddy current damped (Figure 8.13).

The formulas for the sensitivity of the various rotor types are taken from the detailed theory developed by Beavitt (11). For torsion wire suspended cylinders the vapor stream density N is given by:

$$N \sim M/Ft = 2C\theta/\bar{c}r^2h \qquad (8.12)$$

where M is the total mass of impinging atoms, F is the area, t is the time, C is the torque constant of the torsion wire, θ is the deflection angle of the cylinder, r is the radius of the cylinder, h is the height of the cylinder, and

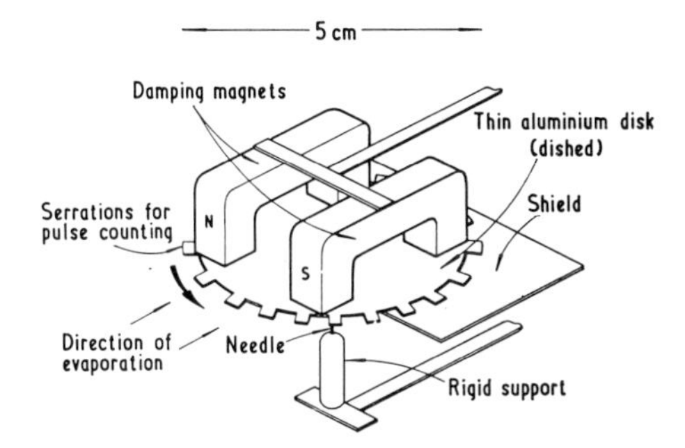

Figure 8.13. Eddy current damping of rotors.

$\bar{c}$ is the mean velocity of the impinging atoms. This equation is modified for the disk rotor to:

$$N \sim M/Ft = 3C\theta/[\bar{c}r^3 \sin(2\varphi)] \qquad (8.13)$$

where φ is the angle between the normal of the disk and the impinging vapor stream. The rotors with pivot bearings are damped by the eddy current method to such a degree that the dynamic friction of the pivot is very much smaller than the electromagnetic damping (Figure 8.13). Since there is no elastic torque in these rotors, they turn as long as atoms are impinging on them. The angular velocity is proportional to the vapor stream density and inversely proportional to the damping constant; the total mass impinging on the rotor is given by the angle of rotation. Beavitt derived the following formulas. For a cylindrical rotor:

$$M/F = 2\pi\omega t 2r\sigma/\bar{c}\tau \qquad (8.14)$$

where M/F is the total mass of impinging atoms per unit area, $2\pi\omega t$ is the angle of rotation during the time t, σ is the mass of the side walls per unit area, and τ is the time constant of damping.

For a disk rotor:

$$M/F = 2\pi\omega t 3r\sigma/[4\bar{c}\tau \sin(2\varphi)] \qquad (8.15)$$

The pivot bearings have a static friction that is appreciably greater than the dynamic friction. A minimum vapor stream density of about 1 Å/sec is required to overcome the static friction. To prove the linearity of thickness measurement with pivotal rotors, Beavitt (11) evaporated weighed amounts of aluminum from a tungsten wire. Figure 8.14 shows the rotation angle of a cylindrical rotor (h = 2.5 cm, r = 1.6 cm, $\tau \ll 1$ sec) as a function of the total mass of evaporated aluminum. The maximum sensitivity of pivotal

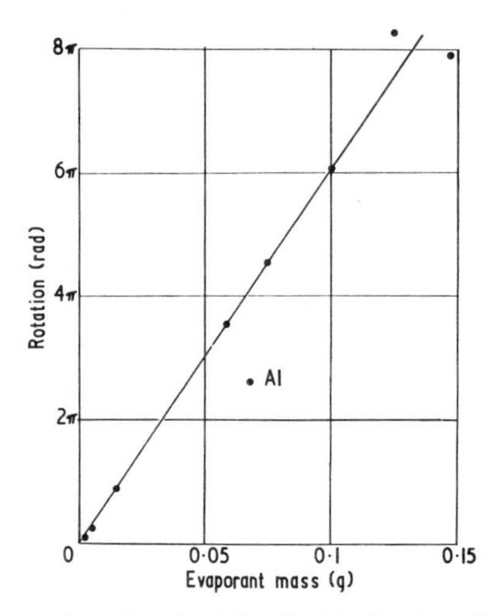

Figure 8.14. Rotation angles of a pivotal cylindrical rotor plotted against mass of evaporated aluminum (source distance 10 cm).

rotors is about 10 Å of Al per revolution and for suspension ratemeters 0.042 Å of Al per minute per degree. The time constant for this latter rotor was about 30 sec.

The type of microbalance described here has recently been of considerable interest. It has high mechanical strength, simple construction, and high sensitivity. The choice of a proper material for magnetic damping could be a problem for UHV use because the Curie temperature must be higher than the bakeout temperature and the outgassing of the magnetic material must be low. Further improvements are required before these rotors can be used for automatic evaporation rate control.

E. Quartz Oscillator Microbalances

A very elegant method for the determination of mass thickness consists of the measurement of the frequency shift of a quartz oscillator loaded homogeneously with additional mass. A detailed discussion of the quartz crystal microbalance is given in Chapter 5. It has been shown that a linear dependence exists between the frequency shift Δf and the mass loading $\Delta m/F$ (F is that part of the area of the quartz crystal on to which mass is deposited) as shown in equation 8.16:

$$\Delta f = -C_f(\Delta m/F) \tag{8.16}$$

where C_f, the proportionality constant, is given by:

$$C_f = f^2/K\rho_Q = f/\rho_Q t_Q \tag{8.17}$$

Here f is the resonance frequency, ρ_Q the density, K the frequency constant, and t_Q the thickness of an AT cut quartz crystal.

For an exact determination of mass thickness it is necessary to take into account the temperature dependence of the resonant frequency. This can be accomplished either by maintaining the quartz at a constant temperature or preferably by measuring the temperature of the quartz oscillator during deposition. The effect of temperature on the determination of mass thickness and the sensitivity and linearity of this method have been investigated by various workers (12–15) and are discussed in detail by Warner in Chapter 5.

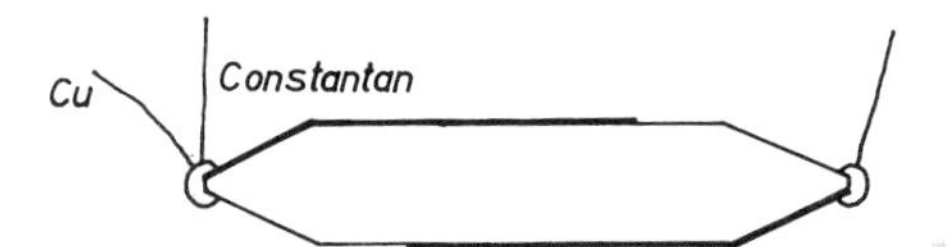

Figure 8.15. Attachment of thermocouple to quartz oscillator with "Einbrennsilber."

The authors have determined the sensitivity of a quartz resonator and its dependence on temperature for small film thicknesses and low evaporation rates. This was accomplished by comparison of the mass changes recorded during simultaneous deposition onto a quartz crystal resonator and a silver substrate suspended from a torsion balance (Figure 8.5). The quartz oscillator had been plated with silver electrodes so that the sticking coefficient could be assumed equal and constant on both substrates. Temperature measurements were made with a thermocouple fixed to a facet of the quartz oscillator with "Einbrennsilber" (Figure 8.15). The experimental arrangement is shown in Figure 8.16. First, the dependence of frequency on temperature was determined by *in situ* radiation heating of the quartz (Figure 8.17). A film was then evaporated onto the quartz resonator and the substrate on the microbalance. Both substrates were at approximately room temperature. Using the geometrical factor between the positions of the quartz resonator and the microbalance pan, two time dependent mass thicknesses were obtained for the resonator position, that is one that has been measured directly by the frequency shift of the quartz and the other that has been calculated from the microbalance measurements. Figure 8.18 shows that these two curves did not agree. In the same figure the temperature variation of the quartz oscillator during the experiment is also plotted.

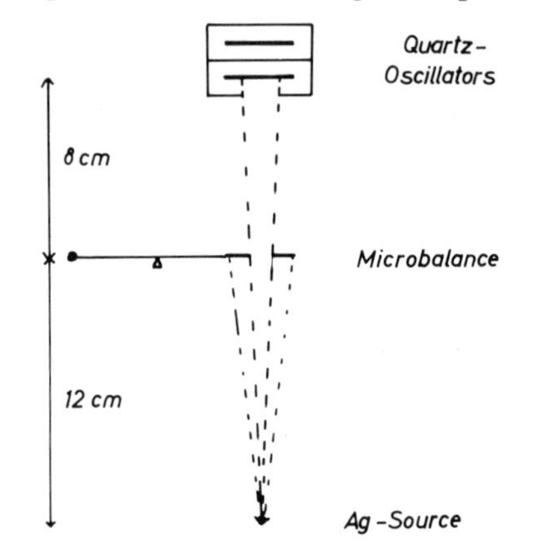

Figure 8.16. Schematic view of experiment arrangement for the comparison of quartz oscillator and microbalance.

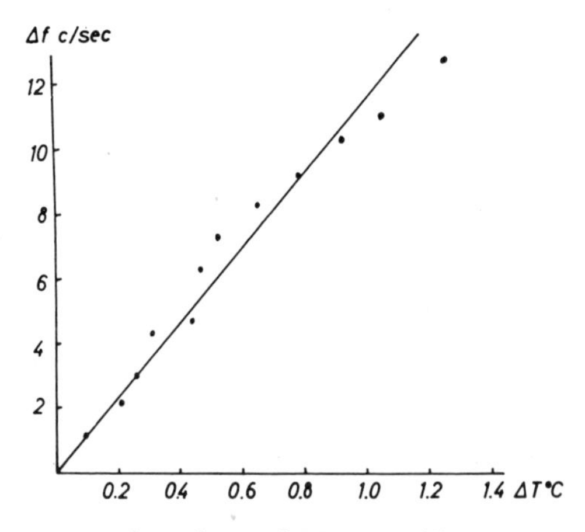

Figure 8.17. Temperature dependence of 5.5 Mc oscillator near room temperature

It is seen that within 10 min there was a temperature increase of about 0.7°C. If the temperature dependence, as shown in Figure 8.17, is used to correct the resonator data for the measured temperature shift, the microbalance and quartz oscillator measurements agree exactly (Figure 8.19). The ratio of the corrected frequency variation to mass increment gave an experimental mass thickness sensitivity, $C_{f,\,\mathrm{ex}}$ of 67.85 ± 1% cps-cm^2/μg

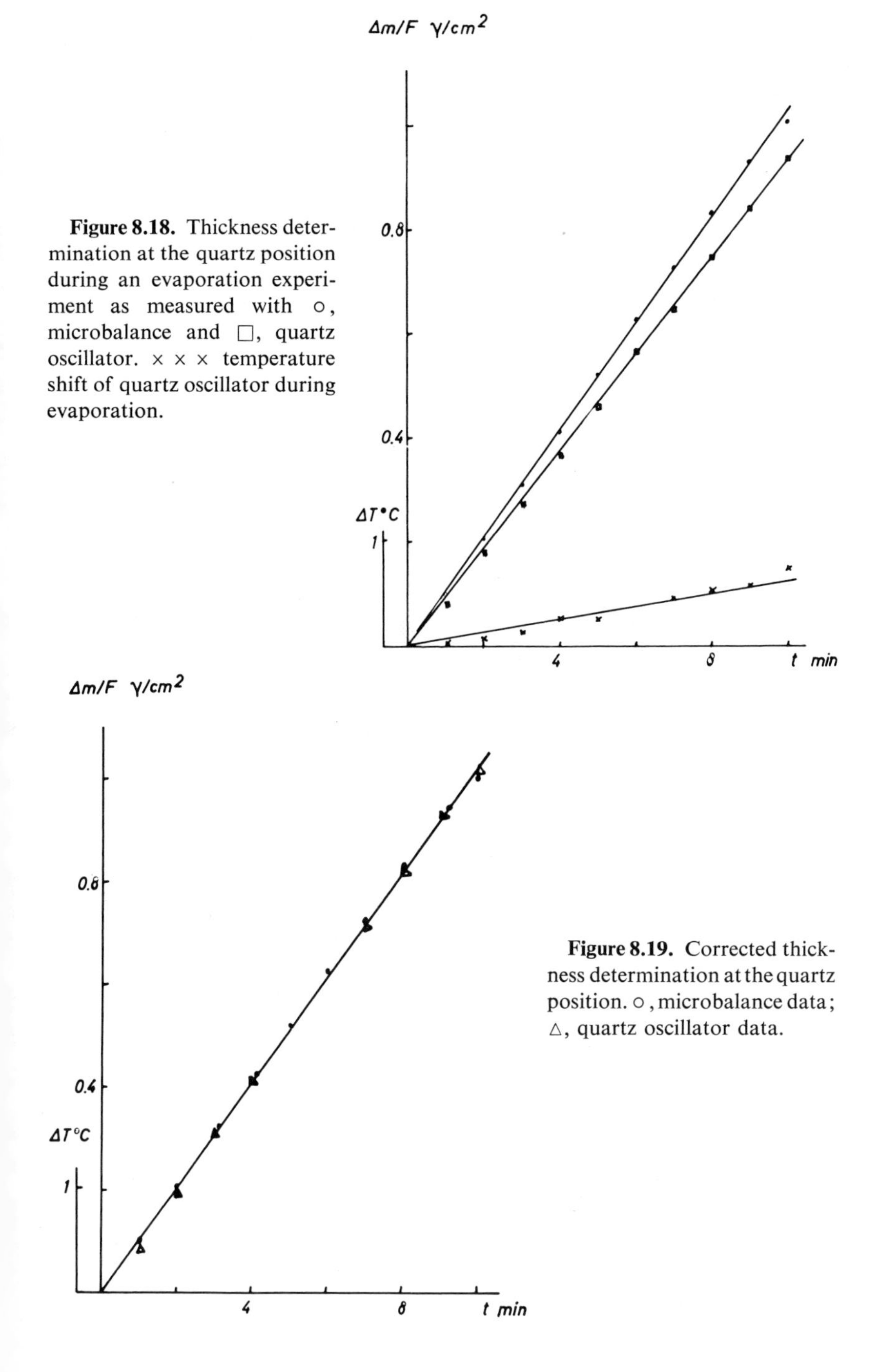

Figure 8.18. Thickness determination at the quartz position during an evaporation experiment as measured with ○, microbalance and □, quartz oscillator. × × × temperature shift of quartz oscillator during evaporation.

Figure 8.19. Corrected thickness determination at the quartz position. ○, microbalance data; △, quartz oscillator data.

which agreed excellently with the calculated sensitivity $C_{f,\,\mathrm{th}}$, of 67.96 cps-cm²/μg for a 5.5 Mc oscillator.

Sauerbrey (13) made a comparison of the quartz oscillator and the microbalance for film thicknesses greater than 10^{-5} g/cm². A similar method was used recently for film thicknesses between 100 and 1000 μg/cm² by Eschbach and Kruidhof (15). In the latter method the quartz oscillator also served simultaneously as the plate of a microbalance (Figure 8.20). Their

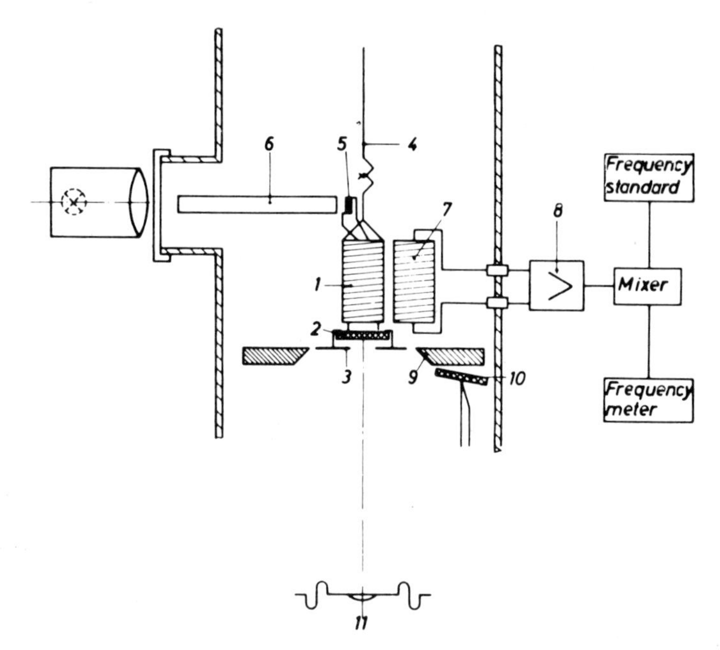

Figure 8.20. Pan of microbalance constructed as a quartz oscillator; (*1*) quartz crystal, (*2*) transmitting coil, (*3*) aluminum mask, (*4*) suspension wire, (*5*) solar cell, (*6*) light guide, (*7*) pickup coil, (*8*) amplifier, (*9*) shield, (*10*) second quartz with thermocouple, (*11*) resistance-heated source.

calibration curve (Figure 8.21), shows the excellent linearity of a 5 Mc quartz oscillator to a weight of 1 mg/cm². The sensitivity of the quartz oscillator, $C_{f,\,\mathrm{ex}}$, was 57.8 cps-cm²/μg which agrees with the calculated value of $C_{f,\,\mathrm{th}}$, of 57.6 cps-cm²/μg. These results have been confirmed by Pulker (14), who checked the results of the quartz oscillator by means of a potentiometric microtitration of the deposit.

These comparisons show that the quartz crystal oscillator can be used for mass thickness determinations without detailed calibration if proper precautions are observed. The effect of temperature on frequency imposes the major limitation of the method. The problem of temperature dependence over a broad temperature range has been investigated by Hillecke (16). Figure 8.22 shows the frequency shift and Table 8.1 the change of the temperature coefficient with temperature for three AT cut 5.5 Mc quartz crystals (AT cut having a zero temperature coefficient at room temperature) in the temperature range between 2 and 360°K. The curve has the general form of a third order parabola. Since the temperature coefficients of each quartz crystal differed slightly in the room temperature range, they were

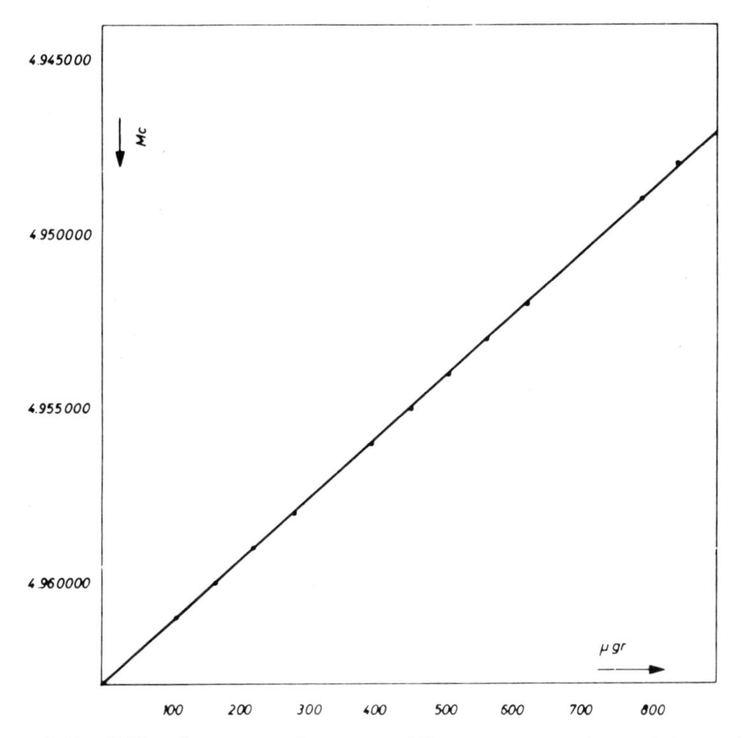

Figure 8.21. Calibration curve: frequency shift versus mass for a 5 Mc quartz.

determined separately for each quartz crystal. At low temperatures, e.g., -100°C, the temperature coefficient appeared to have a uniform value of about $+4.85 \times 10^{-6}$ 1/°C, that is 27 cps/°C. To attain an accuracy of about $\frac{1}{2}$ a monolayer, the quartz resonator temperature must be determined to 0.1°C. Bechmann (17) has studied the influence of the crystal orientation angle on the temperature dependence of the frequency. (Also see Chapter 5 by Warner.) He showed that AT crystals can be cut so that they have a zero temperature coefficient region in the temperature range of 20–500°K. If one defines a change of less than 0.5×10^{-6} 1/°C as a zero temperature coefficient, the width of the zero temperature coefficient plateau for various cut angles can be determined from Bechmann's work. At a temperature of $-20°C$ a 35.5° cut crystal has a temperature plateau of $\pm 25°C$; at $-120°C$ a 36.8° cut crystal has a range of $\pm 10°C$; and finally at $-230°C$ a 41.0° cut crystal shows a plateau of $\pm 5°C$. It is not practical,

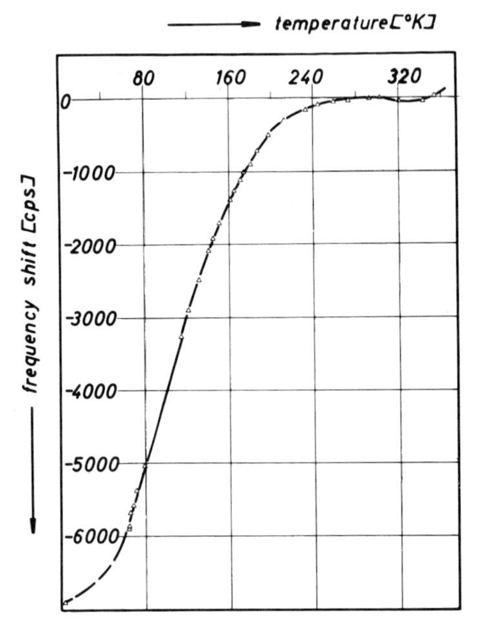

Figure 8.22. Temperature dependence of resonance frequency in the range from 2 to 360°K for a 5.5 Mc AT quartz oscillator.

however, to use a separate quartz crystal for every temperature range. Additionally, as has been shown (12) even with a temperature coefficient of 2×10^{-6} 1/°C, the temperature of the quartz crystal should be monitored for best results.

Thermal conduction obtained by simply pressing the quartz crystal to a temperature stabilized substrate is not very reproducible. In the worst case there is not more than radiation equilibrium.

III. APPLICATIONS

A. Determination of the Density of Thin Films

It has been reported that the density of thin films is smaller than the density of bulk material (see, for instance, Kiessig (18)). More recent measurements have been made by Blois and Rieser (19), who compared the mass thickness as determined by a microbalance with the geometric thickness measured by Tolansky multiple beam interferometry (20). Their silver and copper films of 200 Å thickness had a density that was 30% lower than that of the compact material.

The following is some recent work concerning this problem (21–23). All these authors simultaneously covered a quartz oscillator and a microscope slide with the material investigated. All experiments were done at about 10^{-6} torr. The mass thicknesses were measured with the quartz oscillator and the geometric thickness by the Tolansky method. Hartman (22) additionally polished the surface of his quartz oscillator and determined the geometrical thickness on this substrate directly. For aluminum, gold,

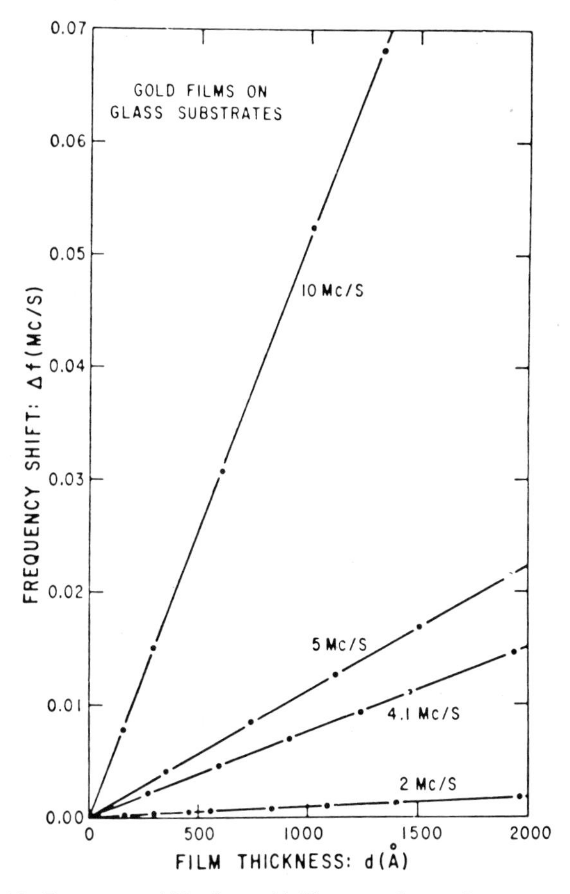

Figure 8.23. Frequency shifts for gold films on glass substrates as a function of interferometrically determined film thicknesses. The various curves were taken with four different quartz crystals.

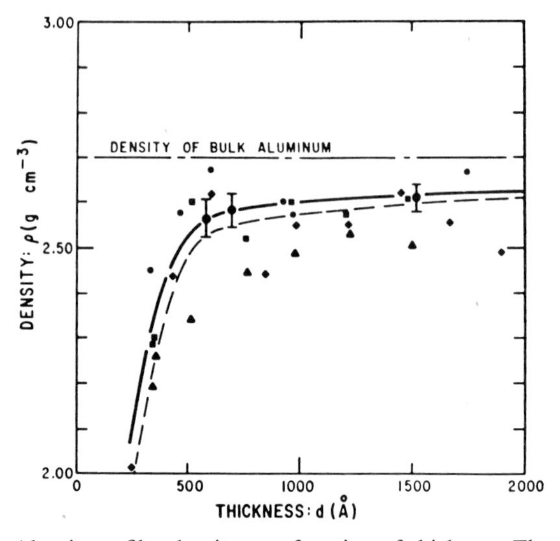

Figure 8.24. Aluminum film density as a function of thickness. The various small dots represent measurements with different quartz crystals; the large dots with arrow flags originate from interferometric thickness measurements taken directly on polished crystal substrates. The dashed line represents the best fits to all points, the solid line is corrected for oxide formation.

silver, and copper, Wolter (21) has found that the density was independent of film thickness and agreed with the density of the bulk material for film thicknesses between 200 and 5000 Å. The density of chromium was similarly independent of thickness. However, the density differed from the bulk density and was dependent on the oxygen partial pressure. The results for gold and silver were confirmed by Hartman (22) (Figure 8.23) who found in contrast to Wolter, in the case of aluminum, a strong dependence of the density on thickness in the range between 200 and 500 Å and a smaller, but reproducible, thickness dependence up to 2000 Å. The deviation from the bulk density for the thicker films was about 4% (Figure 8.24). Wolter's results have been confirmed by Edgecumbe (23) within 5% for aluminum, gold, KCl, and CsI.

These studies show that films thicker than 500 Å have densities within 5% of the bulk value. Variations greater than 5% could be explained by systematic errors in the interferometric thickness determinations. A film with a rough surface will produce a phase shift similar to that resulting from a thickness greater than the real geometric thickness (24). Different film roughnesses resulting from the various film preparation conditions could very well account for the differences in the film density determined by the above authors. For example, the dependence of density on thickness found by Hartman could be explained by a roughness of the silver coating of 150 Å.

B. Investigation of Condensation Phenomena with the Microbalance and with the Combination of Leed and the Quartz Oscillator

Although the microbalance is one of the simplest tools for investigating condensation coefficients in ultrahigh vacuum, it has generally been often ignored in favor of other experimental techniques. The percentage of impinging atoms that stick on a substrate, that is the condensation coefficient,

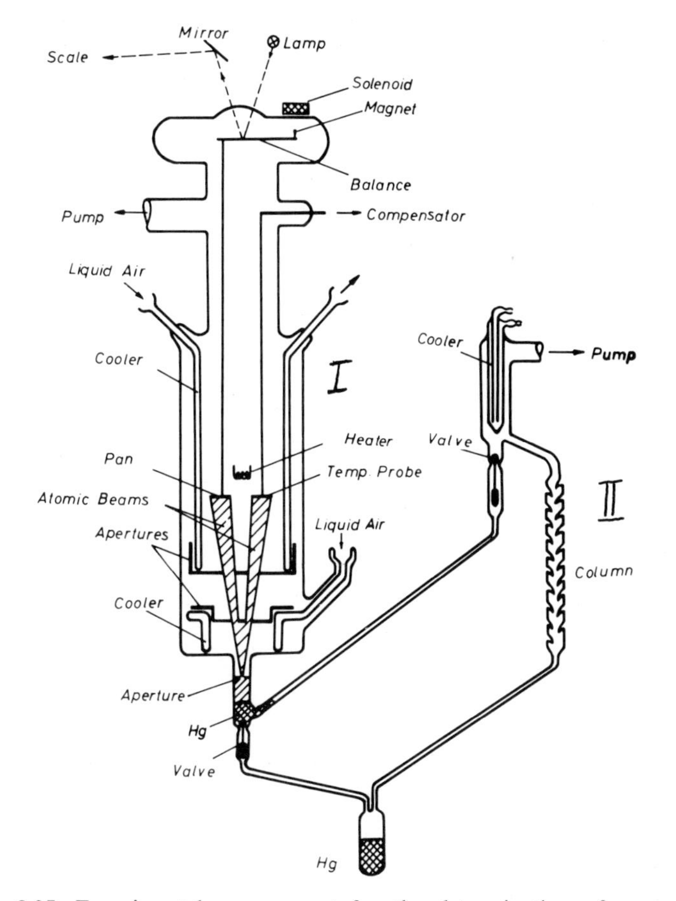

Figure 8.25. Experimental arrangement for the determination of mercury condensation coefficients: I condensation cell, II mercury distillation apparatus.

is of great importance in understanding the initial growth processes of thin films. Göhre (3,5) has used a microbalance (Figure 8.2) to study the condensation of mercury on quartz substrates at very low pressures. The experimental arrangement is shown in Figure 8.25. The pan of the balance was a platelet of fused quartz (area = 0.502 cm²). The pan was cooled with liquid air and the temperature was measured with a probe hanging symmetrically with the pan. The lowest temperature attained was −133°C. Elevated temperatures up to 50°C could be obtained with a heating coil to an accuracy of 0.1°C, while the walls were at liquid air temperature. A well-defined mercury beam, whose intensity was determined by the geometry of the arrangement and the temperature of the furnace, was formed by a system of apertures. The deposited mass was measured as a function of the evaporation time. The excellent reproducibility of the microbalance data shown in Figure 8.3 allowed the continuous plotting of the increased mass as a function of time and as a function of the total amount of condensed mercury over a period of several hours.

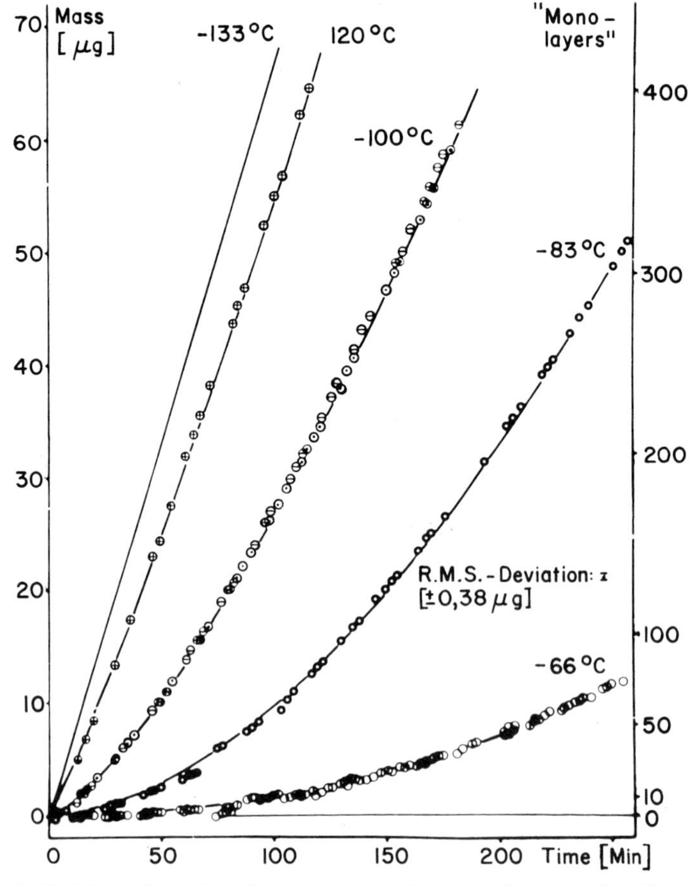

Figure 8.26. Mass of condensed mercury as a function of evaporation time; parameter is the substrate temperature.

These experiments established that the mass increase did not depend linearly on the evaporation time as would be expected if the condensation coefficient was constant. The deviation from linearity was a function of the temperature of the substrate. In Figure 8.26 the weight of the pan is plotted as a function of deposition time for various substrate temperatures. The evaporation rate was 1.33 μg/cm²-min. The condensation coefficient

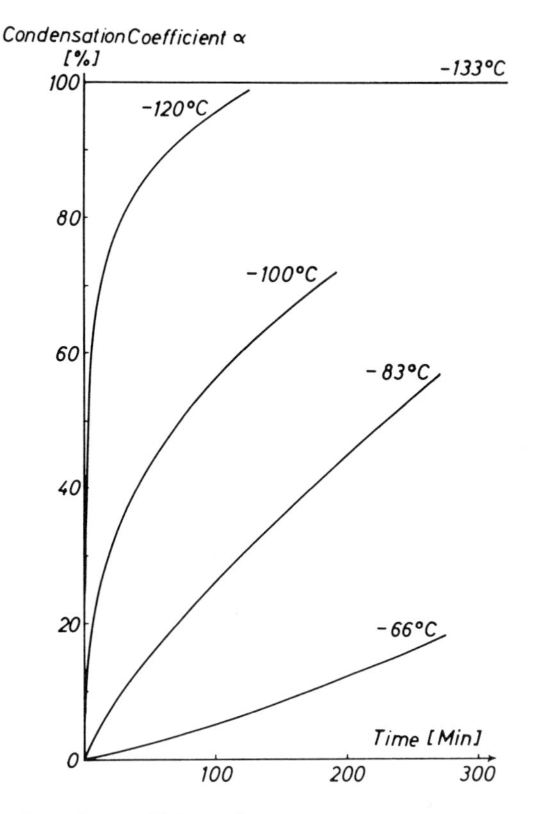

Figure 8.27. Condensation coefficient of mercury on quartz versus evaporation time for various substrate temperatures. Evaporation rate was 1.33 μg/cm² min.

had a constant value assumed to be 1, only at −133°C. At elevated temperatures the condensation coefficient increased slowly from 0 to 1 as a function of the deposition time as seen in Figure 8.27. Apparently the condensation coefficient is also dependent on the deposition rate. From this study it can be concluded that the condensation coefficient on clean surfaces depends on supersaturation in the usual manner and that, in addition, the supersaturation itself varies with the degree of surface coverage.

The influence of precoverage on the condensation coefficient has been clearly demonstrated in the above work for the first time with a microbalance. Similar observations have been made, using other experimental techniques, by Palatnik and Komnik (25) for bismuth, lead, tin, and antimony on glass and by Campbell and Blackburn (8) for zinc sulfide on lithium fluoride.

A more detailed understanding of condensation phenomenon was gained from the work of Spiegel et al. (26–28) who evaporated silver onto silicon

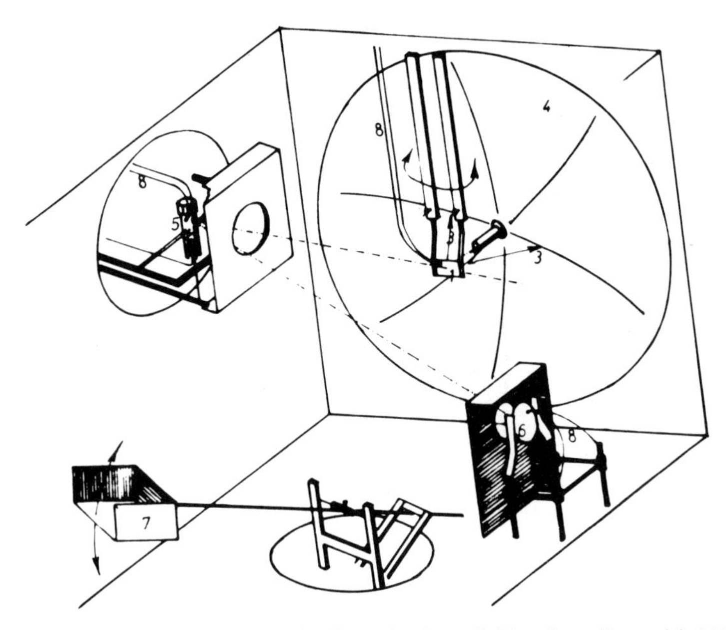

Figure 8.28. Experimental setup for investigation of thin silver films with LEED: (*1*) silicon substrate, (*2*) primary electrons, (*3*) diffracted electrons, (*4*) screen, (*5*) evaporation crucible, (*6*) quartz oscillator, (*7*) shutter, (*8*) thermocouples.

inside a low energy electron diffraction (Leed) apparatus. The experiments were carried out in a vacuum of 10^{-10} torr. The interior of their Leed diffraction chamber is shown in Figure 8.28. The substrate temperature was measured with a thermocouple and the vapor stream density with a quartz oscillator. The structures of the condensed phase were interpreted from the low energy electron diffraction patterns.

Two stages could be distinguished during the film growth:

1. A two-dimensional adsorption layer where the separation of the silver atoms was completely determined by the substrate lattice.

2. The growth of three-dimensional silver nuclei which had the regular silver lattice.

Correlation of the electron diffraction patterns with measurements of the vapor density with a quartz crystal oscillator proved that the adsorption layer, up to a maximum coverage of $\frac{1}{3}$ of a monolayer, was deposited with a condensation coefficient of 1 in the temperature range of 200–400°C. Further condensation occurred at a reduced rate in a manner similar to that observed by Göhre for mercury condensation. Figure 8.29 is a plot of

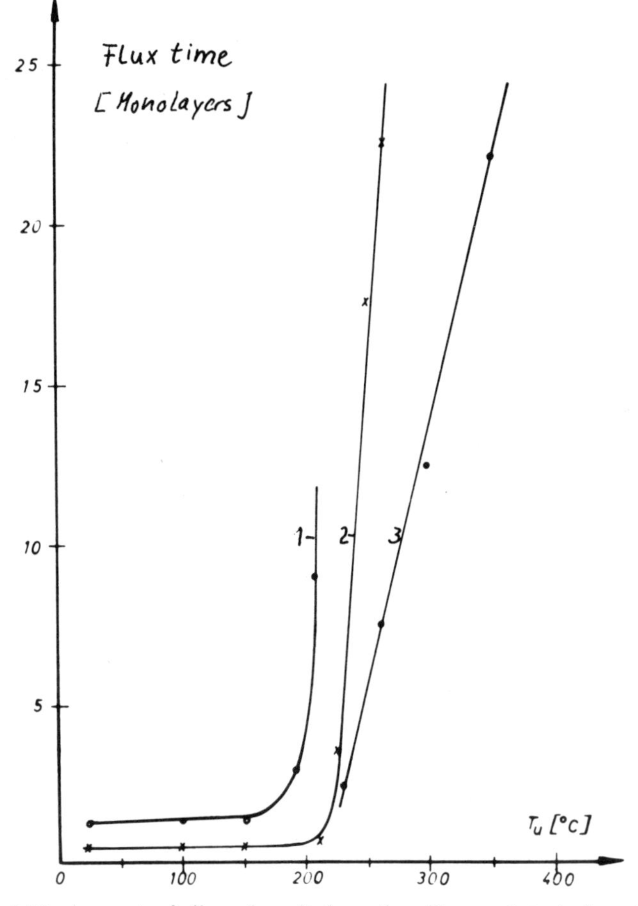

Figure 8.29. Amount of silver deposited on the silicon substrate in order to get three-dimensional silver nuclei as a function of substrate temperature: parameter is the evaporation rate R. (*1*) $R = 0.25$ monolayers per min, (*2*) $R = 1{,}5$ monolayers per min, (*3*) $R = 15$ monolayers per min.

the number of monolayers as a function of substrate temperature that must impinge on the substrate in order to obtain a diffraction pattern of three-dimensional silver nuclei. The three curves correspond to three different vapor stream densities. It is seen that the condensation rate decreased with lower evaporation rates and higher substrate temperatures.

Schmider (29) of this laboratory, is extending the investigations of Göhre by simultaneously measuring the substrate temperature and the resistance of the condensed film on the microbalance plate. Since four additional electrical leads are required, some modifications of the microbalance were made as described in Section I.A. of this chapter.

C. The Thickness Dependence of the Melting Point of Thin Films

Flux measurements with a quartz oscillator combined with high energy electron diffraction provided a means of investigating the relationships between the thickness and melting point of thin films (30,31).

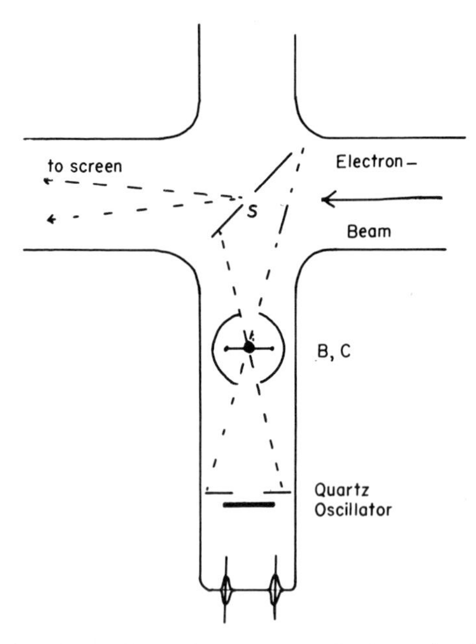

Figure 8.30. Top view of experimental arrangement for determination of melting point of thin films. *S* substrate, *B* evaporation source, *C* rotatable shutter.

For these experiments the apparatus, shown in Figure 8.30 was used. Silver was evaporated from a horizontal tungsten wire onto a thin carbon film on one side and investigating a quartz oscillator on the opposite side. The substrate, *S*, could be heated to temperatures of 900°C with the quartz oscillator remaining at room temperature. With a symmetrical arrangement of the quartz crystal and the substrate the oscillator accurately determined the total flux striking the substrate. Since the condensation coefficient was unknown, the amount of material deposited could not be determined. The structure of the silver film was examined *in situ* by transmission electron diffraction. In this study the vacuum was better than 5×10^{-9} torr.

Two different kinds of experiments have been performed. First the film structure during the growth was investigated. A certain amount of silver was evaporated at a fixed substrate temperature, the shutter C was closed, the flux was determined from the quartz oscillator and an ED picture taken within about 30 sec. Consecutively an additional amount was evaporated and a new picture was taken in the same way. This procedure was

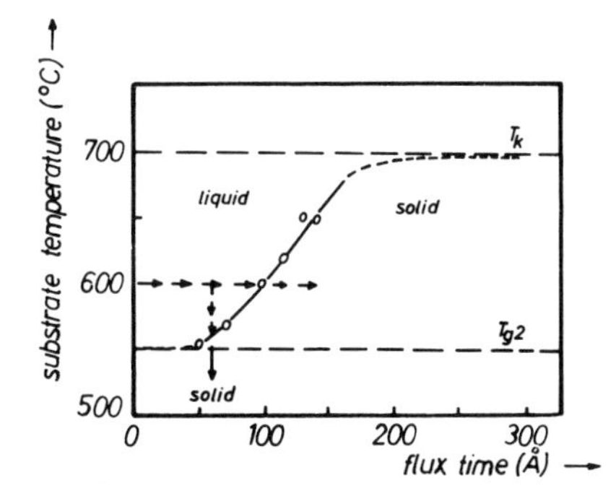

Figure 8.31. Change of the state of matter during deposition of Ag on carbon films.

repeated until about 200 Å silver had been deposited on the quartz crystal. If the substrate temperature was above 550°C the ED patterns for very thin films were those of the liquid phase. A sharp transition from liquid to solid could be observed when a certain temperature dependent flux was exceeded. Figure 8.31 shows the relationship between this transition thickness and substrate temperature. To establish the consistency of the interpretation of the diffuse patterns evaporations were made at 600°C to a flux time of 60 Å where the evaporation was interrupted. The patterns were diffuse at 600°C; however, after cooling to 500°C sharp rings of solid silver were obtained. This experiment is indicated by the vertical arrows in Figure 8.31.

The above observed phenomena could be interpreted in two different ways. First, the liquid phase could be a transitional state as described by the Ostwald step rule. Second, the phase equilibrium could depend on film thickness. To decide which of these interpretations is correct, another experiment was performed. Films of known thickness were formed by evaporation at a substrate temperature of 200°C. Subsequently these films were heated within 2 min to their melting point. During this temperature increase the film structure was observed with the ED so that the transition temperature from the sharp rings of polycrystalline material to the diffuse liquid pattern could be determined. The dependence of the transition temperature on film thickness for silver and copper is plotted in Figure 8.32.

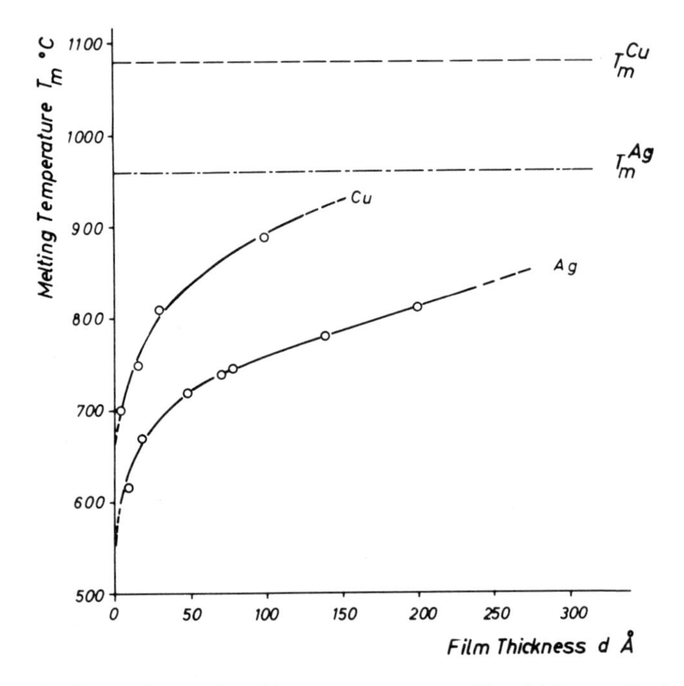

Figure 8.32. Dependence of melting temperature on film thickness: T_mAg melting temperature of bulk silver, T_mCu melting temperature of bulk copper.

The melting point for very thin films was ⅔ of that of the bulk material. Although similar observations have been made qualitatively by other authors (32), the quantitative measurement of the thickness dependence could only be established with use of the quartz oscillator.

D. Magnetization of Thin Films

Measurement of the magnetization of thin films is an excellent example of the utilization of the vacuum microbalance. Sometime ago several theories existed on the magnetization of thin films (see review by Stünkel

(6)). Insufficient experimental data, however, were available to determine which theory provided the most correct description of the phenomena.

To this end Neugebauer (33) measured the magnetization of thin nickel films with a vacuum torsion magnetometer and Stünkel (6) has investigated iron films with the microbalance shown in Figure 8.4. Residual pressures in both experiments were lower than 10^{-9} torr.

The design of the torsion magnetometer used by Neugebauer is shown in Figure 8.33. The substrate was suspended by a thin tungsten wire attached

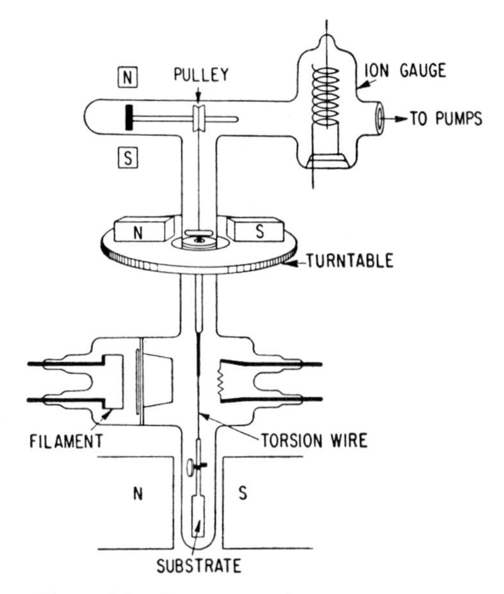

Figure 8.33. Vacuum torsion magnetometer.

to a magnetically rotatable turntable. With a magnetic field acting on the film the turntable was rotated until the film normal made an angle of 45° with the external magnetic field. The measurements were evaluated using a theory based on the following expression for the energy E of a plain magnetic film in an external magnetic field:

$$E = -M_s H \cos \theta + (\tfrac{1}{2}) N M_s^2 \sin^2 (\theta^\circ - \theta) + k \sin^2 (\theta^\circ - \theta) \tag{8.18}$$

In this expression M_s is the saturation magnetization, H is the external field strength, $N = 4\pi$ is the demagnetization factor, k is the anisotropy constant, θ is the angle between H and M_s, and θ° is the angle between H and the film. An expression for the torque can be derived from this formula.

The saturation magnetization and the anisotropy constant can be determined separately. The film thickness was measured by x-ray fluorescence analysis.

The main result derived from this investigation was that clean nickel films had the same fundamental magnetic properties as those of the bulk material. This is demonstrated clearly in Figure 8.34 where the relative magnetization $M_s(T)/M_s(0)$ is plotted as a function of temperature for bulk material (solid line) and for a 27 Å film (circles).

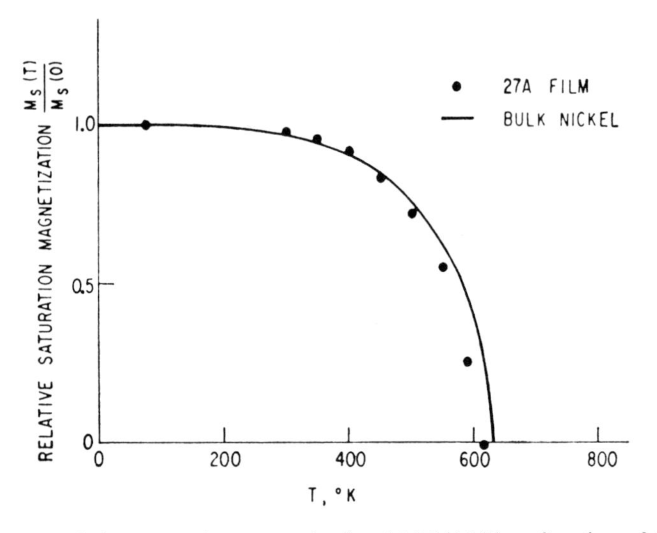

Figure 8.34. Relative saturation magnetization $M_s(T)/M_s(0)$ as function of temperature T: solid line represents bulk nickel, dots a 27 Å nickel film.

Stünkel (6) investigated both the mass thickness and the dipole moment of iron films using a microbalance. The experimental arrangement is shown in Figure 8.35. The microbalance was in the upper part of a Pyrex glass cell. The hanging quartz substrate was positioned so that iron could be evaporated onto it and it would be in an inhomogeneous magnetic field. It was possible to obtain substrate temperatures between 183°K (by liquid air cooling) and 600°K (by radiation heating). The gradient of the vertical magnetic field, $\text{grad}_z H$ was measured. Under these conditions, the force F_1, on a magnetic dipole moment, P, developed by an inhomogeneous magnetic field is given by:

$$F_1 = P \text{ grad}_z H \tag{8.19}$$

The magnetization, M, can be expressed by:

$$M = P/V = F_1/\text{grad}_z HV \tag{8.20}$$

The volume of the film V is related to the weight F_2 by the expression:

$$F_2 = V\rho g \tag{8.21}$$

The weight was also measured with the same microbalance. The magnetization, M, could be determined from equation 8.22 derived by combining equations 8.20 and 8.21:

$$M = (F_1 \rho g / F_2 \,\mathrm{grad}_z\, H) \tag{8.22}$$

Calibration errors of the microbalance cancel out in these determinations.

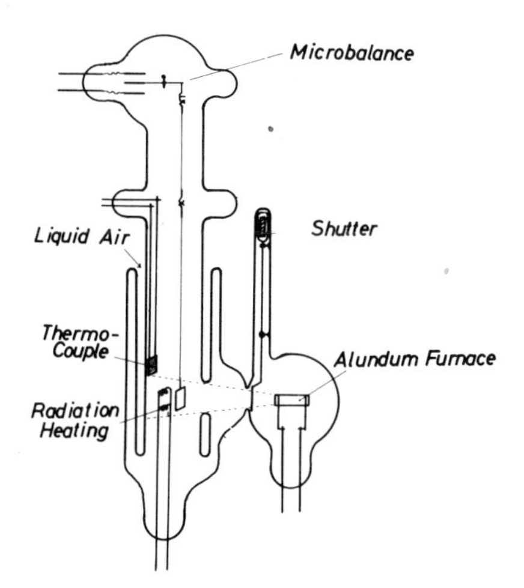

Figure 8.35. Experimental setup for magnetization measurements with a microbalance.

The magnetization data of Stünkel for different film thicknesses are plotted in Figure 8.36 as a function of temperature. Films thicker than 140 Å showed bulk behavior in the whole temperature range between 183 and 600°K. Thinner films showed an irreversible change during heating and consequently a deviation of the temperature dependence of saturation magnetization from bulk characteristics. Stünkel interpreted this deviation as indicating a roughening of the film structure during annealing. As in the work of Neugebauer, and contrary to the theory of Glass and Klein (34), there was no indication of a thickness dependence of magnetization for films thicker than 40 Å. The two different experiments are supplementary. Since Neugebauer assumed plane films in his evaluaton, he was able to determine the anisotropy constant and therefore the direction of magnetization vector within the film. Stünkel, on the other hand, did not assume plane films in his studies. His work was, however, limited to the measurement of that component of magnetization that is parallel to the field gradient.

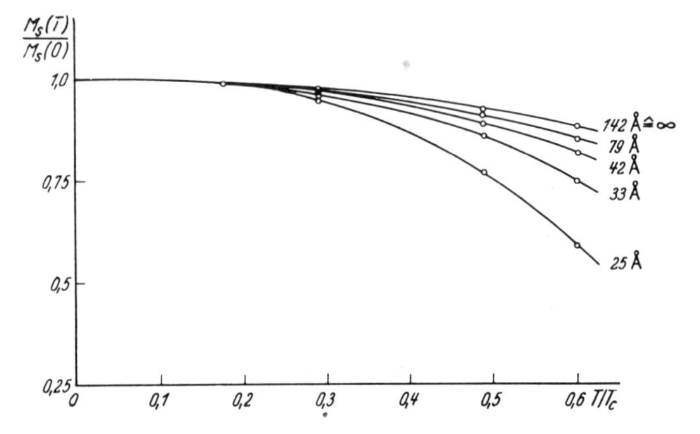

Figure 8.36. Relative saturation magnetization as function of the reduced temperature for different film thicknesses; T_c is the Curie temperature of bulk iron.

E. Technical Applications

The preparation of thin films for devices requires in most cases an exact control of film thickness and evaporation rate. This is particularly important for micro circuitry applications where capacitors, resistances, and inductances have to be evaporated within narrow tolerances. Rate control is also important in the preparation of alloy films.

There appears to be no published data on the application of microbalances and rotor ratemeters to thin film production. However, the Cahn Instrument Company advertises a microbalance for thin film monitoring (see their Bulletin 126, Techniques with the Cahn Electrobalances #9). The quartz oscillator is being used increasingly in all technical applications involving thin films.

Examples of the applications of the quartz oscillator are found in the work of Steckelmacher et al. (35), Bath (36), and Langer and Patton (37). Normally in the laboratory the frequency of the monitoring quartz oscillator is mixed with a fixed frequency of a second oscillator resulting in a beat frequency which is counted by a frequency meter. The above authors (37), however, mix the beat frequency a second time with a variable frequency of an audio oscillator and produce a voltage proportional to the second beat frequency. This method allows a direct measurement of the frequency change, and therefore the evaporation rate by the insertion of an electronic differentiating circuit. Thus, the same instrument can be used as a thickness monitor and a ratemeter. Feedback from the output voltage of the rate meter can be used to regulate the temperature of the crucible as demonstrated by Bath (Figure 8.37). The regulating circuit can be used either for the stabilization of evaporation rates or for evaporation with a

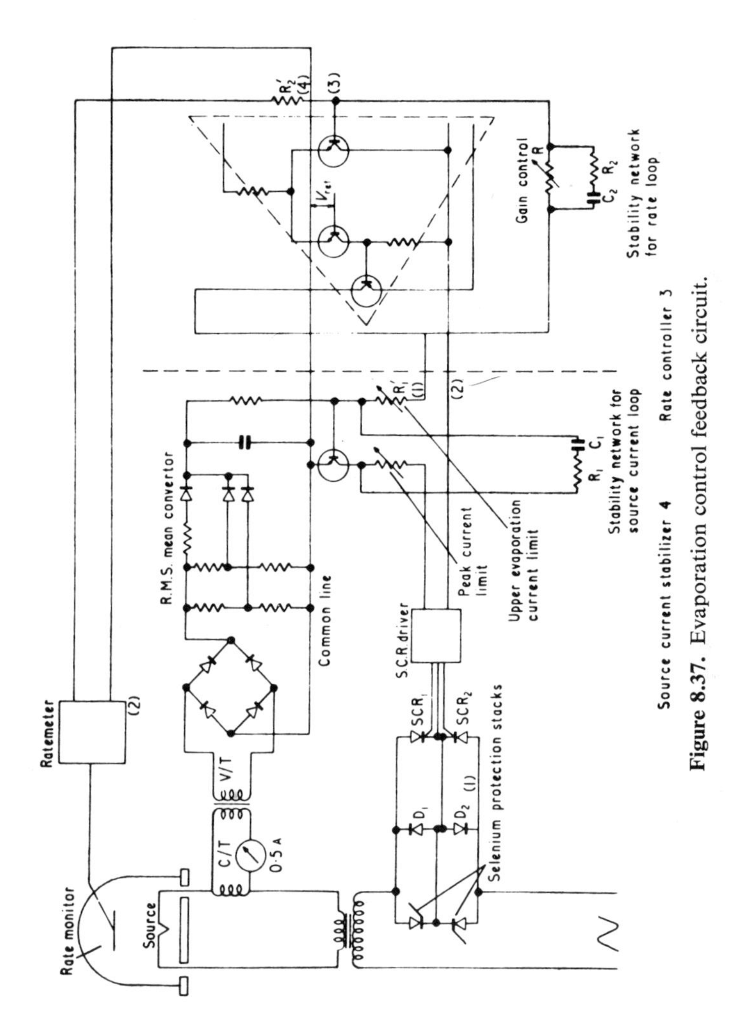

Figure 8.37. Evaporation control feedback circuit.

predetermined rate program. The importance of the stabilization of the evaporation rate in the case of the dielectric properties of isolating films is seen in Figure 8.38. The plotted points represent the averages of six different measurements for each evaporation rate.

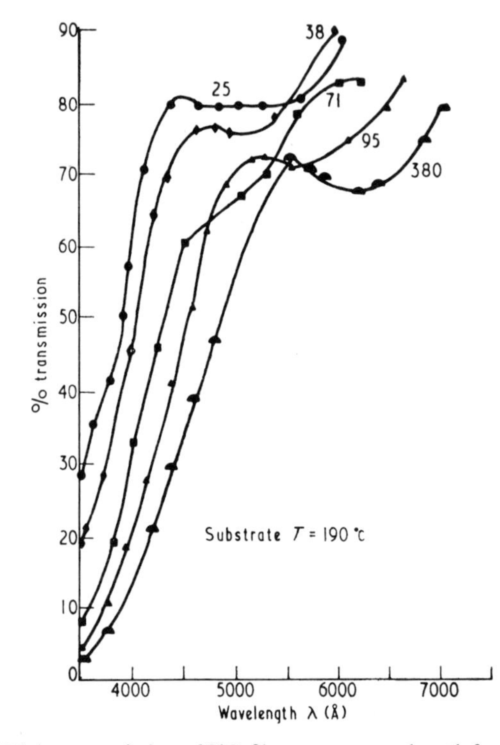

Figure 8.38. Light transmission of SiO-films versus wavelength for several evaporation rates. The numbers at the curves denote evaporation rates in ng cm^{-2} sec^{-1}.

The control of the evaporation rate is equally important in the fabrication of thin film resistors. The thin film resistors are often made from nickel–chromium, tantalum–nickel, chromium–titanium, or similar alloys. The composition of these materials can be controlled by regulating the evaporation rate of the vapor sources. Behrndt and Love (38–40) prepared a great number of permalloy films, the composition of which could be predetermined within 0.5% and the film thickness within 0.2%, using a control circuit similar to that of Bath. The mean thickness of these films was about 500 Å.

The thickness of thin films for optical applications, such as laser mirrors, must be known very accurately. It is not possible, however, to use the mass thickness without corrections because the interferometer thickness may differ from the mass thickness by the roughness factor mentioned earlier in this chapter. It is necessary, therefore, to determine the effective thickness of the multiple layers by optical interference methods. It is convenient however, in the preparation of these films to preset the evaporation rate with a quartz oscillator before starting the deposition. Procedures of this type have been performed by Behrndt and Doughty (41). These authors simultaneously used a quartz oscillator and the reflectance from the deposited film of the monochromatic light of a helium–neon laser. They were able to fabricate mirrors of 15 cm diam consisting of 21 double layers of zinc sulfide and $ThOF_2$. These mirrors had a transmittance of 0.05% and a reflectance of 99.8% between 5600 and 6900 Å.

IV. CONCLUSIONS

In view of the numerous investigations of thin films where film thickness is a very important parameter, it is surprising that the classical microbalances are seldom used for thickness determinations. However, the quartz oscillator is expected to gain a wider appreciation as a convenient tool for thin film work in the future. The limitations originating from the position of the zero coefficient plateau near room temperature can be surmounted without too much effort by careful temperature measurements and the determination of the temperature coefficient of the frequency. The metal electrodes of the polished quartz crystal can be covered with any desired material making available a wide variety of substrates for study.

The application of the classical microbalances has been limited in their use by certain experimental difficulties. It has been shown here and in other chapters, however, that it is very possible to construct sensitive bakeable systems.

Finally, it should be mentioned that in geophysics very sensitive and rugged microbalances are used.

An example is the Worden Gravimeter (42) shown in Figure 8.39. This gravimeter is in principle a spring microbalance (see Chapters 2 and 3). The restoring directional force of the spring is partly compensated by the directional force of the weight arm, caused by its gravity, so that the gravimeter is made astatic to a certain degree

The interesting features of this gravimeter are:

1. Very small weight of the weight arm, i.e., 5 mg
2. Very high sensitivity of about 10^{-5} cm/sec^2
3. Excellent zero stability and ruggedness

The small weight and the high sensitivity to gravity changes indicate that mass changes of 5×10^{-11} g could be detected. The zero stability is achieved by careful adjustment of the various springs (Figure 8.39) so that the temperature drift is compensated to 10^{-4} cm/sec^2-°C. The short term zero drift is 3×10^{-5} cm/sec^2 and the long term drift 10^{-3} cm/sec^2 per

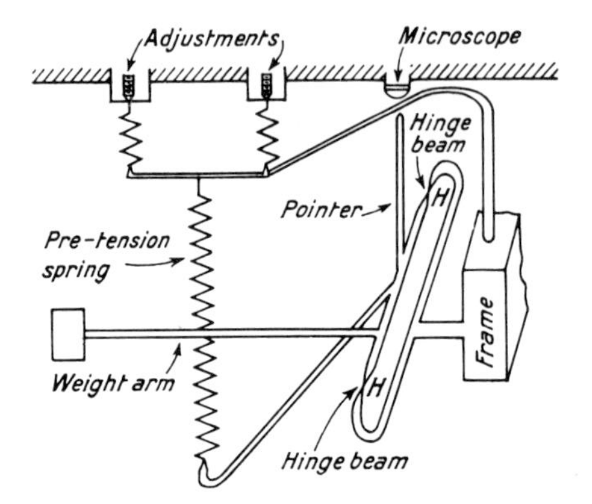

Figure 8.39. Sketch of a Worden Gravimeter. The gravimeter is made completely out of quartz.

day. This would correspond to 5×10^{-9} g per day. The gravimeters exhibit these data although they are submitted to comparatively rough transport conditions.

It can be expected from these experiences, that there are still possibilities in the construction of microbalances to increase the ease of use and the sensitivity.

ACKNOWLEDGMENTS

The authors are deeply indebted to Prof. Mayer whose collection of the literature on thin films facilitated our task a great deal. We also wish to thank P. Schmider for permitting us to use his not yet published work in this chapter.

Finally, we wish to express our gratitude to the editors. They have not only improved the language of this chapter but they have also done a considerable part to clarify some of its aspects by their expert comments.

REFERENCES

1. Th. Gast, *Z. Instr.*, **68**, 30, 1960; *Feinwerktechnik*, **53**, 167 (1949); *Z, Angew. Phys.*, **8**, 164 (1956).
2. L. Cahn and H. R. Schultz, *Vacuum Microbalance Tech.*, **3**, 29 (1963); **2**, 7 (1962).

3. H. Mayer, R. Niedermayer, W Schroen, D. Stünkel, and H. Göhre, *Vacuum Microbalance Tech.*, **3**, 75 (1963).
4. R. Niedermayer, Thesis, Clausthal, 1963.
5. H. Göhre, Thesis, Clausthal, 1962; Verzögerte Kondedsation: Ein Beitrag zum Mechanismus der Kondensation an gekühlten Oberflächen. Proceedings of the 2nd European Symposium Vacuum, Frankfurt, June 1963, R. A. Lang-Verlag, Esch-Idstein 1964.
6. D. Stünkel, Thesis, Clausthal, 1962; *Z. Phys.*, **176**, 207 (1963).
7. R. Niedermayer and W. Schroen, *Vakuum-Technik*, **11**, 36 (1962).
8. D. S. Campbell and H. Blackburn, *Transactions of the 7th National Vacuum Symposium, London, 1960*, Pergamon Press, New York, 1961, p. 313.
9. C. A. Neugebauer, *J. Appl. Phys.*, **35**, 3599 (1964).
10. F. C. Hurlbut, *J. Appl. Phys.*, **28**, 844 (1957); *Recent Research in Molecular Beams*, Academic Press, New York, 1959, Chapter 8.
11. A. R. Beavitt, *J. Sci. Instr.*, **43**, 182 (1966).
12. D. Hillecke and R. Niedermayer, *Vakuumtechnik*, **3**, 69 (1965).
13. G. Sauerbrey, *Z. Physik.*, **155**, 206 (1959).
14. H. K. Pulker, *Z. Angew. Phys.*, **20**, 537 (1966).
15. H. L. Eschbach and E. W. Kruidhof, *Vacuum Microbalance Tech.*, **5**, 207 (1966).
16. D. Hillecke, *Vakuum-Technik*, **17**, 167 (1968).
17. Bechmann, R., *Proc. IRE*, **48**, 1494 (1960).
18. H. Kiessig, *Ann. Phys. (Leipzig)*, **10**, 715 (1931).
19. M. S. Blois and C. M. Rieser, *J. Appl. Phys.*, **25**, 338 (1954).
20. S. Tolansky, *Multiple Beam Interferometry*, Oxford University Press, New York, 1948.
21. A. R. Wolter, A. R., *J. Appl. Phys.*, **36**, 2377 (1965).
22. Th. E. Hartman, *J. Vacuum Sci. Technol.*, **2**, 239 (1965).
23. J. Edgecumbe, J., *J. Vacuum Sci. Technol.*, **3**, 28 (1966).
24. R. Niedermayer, *Basic Problems in Thin Film Physics* (Proc. Intern. Symp. Clausthal-Göttingen 1965), Vandenhoek & Ruprecht, Göttingen 1966, p. 249.
25. L. S. Palatnik and Yu. F. Komnik, *Soviet Phys. Doklady (English Transl.)* **4**, 663 (1959).
26. K. Spiegel, Thesis, Clausthal, 1966.
27. K. Spiegel, R. Niedermayer, and H. Mayer, *Kurznach. Akad. Wissen. Göttingen*, **22**, 103 (1966).
28. K. Spiegel, *Surface Sci.*, **7**, 125 (1967).
29. P. Schmider, to be published.
30. N. T. Gladkich, R. Niedermayer, and K. Spiegel, *Phys. Status Solidi*, **15**, 181, 1966.
31. R. Niedermayer, N. T. Gladkich, and D. Hillecke, *Vacuum Microbalance Tech.*, **5**, 217 (1966).
32. M. Takagi, *J. Phys. Soc. Japan*, **9**, 359 (1954).
33. C. A. Neugebauer, *Phys. Rev.*, **116**, 1441 (1959); *Z. Angew. Phys.*, **14**, 182 (1962).
34. S. J. Glass and M. J. Klein, *Phys. Rev.*, **109**, 288 (1958).
35. W. Steckelmacher, J. English, H. H. A. Bath, D. Haynes, J. T. Holden, and L. Holland, *Transactions of the 10th National Vacuum Symposium 1963*, Macmillan, New York, 1963, p. 415; *Semiconductor Products and Solid State Technology*, p. 17, Dec. 1964.
36. H. H. A. Bath, *J. Sci. Instr.*, **43**, 374 (1966).
37. A. Langer and J. T. Patton, *Vacuum Microbalance Tech.*, **5**, 231 (1966).

38. K. H. Behrndt, *J. Metals*, **14**, 208 (1962).
39. K. H. Behrndt, *Transactions of the 6th National Vacuum Symposium, 1959*, Pergamon Press, London, 1960, p. 242.
40. K. H. Behrndt and R. W. Love, *Vacuum*, **12**, 1 (1962).
41. K. H. Behrndt and D. W. Doughty, *J. Vacuum Sci. Technol.*, **3**, 264, 1966.
42. M. B. Dobrin, *Introduction to Geophysical Prospecting*, McGraw-Hill, New York, 1952. A. Graf, *Gravimetrische Instrumente und Messmethoden* (Jordan-Eggert-Kneissl, *Handbuch der Vermessungskunde*, Vol. Va), J. B. Metzler, Stuttgart, 1967.

Chapter 9

The Simultaneous Measurement of Mass Change and Optical Spectra

C. L. ANGELL

Union Carbide Research Institute
Union Carbide Corporation
Tarrytown, New York
and

A. W. CZANDERNA

Department of Physics
Institute of Colloid and Surface Science
Clarkson College of Technology
Potsdam, New York

I. INTRODUCTION

The importance of carrying out simultaneous measurement of mass and other physical parameters was reviewed in Chapter 2, Section VIII B. In this chapter, the simultaneous measurement of mass change and optical spectra will be described to illustrate the value of simultaneous determination of more than one parameter. In the applications to be discussed, a substrate, which was nonabsorbing in the visible or infrared, was suspended from one arm of an automatic recording microbalance. For the measurement of the absorption of radiation in the infrared, the substrate was suspended in a vertical plane and the light beam was incident in a horizontal plane. In this application, described in Section II, it has been possible to correlate the absorption of light with the surface coverage of adsorbed species on a high surface area zeolite material. For the measurement of the transmittance of thin metal and metal oxide films in the visible region of the spectrum, the substrate was suspended in a horizontal plane and the light beam was incident in a vertical plane. In this work, described in Section III, fundamental physical information about the film has been obtained both from the spectra and from measurements carried out on samples prepared in auxiliary systems. In the latter experiments, the spectra were used to identify the thickness of the original copper film and the extent of its oxidation or reduction.

II. SIMULTANEOUS STUDIES OF THE ADSORPTION OF GASES AND THE INFRARED ABSORPTION SPECTRA

A. Introduction

The importance of infrared spectroscopic measurements for the determination of molecular structure has long been established. During the last twenty years the progress in infrared instrumentation has been tremendous, resulting in the availability of high-quality low-price spectrometers. As a consequence, a storage of thousands of infrared spectra has been accumulated (1) and can be easily reached for reference and comparison purposes.

The infrared region of the electromagnetic spectrum (usually 2–40 μ) corresponds to energies in the range of molecular vibrations and rotations.

The frequency of vibration of a diatomic molecule in the harmonic oscillator approximation is

$$\nu \text{ cm}^{-1} = I/2\pi \, (k/\mu)^{1/2} \tag{9.1}$$

where k is the force constant and μ the reduced mass of the atoms. In polyatomic molecules more complicated expressions are required to account for the observed frequencies in terms of the various stretching and bending modes of the molecules. While there are only a small number of molecules where a complete vibrational analysis of all the modes have been carried out, it has been found that motions between certain atoms are comparatively independent of the rest of the molecule and give rise to characteristic frequencies, so called "group frequencies." For example, the C=O stretching frequencies are in the 1700 cm^{-1} region, the C—O stretching frequencies in the 1000 cm^{-1} region, and the C—H stretching frequencies in the 3000 cm^{-1} region. With these group frequencies the infrared spectrum of a compound has become a highly efficient tool for the identification of various atomic groupings making up the molecule. On the other hand, the actual frequency (and force constant) of a group vibration is a very sensitive measure of the structural and electronic environment of the group and has been used to characterize intramolecular effects like electronegativity of groups, conjugation and induction effects, as well as intermolecular forces, like hydrogen bonding and solvent effects.

While the frequency of an infrared band is a measure of the strength of a part of a molecule, the intensity of the band is related to the dipole moment of the vibrating unit through $A = (\partial\mu/\partial q)^2$, where μ is the bond moment and q is the normal coordinate of the vibration. The band intensity is usually measured as the area of the band, A, and is calculated from

$$A = \frac{1}{cl} \int \log I/I_0 \, d\nu \tag{9.2}$$

where c is the concentration, l the optical path length, I_0 and I the intensity of light entering and leaving the sample and the practical integration limits are the observable measurable wings of the bands. While band intensity measurements are much more tedious than the measurement of band frequencies, they provide useful information about the mechanical and electrical forces within a molecule. Variation of band intensities within a series of molecules is a useful corollary to the frequencies in establishing the various factors mentioned above, affecting the vibrating unit in question.

Infrared spectroscopy as a diagnostic tool can be applied in nearly all branches of science due to its great versatility; it has been applied to gases, liquids, and solids, from very low to high temperatures. About ten years ago it was also demonstrated that it can provide useful information about molecules adsorbed on surfaces (2). Since then a formidable number of articles on this subject has been published (for recent reviews see reference 3). The present publication rate is ample evidence of the activity and versatility of infrared studies.

The main advantage of infrared spectroscopy in the study of adsorbed molecules is that the investigation can be carried out on the molecules *in situ*, viz; when they are adsorbed on the surface. Changes in adsorbate species because of changes in the pressure, temperature, or the presence of other gases can be followed even when no changes occur in the molecular species of the gas phase. Infrared spectroscopy can provide information about (*1*) the nature of surface groups, their stability, their possible exchange with gaseous constituents, and their interaction with adsorbed molecules; (*2*) the identity of the adsorbed molecules, thus making it possible to detect chemical reactions such as polymerization, dissociation resulting in the loss of hydrogen, or uptake of hydrogen by molecules in the adsorbed phase; (*3*) changes in the characteristic frequencies of the molecules, indicating changes of the forces within the molecule; and (*4*) the strength of bonding to the surface as determined from the pressure dependence of characteristic bands, from the changes of force constants within the molecule, and, in some favorable cases, from the observation of the vibration of the bond between the surface and the adsorbate.

The information, *1–4*, can be obtained from the frequency of the bands, that is from the measured infrared spectrum of the system alone, without any knowledge of the amount of adsorbed material. Several studies have been described in the literature in which the intensities of bands in the infrared spectrum of adsorbed molecules have been measured (4–7). However, to calculate the intensity of a band it is necessary *to know* the *concentration* of the material responsible for the infrared absorption, that is the number of molecules adsorbed per unit volume of adsorbent. Thus, Seanor and Amberg (6) used the formula

$$cl = w/Ma \tag{9.3}$$

to calculate the band intensity, where c is the concentration, l is the thickness, a the geometric area of the pressed pellet, w the weight, and M the molecular weight of the adsorbed gas. It can be seen that an isotherm for the adsorption of the compound under investigation is needed at the same temperature at which the infrared spectra are taken.

The isotherms can be obtained from either separate measurements on a sample similar to the one used for the spectroscopic study, or on the same sample simultaneously with the spectroscopic measurements. If the adsorption measurements are made on similar samples, the isotherm may be determined by either volumetric or gravimetric methods. There is, however, one danger, namely the lack of certainty that the two different samples receive exactly the same pretreatment, are subjected to the same experimental conditions, and are, therefore, really equivalent samples. For example, in our studies on zeolites (9) it was found that the nature of the surface depended on the rapidity of the removal of water during activation. Since the infrared samples usually are quite small (20–70 mg), and the samples on which adsorption measurements are carried out usually much larger (1–5 g), it is not difficult to perceive that the rate of removal of water from the same material can be very different in the two experiments even when the pumping conditions are identical.

The conditions of the two types of measurements, therefore, can only be unquestionably identical when they are carried out simultaneously on the same sample. The smallness of infrared samples has already been mentioned; in addition, the volume of the cells used in these measurements is usually large. These factors make volumetric adsorption measurements highly impractical and gravimetric methods desirable.

Peri (10) has described an infrared cell with quartz windows, where the weight changes could be measured by the elongation of a quartz spring on which the sample was hanging. Seanor and Amberg (11) have described a cell in which weight changes were measured by a null-reading quartz spring balance simultaneously with the infrared measurements (Figure 1.2, Chapter 1, Section IV.B.2).

The measurement of infrared spectra of molecular species adsorbed on surfaces, and especially simultaneous infrared and weight measurements present a number of experimental requirements, which are discussed in the following.

1. Since the infrared samples are small, the weight changes due to the adsorbed gas will be small. For example, on a 100 mg sample the weight corresponding to 1 cc/g adsorption is 0.125 mg for CO, 0.080 mg for H_2O, and 0.009 mg for hydrogen. The thickness of the sample cannot be increased much beyond the usual 12–20 mg/cm^2 due to the limited transmission of the solids usually studied. Thus, only the diameter of the sample pellet can be increased to increase the weight. This means that the measurements must be carried out with a microbalance with a sensitivity of 1 μg or better, to study the adsorption process for these gases to very low coverages.

2. Since we are dealing with solids with highly reactive surfaces, it is desirable to carry out measurements in a vacuum system capable of maintaining a pressure of less than 10^{-6} torr, to minimize the amount of adsorbate contamination. Hence, a microbalance capable of operating in a vacuum is needed.

3. Any up and down movement of the sample during the weighing process might mean that different parts of the sample will be exposed to the infrared beam and any variation in the sample either in composition or in thickness will interfere with the accuracy of the optical measurements. To avoid any such movement a null type balance should be used.

4. For the measurement of the spectra of adsorbed species, most infrared cells described in the literature have a long path length of the beam in the gas. Therefore, undesirable gas-phase spectra are produced in addition to the bands of the adsorbate whenever it is necessary to maintain an adsorbate gas pressure. In chemisorption experiments, the system can be evacuated after exposure to the gas for a given time interval, but whenever equilibrium conditions are studied, as in the determination of an isotherm, the gas-phase spectrum will interfere. Systems have been described where the gas spectrum is compensated with another cell in the reference beam of the spectrometer. However, these systems are never quite satisfactory, especially when concerned with the fine structure of the rotational spectrum of small molecules, like CO, CO_2, H_2O, CH_4, ethylene, etc. For example, Figure 9.4 shows that even with the small (10 mm) path-length cell gaseous CO causes serious interference in the spectrum of the adsorbed species above about 50 torr pressure. The best results can be obtained if the optical path length is actually decreased as much as possible. Angell (9) described an improved cell for general purposes with a path length of only 8 mm; for the simultaneous measurements described below a cell with a 10 mm path length was used.

5. In all adsorption experiments it is necessary to outgas the sample at high temperature to activate the surface before experiments by removing various adsorbed impurities that cannot be removed at lower temperatures. On the other hand, the salt crystals (NaCl, CaF_2, KBr, etc.) used to make the infrared transmitting windows cannot be heated much above 150°C. When the sample compartment is large, there is enough room to heat the central portion with the sample and keep the windows cool. This becomes difficult for a narrow sample section. Peri's (10) system was only 30 mm wide, but the windows were quartz so they could be heated to high temperatures with the rest of the system. In systems where the sample is not attached to a balance, it can be positioned in a furnace section for heating and moved between the windows for measuring the spectrum by using mechanical or magnetic devices. However, a sample attached to a microbalance can be moved only with seemingly insurmountable difficulty. A unique method, therefore, was devised in this work to allow the sample to be heated.

Fortunately, the commercially available Cahn RG automatic vacuum microbalance satisfies the conditions set out above in *1–3*. Using this balance a system has been designed that includes a sample compartment which satisfies the condition of small gas path length (*4*). To solve the heating problem the sample was allowed to hang freely on the balance and the envelope of the hangdown tube was moved up and down so that the sample could be positioned either in an all-glass furnace section or between the NaCl windows in the spectrometer. The hangdown tube, which has been described in detail (12), could be moved up or down nearly 5 in. because of a flexible stainless steel bellows while maintaining a high vacuum.

B. Experimental

The experimental apparatus is shown schematically in Figure 9.1 and in a photograph in Figure 9.2. The balance was a Cahn Model RG Electrobalance placed in a 2505 glass vacuum bottle equipped with 75/50 O-ring

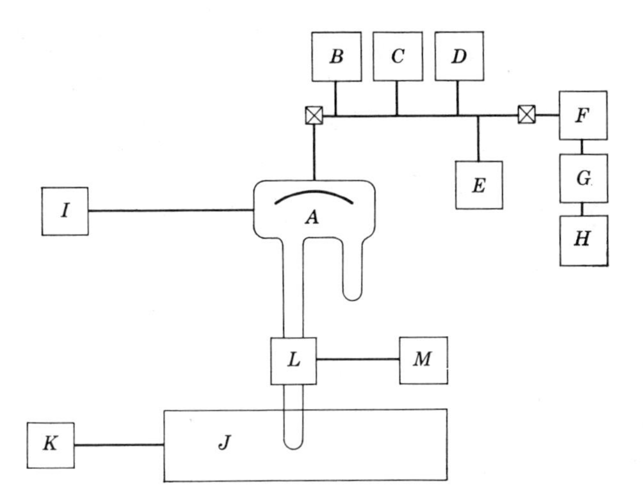

Figure 9.1. General layout of apparatus. *A*, microbalance; *B*, Phillips gauge; *C*, mercury manometer; *D*, McLeod gauge; E, sample inlet; *F*, liquid N_2 trap; *G*, diffusion pump; *H*, mechanical pump; *I*, recorder for balance; *J*, IR spectrometer; *K*, recorder for spectrometer; *L*, heating tape; *M*, heat control.

ball joints. The vacuum was maintained by an oil mechanical pump, a high capacity (104 liters/sec) oil diffusion pump and a metal liquid nitrogen trap. The vacuum line to the bottle was at least $1\frac{1}{2}$ in. in diameter, including the two metal valves. Pressure measurements covering the range of 10^{-7} torr to atmospheric pressure were made with a Phillips gauge, a McLeod gauge, and a mercury manometer in their respective ranges.

For the tare side a short hangdown tube, an aluminum pan, and the tare weights supplied by the manufacturer as standard accessories were used. The sample hangdown tube is illustrated in Figure 9.3. The metal bellows was welded to a glass-Kovar seal on top and to a rotatable Conflat

Figure 9.2. Photograph of apparatus.

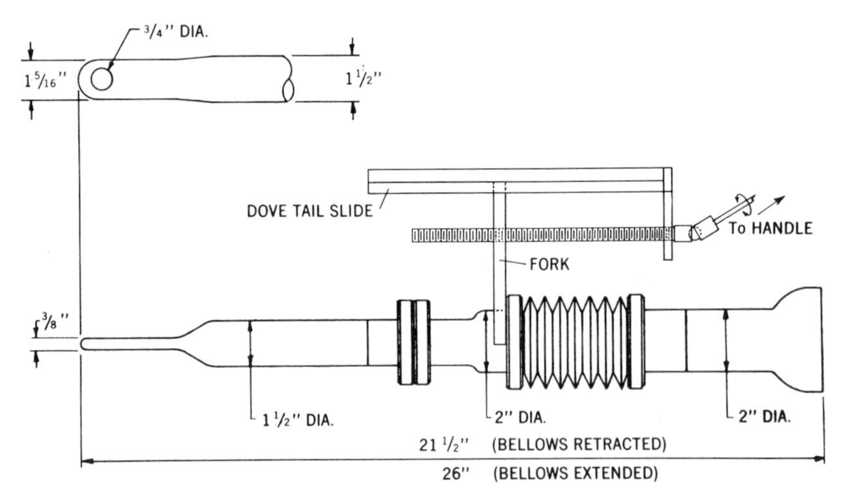

Figure 9.3. Sample hangdown tube.

flange on the bottom. The flange was included to facilitate the changing of samples; it would have been extremely inconvenient to open the system at the ball joint and remove the whole bellows assembly every time a sample had to be changed. A $1\frac{1}{2}$ in. diam glass tube, flattened to a $\frac{1}{2}$ in. by $1\frac{1}{4}$ in. internal dimensions section at the lower part, was attached to the bottom part of the flange, and $\frac{3}{4}$ in. holes drilled in the glass. One inch diam NaCl windows provided the optical path. A section of the round glass tube was wrapped with heating tape. The distance between this section and the NaCl windows of about $4\frac{1}{2}$ in. could be traversed by the up and down movement of the bellows.

Samples were most often in a pellet form, 1 in. in diam, about 0.2 mm thick, weighing between 70 and 100 mg, and held vertically in a tantalum sample holder described previously (12). Since rotation of the sample was not a problem, it was not necessary to use nonrotating hooks as described by Seanor and Amberg (11). On the other hand, if the sample started swinging and bumping into the wall when it was in the narrow part of the tube, it was practically impossible to stop this motion. In such cases, it was found convenient to lower the tube until the sample was in the wide part of the tube where it could swing freely and the motion died out. For calibration, substitution weights made from wire were hung on the same hook as the sample holder.

The changes in weight were recorded on a Photovolt Microcord recorder. The infrared spectrometer was a Perkin-Elmer Model 112 converted to filter-grating operation. The balance and the vacuum system were mounted on a fixed stand, thus, making it necessary to move the spectrometer to the sample. The spectrometer was mounted on a lift table with wheels. The fine adjustment was made by two perpendicular cross slides under the spectrometer.

After activation of the sample and degassing most of the system, dynamic vacua of the order of 5×10^{-7} torr could be achieved. The balance housing and the NaCl windows could not be heated above 100°C. For most measurements, a Recorder Range of 200 μg was chosen to permit ± 1 μg to be read on the chart without difficulty.

C. Results

Peri (10) has used simultaneous weight and spectral measurements in an all-quartz cell in studies of the isomerization of butene on gamma-alumina. Due to the limited transmission of quartz in the infrared only the OH and CH stretching regions were examined. After allowing 1-butene to isomerize at room temperature on alumina, a "desorption isotherm" (both spectral and weight) was determined for the resulting butene mixture by reducing the pressure from 8.5 torr to less than 10^{-3} torr. From the observed weight changes between 8 and 2 torr pressure, a specific absorbance was calculated for adsorbed butene at the 2927 cm^{-1} band. The weight of adsorbed butene was calculated from the infrared absorbance for the entire pressure change. Due to the 30 mm optical path length in the cell, a correction had to be applied for the absorbance of gaseous butene. It was concluded from the calculation that about 1% of the alumina surface remained covered with strongly adsorbed butene, even after evacuation for 15 min. The nature of this strongly adsorbed butene was established by comparing its spectrum with the liquid-phase spectra of a number of butenes.

A more detailed study of both isomerization and polymerization of all four butenes on porous Vycor glass was carried out by Little et al. (5) also using simultaneous measurements to determine band intensities. They obtained the amounts of adsorbed material by the volumetric method. Since when working at low coverages (0.001 surface layer) the amount of butene adsorbed was very low (2×10^{-6} moles per cm^3 sample), and the corresponding pressure also had to be low, large uncertainties were introduced in the intensity measurements. It was believed that optimum conditions might be obtained by working at low initial pressures, thus eliminating interferences from strong bands due to physically adsorbed material at the beginning of the adsorption process and minimizing the effect of the intense bands of the polymeric species when polymerization assumed significant proportions. Indeed, under these conditions chemisorption predominated; the pressure dropped to below 0.01 torr within 1–2 min of admitting the butene samples. Five minutes later, when the spectra were taken, it was found that the surface species were identical for the three *n*-butenes resulting presumably from rapid isomerization. The total band intensity for all the C—H vibrations was determined in each case as well as the percentage intensity of the olefinic band. These figures were compared with intensity measurements for the four butenes and related molecules in solution. While most of the conclusions drawn by the authors (5) were based on frequency correlations, the band intensities were also used to obtain information about the nature of the surface species. When the *n*-butenes were adsorbed on acid-leached Vycor glass, the isomerization was slowed down considerably and no polymerization occurred. The spectra were very similar to those of the same compounds in a solution of CCl_4. The molar intensities of the surface material, however, were considerably less than the corresponding values in solution. The ratios of olefinic to total C—H intensities, however, agreed quite well with the values taken in solution. The following conclusions were drawn about the possible structure of the surface species formed under low-coverage conditions: (*1*) since a band appeared at 3020 cm^{-1}, the species must contain a double bond with at least one hydrogen on it; (*2*) the 10–12% intensity of the olefinic band would make a polymeric species unlikely; (*3*) of the two stable forms frequently postulated for the carbonium ion intermediate in double bond isomerization and polymerization reactions the σ complex

$$\begin{array}{c} \quad\ \mathrm{H}\ \ \ \mathrm{H} \\ \quad\ \ |\ \ \ \ | \\ \mathrm{CH_3{-}C{-}\overset{}{\underset{+}{C}}{-}CH_3} \\ \quad\ \ | \\ \quad\ \mathrm{H}\ \ \ \mathrm{S^-} \end{array} \tag{9.4}$$

is not consistent with the obtained spectra, while the π complex

$$\begin{array}{c} \quad\ \mathrm{H}\ \ \ \mathrm{H} \\ \quad\ \ |\ \ \ \ | \\ \mathrm{[CH_3{-}C{-}C{-}CH_3]^+S^-} \\ \quad\ \ \vdots \\ \quad\ \mathrm{H} \end{array} \tag{9.5}$$

has the olefinic character essentially preserved and might be expected to give rise to absorption above 3000 cm^{-1} and be compatible with the spectra of the surface species.

Seanor and Amberg (6) observed that in the case of CO weakly bonded to various zinc oxide surfaces, the absolute intensities varied markedly with frequency. Using the apparatus mentioned earlier they extended these studies to CO adsorbed on other oxide and metal surfaces. From the equation

$$A = \frac{2.303Ma}{w} \int D\,d\nu \tag{9.6}$$

where D is the optical density at ν, a is the geometric area of the sample and M is the molecular weight of the adsorbed gas, A could be determined from the slope of the plot of the measured values of the integral against w, the weight of gas adsorbed. This method is limited to cases in which the band under study is associated with one particular surface species only.

Intensity measurements were carried out on the bands of CO adsorbed on pure and doped ZnO, on Pt, PtO, Cu, and CuO supported on Cabosil and compared with band intensities of CO trapped in nitrogen and argon matrices. In the region of weak, reversible adsorption (2132–2212 cm^{-1}) a 25-fold change in intensity was noted, and the intensity-frequency correlation did not appear to be sensitive to the chemical nature of the substrate. As an explanation of the intensity changes, the authors (6) have suggested that the weakly and reversibly adsorbed CO molecules obey a dipole moment–internuclear distance function slightly perturbed from that of the free gas. Using the equation

$$A = C\left(\frac{d\mu}{dr}\right)^2 \tag{9.7}$$

where C is constant, μ is the dipole moment, and r is the internuclear distance, the dipole moment gradients were calculated from the measured intensities. The internuclear separations were estimated from the absorption frequencies using Badger's rule. In this way, a dipole moment against internuclear distance curve could be constructed about the known values of μ and $d\mu/dr$ for gaseous CO. This, however, involves the crude assumption that when CO is perturbed by the various surface force fields, a new equilibrium internuclear distance is established, but that μ itself changes very little. Irrespective of whether the simple model is ultimately found to be acceptable, the behavior of infrared band intensities as a function of frequency must originate from changes in the solid substrate. The authors ascribe the changes to the effect of the electrostatic field arising from the structure of the surface.

Further insight into the mechanism of CO adsorption was gained by the work of Angell and Schaffer (13), who studied CO adsorbed on a number of cation exchanged Linde X and Y zeolites. These synthetic zeolites, structurally related to the mineral faujasite, are a three-dimensional highly rigid framework of SiO_4 and AlO_4 tetrahedra linked together by shared oxygen atoms. They have a large intracrystalline surface associated with a crystal-lographically well-defined system of pores and cavities. The negative charge of this framework, arising from the tetrahedrally bonded Al atoms, is counterbalanced by an equal number of cations. As synthesized, these materials contain Na^+ ions but these can be exchanged with other mono-, bi-, and tervalent cations. The possible cation positions in anhydrous zeolites X and Y are essentially the sites hidden from the zeolite surface, being closely surrounded by the atoms of the framework (S_I) and the sites on the zeolite surface (S_{II} and S_{III}) (7). The catalytic activity of the cations in these surface sites give the zeolites their properties of great interest. The cations are only semihedrally shielded by the framework atoms. On their "bare" side they give rise to unusually strong, accessible electrostatic fields and show carboniogenic catalytic activities due to the polarization of the substrate molecules in these fields. These surface cations also present excellent opportunities for the study of interactions between various molecules and cations, because they can be exchanged to practically any desired other cations without changing the structure supporting them. In this way, since all other variations in the system have been eliminated, any effects observed on a series of different cation containing zeolites can be strictly associated with the nature (charge, size, coordinating power) of the cations. For example, it has been shown that the carboniogenic activity in-increases with increase in cation charge and decrease in cation size, and that this is why the electrostatic fields of the cations have been thought to be responsible for the activity.

Very little is, however, known about the distribution of the bivalent cations between the different sites, since only a Ca-exchanged faujasite sample has been examined by x-ray diffraction (8). In this case, the Ca ions showed great preference for the hidden sites. This work on CO adsorbed on various zeolites has not only confirmed the existence and magnitude of the accessible electrostatic field of the cations, through the frequency shifts of the CO band, but, in a more quantitative way, has allowed us to measure the number of surface sites (cations) responsible for the CO adsorption, and thus establish the distribution of bivalent cations between hidden and surface sites.

It has been observed (13) that when carbon monoxide is adsorbed on X- and Y-type zeolites containing a variety of cations, two types of adsorption can be observed by infrared spectroscopy. The first was seen on all the zeolites tested, including X and Y zeolites that contained uni-, bi-, or ter-valent cations and Y zeolites that were partly or almost fully decationized. This type of adsorption produces bands at 2170 and 2120 cm^{-1} unchanged in frequency, intensity, and shape. It was shown that the two bands are associated with two separate adsorption sites, but since they both are independent of the cationic composition of the zeolites, they must be associated with the zeolite framework. In addition to these ubiquitous bands, the zeolites containing bivalent cations exhibited a band at a higher frequency than in the gas phase indicating a second type of adsorption (Figure 9.4). This is a weak adsorption, since the band disappears on pumping at room temperature, although not as readily as the other two bands. Its actual position depends on which bivalent cation is present and to a smaller extent on the type (X or Y) of zeolite, and could be correlated with the radius of the bivalent cation. The shift of the band from the gas-phase frequency was shown to be directly proportional to the calculated field at the expected position of the carbon atom in the M....CO complex for a wide selection of bivalent cations. It was inferred that the frequency shift was primarily due to a distortion of the CO molecule in the electrostatic field, and that the distortion and frequency shift were a measure of the strength of the electrostatic field at the adsorption site.

The dependence of the CO frequency on the cation clearly establishes that the CO molecule is attached to the cations. The specific bands thus have possibilities for the study of the cation–CO complexes as well as the cation distribution and environment in the zeolite. Since it has been shown that univalent cations do not give rise to the high frequency CO band, and it is assumed that in S_I a cation could not be reached by any adsorbed molecule, the appearance of this band is an indication that there are bivalent cations in the surface positions. For example, for CoY and NiY the specific band appeared even when the degree of exchange of added ion for original sodium ion was less than enough to fill all the S_I, showing that these cations did not realize the full expected preference for the hidden site.

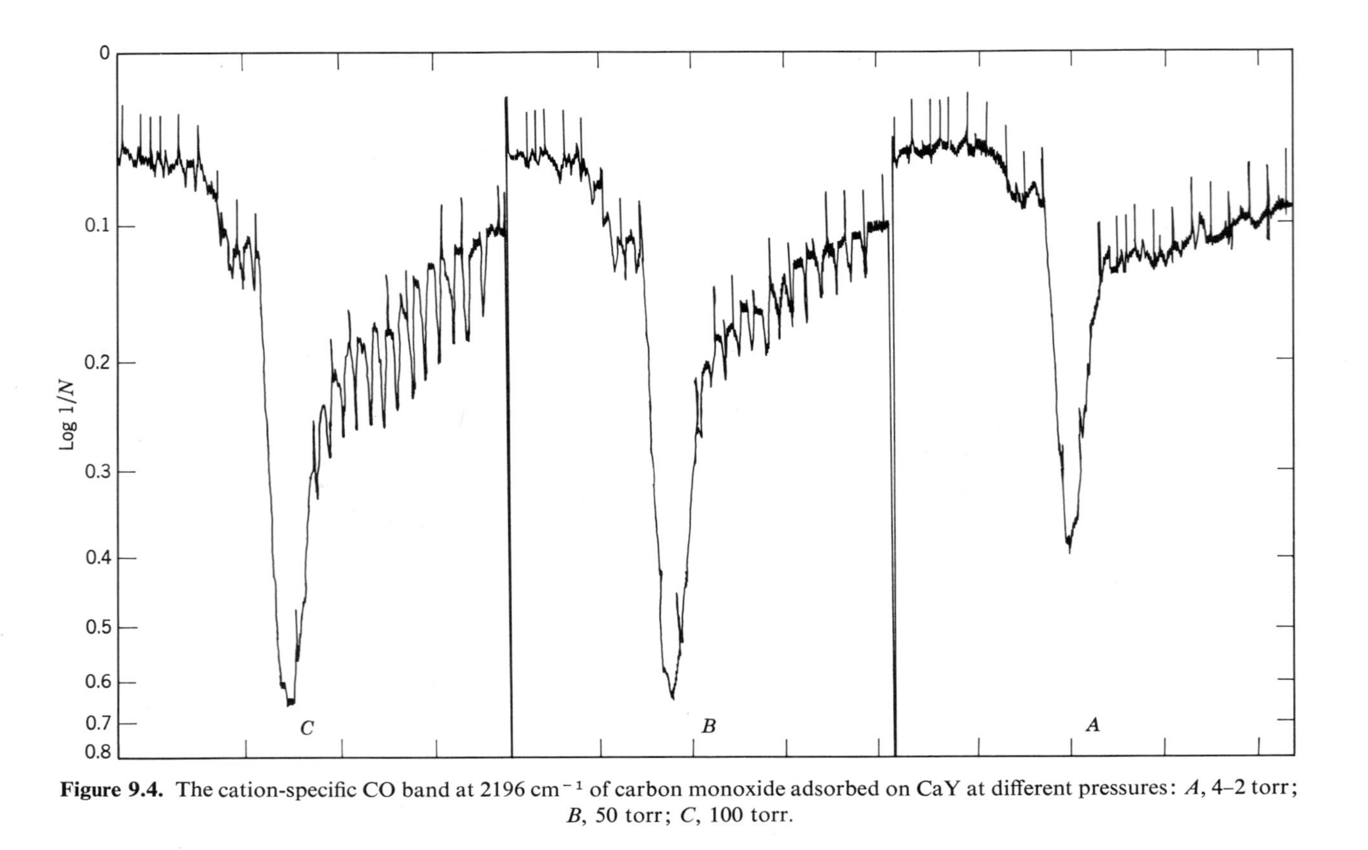

Figure 9.4. The cation-specific CO band at 2196 cm^{-1} of carbon monoxide adsorbed on CaY at different pressures: *A*, 4–2 torr; *B*, 50 torr; *C*, 100 torr.

On the other hand, a 35% exchanged CaX sample (15 bivalent cations for the 16 S_I per unit cell) did not show any trace of the corresponding high-frequency band.

When attempts were made to use the observations described in the previous paragraph for a quantitative measure of the number of surface cations, it was realized that the size of the specific CO band could not be immediately correlated with the amount of CO adsorbed because of the presence of the non-specific, more general adsorption. When the pressure dependence of the peak optical densities (OD) of the specific bands for a number of samples was determined, it was found that the extinction data fitted Langmuir curves well, implying adsorption on approximately equivalent, noninteracting sites. In addition, saturation of the specific bands was reached at pressures where the two nonspecific bands had just barely begun to grow. It was this great difference in pressure dependence that permitted the two types of adsorption to be separated.

For this purpose the optical density (peak intensity) of the specific band and the amount of CO adsorbed were measured over a large range of pressures (7). The optical density, OD, plotted against pressure was represented by the equation

$$\mathrm{OD} = [(\mathrm{OD})_0 n_0 kp]/(1 + kp) \tag{9.8}$$

where n_0 is the number of effective sites, k is the Langmuir combining constant and p is the pressure (see Figure 9.5). Thus, plotting p/OD against p permitted evaluation of k and the quantity $(\mathrm{OD})_0 n_0$.

The room temperature adsorption isotherms showed a sharp rise in the curve at fairly low pressures, followed by a steady, nearly linear rise of the amount adsorbed with increasing pressure (see Figure 9.6). Since the infrared observations were consistent with two different kinds of adsorption, it is reasonable to explain the adsorption isotherm as follows. The sharp increase at the beginning is due to the specific adsorption on the cations and the break in the curve represents saturation of this type of adsorption. The slow steady rise of the curve results from the nonspecific adsorption. This is confirmed by the case of NaY which did not exhibit the sharp initial rise associated with a specific CO band, but revealed a steady increase with pressure. The slope of the linear curve for NaY was practically the same as the slope of the second part of the curves for the bivalent cation zeolites.

Thus, the gas adsorption was represented by the equation

$$n = \frac{n_0 kp}{1 + kp} + Cp = \frac{Bp}{1 + kp} + Cp \tag{9.9}$$

where the first term describes a Langmuir type specific adsorption and the

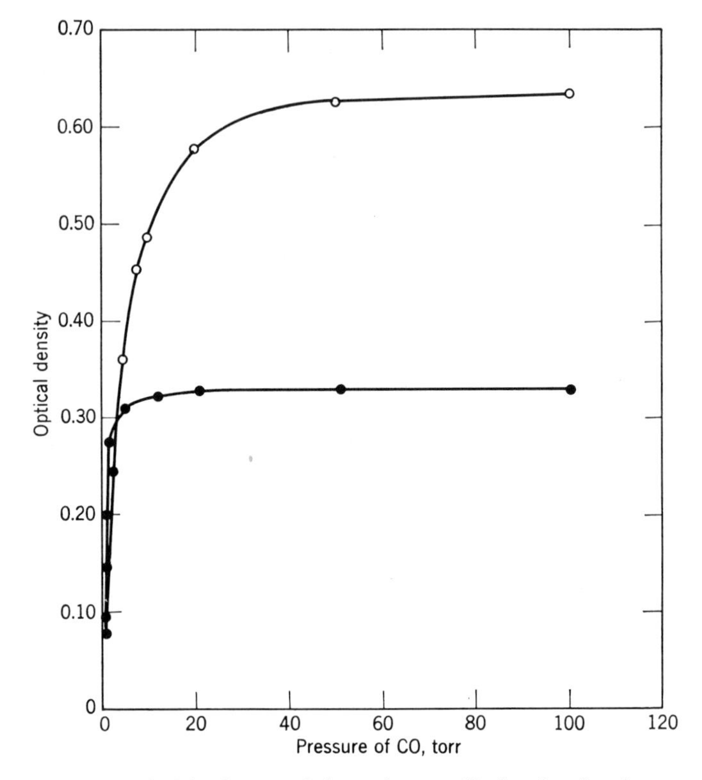

Figure 9.5. The optical isotherms of the cation-specific bands of carbon monoxide adsorbed on CaY (○) and MgY (●).

second term a linear nonspecific adsorption with slope C. For the Langmuir part of these curves the same value of the constant, k, was used as determined from the optical density curves (equation 9.8).

This equation can be rearranged to

$$n(1 + kp)/p = B + C + Ckp \tag{9.10}$$

Thus, a plot of $n(1 + kp)/p$ against pressure, should yield a straight line. The slope Ck, yields C, and the intercept, which is $B + C$, yields B. Since the value of n_0 can be calculated from B, the value for $(OD)_0$, the relative molecular optical density is obtained.

These results are collected in Table 9.1. The combining constants (k) are generally consistent with the frequency shifts, for the closed-shell ions, but for NiY and CoY they are very much larger. It seems that for these transition metal ions in addition to a simple charge polarizability bonding there is also a special covalent or ligand–field bonding. The slopes calculated for the straight line part of the adsorption curves (C) are in excellent agreement with the slope for the NaY adsorption, confirming the assumption that the nonspecific adsorption is taking place on the zeolitic surface and is unaffected by changes in the cations.

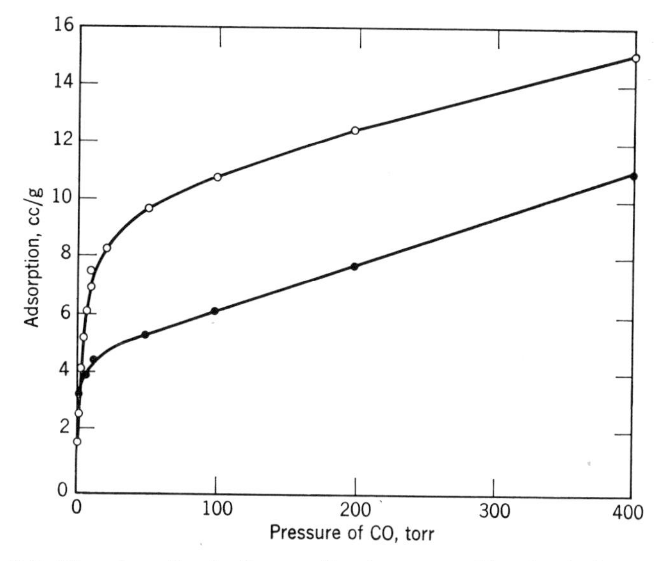

Figure 9.6. The adsorption isotherms of carbon monoxide adsorbed on CaY (o) and MgY (●).

The most revealing results of these quantitative studies are the numbers of effective sites (n_0). These are compared with the number of surface cations calculated from the degree of exchange and the postulated structures, i.e. it is assumed that the bivalent cations fill the 16 hidden sites (S_I) first and only ions in excess of 16 per unit cell go into surface sites. The correspondence is good for CaY which is the only case where there is independent comparable structural information, viz. the x-ray study of Ca^{2+}-exchanged faujasite (8), MgY, MnY, and ZnY. For BaY the number of effective sites appears to be much greater than the number of Ba ions in excess of 16 per unit cell, but this was not unexpected, since the large Ba radius might well keep the barium ions from penetrating the hexagonal prisms that enclose the hidden sites. For CaX, CoY, and NiY the reverse is true. The number of observed effective sites is much smaller than the number of surface cations expected after all the hidden sites are filled. According to our present understanding about the distribution of cations in the zeolite structure the location of the missing cations which makes them unavailable for interaction with the CO molecules is unknown.

Table 9.1. Adsorption and Optical Properties of CO Adsorbed on Various Zeolites

Zeolite	Exchange, %	K, mm^{-1}	C, cc/g mm	n_0, cc/g	n_0 sites/unit cell	M^{2+} in site S_{II}	Frequency, cm^{-1}	Half-width, cm^{-1}	$(OD)_0$	Intensity A
MgY	72	1.81	0.0172	4.40	4.8	4.5	2213	—	0.101	—
CaY	84	0.35	0.0144	13.7	7.6	8	2196	9.5	0.107	3.50
CaY	77	0.29	0.0166	9.35	5.3	5.9	2196	—	—	—
BaY	80	0.00308	0.0130	24.6	16.4	6.8	2178	—	0.186	—
MnY	63	0.56	0.0133	4.90	2.8	2.0	2208	—	0.139	—
CoY	78	103	0.0135	2.36	1.6	6.2	2208	10.0	0.432	19
NiY	76	18.7	0.0178	3.86	2.7	5.6	2217	9.0	0.331	10.3
ZnY	66	0.50	0.0142	4.81	2.8	3.0	2214	10.5	0.149	~5
CaX	98	0.108	0.0206	20.3	12.8	26.6	2192	9.2	0.233	~5

Now that the effective number of CO molecules involved in the cation-specific adsorption is known, it is possible to calculate the molecular band intensity for this absorption. Instead of measuring the area under each band, they were approximated by

$$A = (\pi/2)\Delta\nu_{1/2}(\mathrm{OD})_{\max} \tag{9.11}$$

where $\Delta\nu_{1/2}$ is the half-width and $(\mathrm{OD})_{\max}$ the peak height of the band. Intensities were calculated from equations 9.3 and 9.11, and are also given in Table 9.1. It can be seen that the values for CaY, CaX, and ZnY are in good agreement with the intensities reported by Seanor and Amberg for reversible CO adsorption in the 2212–2132 cm^{-1} range (6). However, in the case of CoY and NiY the intensities are considerably larger, again confirming that the CO bonding to Co and Ni ions is very different from a simple electrostatic polarization effect.

D. Conclusion

Through simultaneous measurements of weight changes and infrared spectra it was possible to ensure unequivocally that the two types of measurements were carried out on the same sample under identical conditions. In the case of zeolites, by the combination of gas adsorption and spectral measurements for CO, it was possible to calculate the strength of bonding, the number of effective sites and the molecular band intensities of the adsorbed CO.

III. SIMULTANEOUS STUDIES OF THE MASS CHANGE AND OPTICAL TRANSMITTANCE OF THIN FILMS IN THE VISIBLE

A. Introduction

In this section, a technique for measuring simultaneously the mass change and optical transmittance of supported thin films will be described (14). In the principal experiment, a thin metal film is evaporated onto a transparent substrate suspended from a microbalance (15). During subsequent oxidation (15) or reduction (16,17) of the film, the mass change of the film and its optical transmittance are continuously and simultaneously measured. The mass measurements were used to determine the gain or loss of mass by the film; the optical measurements provided both fundamental physical information about the film and a unique indication of the extent of oxidation or reduction. Exemplary results obtained during studies of the oxidation of evaporated copper films will be presented. The use of the optical transmittance as a monitor for auxiliary studies of the magnetic susceptibility of partially oxidized copper films will be illustrated (18). A method for determining the optical constants of the thin metal oxide film formed by the oxidation will be outlined (19), and the use of the optical constants as a means of studying the oxidation of copper will be discussed (20,21).

B. Theory

A treatise on the theory of optical thin films is not only beyond the scope of this chapter but unnecessary because of recent publications (22–24). However, the assumptions used for the development of the relatively simple expressions for computing the transmittance of a thin film can be introduced in a skeletal theoretical development that is important to the system air–copper oxide–copper–transparent substrate.

Mathematically, a thin film is regarded as an infinite layer between parallel planes the thickness of which is comparable to the wavelength of incident light. Following the notation used by Heavens (22), the film is characterized by optical constants, n, the refractive index, and k, the absorption coefficient to form a complex index

$$\mathbf{n} = n - ki \tag{9.12}$$

If absorption of light occurs, k is non zero.

If a wave of circular frequency ω is traveling in the direction (λ,μ,ν) in an absorbing medium of refractive index, n, the electric vector may be written

$$E = E_0 \exp i\omega\{t - [(n - ik)(\lambda x + \mu y + \nu z)/c]\} \tag{9.13}$$

for *normal* incidence. In equation 9.13, n is the ratio of the velocity of the wave in vacuum to that of the wave in the medium; k represents the energy absorption. As defined in equation 9.12, k represents the attenuation of the amplitude of the wave for a path of one vacuum wavelength, viz., $\exp(-2\pi k)$.

The laws of electromagnetic theory for an isotropic medium represented by the usual divergence and curl relationships for electric and magnetic fields lead directly to Maxwell's relations for the propagation of electromagnetic disturbances in a medium with zero space charge. The problem of determining the photon energy reflected and transmitted at a boundary separating two media is dealt with by applying boundary conditions to the solutions of Maxwell's equations. While this is tedious for many systems the development of theoretical expressions for the transmittance and reflectance of light has been reported by Heavens (22). By applying suitable boundary conditions to Maxwell's relations, the ratio of the amplitudes of the transmitted and reflected vectors in terms of the incident vectors can be obtained as components in the x and y directions and expressed as Fresnell coefficients. The latter can be conveniently treated by a matrix method.

During the oxidation of copper films, one or two absorbing films are always present on the transparent substrate. Thus, in Figure 9.7, light incident from vacuum onto an oxide growing on a thin metal film supported on a transparent substrate will be reflected or transmitted. The oxide and metal have absorption coefficients, k_1 and k_2, and absorbing thicknesses, d_1 and d_2.

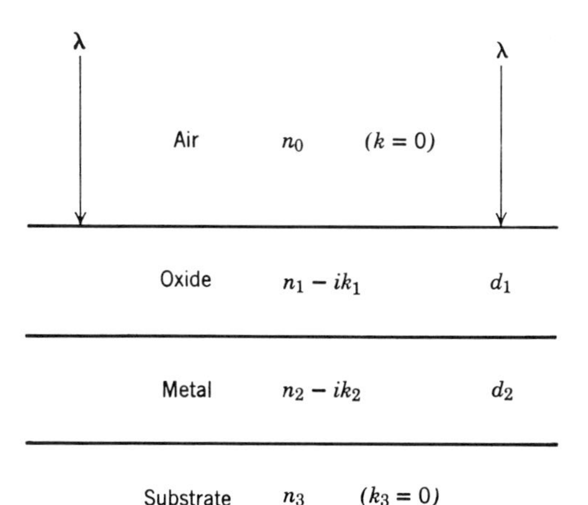

Figure 9.7. Two absorbing films on a nonabsorbing substrate.

For calculating the transmittance of light at normal incidence to a single or double layer on a transparent substrate, the equations given in Appendix I have been cited by Heavens (22). The expressions are readily programmed for calculation of the transmittance by a high speed computer.

C. Experimental

The apparatus used for carrying out simultaneous mass and optical transmittance measurements is shown schematically in Figure 9.8 and in a photograph in Figure 9.9. The apparatus consists of an automated bake-able pivot type beam microbalance, a vacuum station for producing very high vacuum and for manipulating and measuring gas pressures, an evaporator assembly for the deposition of a metal onto a substrate suspended from the microbalance and an optical system for measuring the transmittance of visible light through the sample and substrate.

1. Mass Measurements

Mass changes were determined with an automated pivot type beam ultramicrobalance. This type of balance has been discussed in Chapters 2

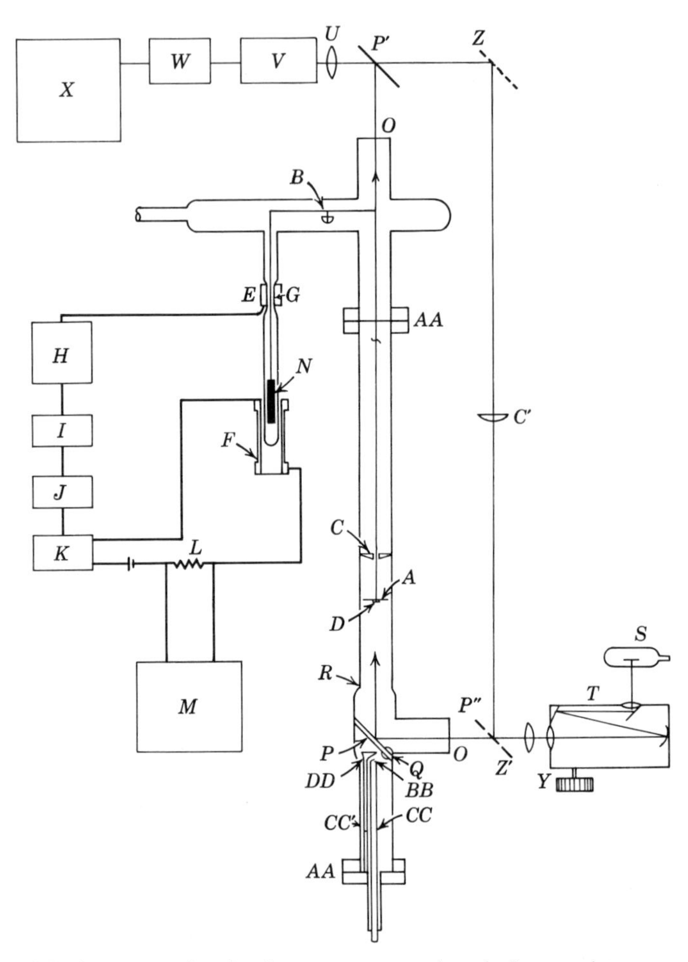

Figure 9.8. Apparatus for simultaneous mass and optical transmittance measurements. *A*, Pyrex glass substrate; *B*, beam microbalance; *C*, C′, field lens; *D*, quartz disk; *E*, transducer; *F*, magnetic compensation solenoid; *G*, piano wire probe; *H*, translator; *I*, amplifier; *J*, servo motor; *K*, helipot; *L*, standard resistor; *M*, bucking voltage sources and recorder; *N*, Cunife compensation magnet; *O*, quartz windows; *P*, *P′*, *P″*, silvered mirrors; *Q*, slotted Pyrex rod; *R*, square Pyrex tubing; *S*, light sources; *T*, monochromator; *U*, opal glass disk; *V*, photomultiplier; *W*, photometer; *X*, recorder; *Y*, motor drive; *Z*, *Z′*, mirror positions; *AA*, stainless steel flanges; *BB*, Mo boat; *CC*, *CC′*, copper rods; *DD*, nickel shield.

and 3 of this book and a very detailed description of the construction, operation, calibration, and limitation of the balance has been published elsewhere (14). In the application for fundamental studies of thin films, the sensibility of the balance was ± 0.1 μg in high vacuum and ± 1.0 μg at the

Figure 9.9. The apparatus for the simultaneous measurement of mass change and optical transmittance. (After A. W. Czanderna and H. Wieder, in R. F. Walker, Ed., *Vacuum Microbalance Tech.*, **2**, 147 (1962), Plenum Press).

oxygen and hydrogen pressures utilized. The decrease in the sensibility in the gaseous environment resulted from the influence of buoyancy and thermomolecular flow effects. During automatic operation, the balance was maintained at a null position. The mass change was monitored by measuring the change in potential difference across a standard resistor (*L*, Figure 9.8).

The copper film was deposited onto the Pyrex glass plate substrate ((*A*) 0.25 mm thick, 40 mm in diameter, and weighing 0.7 g) that was suspended from the balance (*B*). A 6-mm-diameter hole was drilled in the field lens (*C*) to allow adequate clearance of the hangdown suspension fiber. The substrate was held in a horizontal plane by a small quartz disc (*D*).

2. *Optical Measurements*

The optical transmittance of the films was measured using the arrangement shown in Figure 9.8. Light from a ribbon-filament tungsten source (*S*) was incident on a Bausch and Lomb grating monochromator (*T*). The monochromatic light from the exit slit was focused onto the field lens (C,C') and refocused onto an opal glass disc (U) located in front of an RCA 6217 photomultiplier tube (V). An Aminco photometer (W) served as the photomultiplier power supply; the photovoltage was displayed on a 10-mV recorder. The wavelength region from 400 to 800 mμ was scanned by means of a motor drive (Y) attached to the wavelength drum of the monochromator.

To measure the light transmitted by the sample, mirrors P and P' were located in the light path as shown by the solid lines in Figure 9.8. To measure the intensity distribution of the source as a function of wavelength, mirror P' was moved on an optical bench to Z and a mirror (P'') was inserted at Z'. The transmittance of the sample at any wavelength then is given by equation 9.14

$$T = SF_\lambda PS_\lambda / PB_\lambda \qquad (9.14)$$

where PS_λ is the photovoltage via the sample path ($SOPACOP'UV$), PB_λ is the photovoltage via the blank path ($SP''C'P'UV$), and SF_λ is the scattering factor.

The scattering factor is necessary because scattering of light by the microbalance beam and suspension cannot be avoided. Gross losses, which could be caused by the hole in the field lens, were avoided by masking the center portion of the slit. The magnitude of the scattering factor was determined from runs made prior to deposition of any material on the substrate. Since T should be 100%, SF_λ is simply PB_λ/PS_λ. In Figure 9.10, the dependence of the scattering factor on wavelength is plotted for several typical runs. The variation in the amount of scattering from one run to the next resulted from a slightly different alignment of the sample hangdown tube. Thus, determination of SF_λ at the start of each film preparation was absolutely necessary. SF_λ did not change during the study of any given film. For a completely oxidized film, the points on curve 180x8 in Figure 9.10 were obtained using SF_λ (curve 18); the solid line was obtained by removing the sample from the balance and inserting it in the blank path at C' to obtain "PS_λ". It is evident that the initial determination of SF_λ was adequate and indicates there was no change in the optical alignment.

3. *Vacuum Technique and Temperature Control*

Vacuum in the system was produced with a glass three-stage oil diffusion pump and a mechanical forepump. Oil vapor from the diffusion pump was condensed in a zeolite trap. Gold gaskets were used for the seals at the stainless steel flanges (AA); Teflon gaskets were used in the stainless steel bellows valves in the gas handling system. The hangdown tube was heated with a hinged tube furnace and the remainder of the vacuum system was

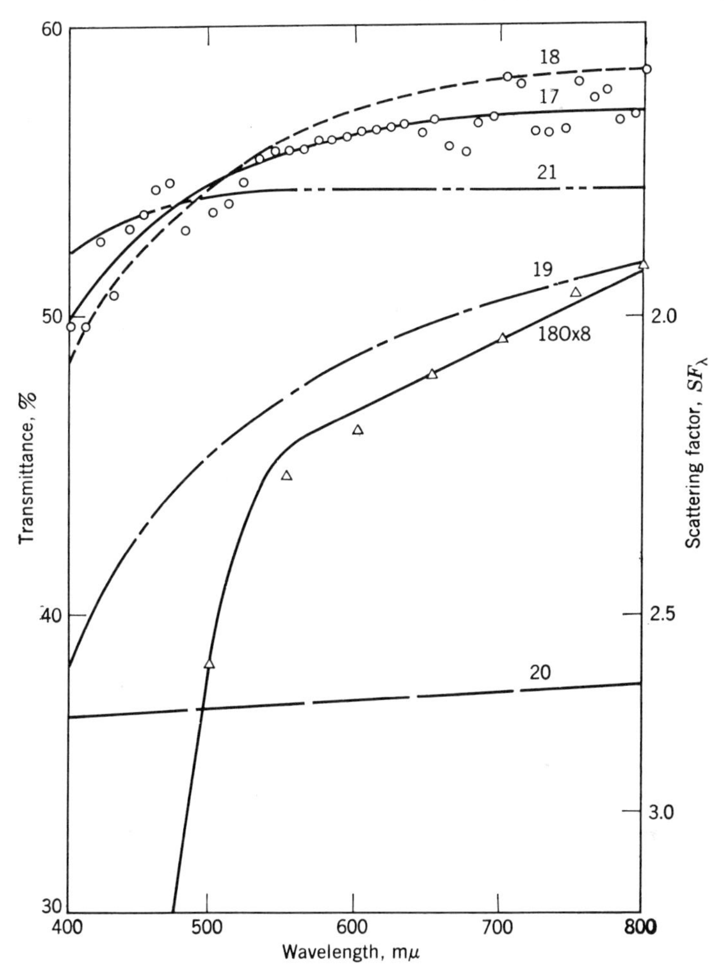

Figure 9.10. The dependence of the scattering factor on wavelength for five different film studies (curves 17–21). In curve 180x8, the transmittance determined by the use of the scattering factor (Δ's) is compared with the transmittance determined by removing the sample from the balance and placing it at C' in the blank path. In the latter case, no scattering occurs.

heated with tapes. Pressure below 1 m torr was read with a Veeco RG-75 ionization gauge and above one Torr with a Wallace and Tiernan dial manometer type FA-145.

Oxygen prepared by the thermal decomposition of chemically pure potassium permanganate and nitrogen prepared by the thermal decomposition of sodium azide were stored in bulbs. Reagent grade hydrogen was obtained in 1-liter glass break seal flasks from the Air Reduction Company.

The temperature of the hinged tube furnace was controlled to $\pm 0.5°$. The furnace temperature was determined with a chromel–alumel thermopile cemented to the furnace as near as possible to the sample plate. Prior to the film studies, the temperature of the sample was calibrated as a function of the thermopile temperature. Thereafter, the thermopile emf was measured and the sample temperature was obtained from the calibration curve.

4. Preparation of Films

For deposition of a film on the substrate (14), the silver mirror (*P*) was rotated to a vertical position using the soft iron core enclosed in glass at the end of the mirror holder (*Q*). The copper charge, which was formed from 99.997% copper sheet, was evaporated from a molybdenum boat.

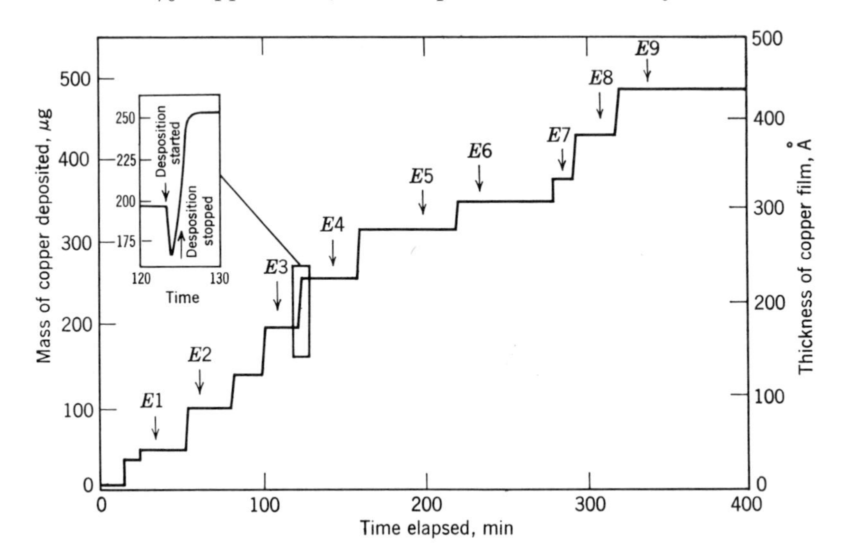

Figure 9.11. The mass gain by the substrate during the incremental deposition of a copper film.

The latter was heated by using a low voltage, high ac current obtained from a saturable core transformer. The copper charge was exposed by moving the nickel shield when sufficient outgassing of the charge was evident. The 25-cm distance from the boat to the substrate resulted in a film uniformity of about 1%. The copper deposited on the walls of the hangdown tube, which served as a getter during the preparation of each film, was removed with nitric acid after completing the study of each film.

The preparation of a copper film, 422 Å thick, which is typical for the technique used, was accomplished by employing several evaporation intervals. When the charge was heated, onset of deposition was observed as an apparent mass loss resulting from the momentum transfer of the impinging evaporant (Figure 9.11, inset). When the pressure reached 1 μ torr, the deposition was interrupted to allow the components near the evaporator assembly to cool before resuming evaporation. Data obtained during the preparation of the 422 Å film are summarized in Table 9.2 and Figure 9.11. The arrows labeled *E1* to *E9* indicate the time during preparation of the film when the transmittance of the copper film was measured. It is evident by inspection of Table 9.2 that it is possible to prepare a film to a desired thickness to within a few Angstroms by monitoring the mass gain during deposition. However, it is essential to establish that the thickness is related to the mass of material deposited by the bulk density or some other suitable constant in independent auxiliary experiments.

Table 9.2. Typical Data Obtained during the Incremental Deposition of a Copper Film

Evaporation no.	Deposition time, sec	Pressure, μT		Mass deposited, μg	Cumulative film		Deposition rate	
		Initial	Final		Mass, μg	Thickness, Å[a]	μg/min	Å/sec
1	19	0.07	1.0	3	3	2.67	9	0.13
2	94	0.5	0.6	30	33	29.3	20	0.30
3	37	0.1	0.5	16	49	43.6	43	0.64
4	122	0.5	1.0	51	100	88.9	30	0.45
5	115	0.3	1.0	39	139	125	20	0.30
6	140	0.3	0.7	58	197	175	25	0.37
7	105	0.3	0.5	57	254	226	33	0.49
8	90	0.2	0.4	58	312	278	39	0.58
9	50	0.1	0.2	18	330	293	21	0.31
10	45	0.1	0.3	18	348	310	25	0.37
11	50	0.1	0.3	27	375	334	33	0.49
12	80	0.1	0.4	54	429	381	40	0.59
13	65	0.1	0.3	46	475	422	42	0.62

[a] Calculated assuming bulk density is valid at all thicknesses.

D. Results and Discussion

1. The Transmittance of Copper Films during Incremental Deposition

The transmittance curves obtained at the notations *E1* and *E9* can be seen in Figure 9.12. This experiment, while quite simple, is nevertheless

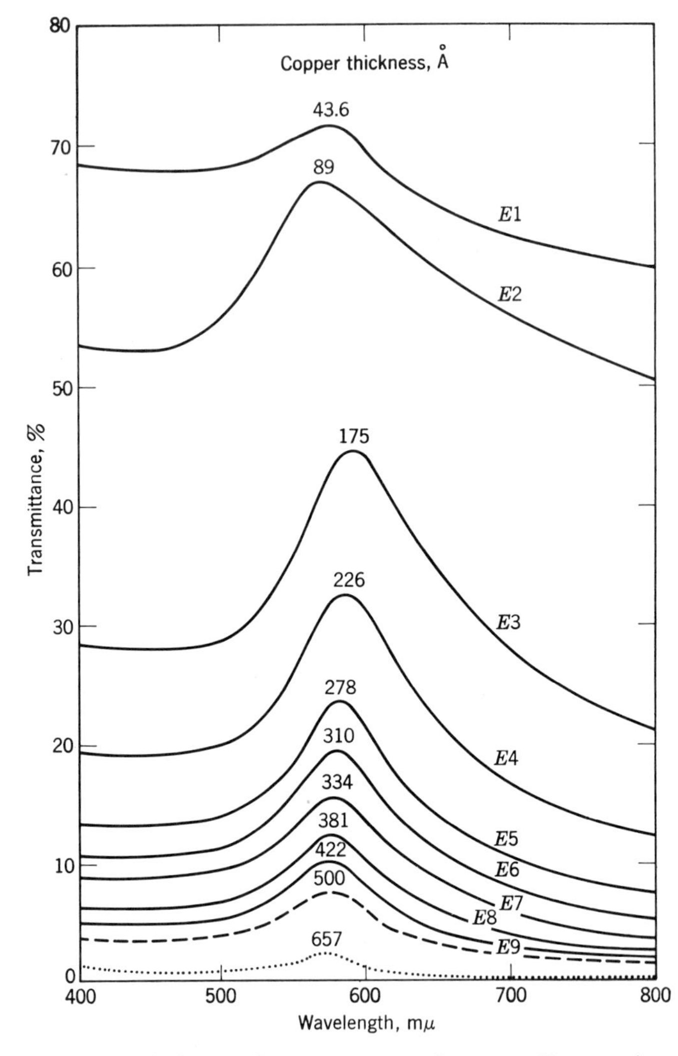

Figure 9.12. The optical transmittance spectrum of a copper film at various stages of preparation by the incremental deposition technique.

very revealing. The transmittance of the copper film at the blue end of the spectrum is a sensitive function of the thickness, showing that most of the light lost is absorbed. A plot of the logarithm of the transmittance versus thickness at 450 mμ, shown in Figure 9.13, is linear above 100 μg Cu (89 Å). The transmittance at 450 mμ measured for several other copper films evaporated onto substrates of the same size are plotted in this figure to show the reproducibility of the relationship between transmittance and thickness of the copper films. This result was achieved by maintaining the conditions of film preparation as constant as possible. This included the substrate temperature, evaporation rate, bakeout cycle, and degassing cycle prior to deposition, etc. When deviations in the pressure of deposition occurred, the transmittance as shown for film 20, would not fall on the line. The probable explanation is that the pressure during deposition of the latter film was allowed to reach 7 μ torr which was nearly an order of

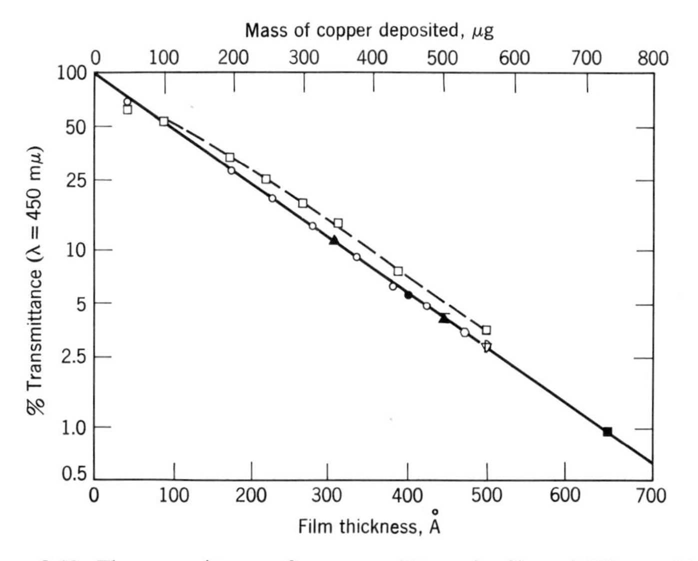

Figure 9.13. The transmittance of copper at 450 mμ for films of different thickness. The data for film 20 deviate from those for all other films because evaporation was carried out at higher pressures. ⬡ film = 11; ᗞ film = 14; ● film = 18; ■ film = 19; □ film = 20; ▼ film = 21; ▲ film = 22; ⊙ film = 17 (Figure 9); ▲̄ film = 3.

magnitude more than normally employed. Although the error in the thickness that would be read from such a point is only about 8%, it is illustrative of the sensitivity of the properties of the film to the preparation cycle. When films were prepared in auxiliary chambers, as discussed in Section III.D.2.b, the conditions used for preparation of the films were controlled the same way as during the *in situ* studies.

2. *The Oxidation of Films*

a. Determination of the Composition of the Film. The region of thickness where simultaneous measurements are useful is limited by the sensibility of the microbalance in the very thin film range and by the detection of transmitted light in the thick film region (14). For studies of the oxidation of copper, this range is 200–1500 Å (15). The optimum region, 400–600 Å, was explored extensively.

In the discussion which follows, the amount of oxygen incorporated into the film will be presented as a value of x in CuO_x. This does not imply that the oxygen is homogeneously distributed throughout the film, except for single phase regions. The actual distribution of oxygen may very well be in an oxide layer of a unique stoichiometry on top of an underlying layer of pure copper.

In previous work (15), it was shown that a gross defect phase of composition $CuO_{0.67}$ can be formed for films 200–1085 Å thick at temperatures of 110–200°C. It was also shown that the same composition and similar transmittance spectra were obtained by oxidation at progressively higher temperatures until $CuO_{0.67}$ was reached or by direct oxidation in the temperature region of metastable existence. Thus, the oxidation of the film described in the previous section by the direct oxidation technique will serve as a convenient example.

For oxidation of the 422 Å film, it was convenient to admit oxygen at room temperature to 100 torr pressure and heat the film to a higher temperature to oxidize it completely. During heating, simultaneous measurements of the mass gain of the film and its optical transmittance were carried out. The oxidation curve, $\Delta m(t)$, was obtained by heating the film to 155° in oxygen until no further oxidation was detectable, cooling to room temperature, and reheating to the same temperature *at the same rate*. The data obtained are plotted in Figure 9.14. The oxidation curve, $\Delta m(t)$, is seen to be the difference between the mass data obtained during the first and second heating cycles (curves I and II, respectively). During the first heating cycle, the recorded mass change resulted from the film oxidation and "other causes," e.g., buoyancy, TMF, etc. During the second and successive heating cycles to the same temperature at the same heating rate, the mass change resulting from "other causes" is recorded. While numerous studies have verified the validity of this technique for copper oxide films, it is evident that if decomposition occurs on cooling, this method of obtaining an oxidation curve is not accurate. It is also clear that in this case the data are of limited value for analysis of the kinetics of oxidation. However, where decomposition occurs, "thermooxidation," in which the temperature would increase linearly with time until oxidation is complete, could be employed rather than allowing the temperature to reach a maximum value asymptotically as was done in this work. Direct oxidation at a constant elevated temperature was not employed because the reaction was too rapid to allow the transmittance data to be taken in detail.

The transmittance of the film at various stages of oxidation can be seen in Figure 9.15. The arrows labeled in Figure 9.14 indicate the stages of oxidation where the corresponding transmittance data were taken. The base of the arrow indicates the time required to determine PS_λ from 400 to 800 mμ. Thus, even for this relatively slow oxidation, the change in x in

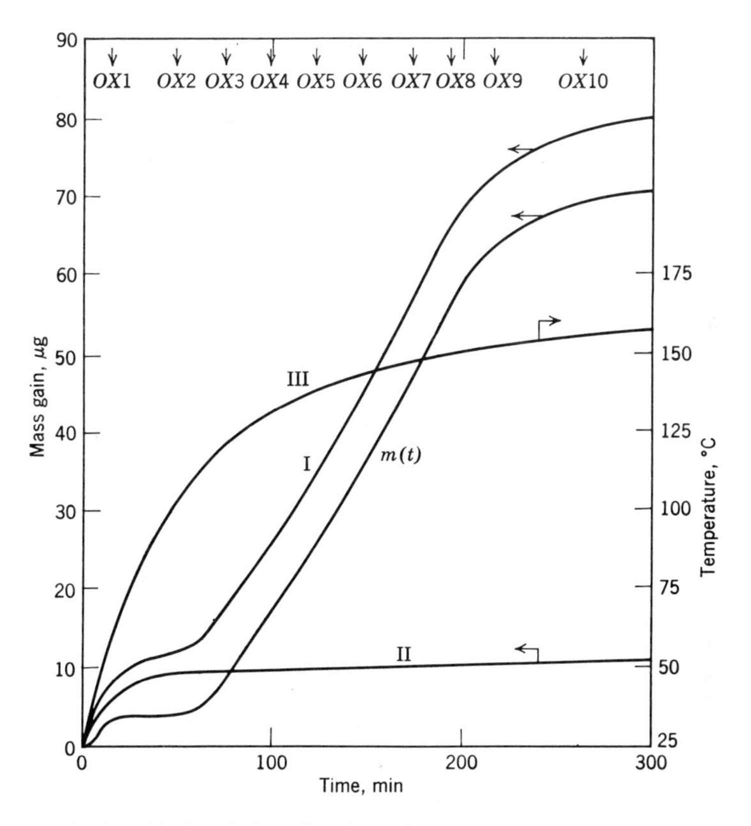

Figure 9.14. Graphical technique for determination of the mass gained during the direct oxidation of a copper film.

CuO_x in the steep portion of the curve was 0.015 from the low to the high wavelength end of the scan. Appropriate corrections were applied for detailed analysis of the transmittance spectra, although the mean values of x were used for labeling Figure 9.15.

b. Use of Transmittance Spectra for Auxiliary Studies. During the oxidation of a number of films of different thicknesses, hundreds of transmittance spectra were obtained at various stages of oxidation (14–20). It was found that for a given film thickness, there was a unique relationship between the transmittance and the stage of oxidation of the film. Thus, the transmittance measurement could be used to "fingerprint" a dynamic process

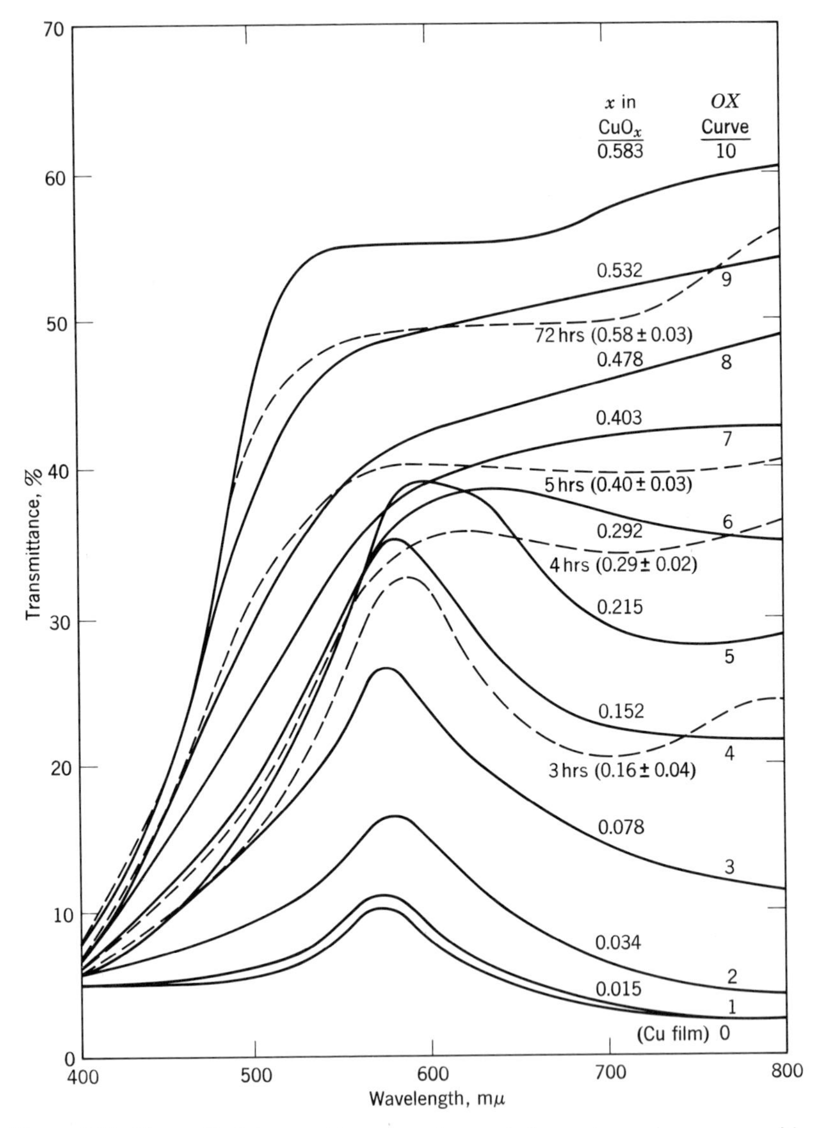

Figure 9.15. The optical transmittance spectrum of the system, air–copper oxide–copper–Pyrex glass at various stages of oxidation.

and allow other parameters to be measured with the convenience of a quasi-static process.

Data obtained during studies of the magnetic susceptibility of partially oxidized copper films serve as a convenient example of the utility of the transmittance spectra. In this study (18), eighteen copper films ranging from 400 to 500 Å in thickness were deposited on Pyrex glass substrates. Partial or complete oxidation of each film was carried out by heating to 140° in 100 torr of oxygen for various time intervals. After evaporation the optical transmittance of the copper film, corrected for room temperature oxidation, was used to determine the thickness of the copper. After partial or complete oxidation, the transmittance was again measured. The latter curve was used to estimate the x in CuO_x by superimposing it on the appropriate set of curves obtained from oxidation studies carried out using the

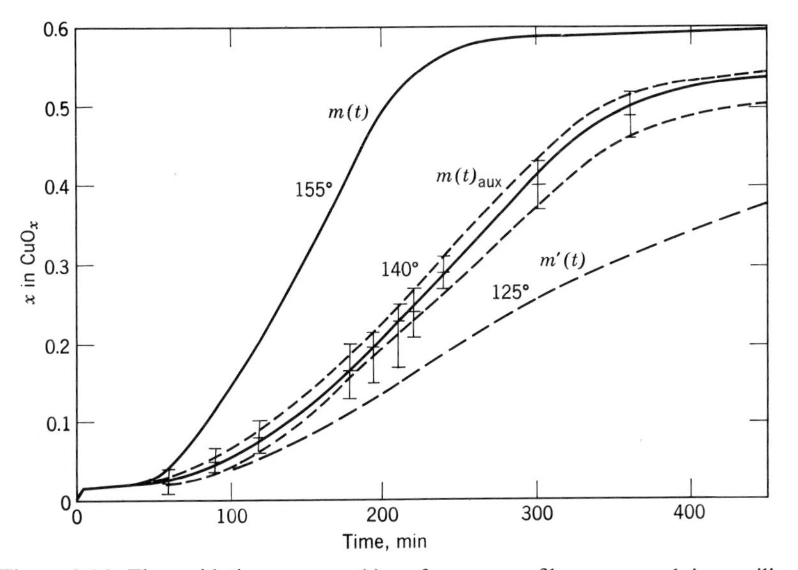

Figure 9.16. The oxidation curve $m(t)_{aux}$ for copper films prepared in auxiliary chambers as constructed from transmittance data. The curves, $m(t)$ and $m'(t)$ were obtained by the direct oxidation of copper films on the microbalance.

microbalance. The curves shown in Figure 9.15 represent one set of a series of film thicknesses, 400, 422, 450, 480, and 500 Å obtained in detail on the microbalance-transmittance apparatus (14) for the susceptibility comparison. Curves obtained during the partial oxidation of films in the auxiliary chambers close to 422 Å are plotted as broken lines in Figure 9.15 with the corresponding times of oxidation in the auxiliary chamber. By estimating the value of x in CuO_x for each film with the transmittance curve and using the time of oxidation on heating from room temperature, it was possible to construct the oxidation curve, $m(t)_{aux}$ for the auxiliary films as shown in Figure 9.16. The similarity between $m(t)_{aux}$ and $m(t)$, which was replotted from Figure 9.14, and $m''(t)$, replotted from previously published data (15), can be seen. The composition-susceptibility isotherm constructed from the measured values of the susceptibility and x in CuO_x is shown in Figure 9.17. The physical significance of this isotherm has been discussed (18).

The transmittance data were used in a similar manner for analysis of the structure of thin oxide films by using x-ray and electron diffraction techniques. This correlation was essential to the conclusion that $CuO_{0.67}$ is a

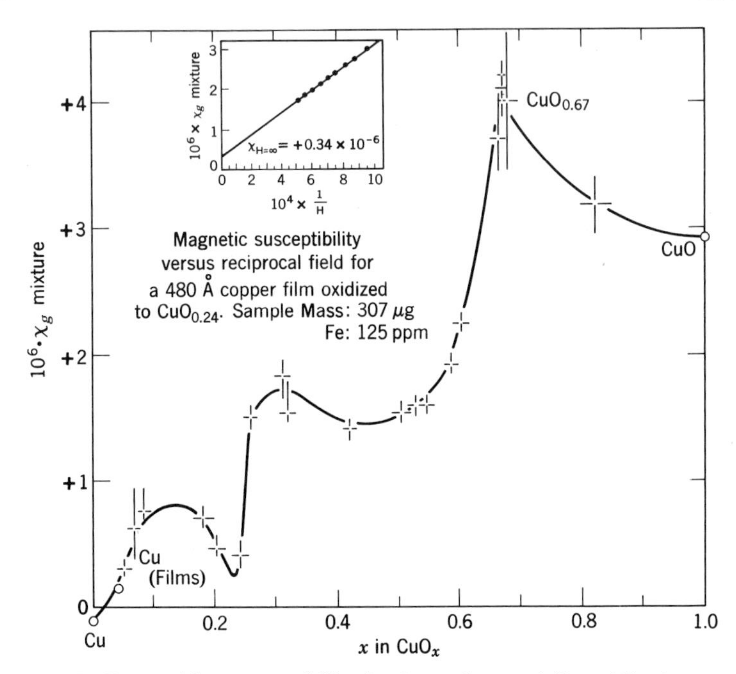

Figure 9.17. Composition susceptibility isotherm for partially oxidized evaporated copper films (400–500 Å). The range of error is indicated by the size of the cross. After A. W. Czanderna and H. Wieder, *J. Chem Phys.*, **39**, 489 (1963), American Institute of Physics.)

gross defect structure of Cu_2O or possibly an ordered defect phase Cu_3O_2 (15). Use of the transmittance data was also made for studies of the density of copper and copper oxide films (15) and for studies of the optical properties below room temperature (25).

c. Use of Transmittance Spectra for the Determination of the Optical Constants of Films. The transmittance spectra are of considerable value for determining the physical properties of thin films. For example, the study of the oxidation of copper films yielded spectra for a broad range of thicknesses. These were analyzed by a computerized trial and error process to evaluate the coefficients n and k of the complex index $\mathbf{n} = n - ik$. The measured transmittance spectra for several of these films are shown in Figure 9.18. The plotted points were calculated from the values deduced for n and k (19). The latter values may be useful in elucidating the mechanism of oxidation of copper in thin films (20,21). Since the oxidation to $CuO_{0.67}$ and then to CuO can be separated by careful control of the oxidation temperature, one of the two layer systems, $CuO_{0.67}$–Cu or CuO–$CuO_{0.67}$ are present. Equations have been published for the transmittance of light at normal incidence for the two layer systems (22). It must be assumed that the films are homogeneous and that the boundary between the layers remains parallel to the surface and the substrate but perpendicular to the incident light. From the optical constants of CuO, $CuO_{0.67}$ and Cu, the thicknesses of the layers, and the wavelength of the incident light, it is possible to calculate the transmittance and reflectance and compare these with experimental values obtained during the oxidation of copper films (15,20). Here, the simultaneous measurements are essential to establishing the value of x in CuO_x for each measured transmittance spectrum. From the assumption that the oxidizing film consists of two stoichiometric layers, the values of the layer thicknesses can be calculated and inserted into the equations. Thus, details of the mechanism of oxidation can be found (20) that would otherwise be difficult to assess in such quantitative detail. This type of study of the oxidation of copper films is under intensive study in the Department of Physics and the Institute of Colloid and Surface Science at Clarkson College of Technology, Potsdam, New York.

3. *Reduction of Films*

a. Mass Loss Determination. It has been shown (18) that it is possible to reduce $CuO_{0.67}$ films in 100 torr of hydrogen at temperatures ranging from 25 to 125°. Typical mass loss data obtained during reduction are plotted as a function of time in Figure 9.19. The sigmoidal shape obtained is typical for a nucleation, growth, and depletion mechanism. In the induction period, the average thickness of the copper nuclei formed on the oxide was 20–30 Å. The transition to growth of the nuclei was abrupt when the reduction temperatures exceeded 70°. The rate of mass loss was constant for compositions of copper mole fractions of 0.2–0.6 at all temperatures. The activation energy calculated for this mole fraction range was 12 ± 2 kcal/mol. The rate determining step during the growth of the nuclei could be the dissociation of a copper–oxygen–hydrogen surface complex or the diffusion of copper on a $CuO_{0.67}$ surface.

b. Use of the Transmittance Spectra. The transmittance data obtained during a typical reduction cycle from $CuO_{0.67}$ to $CuO_{0.04}$ are shown in

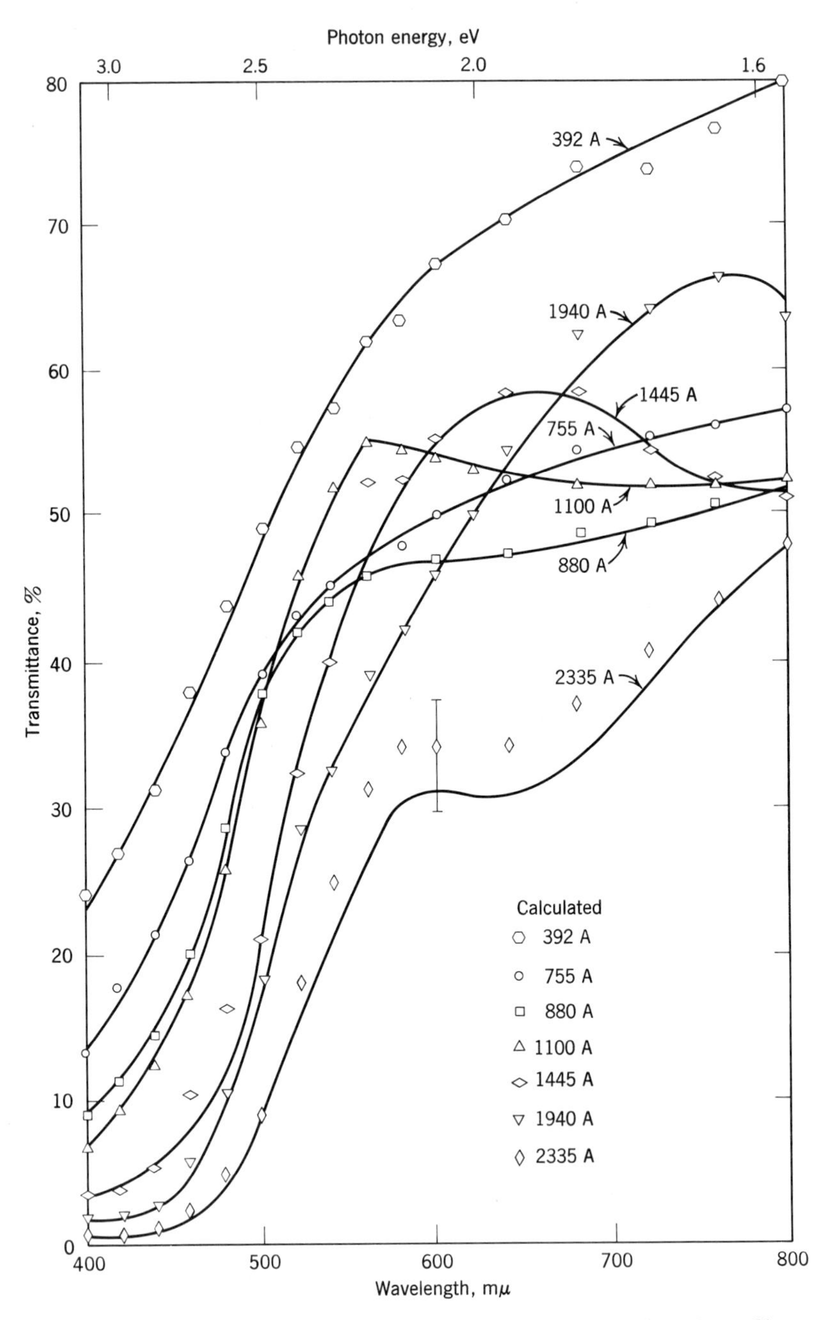

Figure 9.18. Wavelength dependence of the transmittance of $CuO_{0.67}$ films of different thicknesses. The solid lines are experimental curves; the "points" were calculated from the optical constants, n and k. (After H. Wieder and A. W. Czanderna, *J. Appl. Phys.*, **37**, 184 (1966), American Institute of Physics.)

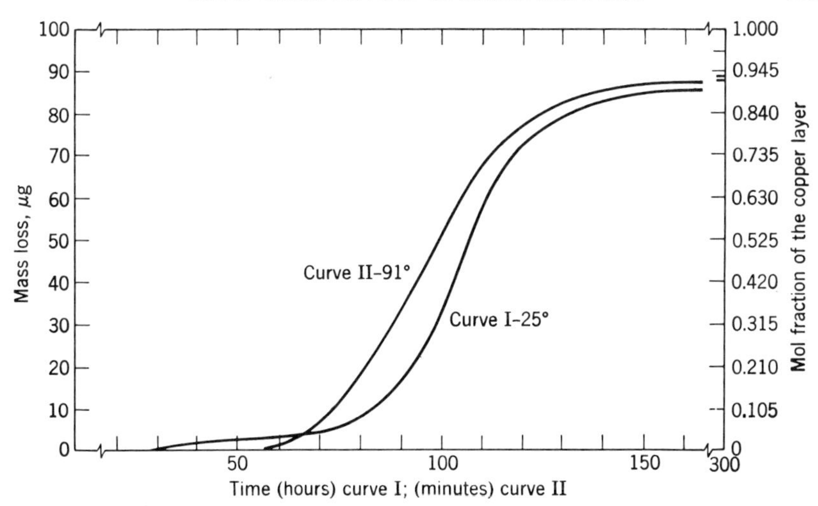

Figure 9.19. The reduction of $CuO_{0.67}$ in hydrogen. (Redrawn from *J. Phys. Chem.*) **69,** 3608 (1965), American Chemical Society.

Figure 9.20. These curves were analyzed as though the reduction process consisted of the formation of parallel layers of copper and $CuO_{0.67}$. Using the equations and optical constants for copper and the oxide, the transmittance of the system air–Cu–$CuO_{0.67}$–Pyrex glass was calculated. The transmittance shown by the experimental curves in Figure 9.20 decreases more rapidly with the formation of a small amount of copper than is calculated. This is not unexpected because each of the small nuclei formed on the surface serves as a scattering center and therefore has a different optical behavior than a smooth metal surface. The calculated curves follow qualitatively the experimental curves shown in Figure 9.20. It was possible to fit all the curves by using the expression

$$T = \exp\left(-\alpha_1 t_1 + \alpha_2 t_2\right) \tag{9.15}$$

where T is the transmittance at a given wavelength, α_1 and α_2 are the absorption coefficients of copper and of $CuO_{0.67}$ at that wavelength, and t_1 and t_2 are the respective thicknesses. In equation 9.15, the entire loss of light intensity is assumed to result from absorption by the film. There are initial effects that must be considered which make the α's effective absorption coefficients (16,17). The good fit of the curves of Figure 9.20 is an indication that no gross deviations from the simplified treatment existed at any stage of the reduction. This seems to indicate that the uniformity of the original film had not been drastically altered by the reduction of the $CuO_{0.67}$ film, in contrast to the reduction of CuO (16). The implication is that the number

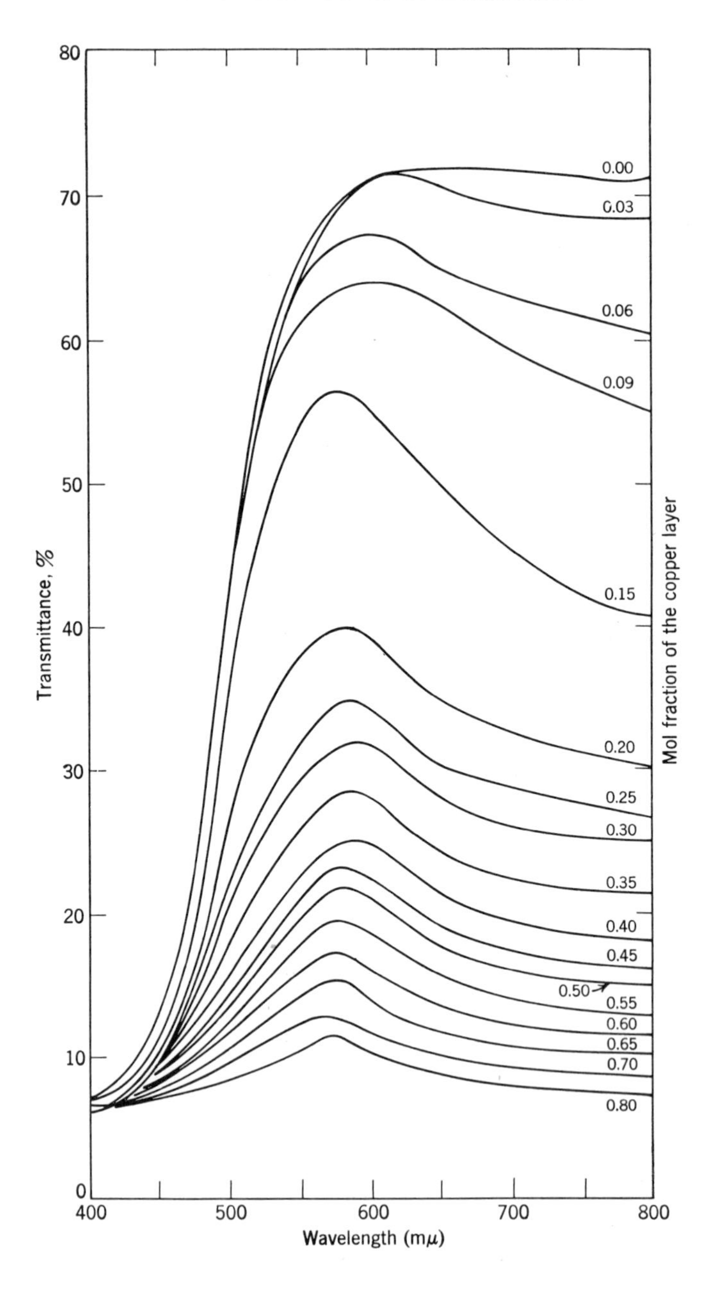

Figure 9.20. The optical transmittance spectrum obtained at various stages of reduction of $CuO_{0.67}$. Initial temperature, 85°, final, 50°. (Redrawn from *J. Phys. Chem.*, **69**, 3608 (1965), American Chemical Society.)

of copper nuclei formed was large and the size of each nuclide was small. It is also possible that the film uniformity was maintained because of epitaxial relationships between Cu and $CuO_{0.67}$.

In the same study (17), the effect of several cycles of oxidation to $CuO_{0.67}$ and reduction to $CuO_{0.04}$ on the transmittance of a copper film was investigated. It was not possible to reduce the oxygen content below $CuO_{0.04}$ at the reduction temperatures employed. However, transmittance spectra could be compared with those for partially oxidized film (20). The spectra obtained for the evaporated copper film, *E*, and the reduced film after several oxidation–reduction cycles are shown in Figure 9.21. The changes in transmittance occurred essentially in the first two cycles. It is likely that the differences result from a change in the extent of annealing of the film (25). The values of n and k deduced for the evaporated copper films (26) differ from the n for bulk copper. Greater values of n are indicative that the copper film is in a stressed condition (27,28). It is possible that the differences in n and k result from a change in the distribution of the small oxygen content of the incompletely reduced films but this is difficult to verify.

IV. CONCLUSIONS

The simultaneous measurements of mass change and optical spectra have been of great value for studies of the adsorption of gases on zeolites and the oxidation of thin copper films. With the modern capabilities of automatic recording microbalances, vacuum technique, and optical instrumentation, future studies of this type should prove fruitful. Other systems should be chosen carefully with consideration given to carrying out the optical measurements in the ultraviolet region of the electromagnetic spectrum.

Through simultaneous measurements of weight changes and infrared spectra it was possible to ensure unequivocally that the two types of measurements were carried out on the same sample under identical conditions. In the case of zeolites, by the combination of gas adsorption and spectral measurements for CO it was possible to calculate the strength of bonding, the number of effective sites and the molecular band intensities of the adsorbed CO.

ACKNOWLEDGMENTS

It is a pleasure for one of us (A.W.C.) to express gratitude to Dr. Harold Wieder with whom most of the work presented in Section III was carried out. The partial support by the Atomic Energy Commission for

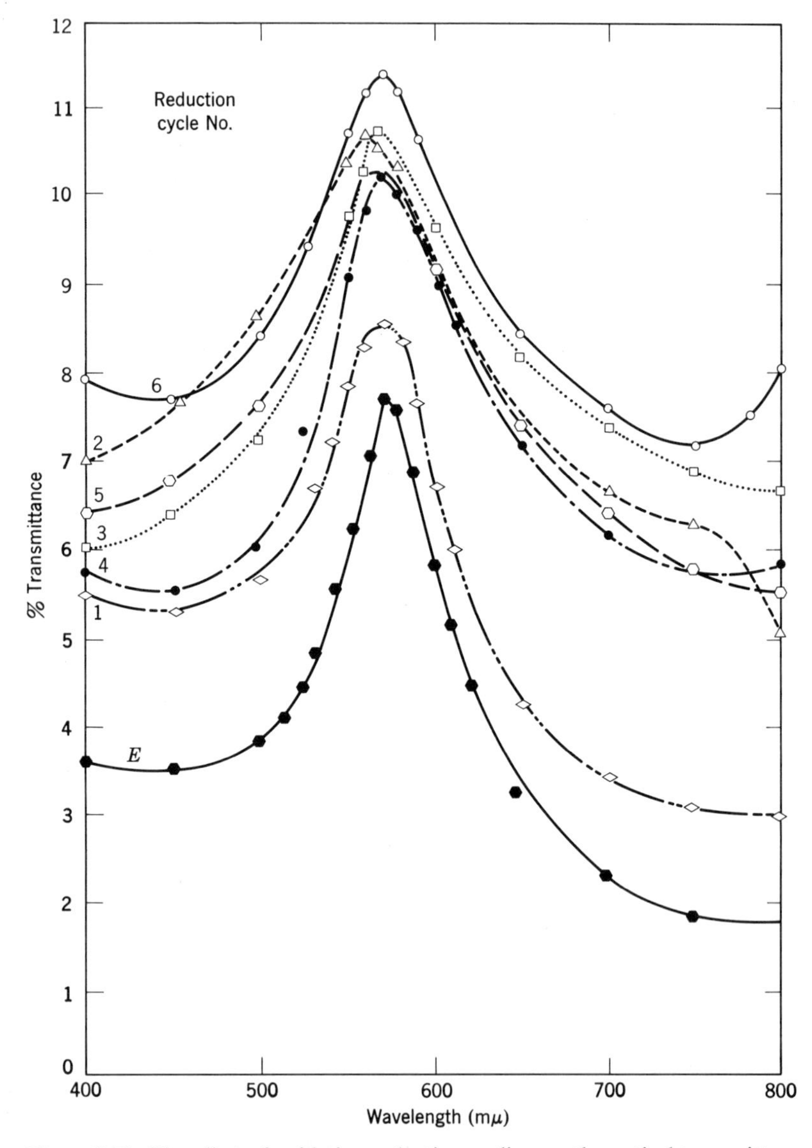

Figure 9.21. The effect of oxidation reduction cycling on the optical transmittance of a 500 Å copper film. *E* is for the film after it was evaporated; 1–6 corresponds to the spectrum obtained at the end of each oxidation reduction cycle. (Redrawn from *J. Phys. Chem.*, **69**, 3609 (1965), American Chemical Society.)

continuing this work, under contract AT(30–1)–3817, is also gratefully acknowledged.

V. APPENDIX 1

THEORETICAL EQUATIONS FOR COMPUTATION OF TRANSMITTANCE AND REFLECTANCE

The equations for computing the transmittance (T) and reflectance (R) for light incident from vacuum (n_0) onto an oxide film (n_1) on a metal film (n_2) supported on a transparent substrate (n_3) with absorption coefficients k_1 and k_2, ($k_0 = k_3 = 0$) and absorbing layer thicknesses of d_1 and d_2 at any wavelength (λ); i.e.

$\lambda \downarrow$		$\lambda \downarrow$		$\lambda \downarrow$
	air		$n_0, k_0 = 0$	
	oxide film	layer 1	n_1, k_1, d_1	
	oxide or metal film	layer 2	n_2, k_2, d_2	
	substrate		$n_3, k_3 = 0$	

are given by:

$$T = \frac{n_3(l_{13}^2 + m_{13}^2)}{n_0(p_{13}^2 + q_{13}^2)} \qquad R = \frac{t_{13}^2 + u_{13}^2}{p_{13}^2 + q_{13}^2}$$

where

$$l_{13} = (1 + g_1)(1 + g_2)(1 + g_3) - h_2h_3(1 + g_1) - h_3h_1(1 + g_2) - h_1h_2(1 + g_3)$$

$$m_{13} = h_1(1 + g_2)(1 + g_3) + h_2(1 + g_3)(1 + g_1) + h_3(1 + g_1)(1 + g_2) - h_1h_2h_3$$

$$p_{13} = p_{12}p_3 - q_{12}q_3 + r_{12}t_3 - s_{12}u_3 \qquad t_{13} = t_{12}p_3 - u_{12}q_3 + v_{12}t_3 - w_{12}u_3$$

$$q_{13} = q_{12}p_3 + p_{12}q_3 + s_{12}t_3 + r_{12}u_3 \qquad u_{13} = u_{12}p_3 + t_{12}q_3 + w_{12}t_3 + v_{12}u_3$$

$$\left.\begin{aligned}
p_{12} &= p_2 + g_1t_2 - h_1u_2 & w_{12} &= w_2 + h_1r_2 + g_1s_2 \\
q_{12} &= q_2 + h_1t_2 + g_1u_2 & v_{12} &= v_2 + g_1r_2 - h_1s_2 \\
r_{12} &= r_2 + g_1v_2 - h_1w_2 & u_{12} &= u_2 + h_1p_2 - g_1q_2 \\
s_{12} &= s_2 + h_1v_2 + g_1w_2 & t_{12} &= t_2 + g_1p_2 - h_1q_2 \\
p_2 &= \cos\gamma_1 \exp(a_1) & q_2 &= \sin\gamma_1 \exp(a_1) \\
p_3 &= \cos\gamma_2 \exp(a_2) & q_3 &= \sin\gamma_2 \exp(a_2) \\
r_2 &= (g_2\cos\gamma_1 - h_2\sin\gamma_1)\exp(a_1) & s_2 &= (h_2\cos\gamma_1 + g_2\sin\gamma_1)\exp(a_1)
\end{aligned}\right\} \quad (1)$$

$$\left.\begin{aligned}
t_2 &= (g_2 \cos \gamma_1 + h_2 \sin \gamma_1) \exp(-a_1)\\
t_3 &= (g_3 \cos \gamma_2 + h_3 \sin \gamma_2) \exp(-a_2)\\
u_2 &= (h_2 \cos \gamma_1 - g_2 \sin \gamma_1) \exp(-a_1)\\
u_3 &= (h_3 \cos \gamma_2 - g_3 \sin \gamma_2) \exp(-a_2)\\
v_2 &= \cos \gamma_1 \exp(-a_1)\\
w_2 &= -\sin \gamma_1 \exp(-a_1)
\end{aligned}\right\} \quad (2)$$

$$\left.\begin{aligned}
h_1 &= \frac{2n_0k_1}{(n_0 + n_1)^2 + k_1^2} & g_1 &= \frac{n_0^2 - n_1^2 - k_1^2}{(n_0 + n_1)^2 + k_1^2}\\
h_2 &= \frac{2(n_1k_2 - n_2k_1)}{(n_1 + n_2)^2 + (k_1 + k_2)^2} & g_2 &= \frac{n_1^2 - n_2^2 + k_1^2 - k_2^2}{(n_1 + n_2)^2 + (k_1 + k_2)^2}\\
h_3 &= \frac{-2n_3k_2}{(n_2 + n_3)^2 + k_2^2} & g_3 &= \frac{n_2^2 - n_3^2 + k_2^2}{(n_2 + n_3)^2 + k_2^2}
\end{aligned}\right\} \quad (3)$$

$$\left.\begin{aligned}
\gamma_1 &= \frac{2\pi n_1 d_1}{\lambda} & a_1 &= \frac{2\pi k_1 d_1}{\lambda}\\
\gamma_2 &= \frac{2\pi n_2 d_2}{\lambda} & a_2 &= \frac{2\pi k_2 d_2}{\lambda}
\end{aligned}\right\} \quad (4)$$

REFERENCES

1. "Sadtler Standard Spectra" collection, published by Sadtler Research Laboratories, Philadelphia, now contains over 30 thousand spectra.
2. (*a*) N. G. Yaroslavsky and A. N. Terenin, *Dokl. Akad. Nauk. SSSR*, **66**, 885 (1949); (*b*) D. J. C. Yates, N. Sheppard, and C. L. Angell, *J. Chem. Phys.*, **23**, 1980 (1955); (*c*) W. A. Pliskin and R. P. Eischens, *J. Chem. Phys.*, **24**, 482 (1956).
3. (*a*) L. H. Little, *Infrared Spectra of Adsorbed Species*, Academic Press, New York, 1966; (*b*) C. H. Amberg, *Infrared Spectroscopy and the Solid-Gas Interface*," in *The Solid–Gas Interface*, Vol. II, E. A. Flood, Ed., Dekker, New York, 1967; (*c*) M. L. Hair, *Infrared Spectroscopy in Surface Chemistry*, Dekker, New York, 1967.
4. L. H. Little, N. Sheppard, and D. J. C. Yates, *Proc. Roy. Soc.* (*London*), **A259**, 242 (1960).
5. L. H. Little, H. E. Klauser, and C. H. Amberg, *Can. J. Chem.* **39**, 42 (1961)
6. D. A. Seanor and C. H. Amberg, *J. Chem. Phys.*, **42**, 2967 (1965).
7. J. A. Rabo, C. L. Angell, P. H. Kasai, and V. Schomaker, *Discussions Faraday Soc.*, **41**, 328 (1966).
8. J. M. Bennett and J. V. Smith, *Materials Research Bulletin*, **3**, 633 (1968).
9. C. L. Angell and P. C. Schaffer, *J. Phys. Chem.*, **69**, 3463 (1965).
10. J. B. Peri, *Proc. 2nd Intern. Congr. Catalysis, Paris*, 1960, p. 1333. Edition Technip, Paris, 1961.
11. D. A. Seanor and C. H. Amberg, *Rev. Sci. Instr.*, **34**, 917 (1963).
12. C. L. Angell, in *Vacuum Microbalance Tech.*, **6**, 73 (1967).
13. C. L. Angell and P. C. Schaffer, *J. Phys. Chem.*, **70**, 1413 (1966).
14. A. W. Czanderna and H. Wieder, *Vacuum Microbalance Tech.*, **2**, 147 (1962).
15. H. Wieder and A. W. Czanderna, *J. Phys. Chem.*, **66**, 816 (1962).

16. H. Wieder and A. W. Czanderna, *J. Chem. Phys.*, **35**, 2269 (1961).
17. A. W. Czanderna, *J. Phys. Chem.*, **69**, 3607 (1965).
18. A. W. Czanderna and H. Wieder, *J. Chem. Phys.*, **39**, 489 (1963).
19. H. Wieder and A. W. Czanderna, *J. Appl. Phys.*, **37**, 184 (1966).
20. A. W. Czanderna and H. Wieder, unpublished.
21. J. R. Biegen, M.S. Thesis, Clarkson College of Technology, Potsdam, New York, 13676, 1967.
22. O. S. Heavens, *Optical Properties of Thin Solid Films*, Dover, New York, 1965, p. 46.
23. P. H. Berning, in *Physics of Thin Films*, Vol. 1, G. Hass, Ed., Academic Press, New York, 1963, p. 69.
24. O. S. Heavens, in *Physics of Thin Films*, Vol. 2, G. Hass and R. Thun, Eds. Academic Press, New York, 1964, p. 187 ff.
25. H. Wieder, unpublished; F. L. Boyko, M.S. Thesis, Clarkson College of Technology, Potsdam, New York, 13676, 1968.
26. A. W. Czanderna and F. L. Boyko, to be published.
27. O. S. Heavens, *Optical Properties of Thin Solid Films*, Dover, New York, 1965, f. 199.
28. L. G. Schultz, *J. Opt. Soc. Am.*, **44**, 357 (1954).

Chapter 10

High Temperature Reaction Studies

EARL A. GULBRANSEN AND FRED A. BRASSART

Westinghouse Research Laboratories
Pittsburgh, Pennsylvania

I. INTRODUCTION

A. Types of Chemical Reactions

One of the important and interesting properties of high temperature materials is the chemical reactivity. The chemical reactivity is important in the winning, pretreatment, and use of the material. It is not often realized that a metal at high temperature is a zone of many different types of

Table 10.1. Types of Chemical Reactions with Examples

1. Direct oxidation, nitriding or hydriding

$$Zr(s) + O_2(g) \rightleftharpoons ZrO_2(s)$$

2. Direct formation of volatile oxide

$$Mo(s) + \tfrac{3}{2}\, O_2(g) \rightleftharpoons \tfrac{1}{3}(MoO_3)_3(g)$$

3. Direct solution of oxygen, nitrogen, or hydrogen in metal

$$Zr(s) + O_2(g) \rightleftharpoons Zr \text{ (solid solution of } O_2)$$

4. Reaction with H_2O

$$Zr(s) + 2H_2O(g) \rightleftharpoons ZrO_2(s) + 2H_2 \text{ (solid solution or gas)}$$

5. Direct oxidation of lower oxide

$$2Fe_3O_4(s) + \tfrac{1}{2}O_2(g) \rightleftharpoons 3Fe_2O_3(s)$$

6. Formation of complex oxides

$$FeO(s) + Fe_2O_3(s) \rightleftharpoons Fe_3O_4(s)$$

7. Reduction of higher oxide by metal

$$Fe_3O_4(s) + Fe(s) \rightleftharpoons 4FeO(s)$$

8. Reduction of oxide by C or S

$$FeO(s) + C(s) \rightleftharpoons Fe(s) + CO(g)$$

9. Volatility of oxide

$$3WO_3(s) \rightleftharpoons (WO_3)_3(g)$$

chemical reactivity. Figure 10.1 shows some of the physical and chemical phenomena occurring on or near a metal surface. Five zones are shown: metal, metal–oxide interface, oxide, oxide–gas interface, and gas. Table 10.1 shows the reactions in the form of chemical equations.

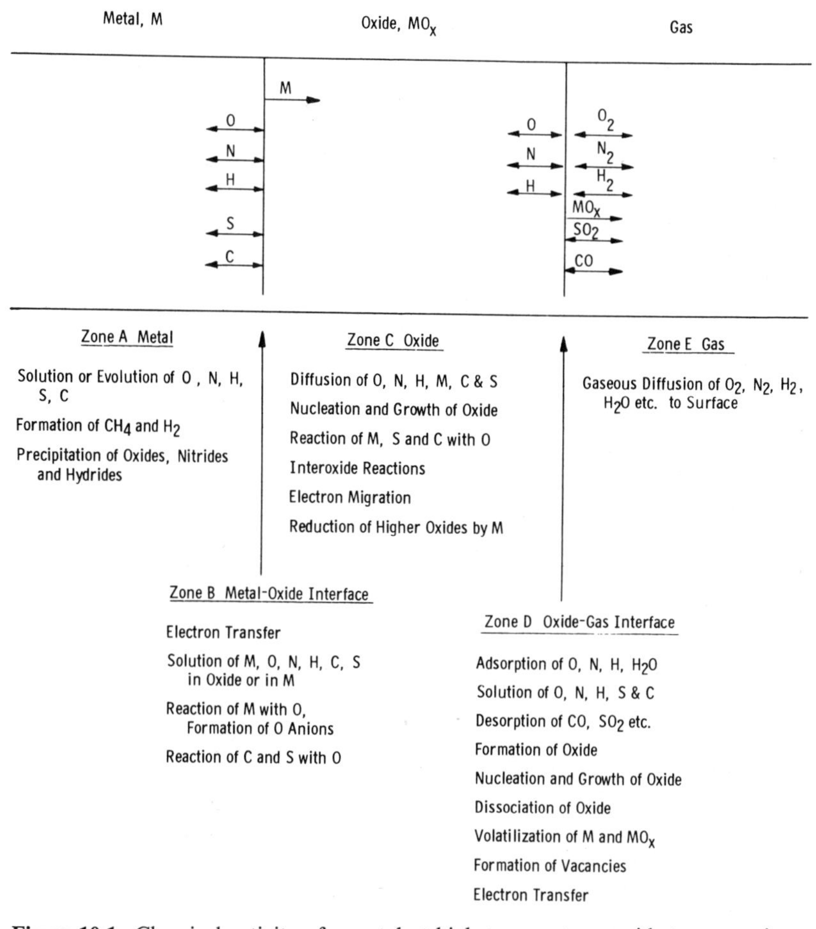

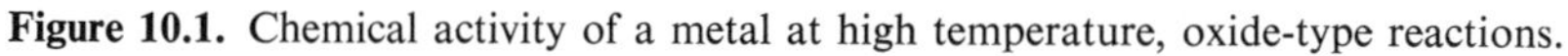
Figure 10.1. Chemical activity of a metal at high temperature, oxide-type reactions.

The occurrence of a particular reaction depends on the free energy of the reaction and to a lesser extent on physical factors such as gas flow and temperature gradients in the reaction system. The free energy of a particular reaction is a function of temperature, pressure, and gas composition. Reactions to form vacancies and other lattice defects are not included in Table 10.1.

From a practical point of view all of the reactions shown in Figure 10.1

and Table 10.1 must be understood if the material is to be used at high temperature. Unfortunately, only a limited amount of information on these reactions is available for most metals.

For the common metals, such as iron, chromium, nickel, cobalt, and aluminum, oxidation occurs by the transport of oxygen atoms or ions or metal atoms or ions and electrons through an oxide film formed on the metal. The general principles for this type of oxidation are understood. The oxidation resistance of these metals is limited by (*1*) the melting points, (*2*) the volatility of the metal, (*3*) the transport of metal or oxygen ions through the oxide, and (*4*) the breakdown of the oxide film.

The oxidation resistance of the highest melting point and lowest volatility elements, tungsten, molybdenum, carbon, and rhenium, is not limited by transport of reaction species through an oxide scale or film since volatile oxides are formed at high temperatures. For some conditions, adsorption of oxygen, desorption of the oxide or chemical reaction of the adsorbed species limits the kinetics of oxidation. For other conditions, the diffusion of oxygen molecules through a barrier layer of volatilized oxide vapors controls the rate of oxidation of these elements. The transition between reactions controlled by chemical processes occurring at the interface and those controlled by diffusion of oxygen through the volatilized gases has recently been studied (1,2).

High temperature materials for many energy producing and space systems are used under severe flow conditions. Flow can accelerate gas–solid reactions by removing the reaction products so that normal equilibrium conditions are not achieved, i.e., by removing the barrier layer of oxide vapors which may normally limit access of oxidizing gases to the surface. Thermochemical calculations on these reactions must be used with caution since it is difficult to define the reaction conditions when the products of reaction are being removed by the gas flow.

The nature of the reacting gases, whether molecular or atomic, is also important. The equilibrium between atomic and molecular oxygen can be calculated from free energy data (3). Above 1500°C, the dissociation of molecular to atomic oxygen becomes important for oxygen pressures below 1 torr. The oxygen atom concentration for temperatures and pressures of interest in this study can be neglected.

B. Purpose of Chapter

It is proposed to discuss vacuum microbalance apparatus and methods used for studying various types of oxidation reactions occurring at temperatures from 500 to 1650°C with special emphasis on the temperatures from 1000 to 1650°C. To establish the problem, it is necessary to discuss the thermochemistry of the reactions and some of thc theoretical analyses of the various reaction mechanisms. Unless the relationship between the experimental kinetics, the thermochemistry, and the mechanisms of reaction can be achieved, each study will stand alone. The high temperature oxidation of carbon, molybdenum, and tungsten will be used to illustrate the value of vacuum microbalance methods for evaluating high temperature reactions.

C. Preliminary Thermochemical Analyses

One of the first questions a chemist should ask before starting a study is what are the thermochemical conditions required for a reaction to occur? The free energy of the particular reaction can be used to answer this question. The free energies of formation of the reaction products of chemical reactions, which vary widely, can be used to classify the various gas–metal and gas–metal oxide reactions.

One of the practical uses of thermochemical analyses is to evaluate the relative stabilities of the various oxides, nitrides, and hydrides of the common metals. Table 10.2 gives the standard free energies of formation ΔG_f° of a number of oxides, nitrides, hydrides, and carbides and values of the equilibrium constants (K_p) at 25°C. The term ΔG_f° is related to K_p by the equation

$$-\Delta G_f^\circ = 4.575T \log K_p \tag{10.1}$$

Equilibrium pressures of oxygen, hydrogen, or nitrogen over the various compounds can be evaluated from log K_p. For example, the equilibrium pressure of oxygen over NiO and Ni at 25°C, as derived from equation 10.2

$$K_p = 1/(p_{O_2})^{1/2} \tag{10.2}$$

is $10^{-75.8}$ atm at 25°C.

The oxides in Table 10.2 are classed as unstable, reducible by H_2 or CO, or refractory. Nitrides are classed as unstable or stable. It will be noted that only a few of the metal oxides can be reduced by H_2 or CO.

We conclude from Table 10.2 and similar data that all metals with the exception of gold react with oxygen at room temperature conditions. Even gold may react at sufficiently high pressure and temperature. Some metals react with nitrogen to form stable nitrides and a few react with hydrogen to form stable hydrides. Zirconium hydride is given as an example in the latter case.

Table 10.2 gives the thermochemical data at a single temperature, 25°C. Many reaction studies are of interest at different temperatures or at more than one temperature; therefore, it is essential to know the free energy

Table 10.2. Free Energies of Formation and Equilibrium Constants of a Number of Oxides, Nitrides, and Hydrides

Compound	$-\Delta G^{\circ}_{52°C}$ kcal/atom of O, N, or H	$\log K_{p(25°C)}$	
	Oxides		
Au_2O_3(s)	−13	−28.6	Unstable
Ag_2O(s)	2.6	1.9	Reducible
CuO(s)	30.4	22.3	with H_2
CoO(s)	51.0	37.4	or CO
NiO(s)	51.7	37.9	
MoO_3(s)	54.0	118.7	
$[H_2O–H_2]$(g)[a]	54.6	40.0	
MnO_2(s)	55.7	81.7	
$FeO_{0.95}$(s)	58.4	42.8	
WO_3(s)	60.8	133.7	
$[CO_2–CO]$(g)[a]	61.5	45.0	
V_2O_5(s)	68.8	252.1	Refractory
Cr_2O_3(s)	83.4	183.4	
Ta_2O_5(s)	94.1	344.9	
TiO_2(rutile, s)	101.9	149.4	
ZrO_2(s)	122.2	179.1	
Al_2O_3 (corundum, s)	125.6	276.2	
	Nitrides		
Fe_4N	−0.89	−0.65	Unstable
Fe_2N	−2.6	−1.9	
CrN	29.8[b]	—	Stable
VN	35	25.7	
AlN	50.1	36.7	
TiN	66.1	48.5	
ZrN	75.4	55.3	
	Hydrides		
$ZrH_{1.5}$[c]	17.1	—	Stable

[a] Oxygen in equilibrium with H_2–H_2O or CO–CO_2 gas of 1–1 mixture.
[b] Value is for $\Delta H^{0}_{f(25°C)}$.
[c] See reference 4.

data or equilibrium constants at the temperatures of interest. Thermochemical data at various temperatures have been compiled by a number of workers (5–7). Thermochemical data as in Table 10.2 and data compiled at higher temperatures can be combined with other thermochemical data to predict equilibrium conditions for a wide variety of reactions related to the system of interest at different temperatures.

Where possible the phase diagram of the gas–metal system should be utilized to establish the reaction conditions. Such a phase diagram should include the solubility limit of the gas in the metal over the temperature range, the range of homogeneity of each gas–metal compound, the range of existence of mixed phases, and the free energy values for each of the phases.

Precautions must be observed in order to apply thermochemical data which are valid for equilibrium conditions. Many reactions involve the formation of gases which (*1*) are removed from the reaction zone as in a dynamic system, (*2*) react with the furnace tubes, and (*3*) condense on cooling or are quenched by cold surfaces. For these conditions, chemical reactions which may otherwise not occur can proceed since equilibrium is never achieved. An interesting example is the reaction of carbon with hydrogen at 1200–1600°C. In a static reaction system, only small quantities of hydrocarbon gases would form since these gases are unstable. However, in a flow system the small quantities of hydrocarbons formed are continually removed thereby allowing the reaction to progress (8).

D. Extent of Reaction

Microbalance methods can be used in a number of ways to evaluate the high temperature reactivity of metals. For many purposes the measurement of the thickness of the reaction product or the extent of metal loss provides sufficient data. The rate of reaction and its time, temperature, and pressure dependence are more difficult to evaluate. However, it is this type of information that can be used to predict the long time reaction behavior at various conditions of pressure and temperature. In addition, an analysis of such data may provide an understanding of the mechanisms of reaction.

Figure 10.2 shows a scheme for classifying the extent of an oxidation reaction, when stable reaction products are formed, by using the thickness of the reaction product. When metal is lost as in the case of a volatile oxide, the thickness of the metal removed can be calculated directly from the change in weight of the metal and its density. For oxidation without volatilization the following equation is used

$$X = (W/\rho d)(\mathrm{M}_x\mathrm{O}_y/y\mathrm{O}) \times 10^8 \qquad (10.3)$$

Here X is the thickness in angstroms, ρ is the surface roughness ratio, d is the oxide density, M_xO_y and yO are the formula weight of the oxide and oxygen, respectively, and W is the weight gain in grams per cm². Weight gain per unit area can also be used to classify the extent of reaction. Six thickness ranges for stable reaction products are shown in Figure 10.2.

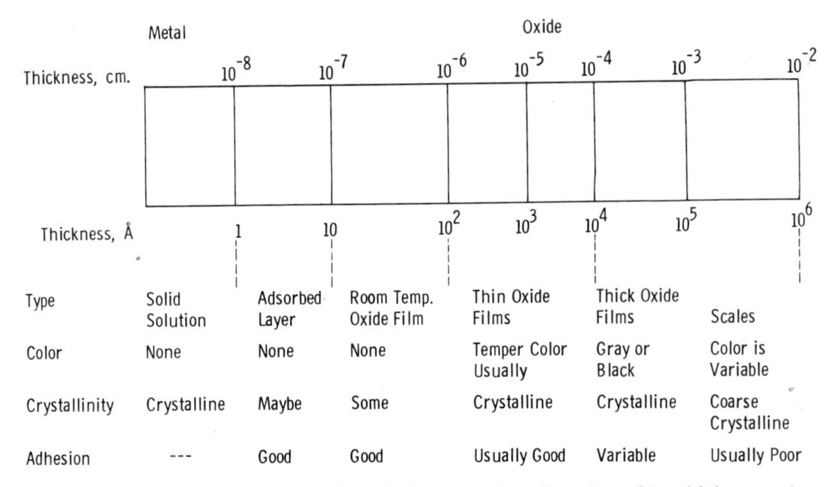

Figure 10.2. Classification of oxidation reactions based on film thickness.

The classification scheme of Figure 10.2 is arbitrary, especially in relation to the thickness limits assigned to thick and thin films and scales. One value of this classification scheme is to point out that oxide films continually change in crystal size and adhesion characteristics as oxidation proceeds. Changes in crystallinity, composition, structure, and adhesion make it difficult to develop empirical and theoretical expressions for the rate of reaction.

1. Units

A wide variety of units are used to describe the velocity of gas–metal reactions. Some of the more commonly used units are based on the rates of change of weight, dw/dt, of thickness, dx/dt, and of the number of atoms reacted, dn'/dt. The weight change, gain or loss, can have the units of grams, milligrams, or micrograms per cm²-sec or per dm²-sec. Many engineers prefer the units of milligrams per dm²-month or day which yield more manageable numerical values. The thickness term, dx/dt, refers to oxide films formed or to metal lost and may be expressed in units of angstroms, centimeters, or inches per second, day, month, or year. The term dn'/dt refers to the number of oxygen or metal atoms reacted per cm²-sec. The term dn'/dt is very useful since it allows a direct comparison with collision theory calculations and the efficiency of conversion of oxygen to oxide in a flow system.

II. EMPIRICAL AND THEORETICAL RATE LAWS

A. General

One of the principal objectives of studies of gas–metal reactions is the development of empirical and, where possible, theoretical rate laws based on physical theory to explain the kinetics of the reaction. Although many workers have attempted to explain the kinetics of gas–metal reactions with one unified rate law, this objective has not been realized.

Empirical rate laws allow one to extrapolate rates of reaction to new experimental conditions while theoretical rate laws, in addition, lead to an understanding of the reaction. A complete understanding allows one to make predictions on other reactions. Four variables are usually considered in the development of either empirical or theoretical rate laws: time, temperature, pressure, and gas flow.

The various rate controlling processes can be classified into four main groups: (*1*) diffusion of the reacting gas or reaction products to and from the reacting surface, (*2*) reactions at the gas–oxide or gas–metal interface, (*3*) diffusion of metal, gas, and electrons in the oxide or metal phases, and (*4*) solid state processes other than diffusion.

Table 10.3 shows details of this classification system including the effects of time, temperature, pressure, and gas flow. The major effects of gas flow are to maintain the supply of reaction gas and to remove any gaseous reaction products. Flow also affects the transition temperature between interface controlled and gas diffusion controlled reactions. Table 10.3 is used to classify the kinetics of a wide variety of chemical reactions where oxide films are formed or where volatile reaction products are formed.

Empirical relationships for chemical reactions can be developed by using log W vs log time, log pressure, or log temperature plots. This enables the development of empirical rate laws, such as the cubic rate law which gives the time dependence of the reaction.

$$W = At^{1/3} + D \tag{10.4}$$

Here W is the weight gain, A and D are constants, and t is the time. The term A, which incorporates the temperature dependence, is given by an exponential equation

$$A = B\exp(-C/T) \tag{10.5}$$

Table 10.3. Rate Controlling Processes

Process: Type	Process: Subtype	Effect of time on rate	Effect of pressure on rate	Effect of temperature on rate	Effect of gas flow on rate
Gas diffusion	*a.* O_2, N_2, or H_2 to interface	None	Linear	Small	Large
	b. Reaction product into gas phase, e.g. $(MoO_3)_3$	None	Linear	Small	Large
Surface interface oxide free	*a.* Adsorption	None	$p^{1/2}$ or p	Exponential $\Delta H < 50$ kcal/mole	None
	b. Desorption	None	None	Exponential $\Delta H > 50$ kcal/mole	None
	c. Chemical reaction	None	$p^{1/2}$ or p	Exponential $\Delta H < 50$ kcal/mole	None
Diffusion in solid	*a.* Oxide	$t^{1/2}$ or $t^{1/n}$	Depends on mechanism	Exponential	None
	b. Metals	$t^{1/2}$ or $t^{1/n}$	Depends on mechanism	Exponential	None
	c. Electrons	$\log t$	Depends on mechanism	Exponential	None
Physical	*a.* Phase transformation	Unknown	Unknown	Unknown	Unknown
	b. Crystal growth	Unknown	Unknown	Unknown	Unknown
	c. Recrystallization	Unknown	Unknown	Unknown	Unknown
	d. Spalling	Unknown	Unknown	Unknown	Unknown

Here B and C are constants and T is the absolutc temperature. If a mechanism can be found for the reaction, C can be expressed as $\Delta H/R$ where ΔH represents the enthalpy of activation of the rate controlling process. The preexponential term, B, may also be related to details of the reaction mechanism. This equation has been found useful in representing the effect of time on the oxidation of zirconium over a wide range of temperatures, pressures, and reaction times.

Theoretical rate laws have been developed for most of the rate controlling processes of Table 10.3. These laws could be the subject of a separate paper and will not be developed here.

Table 10.3 can be used to analyze experimental data in order to choose a possible mechanism for a reaction. This is illustrated in Appendix 1 which contains a theoretical analysis of the oxidation mechanisms of graphite. The oxidation of graphite is a classic example of a surface interface reaction free from oxide film formation. Similar analyses could be made for the oxidation of molybdenum, tungsten, and rhenium.

III. APPARATUS AND METHODS

Static and dynamic or flow types of reaction systems can be used to measure the wide variety of chemical reactions occurring on metal surfaces. A static reaction system is one where no external flow is imposed on the system. Some gas circulation occurs as a result of thermal gradients along the furnace tube. If such a system is operated at constant pressure, the makeup gas as well as the weight change of the sample can be used to evaluate the kinetics of the reaction.

A flow reaction system is one in which external gas flow is imposed on the system which is usually at atmospheric pressure. For the study of fast reactions, a flow-type reaction system has many advantages. Flow systems operating at pressures below atmospheric require control of the reacting gas source and also a pump to exhaust the product gases.

Some of the common methods for following rates of gas–solid reactions involve observations of: (*1*) weight change, (*2*) oxygen consumption, (*3*) thermal conductivity, (*4*) gas composition, (*5*) electrical resistivity of the metal specimen, and (*6*) specimen dimensional change.

Weight change methods which can be used in static and dynamic reaction systems have been developed extensively during the past six years. Oxygen consumption methods have a bright future with the recent development of sensitive capacitance manometers and bakeable quartz pressure gauges. Thermal conductivity and gas analysis methods have also been developed extensively. This discussion will be limited to weight change and oxygen consumption methods.

A. Static Reaction Systems

In a static reaction system operating at constant pressure, it is possible to use both weight change and oxygen consumption methods to follow the course of a gas–solid reaction. The combined use of the two methods in the

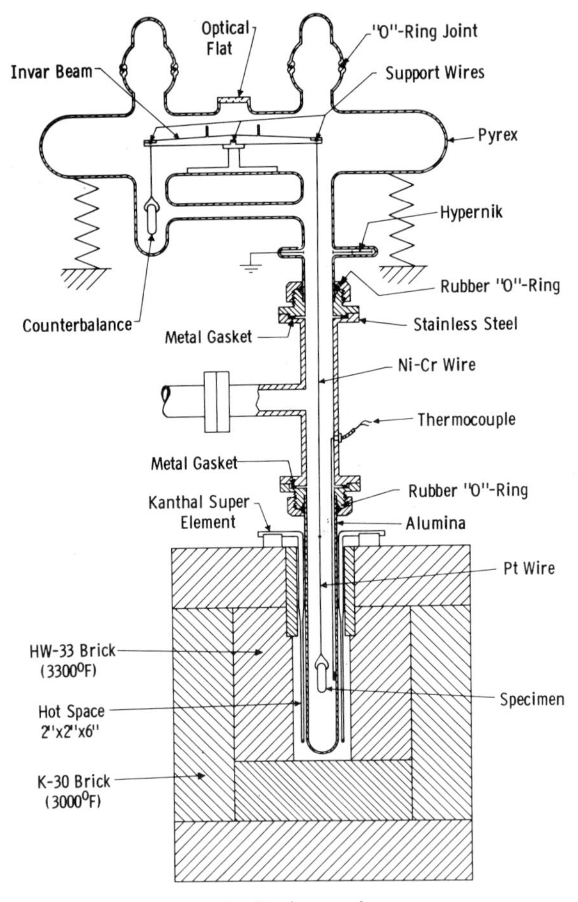

Figure 10.3. Static reaction system.

same system permits an analysis of complex chemical reactions such as those involving the simultaneous formation of oxide films and the volatilization of oxides (3,9–11).

Figure 10.3 shows the balance housing, furnace tube, and associated reaction system. Various balances having an overall beam length of

14.5 cm and sensitivities as low as 1 μg/0.001 cm deflection at 7.25 cm have been used. The above deflection is equivalent to an angular deflection of 0.00014 radians. Figure 10.4 shows a gold-plated Invar beam balance (12) which was designed for the study of fast reactions. The sensitivity and period can be varied by positioning movable beam weights which are not shown in Figure 10.4. The beam weighs 46 g and has a sensitivity of 75 μg/0.001 cm at 7.25 cm distance for a 1.4 g sample weight.

Specimens were suspended from the balance using 8 or 20 mil platinum wire in the hot zone and 2 mil nickel–chromium wire in the cold zone. Above 1500°C, platinum was unsatisfactory for supporting many materials. Depending upon the sample material, silica or alumina fibers were used.

Figure 10.4. Invar balance.

One of the critical problems in the design of practical reaction systems is the furnace design. The furnace tube shown in Figure 10.3 was a 1 in. diam high density mullite or alumina tube. Alumina tubes were used for temperatures above 1200°C. The high temperature properties of molybdenum disilicide and a furnace design based on these elements are shown in Figure 10.3 and described in Appendix 2. The furnace, capable of achieving temperatures up to 1650°C, was used without a temperature controller. Calibrated Pt–Pt + 10% Rh thermocouples mounted adjacent to the specimen as shown in Figure 10.3 were used. The temperature was established by controlling the current in the heater elements.

The combination of weight change and oxygen consumption measurements in the same system permits an analysis of the complex types of chemical reactions such as those involving the simultaneous formation of oxide films and the volatilization of oxides. Figure 10.5 shows a schematic diagram of the vacuum system, gas preparation trains, and the system for measuring the oxygen consumption in the reaction. The vacuum system was not bakeable. As a result, the residual pressure at room temperature was 10^{-6} torr. A calibrated Wallace and Tiernan pressure gauge was used in the control of the oxygen pressure. A precision Granville Philips leak was used

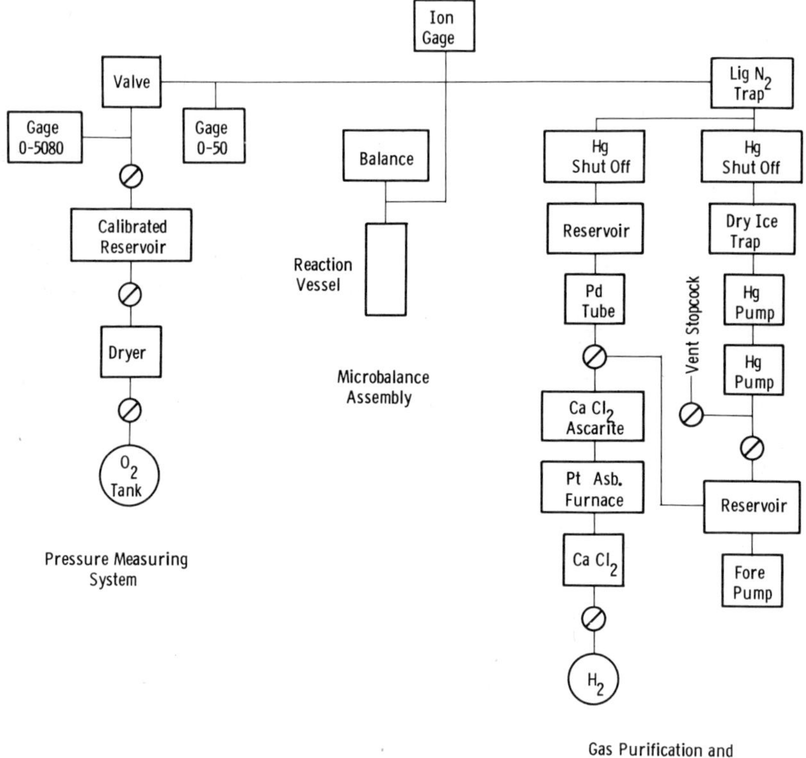

Figure 10.5. Schematic diagram of static reaction system.

to add make up oxygen to the system from a calibrated reservoir having an accurate pressure gauge. Pressure was controlled in the 2.6 liter reaction system to 10^{-1} torr. Oxygen consumption could be measured to $\pm 4 \times 10^{-4}$ g.

The selection, pretreatment, and care of the furnace tubes used in the high temperature reaction system was very critical. A detailed discussion of furnace tubes and support wires and fibers is included in Appendix 3.

B. Flow Reaction Systems

If a flow-type reaction system is operated at constant pressure, oxygen consumption apparatus can be used to provide and measure the amount of oxygen gas flowing over the specimen. Weight change is then used to measure the rates of reaction. As noted previously, the major advantage of a flow type reaction system is that gas flow can be used to diminish or eliminate the effect of the barrier layer of volatilized gases on the rates of oxidation.

Figure 10.6 shows a flow reaction system (13). A Cahn electrobalance with a capacity of about 1 g, a sensitivity of 10^{-6} g, and a period of less than 0.1 sec was used to follow the weight changes occurring during the reaction. The system pressure and flow rate were changed by controlling the inlet gas leak valve and the pumping speed of the exhaust line. Pressures from 2 to 76 torr were used. Flow rates could be varied between 5×10^{18} and 1.8×10^{20} molecules of O_2 per sec.

Furnace temperatures up to 1650°C were achieved with a Kanthal-Super Furnace (14). A vacuum tight alumina furnace tube attached to the stainless steel connecting tubing by a water cooled stainless steel flange and a rubber O-ring enclosed the sample. The furnace tube, balance housing, and associated apparatus were attached to a pumping system of two mercury diffusion pumps in series and a mechanical pump. A liquid nitrogen trap was employed to keep mercury vapor from the reaction volume. A vacuum of lower than 10^{-5} torr was achieved with the furnace tube at 1200°C.

C. Special Problems

1. Temperature

Since large heats of reaction occur during oxidation, the surface temperature of the metal will differ from the furnace temperature. Flowing the reacting gas over the specimen may cool the specimen unless the gas is preheated. Optical pyrometric methods could not be used for temperature measurement due to the furnace system geometry and the use of an opaque alumina furnace tube. A Pt–Pt 10% Rh thermocouple mounted inside the reaction tube adjacent to the sample was used to monitor the temperature of the reacting gases. Since the gases were preheated, no cooling effects were noted at the thermocouple.

The actual surface temperature was estimated from the rate and heat of oxidation, the heat of evaporation, the surface emissivity, and by assuming that at temperatures greater than 1000°C all of the heat is lost by radiation.

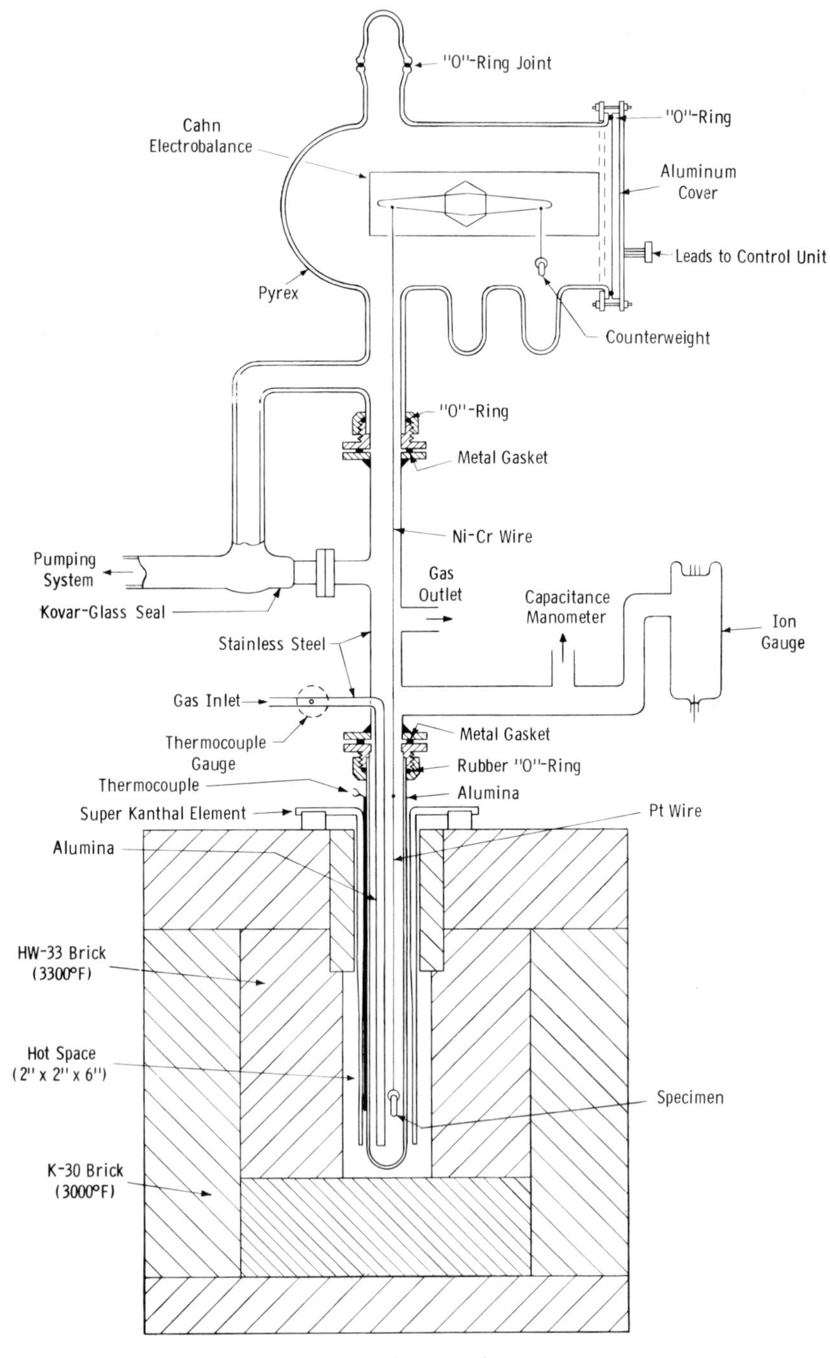

Figure 10.6. Flow reaction system.

The Stefan-Boltzmann equation can be used to calculate the surface temperature T. The equation states

$$h_r = C_r\varepsilon(T^4 - T_a^4) \tag{10.6}$$

Here h_r is the heat rate in cal/cm^2-sec and can be evaluated by knowing the rate of reaction, the heat of reaction, and the heat of vaporization of the oxide, ε is the emissivity, $C_r = 1.36 \times 10^{-12}$ cal cm^{-2} deg^{-4} sec^{-1}, and T_a is the furnace wall temperature. Since the ratio of the tube diameter to the specimen diameter is large, it can be assumed that the emissivity of the walls is unity. An emissivity of unity was assumed for graphite and an emissivity of 0.5 for molybdenum and tungsten.

2. *Gas Flow*

Fast gas flow rates make the specimen and balance system unstable and large buoyancy corrections may be required. Cylindrical samples with hemispherical ends are used in most of the flow studies to minimize such instability. Flow velocities of 500 cm/sec at 1500°C and 19 torr ambient pressure have been used without difficulties.

3. *Pressure*

Measurements of pressure in the range of 2–38 torr in a flow system presents no difficulties. At pressures below 10^{-1} torr, pressure changes can occur in the vicinity of the sample. Here, it is necessary to place pressure gauges close to the sample for reliable readings.

4. *Conclusions*

For kinetic studies of fast reactions it can be concluded that it is important to specify all of the reaction parameters including temperature, pressure, sample size, gas flow, and the reaction system apparatus if useful results are to be obtained. Comparison of kinetic data with that of other workers and the practical application of such data would be difficult unless all of the above parameters were specified. The microbalance is a convenient apparatus for many such studies.

IV. APPLICATIONS

A. Vapor Pressure of Silicon

One of the useful applications of high temperature microbalance methods is the measurement of the vapor pressure of metals and nonmetals. Two general techniques can be used. They are the Langmuir and the

Knudsen methods. In the Langmuir method, the weight loss from rods or strips of the material is determined as a function of time and temperature using the surface area of the specimen. In the Knudsen method, the specimen is supported in a platinum bucket containing a lid with a small hole for the effusion of the vapor. The Knudsen method is the more accurate method; however, higher temperatures must be used to achieve the same rates of weight loss as in the Langmuir method. This may lead to container and support wire problems.

An experimental study of the vapor pressure of high purity silicon using the Langmuir method is presented in this section.

1. Experimental

A Cahn type of microbalance was used to follow the weight loss behavior of silicon. The reaction system consisted of a high-density alumina furnace tube 2 cm in diameter attached to the system by a Viton O-ring. The furnace tube was pretreated by baking in oxygen at 1500°C to remove any residual carbon gases and carbonaceous material which could react to form SiC films on the silicon surface.

The samples were cut from p-type semiconductor grade silicon having a resistivity of 425 ohm-cm and a boron content of 10^{16} atoms/cm^3. The single crystal silicon strip was 0.025–0.030 cm thick, 1 cm wide, and had the (111) crystal plane in the surface. Pieces 0.5 cm wide were prepared by scribing and cracking, and a 0.07 cm diameter hole was made by ultrasonic cavitation to support the specimen. A 0.06 cm quartz fiber was used to suspend the specimen from the balance.

2. Results and Discussion

The specimens were cleaned and pretreated at 1300°C in vacuum before starting the vaporization experiments. Several series of weight loss curves were made at temperatures from 1100 to 1350°C. Typical weight loss data in units of g/cm^2-sec are given in Table 10.4. Assuming the weight loss curves are due exclusively to the volatilization of silicon, the vapor pressures can be evaluated from the Langmuir equation (15)

$$dw'/dt = \alpha P(M/2\pi RT)^{1/2} \tag{10.7}$$

Here M is the molecular weight of the evaporating species, T is the absolute temperature, dw'/dt is the rate of evaporation in g/cm^2-sec, and α is the condensation coefficient.

Table 10.4 and Figure 10.7 show the calculated vapor pressure values assuming monatomic silicon and $\alpha = 1$. Figure 10.7 also includes the

Table 10.4. Rate of Vaporization, Vapor Pressure, and Heats of Sublimation of Silicon

Temp., °K	Rate, g/cm^2-sec × 10^7	log P, atm	$\Delta H^\circ_{298.15}$ [a] kcal/mole
1373	3.76×10^{-2}	−9.20	106.15
1423	9.86×10^{-2}	−8.80	107.39
1423	9.63×10^{-2}	−8.81	107.46
1473	4.34×10^{-1}	−8.15	106.73
1473	4.36×10^{-1}	−8.15	106.73
1523	1.10	−7.74	107.46
1523	1.15	−7.72	107.32
1573	2.76	−7.33	107.96
1573	3.36	−7.24	107.32
1623	14.1	−6.61	105.98

[a] Mean $\Delta H^\circ_{298.15} = 107.1 \pm 0.6$.

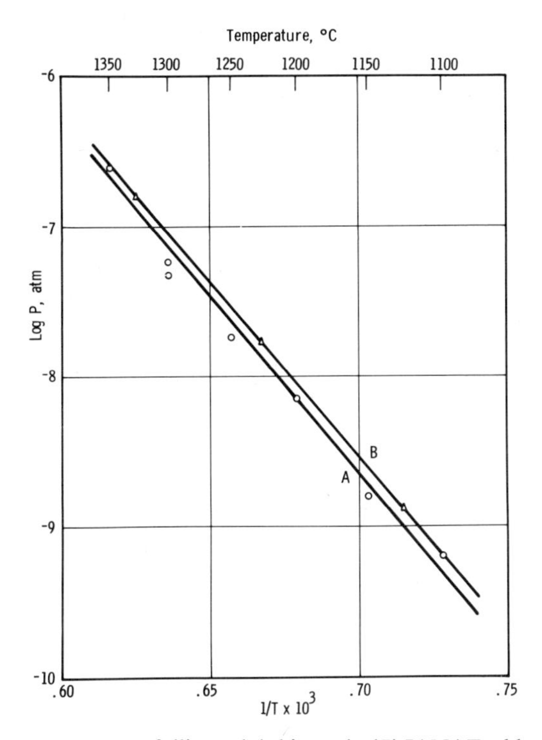

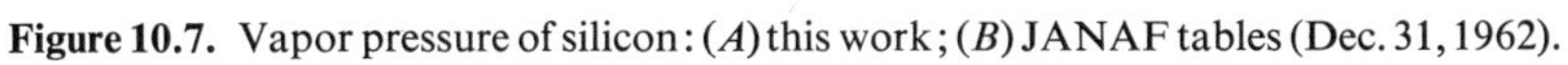
Figure 10.7. Vapor pressure of silicon: (*A*) this work; (*B*) JANAF tables (Dec. 31, 1962).

values given by the JANAF tables (5). An evaluation of some earlier data is made in these tables.

Table 10.4 shows the calculated values for the heats of sublimation $\Delta H^\circ_{298.15}$ using the equation

$$\Delta H^\circ_{298.15}/T = -R \ln P - (F^\circ - H^\circ_{298.15})/T_{\text{gas}} + (F^\circ - H^\circ_{298.15})/T_{\text{solid}} \tag{10.8}$$

where $(F^\circ - H^\circ_{298.15})/T_{\text{gas}}$ and $(F^\circ - H^\circ_{298.15})/T_{\text{solid}}$ are the free energy functions of the gas and solid, respectively. The mean value of $\Delta H^\circ_{298.15}$ is 107.1 $\pm$ 0.6 kcal/mole and may be compared with an averaged value of 106.0 kcal/mole given in the JANAF tables (5).

B. Oxidation of Graphite Using the Static Reaction System

A static system was used in the early studies of this reaction (3,11). One of the important results of these studies was the realization of the importance of the surface area as a variable in the study of oxidation reactions when volatile gases are formed. Specimen surface area can be used in the same manner as gas flow to study details of the mechanisms of fast oxidation reactions.

The importance of surface area was first realized from a comparison of the rates of oxidation of normal graphite samples having an area of 6.2 cm^2 with those of high density pyrolytic graphite samples having a 1 cm^2 area. The smaller pyrolytic carbon specimens oxidized faster at 1000°C than the large graphite specimens. This unexpected observation led to a study of the sample area effect and how it could be used to extend the usefulness of static reaction systems for the investigation of fast reactions. Before discussing the use of the surface area effect in detail, the earlier studies using a Gulbransen-type vacuum microbalance having a sensitivity of about 1 μg/0.001 cm at 7.25 cm will be discussed.

1. Effect of Time and Temperature

A kinetic study of the oxidation of graphite using a sensitive vacuum microbalance method and 6.3 cm^2 plate shaped samples at temperatures between 425 and 575°C and pressures between 1.5 and 98 torr was presented in an earlier paper (11). The purpose of this study was to determine the mechanism of oxidation of graphite in the chemical controlled region.

The weight loss versus time curves were nearly linear. Above 500°C the rates of oxidation increased slightly with time indicating that internal surface area changes were occurring during reaction. Weight losses were given in units of micrograms of carbon reacted per cm^2 of geometrical area.

In a later study (3) the temperature range was extended to 1500°C

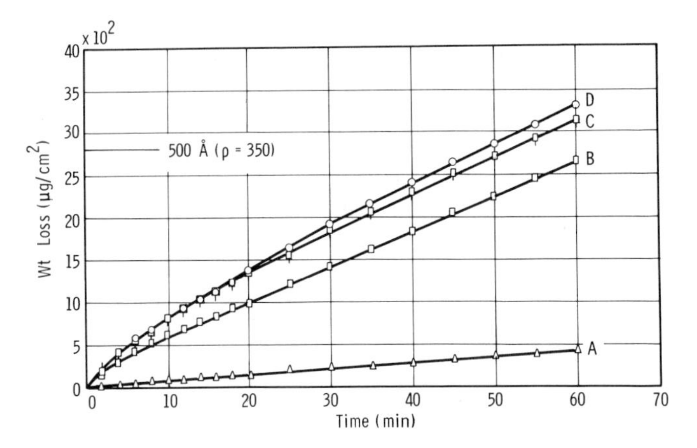

Figure 10.8. Effect of temperature on oxidation of graphite: 600–1000°C, 76 torr. (*A*) 600°C; (*B*) 700°C; (*C*) 800°C; (*D*) 1000°C.

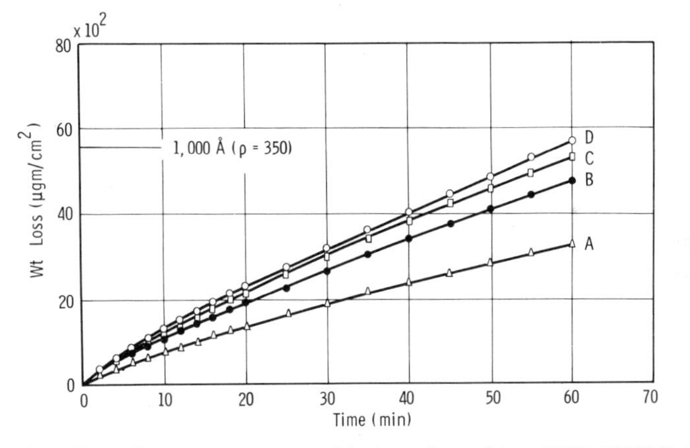

Figure 10.9. Effect of temperature on oxidation of graphite: 1000–1500°C, 76 torr O_2. (*A*) 1000°C; (*B*) 1200°C; (*C*) 1400°C; (*D*) 1500°C.

using 6.3 cm^2 rod-shaped samples. Figures 10.8 and 10.9 show the results at 76 torr oxygen pressure in a static reaction system.

The rates of oxidation, dw/dt, decrease with time as might be expected since oxygen is consumed and carbon dioxide and carbon monoxide were produced in this static system.

Figure 10.8 shows a large increase in the rate of oxidation between 600 and 700°C and much smaller increases above 700°C. Similar results were found in experiments at pressures of 38, 19, 9.5, and 2 torr indicating that a major change occurs in the mechanism near 700°C. Figure 10.9

shows that the rate of oxidation increased slowly between 1000 and 1500°C. The thickness of carbon removed was estimated from the density, 1.6 g/cm^3, and the surface roughness value, ρ, of 350 (11).

The existence of a change in the reaction mechanism at 825°C was reported by Tu *et al.* (16). These authors reported a change from a chemical controlled reaction below 825°C to a diffusion-controlled reaction above 825°C. Kuchta *et al.* (17), Levy (18), and Blyholder and Eyring (19,20) also report a change in mechanism near 825°C. Our results suggest a transition from chemical to diffusion control at about 700°C for a pressure of 76 torr and a sample area of 6.3 cm^2. Using 0.136 cm^2 samples, it has been possible to raise this transition temperature to 1000°C. The fact that earlier workers found a transition temperature near 825°C indicates that their reaction systems, specimen areas, and oxidation conditions were similar to the systems used in this investigation.

2. *Chemical Controlled Region of Oxidation*

a. Temperature Dependence and Heat of Activation. Figure 10.10 shows the data for three pressures on a log dw/dt vs $1/T$ plot for the temperatures from 425 to 1500°C. Although a pressure of 0.026 torr was used, the data of Blyholder and Eyring (19,20) are consistent with the results that are presented. At 725°C, a change is noted in the kinetics of oxidation at all pressures. For this region above 725°C, an energy of activation of 3.6 kcal/mole is calculated which would indicate a diffusion controlled reaction. Below 725°C, the data fall on parallel straight lines. In this region the reaction is chemically controlled since the calculated energy of activation was 39 kcal/mole. This is 2.3 kcal/mole higher than our earlier value.

b. Pressure Dependence Chemical. Figure 10.11 shows the effect of pressure on the initial rates of oxidation of graphite at 700°C for pressures from 2 to 76 torr of oxygen. The initial rates of oxidation fitted the equation

$$dw/dt = (0.06 + 0.01565P)\ 10^{-6} \qquad (10.9)$$

Here dw/dt is the rate in grams per cm^2-sec and P is the pressure in torr. This confirmed the earlier results at 500°C (11) that for the chemical controlled region, the rates of oxidation were nearly proportional to pressure.

3. *Gas Diffusion Region of Oxidation*

The effect of sample area in the diffusion region has been investigated by determining the rate of oxidation, dn'/dt, for a series of samples of different areas at several reaction temperatures. Here the rate of oxidation, dn'/dt, in

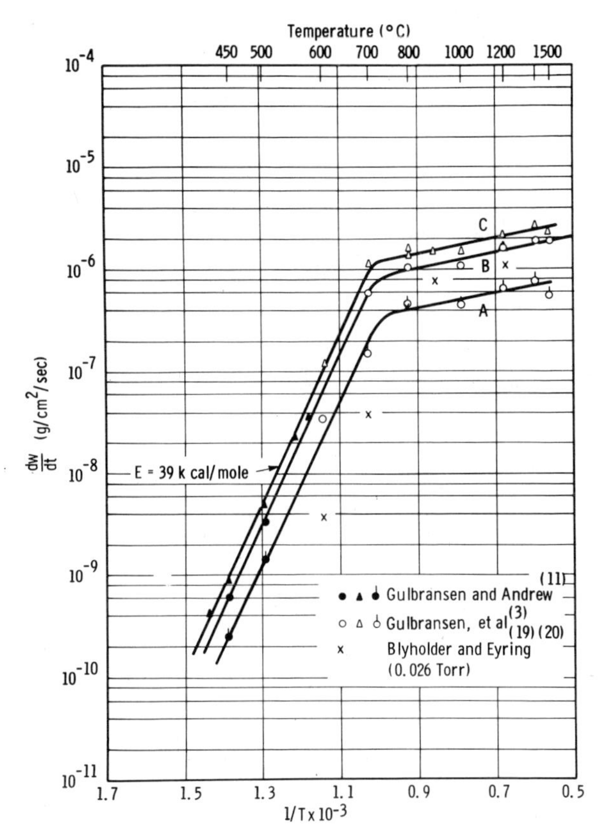

Figure 10.10. Temperature dependence of the rates of oxidation of graphite: (*A*) 2 torr; (*B*) 38 torr; (*C*) 76 torr.

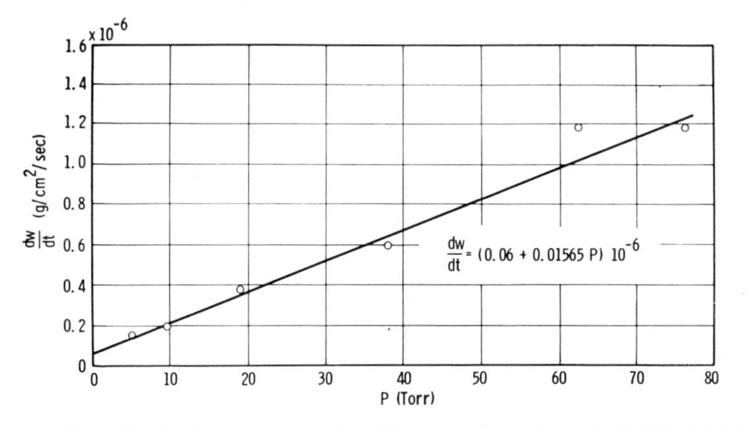

Figure 10.11. Effect of pressure on the oxidation of graphite at 700°C, 0–76 torr.

units of atoms of carbon reacted per cm^2-sec was used. In these experiments, sample areas were 6.3–0.136 cm^2.

Figure 10.12 shows the results of surface area variations for an oxygen pressure of 76 torr together with earlier data at the lower temperatures (11). Line *AB* and its extension represents the data for the chemical controlled region of oxidation. Line *ABD* was obtained using 6.0–6.3 cm^2 samples. Line *BCG* was obtained using 0.136 cm^2 samples. Points *E*, *F*, and *G* were obtained by using samples of different surface areas. Point *H* is an estimated value for a 0.032 cm^2 particle.

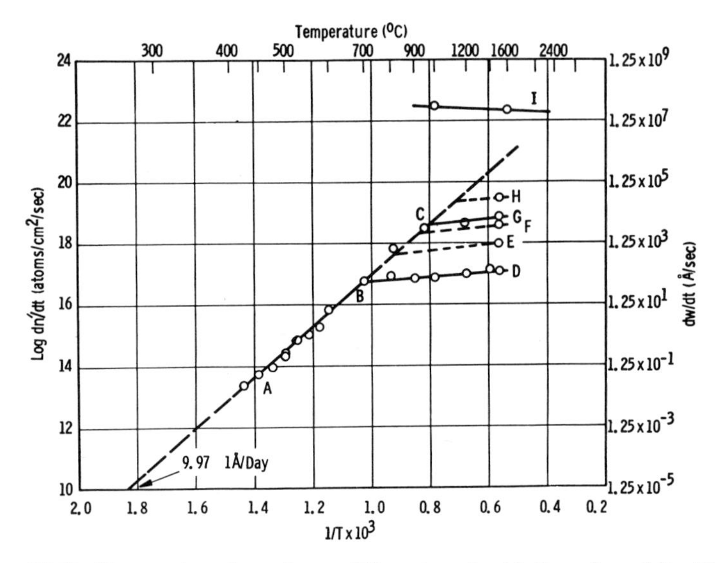

Figure 10.12. Temperature dependence of the rates of oxidation of graphite, 76 torr; line *ABC*, chemical control. Sample areas: (*D*) 6.3 cm^2; (*E*) 0.947 cm^2; (*F*) 0.232 cm^2; (*G*) 0.136 cm^2; (*H*) estimated 0.0316 cm^2; (*I*) theory, $C + O_2$.

Line *ABC* represents chemical-controlled oxidation with an energy of activation of 39 kcal/mole. Reaction rates to the right of line *ABC*, area *BCHD*, represent diffusion-controlled oxidation. In this region, temperature has only a minor effect on the rate of reaction. The important variable is the surface area.

At 1500°C the sample area was varied by a factor of 44 to determine the relationship of the area to the transition temperature between diffusion- and chemical-controlled reaction. Since the value of dn'/dt increased on reducing the specimen size, it was concluded that the oxidation was still under diffusion control for the smallest sample. The oxidation reaction produced appreciable changes in surface temperatures. Observations on the thermocouple near the sample showed temperature changes of up to 7°C for the 0.136 cm^2 sample and calculated temperature increases were less than 13°C except for the smallest samples. Calculated temperatures were used in constructing Figure 10.12.

The measured rate of reaction for a 0.136 cm^2 sample at a furnace temperature of 1500°C and 76 torr pressure was 5.9×10^{18} atoms/cm^2-sec. The ablation rate in angstroms per second is given at the right in Figure 10.12.

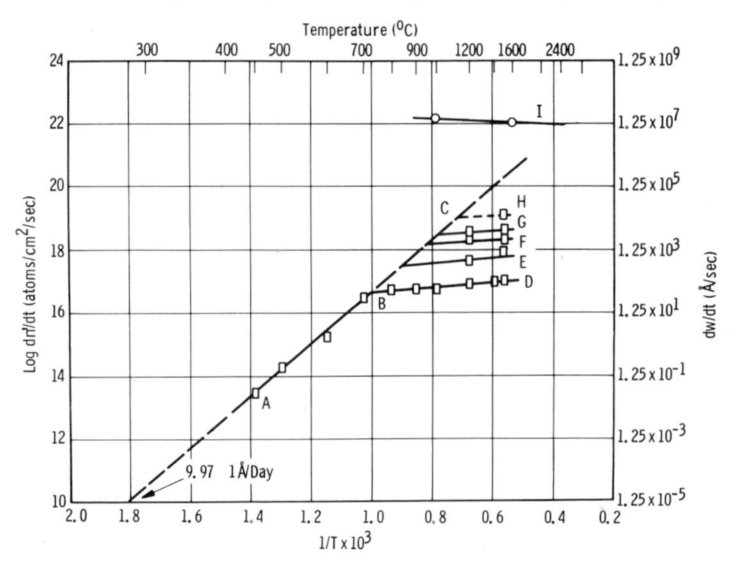

Figure 10.13. Temperature dependence of the rates of oxidation of graphite, 38 torr; line *ABC*, chemical control. Sample areas: (*D*) 6.2 cm^2; (*E*) 1.19 cm^2; (*F*) 0.232 cm^2; (*G*) 0.136 cm^2; (*H*) estimated 0.0316 cm^2 particle; (*I*) theory, $C + O_2$.

Curve I of Figure 10.12 shows the calculated dn'/dt values from kinetic theory for 76 torr pressure. A value of 2.4×10^{22} atoms/cm^2-sec was found at 1600°C.

Figures 10.13 and 10.14 show the rate data for pressures of 38 and 2 torr oxygen pressure. These data support the picture of the oxidation reaction already developed.

a. Pressure Dependence. Figure 10.15 shows a linear plot of the effect of pressure on the initial rates of oxidation for temperatures between 1000 and 1500°C. To establish the pressure relationship, a log dw/dt vs log P plot is given in Figure 10.16 for the data at 800, 1200, and 1500°C. Combining the pressure and temperature coefficient we have the following equation for the data from 800 to 1500°C.

$$dw/dt = 1.86 \times 10^{-6} P^{0.32} \exp(-3600/RT) \qquad (10.10)$$

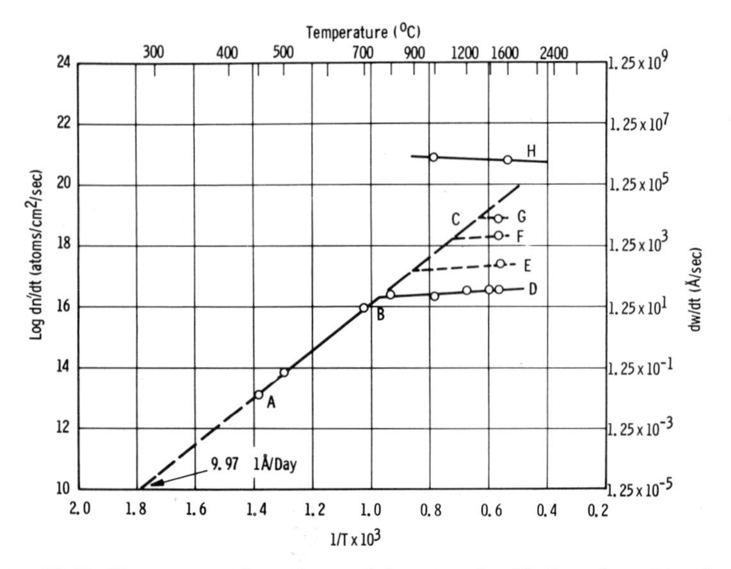

Figure 10.14. Temperature dependence of the rates of oxidation of graphite, 2 torr; line *ABC*, chemical control. Sample areas: (*D*) 6.2 cm²; (*E*) 1.19 cm²; (*F*) 0.136 cm²; (*G*) estimated 0.0316 cm² particle; (*H*) theory, $C + O_2$.

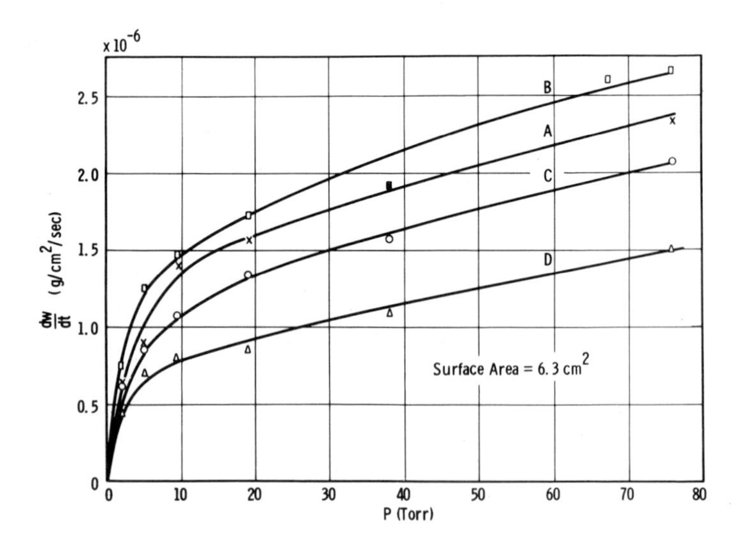

Figure 10.15. Effect of pressure, *dw*/*dt* vs *P* (torr) 1000–1500°C, oxidation of graphite: (*A*) 1500°C; (*B*) 1400°C, (*C*) 1200°C; (*D*) 1000°C.

Here P is the pressure in torr and R is the gas constant. The fractional exponential power suggests a complex mechanism for the reaction.

b. Effect of Pretreatment. In an earlier paper (11) it has been shown that certain pretreatments of the graphite have an important influence on the rate of oxidation below 575°C. This was to be expected since heat treatments and oxidation can affect the surface area and the extent of surface oxide formation. Above 800°C, the surface oxide is unstable and the rate of oxidation should be controlled by transport of reacting gas through the boundary layer of reaction product to the reacting interface.

Table 10.5 shows the results of a series of experiments which confirm this hypothesis since no important effect was found above 800°C.

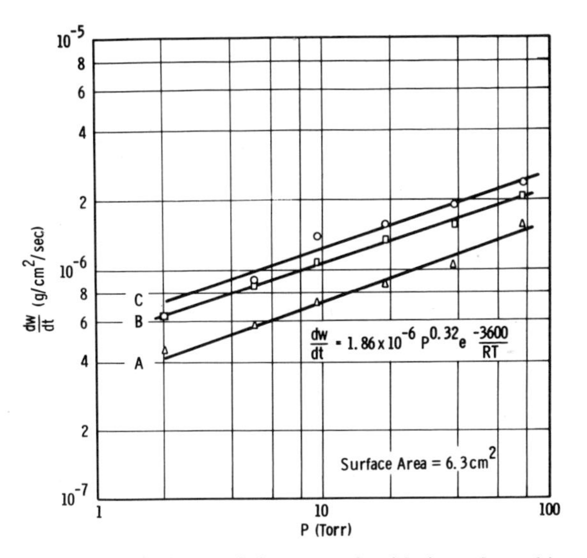

Figure 10.16. Pressure dependence of the rates of oxidation of graphite: (*A*) 800°C; (*B*) 1200°C; (*C*) 1500°C.

Table 10.5. Effect of Pretreatment on Rates of Oxidation of Graphite, 800°C and 76 torr

Treatment	Wt loss, 20 min, g/cm^2	Initial dw/dt, g/cm^2-sec
New sample	1.28×10^{-3}	1.50×10^{-6}
After 1 oxidation	1.33×10^{-3}	1.58×10^{-6}
After 4 oxidations	1.32×10^{-3}	1.53×10^{-6}
Preheat 1300°C, 16 hr	1.29×10^{-3}	1.46×10^{-6}

4. *Capability Factor for Reaction Systems*

The results presented in Section IV.B.3 show that in the gas diffusion region of graphite oxidation the rates of oxidation per unit area increased as the specimen area was decreased. Figure 10.17 shows the results of a study made at 1200°C and a pressure of 38 torr to establish this effect quantitatively. Cylinders and spheres of different areas and mass to area ratios were used.

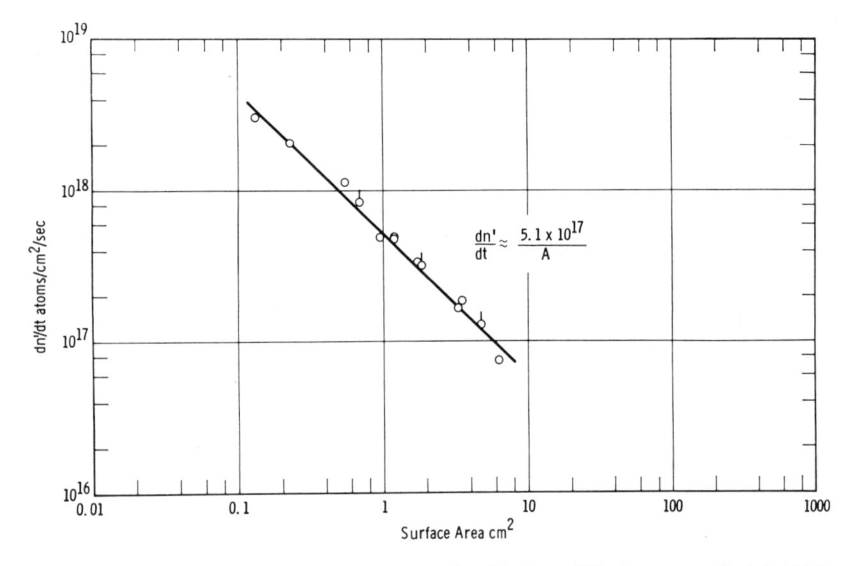

Figure 10.17. Effect of surface area on rate of oxidation, diffusion controlled 1200°C, 38 torr: 6.3–0.136 cm²; (○) cylinders; (♂) spheres oxidation of graphite.

The following equation fitted the rate of oxidation, dn'/dt, versus surface area, A, data.

$$dn'/dt = 5.1 \times 10^{17}/A = K_{(p)}/A \tag{10.11}$$

$K_{(p)}$ is a constant dependent on the pressure and the geometry of the reaction system. Figure 10.17 indicates that very fast rates of oxidation could be measured using small specimens. The term $K_{(p)}$ is defined as the capability constant for the reaction system.

5. *Explanation of Surface Area Effect*

The surface area effect in high temperature oxidation reactions involving volatile reaction products is one of the interesting facts brought out in this study. This effect can be explained on the basis of gas diffusion of oxygen to the graphite surface where reaction occurs. This gas diffusion is a function of the system geometry, pressure, and temperature gradients. For these conditions, the total reactivity, dn/dt, is the important quantity and not the reactivity per unit area, dn'/dt. The total reactivity is given by

$$\frac{dn}{dt} = \frac{dn'}{dt} A \tag{10.12}$$

Here A is the surface area. For small surface areas, large values of dn'/dt can be measured and vice versa.

C. Oxidation of Graphite Using the Flow Reaction System

1. Effect of Gas Flow

The capability of a rate measuring method to measure fast reactions of graphite can be extended at least 100-fold by imposing external flow on the reaction system. Figure 10.18 shows the results of a study of graphite oxidation at 1200°C and 9.5 torr oxygen pressure using a flow rate of

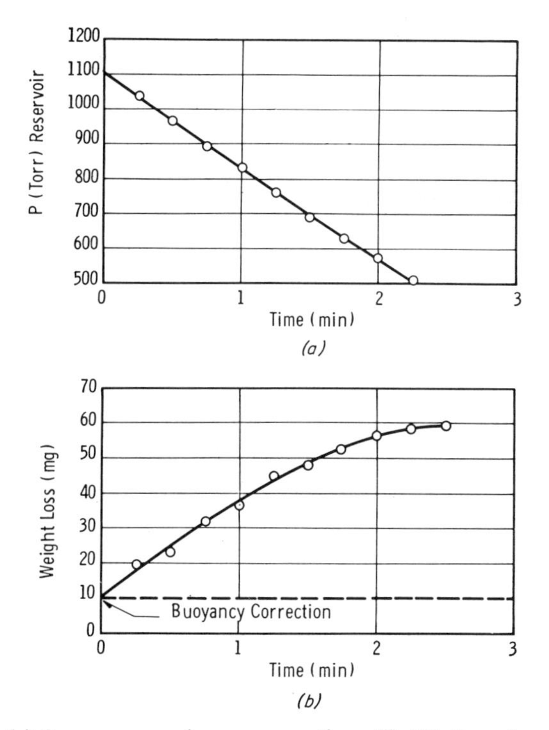

Figure 10.18. (*a*) Oxygen reservoir pressure vs time. (*b*) Ablation of graphite 1200°C, 9.5 torr, flow 8.2×10^{19} atoms O/sec., sample wt, 50.1 mg.

4.1×10^{19} oxygen molecules per sec. On an oxygen atom basis, this is equivalent to 8.2×10^{19} oxygen atoms per sec. Since spalling may occur for some conditions, the term rates of ablation is used to cover the very fast oxidation reactions.

Figure 10.18*a* shows the pressure readings of the gas reservoir as a function of time. Figure 10.18*b* shows the balance readings on the 50 mg graphite sample. An ablation or oxidation rate of 2.7×10^{19} atoms of C per sec was calculated at time $t = 0$ for the 0.55 cm^2 surface area sample.

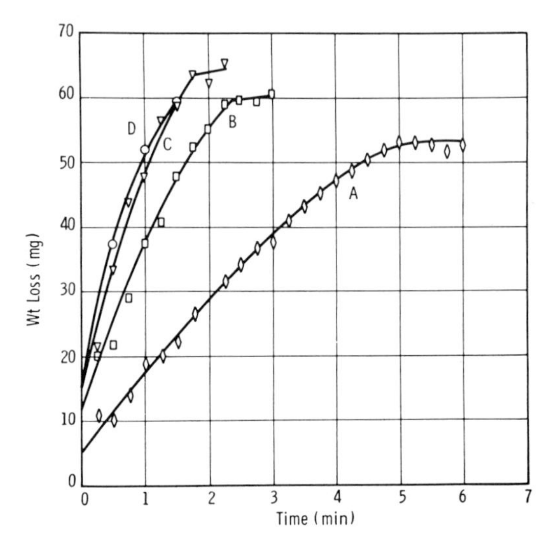

Figure 10.19. Effect of flow on ablation of graphite, 1200°C, 9.5 torr. (*A*) 2.4×10^{19} atoms/sec; (*B*) 7.04×10^{19}; (*C*) 1.30×10^{20}; (*D*) 1.6×10^{20}.

An analysis of Figure 10.18*b* shows the ablation–time curve to be related to the change in sample surface area. The sample was completely reacted after 140 sec.

Figure 10.19 shows the data of a series of experiments on the effect of flow on the rates of oxidation of graphite at 1200°C and 9.5 torr oxygen pressure. The flow rate was varied by a factor of 7.

Figure 10.20 shows the rate data on a log ablation rate vs $1/T$ plot together with results of our earlier studies using a static reaction system. The rate data obtained with the flow reaction system fit well with extrapolations based on studies made for the chemical controlled region of oxidation at temperatures of 425–1000°C (3,11). These experiments show that the chemical controlled region of oxidation can be extended to at least 1332°C using flow rates of 1.6×10^{20} atoms of oxygen per sec. In units of angstroms per sec, a maximum ablation rate of nearly 1.2×10^5 Å/sec was measured based on geometrical area. The above results are an example of the value of the microbalance method for studies of reactions in highly dynamic experimental conditions.

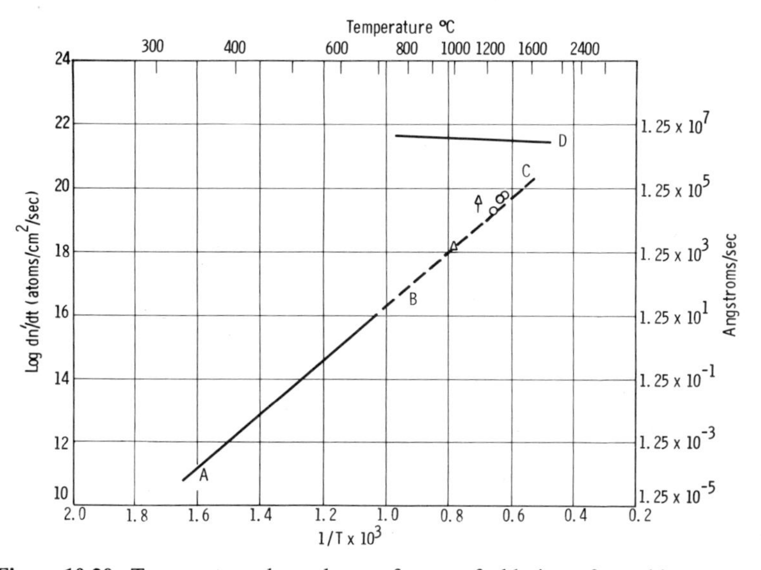

Figure 10.20. Temperature dependence of rates of ablation of graphite, 9.5 torr: (*A–B*) low temp. data; (*B–C*) extrapolated; (*D*) kinetic theory; △-no spalling 1000°C; ⍋-spalling 1000°C initial temp.; (○) some spalling 1200°C initial temp.

2. *Efficiency Calculations*

Two efficiency calculations can be made, flow efficiency and kinetic theory efficiency. The flow efficiency, E, is given by the equation

$$E = (dn'/dt \times A \times 1/F) \times 100 \tag{10.13}$$

Here F is the flow rate in atoms of oxygen per sec and A is the geometrical surface area. A flow efficiency of 26.1% was calculated for the oxidation experiment at 19 torr pressure and 1425°C using a specimen area of 0.55 cm^2 and a flow rate of 3.6×10^{20} atoms of oxygen per sec which corresponds to a flow rate of 563 cm/sec.

The rate data of Table 6 can be compared with kinetic theory by using equation 10.18, Appendix 1. At 1524°C and 19 torr, a theoretical rate of ablation of 5.6×10^{21} atoms of C per cm^2-sec was calculated. This value can be compared to the experimental value of 1.71×10^{20} atoms of C per cm^2-sec and used to calculate a kinetic theory efficiency of 3.0%.

Table 10.6. Ablation Rate vs Pressure and Flow, Oxidation High-Purity Graphite, Initial Temperature, 1200°C

Reaction conditions			Rates of ablation of graphite			
Pressure, torr	Flow rate, atoms/sec	Estimated temp., °C	dw/dt g/sec	dn/dt atoms/sec	dn'/dt atoms/cm²-sec	log dn'/dt
2	1.1×10^{19}	1215	6.6×10^{-5}	3.4×10^{18}	6.2×10^{18}	18.79
2	1.2×10^{19}	1214	6.1×10^{-5}	3.1×10^{18}	5.6×10^{18}	18.75
5	7.6×10^{18}	1211	4.5×10^{-5}	2.2×10^{18}	4.0×10^{18}	18.60
5	2.0×10^{19}	1222	8.7×10^{-5}	4.4×10^{18}	8.0×10^{18}	18.90
5	3.2×10^{19}	1238	1.6×10^{-4}	8.0×10^{18}	1.5×10^{19}	19.18
5	5.2×10^{19}	1276	3.4×10^{-4}	1.7×10^{19}	3.2×10^{19}	19.51
9.5	2.4×10^{19}	1248	2.1×10^{-4}	1.0×10^{19}	1.9×10^{19}	19.28
9.5	7.0×10^{19}	1304	4.8×10^{-4}	2.4×10^{19}	4.4×10^{19}	19.64
9.5	9.1×10^{19}	1323	5.8×10^{-4}	2.9×10^{19}	5.2×10^{19}	19.72
9.5	1.3×10^{20}	1324	5.9×10^{-4}	3.0×10^{19}	5.5×10^{19}	19.74
9.5	1.6×10^{20}	1332	6.3×10^{-4}	3.1×10^{19}	5.7×10^{19}	19.76
12	2.0×10^{20}	1345	7.0×10^{-4}	3.5×10^{19}	6.4×10^{19}	19.81
14	2.6×10^{20}	1420	1.1×10^{-3}	5.7×10^{19}	1.1×10^{20}	20.04
19	3.6×10^{20}	1524	1.9×10^{-3}	9.4×10^{19}	1.7×10^{20}	20.23

3. *Activated State Theory of Surface Reactions*

The above comparison of the experimental rate of oxidation of graphite in the chemical controlled region with the theoretical rate from kinetic theory showed poor agreement. These rates will now be compared with the predictions of the activated state theory of surface reactions which is a more appropiate theoretical approach.

Table 10.7. Comparison of Theoretical and Experimental Oxidation Rates for Graphite

Reaction conditions			Experimental, dn'/dt	Theoretical, dn'/dt	
Press, torr	Calc. temp., °C	Flow, atoms/sec	Ablation rate atoms/ cm²-sec	Mobile adsorption atoms C/cm²-sec	Desorption + diffusion, atoms C/cm²-sec (Blyholder and Eyring)
19	1492	3.6×10^{20}	1.5×10^{20}	1.1×10^{20}	2.9×10^{20}
5	1220	2.0×10^{19}	6.9×10^{18}	4.0×10^{18}	1.9×10^{19}
19	700	static	1.9×10^{16}	2.6×10^{16}	2.4×10^{16}

Table 10.7 shows a comparison of the predictions of two rate-limiting mechanisms presented in Appendix 1 for the ablation of graphite. For the Blyholder and Eyring (19) mechanism, we use a value of 80 kcal/mole for $\Delta H_2^\ddagger$, and for the mobile absorption mechanism, we use a value of 39 kcal/mole for $\Delta H_1^\ddagger$. The surface area for calculating these theoretical rates was evaluated from surface area measurements made in an earlier paper (11) on a similar type of graphite of equal density. That oxidized graphite sample had a surface roughness of 570 or a surface area of 14,200 cm²/g which may be compared with a value of 16,000 cm²/g that was found by Walker and Raats (21) for another type of carbon. The sample size used in this case had an area of 710 cm².

Considering the errors in evaluating the ablation rate, the agreement with the theoretical value based on mobile adsorption of oxygen molecules as the rate limiting process is good over a range of 10^4 in the ablation rate. The agreement of the experimental data with Blyholder and Eyring's mechanism is also reasonable. The nearly linear pressure dependence, however, favors the mobile adsorption mechanism of reaction.

D. Oxidation of Molybdenum–Static Reaction System

1. Oxidation Processes

The oxidation of graphite presented a nearly ideal system for the study of oxidation mechanisms where volatile oxidation products were formed. Molybdenum and tungsten present a more complex picture since stable oxide films and scales can form at low temperatures. At high temperature two volatile oxides are formed for each metal and several molecular species exist for the trioxides, i.e., MoO_3, $(MoO_3)_2$, $(MoO_3)_3$, and $(MoO_3)_4$.

To separate the processes of oxide volatility and the formation of oxide films, it is essential to use both oxygen consumption and weight change measurements.

In an early study (22) using a vacuum microbalance of the Gulbransen type, it was found that adherent oxide films formed on molybdenum below 400°C. The data fitted a parabolic rate law, and an energy of activation of about 36.0 kcal/mole was calculated. Above 400°C deviations from the parabolic rate law occurred. Volatilization of molybdenum trioxide occurred at 475°C under vacuum conditions. These results were confirmed by Gorbounova and Arslambekov (23). In a recent paper (9) the literature was reviewed and an extension of the studies to 1700°C was presented.

The studies on molybdenum had three objectives: (*1*) to characterize the several oxidation processes between 550 and 1500°C, (*2*) to determine the nature of the chemical controlled oxidation processes, and (*3*) to investigate the transition between chemical and gas diffusion controlled oxidation mechanism.

a. 600°C and 76 torr Oxygen Pressure. Both oxygen consumption and weight change measurements were used to evaluate oxide film formation and oxide volatility. Curves *A* and *B* of Figure 10.21 show oxygen consumption and weight change measurements. Both measurements were in units of milligrams per cm^2. The oxygen consumption data show a nearly linear rate law after an initial period of fast reaction while the weight change data show a slow initial reaction rate followed by a period of increasing rate of reaction. The weight change represents the difference between the oxygen reacting to form oxide scale and molybdenum lost as volatilized oxide.

From the chemical equations for the formation of MoO_3, the following equation for the weight of molybdenum lost, W_{Mo}, as $MoO_3(g)$ can be used,

$$W_{Mo} = \tfrac{2}{3}\,(W_O - W_B) \qquad (10.14)$$

Here W_O is the oxygen consumption and W_B is the weight change of the sample during oxidation. The total molybdenum reacted is obtained directly from W_O. From the difference between total molybdenum reacted and the molybdenum volatilized, the weight of oxygen reacting to form oxide scale can be calculated.

Curves *C* and *D* show the weight of molybdenum volatilized and the weight of oxygen reacting to form oxide scale, respectively. These curves show that the rate of loss of molybdenum decreases as oxidation proceeds.

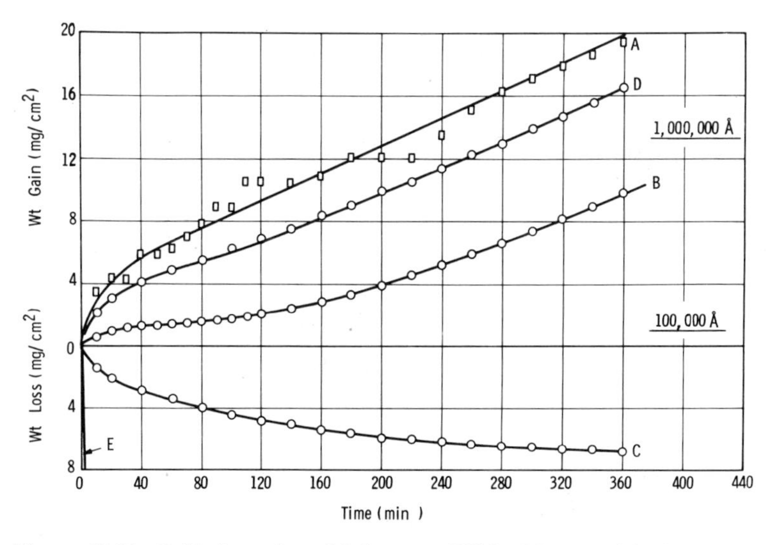

Figure 10.21. Oxidation of molybdenum, 600°C, 76 torr. (*A*) O_2 consumed; (*B*) wt change; (*C*) molybdenum lost; (*D*) O_2 in oxide scale; (*E*) volatility of MoO_3 in vacuum.

After 6 hr of reaction, a very low volatility of molybdenum trioxide is observed. This may be related to a build up of a saturated vapor environment in the static reaction system. A factor of 66,500 is used to relate curve *D* to oxide thickness in angstroms (22,9). Thickness markers are placed on the right of Figure 10.21. This evaluation assumes MoO_3 as the main oxide, a surface roughness ratio of unity, and an oxide which is not porous or full of cracks.

The total weight of molybdenum reacting can be calculated from curve *A* using the stoichiometric weight ratio of 2.00. The surface recession of the metal in angstroms can be calculated from curve *A* using the stoichiometric weight ratio of 2.00 and the density of 10.2. A factor of 19,600 is found.

Figure 10.21 shows several interesting features for the 600°C, 76 torr reaction conditions: (*1*) both oxide scale formation and oxide volatility occur; (*2*) 80% of the oxygen reacted went to oxide scale formation;

(*3*) a nearly linear rate of oxidation was observed; (*4*) loss of molybdenum occurred very rapidly during the initial period of reaction with this rate decreasing as oxidation proceeded. This investigation provided an excellent illustration of the inadequacy of using weight change methods alone to study a reaction in this temperature range.

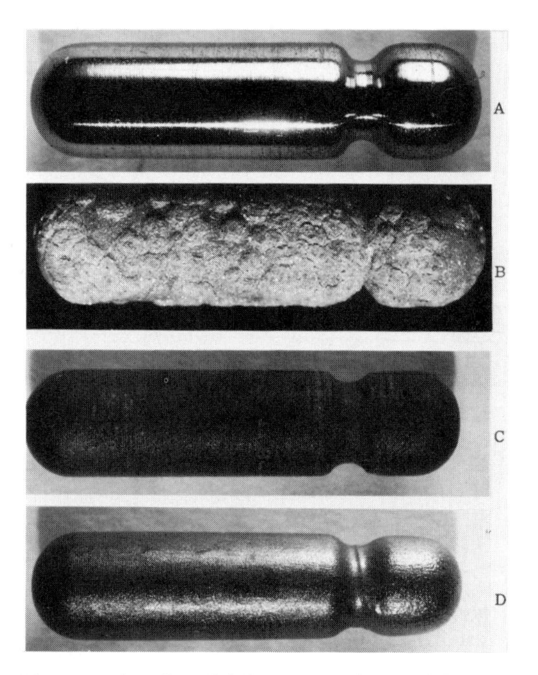

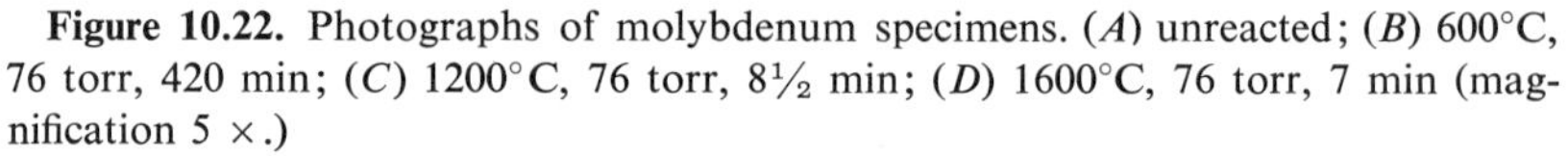
Figure 10.22. Photographs of molybdenum specimens. (*A*) unreacted; (*B*) 600°C, 76 torr, 420 min; (*C*) 1200°C, 76 torr, 8½ min; (*D*) 1600°C, 76 torr, 7 min (magnification 5 ×.)

Figure 10.22*A*, *B* are photographs of the unreacted and oxidized specimens, respectively. The oxidized specimen shows that a poorly adhering oxide scale was formed.

Oxidation studies at 650 and 700°C at 76 torr oxygen pressure show similar phenomena to that observed at 600°C. The percentage of oxygen forming solid oxide decreases as the temperature is raised. At 700°C, only 30% of the oxygen used formed oxide scale. At 800°C all of the oxygen used formed volatile molybdenum trioxide as may be expected since at 795°C the vapor pressure of molybdenum trioxide is 11.7 torr.

b. 1000°C and 76 torr Pressure. At 1000°C only volatile molybdenum trioxide is formed. Curves *A* and *B* of Figure 10.23 represent the oxygen consumption and weight change, respectively. Nearly linear rates of reaction are also found under these conditions. The small decrease in slopes can be related to the decrease in specimen area during reaction.

Using the appropriate stoichiometric weight ratio and assuming all the oxygen forms volatile molybdenum trioxide, curve *A* can be used to calculate the expected weight loss of molybdenum. The good agreement found

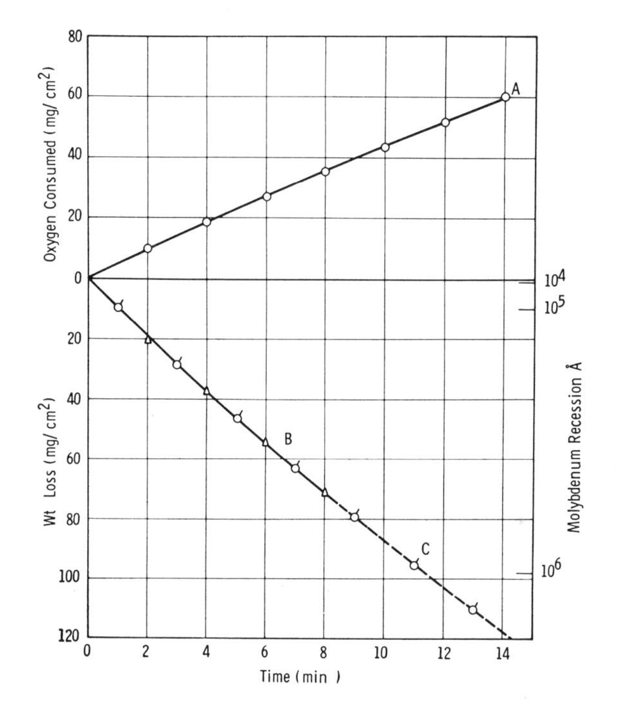

Figure 10.23. Oxidation of molybdenum, 1000°C (1047), 76 torr. (*A*) oxygen consumption; (*B*) wt loss △——△; (*C*) calculated from (*A*) ♂——♂.

with this assumption confirms the nature of the reaction. The recession of molybdenum can be calculated using the previously mentioned relation 1 mg/cm^2 = 19,600 Å.

Although many units are used in presenting reaction rates, the unit of atoms of Mo per cm^2-sec is preferred. The initial rate of reaction of the 1000°C oxidation at 76 torr is 1.08×10^{18} atoms/cm^2-sec. This is in the class of very rapid reactions.

For high rates of reaction, it is essential to have a more realistic value for the surface temperature which takes into account the effects of the oxidation reaction. The surface temperature during reaction can be estimated with assumptions similar to those previously used, namely, (*1*) that radiation is the major source of loss of heat, (*2*) the emissivity of the surface and walls is 0.5, and (*3*) the heat source is the sum of the heat of formation of MoO_3(s, 1) (24) and the heat of vaporization of the oxide (25). For the conditions of the present experiment at 1000°C, a sample surface temperature of 1047°C was estimated. In all of the tables and figures, both the furnace temperature and the calculated temperature are listed.

It has been shown that in the oxidation of molybdenum between 600 and 800°C both oxide film formation and oxide volatility occur. Above 800°C, however, only volatile molybdenum trioxide is formed.

2. *Effect of Temperature*

Figures 10.24 and 10.25 and Table 10.8 are the result of a more detailed study of the effect of temperature on the oxidation of molybdenum at 76 torr oxygen pressure. The weight change in mg/cm² is plotted against

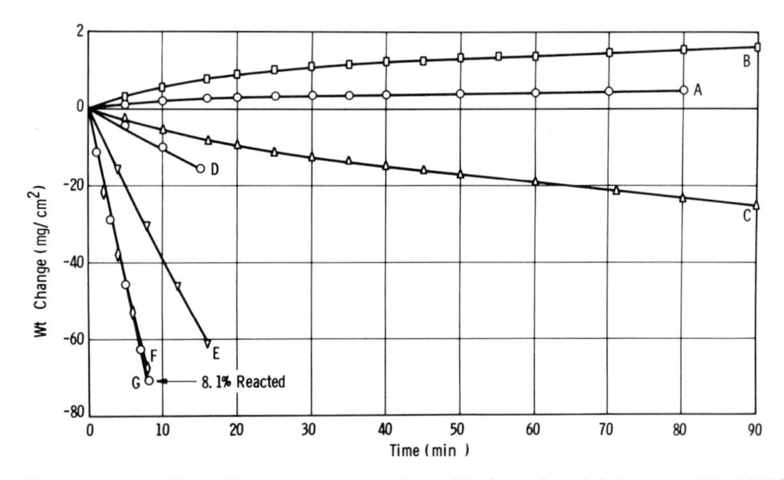

Figure 10.24. Effect of temperature on the oxidation of molybdenum: 550–1000°C, 76 torr O_2. (*A*) 550°C; (*B*) 600°C; (*C*) 650°C; (*D*) 700°C; (*E*) 800°C (829); (*F*) 900°C (957); (*G*) 1000°C (1047).

time in minutes. Figure 10.24 shows experimental results for the temperature range 550–1000°C. Simultaneous oxide scale formation and evaporation occur during oxidation at temperatures between 550 and 700°C. Table 10.8 shows the calculated reaction temperatures and the initial rates of total reaction calculated on the basis of oxygen used in the reaction. Figure 10.24 illustrates the transition in oxidation phenomena between reactions with oxide scale formation and those with complete oxide evaporation.

Table 10.8. Effect of Temperature on Initial Rates of Oxidation of Molybdenum (P = 76 torr; Surface Area = 1.215 cm^2)

Furnace temp., °C	Calculated temp., °C	dn'/dt, atoms/cm^2-sec	log dn'/dt
700	—	1.49×10^{17}	17.17
800	829	4.30×10^{17}	17.63
900	957	1.12×10^{18}	18.05
1000	1047	1.08×10^{18}	18.03
1100	1135	1.10×10^{18}	18.04
1200	1227	1.05×10^{18}	18.02
1400	1418	1.28×10^{18}	18.11
1600	1614	1.11×10^{18}	18.05

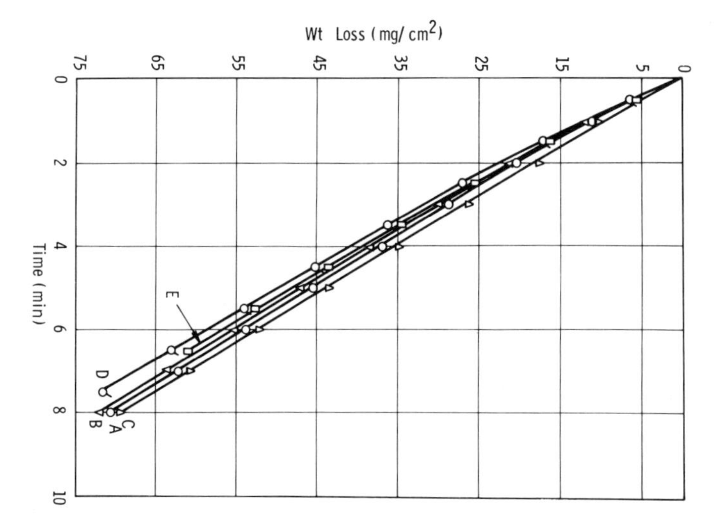

Figure 10.25. Effect of temperature on oxidation of Mo: 1000–1600°C, 76 torr O_2. (*A*) 1000°C (1047); (*B*) 1100°C (1135); (*C*) 1200°C (1227); (*D*) 1400°C (1418); (*E*) 1600°C (1614).

Figure 10.25 presents the results in the higher temperature range of 1000–1600°C. The calculated temperatures are given in brackets. The curves show a small decrease in the rate of reaction resulting from a change in surface area. On the basis of these results alone, it was concluded that temperature has little effect on the rate of oxidation in this temperature range. This could be predicted for oxidation reactions where the rates of reaction are limited by the gaseous diffusion of oxygen (3).

Figure 10.22*C*, *D* are photographs of the oxidized specimens after reaction at 1200 and 1600°C. Figure 10.22*C* indicates the presence of a thin oxide film on the metal. Figure 10.22*D* indicates a bare metal surface to be present during oxidation. This is due to the increased volatility of $(MoO_3)_3$ at 1600°C.

3. *Effect of Pressure*

Table 10.9 is a summary of the data illustrating the effect of pressure at three temperatures. The surface area of 1.215 cm^2 was the same as in the

Table 10.9. Effect of Pressure on Initial Rates of Oxidation of Molybdenum (Surface Area = 1.215 cm^2)

Furnace temp., °C	Calculated temp., °C	Pressure, torr	dw/dt mg/cm²-sec	dn'/dt atoms/cm²-sec	log dn'/dt
800	829	76	0.0684	4.30×10^{17}	17.63
800	803	38	0.0180	1.13×10^{17}	17.05
800	803	19	0.00733	4.60×10^{16}	16.66
800	801	5	0.00327	2.05×10^{16}	16.31
1200	1227	76	0.167	1.05×10^{18}	18.02
1200	1224	38	0.150	9.42×10^{17}	17.97
1200	1219	19	0.116	7.28×10^{17}	17.86
1200	1207	5	0.044	2.76×10^{17}	17.44
1600	1614	76	0.177	1.11×10^{18}	18.05
1600	1612	38	0.156	9.80×10^{17}	17.99
1600	1611	19	0.141	8.85×10^{17}	17.95
1600	1610	5	0.127	7.98×10^{17}	17.90

experimental data shown in Table 10.8. The initial rates of reaction are tabulated in mg/cm²-sec. At 800°C, the effect of pressure on the rate of reaction is large and the rate follows the 1.5 power of the pressure. At 1600°C, the effect of pressure is small and follows the 0.14 power of the pressure.

4. Classification of Oxidation Phenomena

A classification scheme of the phenomena occurring during the oxidation of molybdenum is shown in Table 10.10.

Four temperature regions are proposed. Pressure affects the temperature limits of the several regions with low pressures favoring volatility of the oxide. Three types of rate-controlling processes are given in Table 10.10; (*1*) a Wagner type of diffusion of metal or oxygen through the oxide,

Table 10.10. Classification Scheme: Oxidation of Molybdenum

Class	Reaction conditions	Oxidation phenomena	Rate-controlling process
1	Below 450°C	Adherent oxide films or scales form	Wagner type diffusion of metal or oxygen through oxide
2	500–700°C	Oxide scales form; also oxide volatilizes; low pressure favors volatility of oxide	Oxide scales not protective; probably chemical-type processes on metal interface
3	801°C to transition temperature	Liquid oxide can form; volatilizes as soon as oxide forms	Chemical processes on metal interface
4	Above transition temperature	Oxide volatilizes as fast as it forms	Transport of oxygen to metal interface; turbulence in gas phase important

(*2*) in the intermediate temperature range where oxides are not present, a surface type of chemical reaction; i.e., adsorption, compound formation, and desorption, and (*3*) above a certain transition temperature, a complex type of transport process.

5. Transition Between Chemical Controlled Oxidation and Gas Diffusion Controlled Oxidation

Gas flow methods have been used to study the mechanism of oxidation (26,27). Unfortunately, the gas flow was not varied over a sufficient range to change the mechanism of reaction. If the transport of oxygen to the surface and the transport of reaction products away from the surface to a cold zone of the reaction system controls the rate of oxidation, the total amount of reaction occurring per second is the factor of prime importance. The rate of oxidation per unit area, dn'/dt, can be varied by changing the surface area over a wide range.

Figure 10.25 and Table 10.9 indicate the rate of oxidation to be nearly independent of temperature above 950°C at 76 torr when using 1.2 cm^2 specimens. A similar behavior was noted for 0.947 cm^2 graphite specimen data that is plotted in Figure 10.12. The rates of oxidation were in the same range of about 10^{18} atoms of C or Mo per cm^2-sec. This suggests similar oxidation processes for molybdenum and graphite under these conditions.

The effect of sample area on the initial rates of the oxidation of molybdenum is shown in Table 10.11. Samples having areas of 0.605, 0.304, and 0.12 cm^3 were used.

Table 10.11. Effect of Sample Area on Initial Rates of Oxidation of Molybdenum (P = 76 torr)

Furnace temp., °C	Calculated temp., °C	Sample area, cm^2	dn'/dt atoms Mo/cm^2-sec	log dn'/dt
1000	1047	1.215	1.08×10^{18}	18.03
1000	1124	0.604	2.22×10^{18}	18.35
1000	1159	0.304	3.49×10^{18}	18.54
1200	1227	1.213	1.05×10^{18}	18.02
1200	1262	0.605	2.46×10^{18}	18.39
1200	1296	0.301	3.90×10^{18}	18.59
1200	1410	0.121	7.92×10^{18}	18.90
1400	1418	1.216	1.28×10^{18}	18.11
1400	1451	0.605	2.95×10^{18}	18.47
1400	1509	0.304	6.59×10^{18}	18.82
1600	1614	1.218	1.11×10^{18}	18.05
1600	1634	0.608	2.73×10^{18}	18.44
1600	1660	0.303	4.94×10^{18}	18.69
1650	1704	0.302	4.84×10^{18}	18.68

The rates of oxidation are nearly inversely proportional to the specimen area and nearly independent of temperature. Equation 10.11, which was developed for the oxidation of graphite, can be applied to the oxidation of molybdenum if a small change is made in the constant $K_{(p)}$ to take into account the pressure, the stoichiometry of the reaction, and the nature of the volatilized oxide vapors.

Figure 10.26 shows a log dn'/dt vs $1/T$ plot of the data at 76 torr pressure. The calculated surface temperatures are used. Part of the data fall along a straight line AB. Oxidations along AB are interpreted as being under chemical control. The activation energy is 19.7 kcal/mole. The rate constants falling to the right of AB are interpreted as being in the region of transport process control. Smaller values for dn'/dt at a given temperature than those predicted by AB are found.

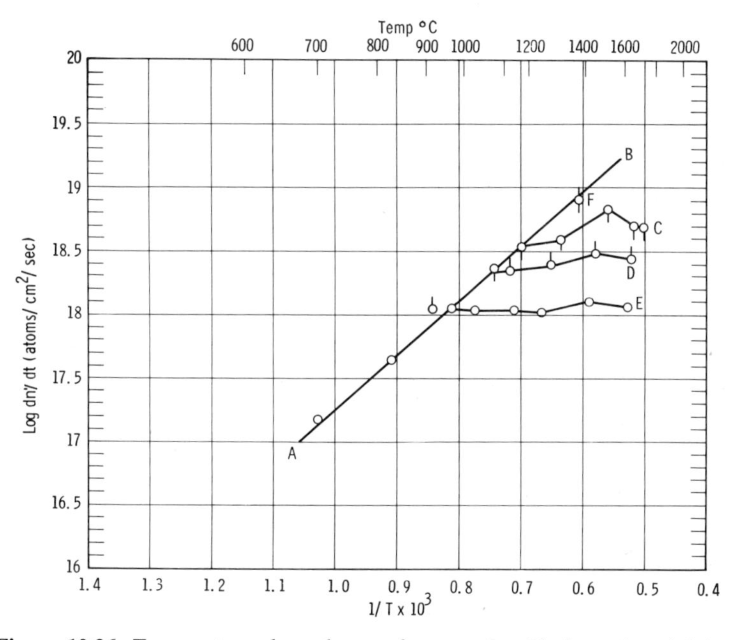

Figure 10.26. Temperature dependence of rates of oxidation of molybdenum, 600–1704°C, 76 torr O_2; line (A–B) chemical control ($\Delta H_{AB} = 19.7$ kcal/mole); area to right of (A–B) diffusion control; areas of samples: (C) 0.304 cm²; (D) 0.604 cm²; (E) 1.215 cm²; (F) 0.12 cm².

The rate data for the several sample areas lie on curves C, D, and E. Point F is that for a 0.1 cm² sample area for which a reaction rate of nearly 10^{19} atoms/cm²-sec was found. Since this point lies on the line AB, it is concluded that the reaction rate at this point is chemically controlled.

The reaction rate of 10^{19} atoms/cm²-sec is the highest rate that has been recorded for a molybdenum oxidation reaction. Using a flow system with air at 1371°C, Semmel (26) found a reaction rate of 1.98×10^{18} atoms/cm²-sec. Modisette and Schryer (27) using a flow system and a 21.5% oxygen–helium mixture found a value of 1.09×10^{18} atoms/cm²-sec. Both values lie close to line E of Figure 10.26.

6. *Kinetic Efficiency Calculations*

The flow efficiency of the oxidation reaction of molybdenum cannot be calculated since flow was not imposed on the reaction system. The kinetic theory efficiency can be calculated. At 1400°C and 76 torr, a collision rate of 1.52×10^{22} molecules of oxygen per cm^2-sec was calculated. The observed reaction rate of point F in Figure 10.26 is 7.9×10^{18} for the 0.121 cm^2 sample. Assuming 3 atoms of oxygen are required to remove 1 atom of molybdenum, a kinetic theory efficiency of 0.08% was calculated.

7. *Calculation of Effective Collision Number,* n

Line AB of Figure 10.26 can be represented by equation 10.19 (Appendix 1). For molybdenum the ratio of the real to geometrical surface area is assumed to be unity.

The experimental value of n from equation 10.19 (Appendix 1) is compared with that calculated from kinetic theory using equation 10.18 (Appendix 1). From the data of Figure 10.26 at 1159°C and 76 torr and using a heat of activation of 19.7 kcal/mole, an experimental value of 3.5×10^{21} was calculated. A value of 8×10^{21} was calculated for n from the kinetic theory, equation 10.18. The agreement of the two values of n is reasonable considering the uncertainty in establishing the true reaction temperature.

8. *Mechanism of Reaction*

Two mechanisms are tested, namely mobile adsorption given by equations 10.20 and 10.21 (Appendix 1) and desorption given by equations 10.22 and 10.23 (Appendix 1). Table 10.12 is a comparison of the experimental oxidation rate data for molybdenum with that calculated on the basis of mobile adsorption and desorption. A desorption process is impossible for a reaction having a ΔH of 19.7 kcal/mole. In contrast to

Table 10.12. Comparison of Theoretical and Experimental Oxidation Rates for Molybdenum, 900°C Furnace Temperature, 76 torr Oxygen

Experimental oxidation rate, atoms/cm^2-sec		Theory	
		Mobile adsorption	Desorption
$\rho = 1$[a]	$\rho = 2$		
1.08×10^{18}	5.4×10^{17}	3.6×10^{17}	1.29×10^{24}

[a] ρ = surface roughness ratio.

graphite, the porosity of the molybdenum presented no problem, since the surface was nearly smooth. The agreement between theory and experiment is good if a surface roughness ratio of 2, which is a reasonable value for a metal surface undergoing reaction, is used.

In Table 10.12 it was assumed that all of the gas was involved in mobile adsorption with an activation energy of 19.7 kcal/mole. Actually it is postulated that a monolayer of oxygen is momentarily adsorbed with a much lower activation energy. The reaction mechanism as illustrated below

$$\mathrm{Mo \cdot Mo{-}O} + \mathrm{O_2} \rightleftharpoons \mathrm{Mo \cdot Mo} \overset{\diagup \mathrm{O}}{\underset{\diagdown \mathrm{O}}{-\mathrm{O}}} \qquad (10.15)$$

involves mobile adsorption of O_2 on Mo–O monolayer.

This complex undergoes the chemical reaction

$$\mathrm{Mo \cdot Mo} \overset{\diagup \mathrm{O}}{\underset{\diagdown \mathrm{O}}{-\mathrm{O}}} \rightleftharpoons \mathrm{Mo{-}MoO_3} \qquad (10.16)$$

and desorption to form gaseous molybdenum trioxide

$$n\mathrm{Mo{-}MoO_3} \rightleftharpoons n\mathrm{Mo} + (\mathrm{MoO_3})_n \qquad (10.17)$$

The experimental values given in Table 10.12 should be reduced by $\frac{1}{3}$ to account for oxygen momentarily adsorbed in the monlayer. The agreement of theory and experimental is still good.

E. Oxidation of Tungsten

A complete analysis of the processes occurring in the oxidation of tungsten will not be given here. Much work needs to be done before the reaction can be understood. In addition several difficulties occur in the study of the chemical and gas diffusion controlled oxidation regions. (*1*) The temperature range for pressures of 2 torr and higher must be above 1350°C. (*2*) Tungsten trioxide vapors dissolve in the ceramic furnace tube. (*3*) High heats of reaction occur which make the surface temperature difficult to evaluate. (*4*) The high rates of oxidation found when oxides are present and the high rates of gas formation, $(WO_3)_n(g)$, at slightly higher temperature bring on gas diffusion controlled oxidation processes. These factors make it very difficult to isolate the chemical controlled region of oxidation for pressures in the range of 2–76 torr.

Some of the recent literature has been reviewed in our earlier papers (28,29), and by Barth and Rengstorff (30). At present only a comparison of the rates of oxidation of tungsten with those for carbon and molybdenum in the gas diffusion region of oxidation can be made.

Table 10.13. Comparison Absolute Values of Rates of Oxidation of Carbon, Molybdenum, and Tungsten: 19 and 38 torr Oxygen Pressures and Surface Areas of 0.6 to 0.7 cm²

Element	Temp., °C	Pressure, torr	dn'/dt atoms/cm²-sec
C	1500	19	4.0×10^{17}
Mo	1400	19	2.5×10^{18}
W	1465	19	2.5×10^{18}
C	1500	38	5.0×10^{17}
Mo	1600	38	2.5×10^{18}
W	1615	38	6.2×10^{18}

Table 10.13 shows such a comparison of the rates of oxidation for two oxygen pressures and for two temperatures. The rates of oxidation are given in absolute units of atoms of C, Mo, or W reacting per cm²-sec. Tungsten at 1465°C is compared to molybdenum at 1400°C and carbon at 1500°C for specimens of about 0.6 cm² surface areas. The rates of oxidation of carbon and molybdenum are in the diffusion controlled region where temperature has only a minor effect on the rate of oxidation. At 19 torr pressure and 1465°C, tungsten and molybdenum oxidize at about the same rate, and carbon oxidizes at a rate $\frac{1}{6}$ as fast. Since the carbon gases do not condense in the reaction system, the barrier for diffusion of oxygen is greater than that found for the diffusion of oxygen through the reaction products of molybdenum and tungsten.

The comparison at 38 torr oxygen pressure and 1615°C shows tungsten oxidizing $12\frac{1}{2}$ times faster than carbon. Again, these results can be interpreted in terms of diffusion of oxygen gas through the gaseous reaction products. The diffusion barrier for oxygen when tungsten is oxidized at 1615°C is small, since the alumina furnace tube absorbs WO_3. The diffusion barrier is greatest for carbon where completely volatile reaction products are formed.

It can be concluded from these studies that in the gas diffusion controlled region of oxidation, tungsten oxidizes somewhat faster than molybdenum and considerably faster than carbon. No information is available on the chemical controlled region of oxidation.

V. GENERAL COMMENTS

The vacuum microbalance method has proven very useful for the study of physical chemical problems involving gases and metals or solids for the following reasons. (*1*) Small microbalances of sensitivity of 1 μg per 1.33×10^{-4} radian deflection and high stability can be built to operate in a 5 cm diam glass tube attached to the vacuum or reaction system. This sensitivity makes possible the study of the formation of chemisorbed monolayers of gases on a 10 cm^2 surface. For a 1-day reaction period this corresponds to a rate of reaction of 1 to 2×10^{-5} Å per sec. Automation of microbalances has made possible in the same apparatus the measurement of fast rates of reaction as high as 10^5 Å per sec. (*2*) Enclosing the microbalance in the vacuum or reaction system makes possible a study of a wide variety of chemical reactions as well as pretreatments and posttreatments of the specimen under precisely controlled conditions of temperature, pressure, gas atmosphere, and gas flow.

Several theoretical and experimental disciplines must be utilized and coordinated in planning, carrying out, and interpreting vacuum microbalance measurements. Sensitive weighing methods must be coordinated with high vacuum techniques and with high temperature materials technology. Thermochemical and kinetic theory calculations must be made on the chemical reaction and all possible side reactions with the specimen support and furnace tube. These calculations help specify the type of vacuum system, furnace tube, specimen support, specimen preparation, and experimental procedure.

VI. APPENDIX 1

THEORY OF SURFACE INTERFACE REACTIONS WITH NO OXIDES PRESENT

A. Processes

During the oxidation of elements under conditions where no oxide films are formed, a barrier zone of volatilized oxide may form around the sample. These vapors can limit the access of oxygen to the sample surface. The overall reaction may be separated into five distinct processes, the slowest of which determines the rate of reaction.

1. Transport of oxygen gas to the surface.
2. Chemisorption of oxygen.
3. Chemical reaction at the surface.
4. Desorption of volatile oxide.
5. Transport of reaction products away from the surface.

Processes *1* and *5* are transport processes and if rate controlling, the temperature dependence of the reaction rate may vary as $T^{1/2}$ or a small power of the absolute temperature. Chemical reactions *2*, *3*, and *4* usually have high activation energies. This factor allows one to distinguish chemical reactions from diffusion process. The chemical processes *2*, *3*, and *4* can be distinguished from each other by the pressure dependence and by the magnitude of the activation energies. Chemisorption usually has much smaller activation energies than desorption. Chemisorption also is pressure sensitive while desorption does not involve pressure.

The following theoretical analysis is for the oxidation of graphite. Similar equations can be derived for other systems such as the oxidation of Mo, W, and Re.

B. Kinetic Theory

In the absence of oxide films and scales and barrier layers of reaction products surrounding the specimen undergoing reaction, we can calculate from kinetic theory the maximum rate of oxidation (31).

The following equation gives the number of collisions, n, of oxygen molecules with a 1 cm^2 surface per sec.

$$n = 3.52 \times 10^{22} \times P/(MT)^{1/2} \tag{10.18}$$

Here M is the molecular weight, T is the absolute temperature, and P is the pressure in torr. At 1683°K and 76 torr, $n = 1.15 \times 10^{22}$ collisions per cm^2-sec or 0.6 g of O_2 per cm^2-sec. For carbon, this corresponds to a rate of oxidation of 2.3×10^{22} atoms of C per cm^2, assuming CO as the reaction product and a stoichiometric reaction coefficient of one.

In fast flow environments where the gas velocity approaches or exceeds the molecular velocity, faster rates of reaction than that given by equation 10.18 would be calculated.

Equation 10.18 assumes all collisions result in reaction. If the reaction is limited by an adsorption process involving a heat of activation, ΔH, then the rate of oxidation is given by

$$dn'/dt = nm \exp(-\Delta H/RT) \tag{10.19}$$

Here m is the stoichiometric factor relating oxygen molecules to metal atoms reacted, n is the number of collisions given by equation 10.18, dn'/dt is the rate of oxidation based on 1 cm^2 of geometrical area. To compare this with rates based on real area, dn'/dt must be divided by the ratio of real area to geometrical area. For graphite this ratio is usually greater than 1000, while for molybdenum, tungsten, and rhenium this ratio is between 1 and 2.

If desorption is rate controlling, the empirical rate law may have the same form as equation 10.19, however, n could not be calculated from the preexponential factor.

Equations 10.18 and 10.19 can be used to check the mechanisms of reaction. The number of collisions, n, can be evaluated from 10.18 and compared with the experimental value calculated by 10.19 if the heat of activation of the reaction is known.

C. Absolute Reaction Rate Theory

The chemical processes of adsorption, desorption, and chemical reaction can be treated using the absolute reaction rate theory of surface reactions (32).

This theory assumes the formation of a complex between the reacting gas and the surface and the chemisorbed reaction product and the surface. The rate of any one of these surface reactions may be considered in terms of reaction complexes passing from one region of configuration space to another. According to Eyring and coworkers (32), the number of reaction complexes crossing the energy barrier is given by the product of the number of complexes in the initial state at time t, the probability that the reaction complex crosses the barrier in any one attempt, and the frequency with which the complexes cross the energy barrier.

Three processes will be considered here: (*1*) mobile adsorption, (*2*) desorption, and (*3*) desorption with diffusion. The latter process applies to porous materials like graphite. Similar equations apply to the oxidation of Mo, Re, and W. However, the porosity factor can be neglected.

1. Mobile Adsorption

Several adsorption equations have been suggested by Eyring and coworkers. Earlier work (11) in our laboratory has suggested mobile adsorption as the rate limiting mechanism for the oxidation of carbon. The applicable absolute reaction rate equation is

$$dn''/dt = 2C_o k_B T/(2\pi m k_B T)^{1/2} \times \exp(-\Delta H_1^\ddagger/RT) \qquad (10.20)$$

where:

dn''/dt = rate of oxidation in atoms of C per cm^2 real area-sec
m = mass of O_2 molecule
$\Delta H_1^\ddagger$ = heat of activation for adsorption
C_o = concentration of O_2 in gas near surface
k_B = Boltzmann constant
T = absolute temperature

The factor 2 takes into account that 2C atoms react for each O_2 molecule adsorbed. The total reaction is given by:

$$dn'/dt = dn''/dt \times S \qquad (10.21)$$

where:

dn'/dt = rate of oxidation in atoms of C per cm^2-sec based on geometrical area

S = surface area cm^2 per cm^2 geometrical area

2. Desorption of CO

The velocity of reaction, dn''/dt, for the desorption of CO on a graphite surface is given by the equation

$$dn''/dt = (KC_a k_B T f^{\ddagger}/f_s) \times 1/h \times \exp(-\Delta H_2^{\ddagger}/RT) \qquad (10.22)$$

where:

k_B = the Boltzmann constant
h = the Planck constant
T = the temperature °K
$f^{\ddagger}$ = the partition function of activated complex
f_s = the partition function of surface with oxygen on it
R = the gas content
C_a = the number of reaction sites/cm^2 $\sim 3.5 \times 10^{15}$
$\Delta H_2^{\ddagger}$ = the activation energy, cal/mole
K = the transmission coefficient
K = is taken as unity
$f^{\ddagger}/f_s$ = is taken as unity

The total reaction is given by equation 10.22 multiplied by the surface area ratio, S.

3. Desorption of CO with Diffusion

Blyholder and Eyring (19) developed a mechanism for the oxidation of carbon based on a combination of desorption and pore diffusion. Without pore diffusion, oxidation is limited by a zero order surface reaction with an 80 kcal/mole activation energy.

The following equation was developed

$$dn'/dt = 1.68 \times 10^{11}\, \theta' (k_1 P T^{1/2})^{1/2} \qquad (10.23)$$

where P is the pressure in torr and θ' is the porosity. $\theta' = P_s V_g$ where P_s is the density of the solid and V_g is the pore volume. When a density of 1.6 and a pore volume of 0.3 g/cm^3 is used, $\theta' = 0.48$. The term k_1 is the intrinsic reactivity of the surface molecules in units of O_2 molecules reacted per cm^2-sec and is given by the equation

$$k_1 = \tfrac{1}{2}(dn''/dt) \quad \text{desorption} \qquad (10.24)$$

where (dn''/dt) desorption is given by equation 10.22.

To evaluate the chemical rate controlling process, a careful study must be made of the dependence of the chemical reaction on temperature, pressure, gas flow, and surface area and the results compared with the equation corresponding to the possible rate controlling steps.

VII. APPENDIX 2

FURNACES FOR TEMPERATURES UP TO 1650°C

Silicon carbide furnace elements and platinum or platinum alloy wire type furnaces are normally used for achieving temperatures between 1200 and 1500°C. Silicon carbide elements can be used at temperatures in excess of 1500°C but their life is short. Platinum–rhodium alloys can be used above 1500°C; however for long life, heavy sections should be used which makes the elements expensive. Molybdenum and tungsten wire or strip furnaces can be used at temperatures above 1500°C; however the heater wires must be enclosed in an inert gas atmosphere or vacuum.

Molybdenum disilicide heating elements can be used to operate furnace elements in air for long periods of time at temperatures up to 1650°C provided care is used in the design and use of the elements. Commercial elements are available. Kanthal-Super, for example, is chiefly molybdenum disilicide with a binder and is produced by powder metallurgical processes in the form of hair pin-type rods.

A. Mechanical and Physical Properties of Molybdenum Disilicide Elements

These elements are essentially vitreous. At 25°C the material is hard and brittle. The impact strength of this material is low although it has high bending and tensile strength. The material should not be subjected to any substantial impact or bending stresses below 1100°C. New elements are ductile above 1100°C; however used elements cannot be subjected to impact or bending stresses even at elevated temperatures.

The specific electrical resistance of the material increases rapidly with temperature. Using a constant voltage, the power consumption decreases as the temperature rises. The resistance at 1650°C is 4 times the resistance at 500°C and 16 times that at 25°C. The average temperature coefficient is 0.0048 per °C between 20 and 1600°C. It has been found that the resistance of the elements does not change with age or useage.

B. Electrical Supply Equipment

Since the resistance of the elements is low at room temperature, heavy current drains would occur as the elements are switched on to the lines.

A starting voltage of 1/3 of the operating voltage is recommended. For the standard 6 mm diameter element, currents of 130 A are required for a 1600°C furnace operating temperature. One way to limit the starting current is to use an adjustable transformer on the input of the low voltage transformer and thus limit the current on the elements. The input of the adjustable transformer should be protected by fuses.

C. Useful Life

The useful life of the elements depends upon many factors including intermittence of operation, element temperature, surface load, furnace atmosphere, refractories used in the element support, and cooling conditions. One set of elements operated for 3 years at temperatures of 1000–1650°C. It is recommended that the furnaces be kept at temperatures above 1000°C or at 1/2 of the operating voltage at all times.

D. Atmospheric Effects

The elements operate most efficiently when freely suspended in oxidizing atmospheres such as air, oxygen, water vapor, or carbon dioxide. If used in reducing atmospheres, the elements should be preoxidized for a few hours in a flowing air atmosphere. Since the oxide layer may scale, the elements should be reoxidized before being used a second time in a reducing atmosphere.

Care should be taken in using pure dry hydrogen since reduction of the surface layer can occur. Sulfur must be avoided; however SO_2 and SO_3 gases are not particularly harmful to the elements. Kanthal-Super can be used to an element temperature of 1600°C in a SO_2 gas with no apparent damage. Metals and enamels which form easily fusible compounds with silicon or silicon dioxide also should be avoided.

E. Furnace Design: Refractories

The elements must be supported so that they are free to expand and contract. If the elements are used above 1600°C, the silicon dioxide surface layer softens and sticking may occur between the elements and the furnace refractories. It is best to mount the furnace elements vertically and out of contact with the surrounding refractories. To avoid chemical reaction when the elements are in contact with refractories, it is essential that sillimanite type of bricks be used. The following compositions are satisfactory Al_2O_3:60–70%, SiO_2:30–40% and Fe_2O_3:max. of 1%. Bricks lower in SiO_2 or silicon carbide bricks can also be used.

Figure 10.27 shows a cross section drawing of the furnace and furnace element while Figure 10.28 shows a photograph of the furnace and furnace

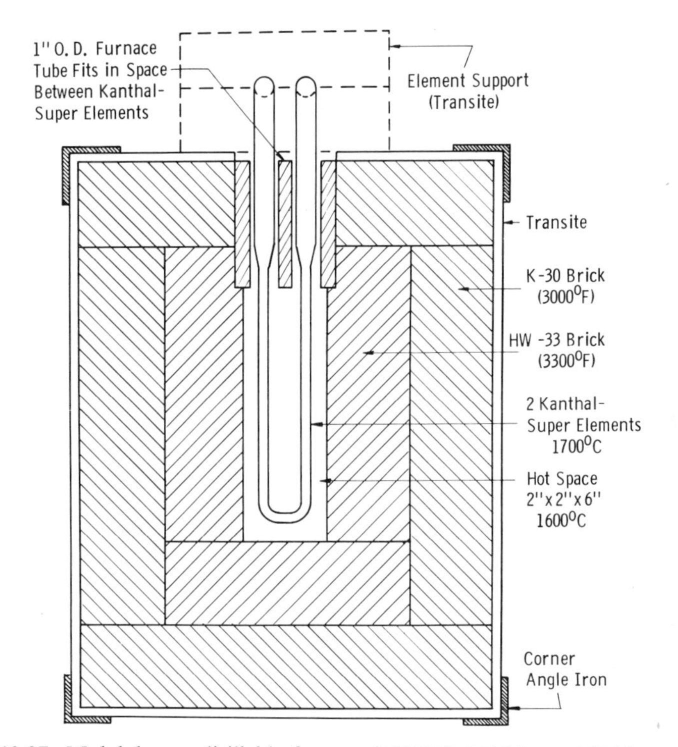

Figure 10.27. Molybdenum disilicide furnace (1600°C), 10½ in. × 10½ in. × 13 in. maximum power, 1760 W.

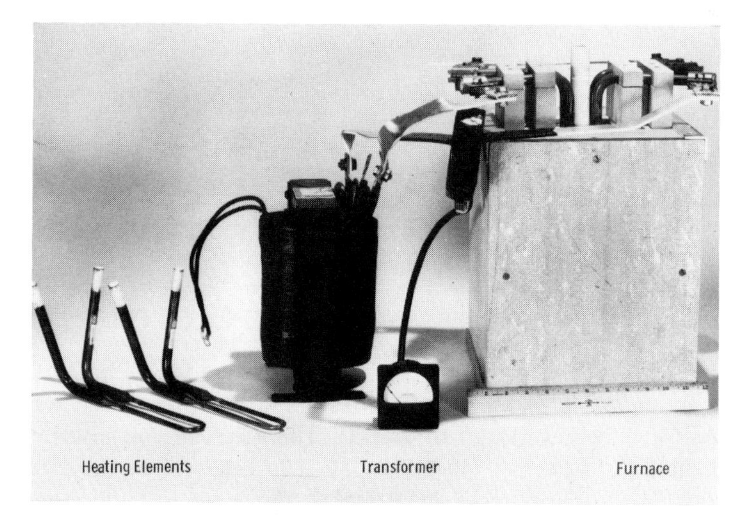

Figure 10.28. Kanthal furnace.

element supports. The furnace is 10½ in. square and 13 in. high. The element supports are 2¾ in. above the top of the furnace. The hot space is 2 in. square and 6 in. long. Two S-306 Kanthal-Super elements operating at temperatures up to 1700°C are used. Details of the furnace elements are given in Table 10.14. K-33 brick is used to enclose the hot space. This brick is good to 1815°C. To lower the heat loss, K-30 brick is used on the outside where lower temperatures are encountered. The furnace is enclosed in ¼ in. thick transite box. The elements are freely suspended into the hot space volume. To minimize radiation and convection losses, the bricks are carefully grooved and fitted together.

Table 10.14. Furnace Element and Transformer Details

Specifications S-306 Element	
Heating length	150 mm
Terminal length	250 mm
Heater diameter	6 mm
Terminal diameter	12 mm
Distance between legs	25 mm
Hot resistance	0.0459 ohms
Power	880 W
Surface heating load	10–20 W/cm^2
Operating voltage	6.3 V
Starting voltage	2.1 V
Current	140 A
Power in leads	18%
Specifications Transformer	
Input	110 V ac
Output	5 and 14 V
Current	140 A max.
Power	1.8 kV-A for 2 elements

VIII. APPENDIX 3

FURNACE TUBES AND REACTION SYSTEMS

Vycor and fused quartz furnace tubes can be used to temperatures of about 900 and 1100°C, respectively. Mullite and zirconia furnace tubes can be used for temperatures up to 1200°C. These ceramic materials can be sealed directly to Pyrex 7740 glass (33). The behavior of these furnace tubes at temperatures up to 1175°C was discussed by Gulbransen and Andrew (33).

A. Criteria of Furnace Tube Performance

Two criteria were used to describe the performance of a furnace tube and the associated vacuum system (33). These were: (*1*) the actual pressure achieved at a given temperature after pumping for a definite period of time, and (*2*) the apparent leak rate of gases into the closed reaction system in

Table 10.15. Effect of Gas Treatment of Tube on the Apparent Leak Rate (900°C Mullite Double-walled Tube, 30-min Test)

Time, min	Apparent leak rate, cc (NTP) per sec
	H_2 at 900°C to 20 torr pressure for 2 hr
10	2.0×10^{-6}
90	5.2×10^{-7}
226	2.2×10^{-7}
305	1.6×10^{-7}
4203	4.0×10^{-9}
	He at 900°C to 76 torr pressure for 2 hr
10	8.8×10^{-6}
90	1.4×10^{-6}
210	5.2×10^{-7}
1402	8.0×10^{-9}
	N_2 at 900°C to 76 torr pressure for 2 hr
12	2.0×10^{-7}
102	2.0×10^{-8}
258	1.0×10^{-8}
1259	4.0×10^{-9}
	O_2 at 900°C to 76 torr pressure for 2 hr
10	4.0×10^{-7}
90	6.0×10^{-8}
226	5.0×10^{-9}
	Ar at 900°C to 76 torr pressure for 2 hr
10	2.0×10^{-7}
90	1.4×10^{-8}
210	4.0×10^{-9}
1208	2.0×10^{-9}

units of cc-atm sec^{-1}. The apparent leak rate gives a very useful measure of the gases accumulating in a closed reaction system without pumping. This criteria is very useful in analyzing the performance of a reaction system where chemical reactions are studied.

Table 10.15 shows the effect of gas treatment of a mullite furnace tube on the apparent leak rate. Here the intermediate chamber in a double walled reaction vessel was maintained under vacuum conditions. The gases were added at 900°C to the reaction system for 2 hr and the apparent leak rates determined after several periods of time. The gas collection time was 30 min. When helium was added for 2 hr at 900°C and at 76 torr pressure, the vacuum system had not recovered after 1402 min. Hydrogen behaved in a similar manner. Argon gas gave the lowest value for the apparent leak rate of the five gases studied.

B. Types of Furnace Tubes

As noted, mullite and zirconia furnace tubes can be used up to 1200°C. Above this temperature, high density recrystallized alumina tubes must be used. These tubes have been used at temperatures up to 1665°C.

Figure 10.29. Reaction of tungsten oxide with alumina and resulting crack.

Care must be used in handling the alumina tubes. At the higher temperatures, dust, dirt, salt, carbon, and oxide vapors react with the alumina. Graphite reacts with aluminum oxide to form carbon monoxide at 1500°C. The reaction of tungsten trioxide gas with alumina was very rapid at temperatures above 1400°C. At 1600°C, alumina furnace tubes crack after exposure to tungsten trioxide vapors for 10–15 min. Figure 10.29 shows cross section photographs of an alumina furnace tube after reaction with tungsten trioxide.

C. Specimen Supporting Wires

The specimen support wires are one of the critical factors in vacuum microbalance technology. Platinum wires can be used to 1500°C with all materials except those which contain silicon or which readily alloy with platinum. Drawn silica fibers 20–100 μ in diameter are good for many materials but not carbon. Sapphire fibers purchased from Linde Company can be used for the most severe service conditions.

Choice of specimen support must be considered carefully in relation to its possible reaction with the specimen and the reaction gases.

REFERENCES

1. D. E. Roxner and H. D. Allendorf, *J. Chem. Phys.*, **40**, 3441 (1964).
2. E. A. Gulbransen, K. F. Andrew, and F. A. Brassart, in *Progress in Astronautics and Aeronautics*, Vol. 15, H. G. Wolfhard, I. Glassman, and L. Green, Jr., Eds., Academic Press, New York, 1964, p. 227.
3. E. A. Gulbransen, K. F. Andrew, and F. A. Brassart, *J. Electrochem. Soc.*, **110**, 476 (1963).
4. E. A. Gulbransen and K. F. Andrew, *J. Electrochem. Soc.*, **101**, 474 (1954).
5. *JANAF Thermochemical Tables*, The Dow Chemical Co., Midland, Michigan.
6. *Selected Values of Chemical Thermodynamic Properties*, Circular 500, National Bureau of Standards, Washington, D.C., 1952.
7. O. Kubaschewski and E. L. Evans, *Metallurgical Thermochemistry*, Pergamon Press, New York, 1956.
8. E. A. Gulbransen, K. F. Andrew, and F. A. Brassart, *J. Electrochem. Soc.*, **112**, 49 (1965).
9. E. A. Gulbransen, K. F. Andrew, and F. A. Brassart, *J. Electrochem. Soc.*, **110**, 952 (1963).
10. E. A. Gulbransen, K. F. Andrew, and F. A. Brassart, *J. Electrochem. Soc.*, **111**, 103 (1964).
11. E. A. Gulbransen and K. F. Andrew, *Ind. Eng. Chem.*, **44**, 1034 (1952).
12. E. A. Gulbransen and K. F. Andrew, *Vacuum Microbalance Tech.*, **2**, 129 (1962).
13. E. A. Gulbransen, K. F. Andrew, and F. A. Brassart, *Vacuum Microbalance Tech.*, **4**, 127 (1965).
14. E. A. Gulbransen, K. F. Andrew, and F. A. Brassart, *Vacuum Microbalance Tech.*, **3**, 179 (1963).
15. I. Langmuir, *Phys. Rev.*, **2**, 329 (1913).
16. C. M. Tu, H. Davis, and M. C. Hottel, *Ind. Eng. Chem.*, **26**, 749 (1934).
17. J. M. Kuchta, A. Kant, and G. H. Damon, *Ind. Eng. Chem.*, **44**, 1559 (1952).
18. M. Levy, *Ind. Eng. Chem., Prod. Res. Devel.*, **1**, 19 (1962).
19. G. Blyholder and H. Eyring, *J. Phys. Chem.*, **61**, 682 (1957).
20. G. Blyholder and H. Eyring, *J. Phys. Chem.*, **63**, 1004 (1959).
21. P. L. Walker, Jr. and E. Raats, *J. Am. Chem. Soc.*, **60**, 364 (1956).
22. E. A. Gulbransen and W. S. Wysong, *Trans. AIME* (*Metals Div.*), **175**, 628 (1948).
23. K. M. Gorbounova and V. A. Arslambekov, in *Reactions Superficielles des Gaz sur les Metaux* 6ᵉ Reunion De La Societe De Chimie Physique, May 29–June 1, 1956, Paris, France, p. 211.

24. E. G. King, W. W. Weller, and A. U. Christensen, *U.S. Dept. of Interior.*, Bureau of Mines, RI, 5664 (1960).
25. E. A. Gulbransen, K. F. Andrew, and F. A. Brassart, *J. Electrochem. Soc.*, **110**, 242 (1963).
26. J. W. Semmel, Jr., *High Temperature Materials*, Wiley, New York, 1959, p. 510.
27. J. L. Modisette and D. R. Schryer, NASA-TN-D-222, March 1960.
28. E. A. Gulbransen and K. F. Andrew, *J. Electrochem. Soc.*, **107**, 619 (1960).
29. E. A. Gulbransen and W. S. Wysong, *Trans. AIME*, **175**, 611 (1948).
30. V. D. Barth and G. W. P. Rengstorff, "Oxidation of Tunsten," (DMIC Report 155) AF33(616)-7747, Defense Metals Information Center, Battelle Memorial Institute, Columbus, July 1961.
31. J. H. De Boer, *The Dynamical Character of Adsorption*, Oxford University Press, London, 1953, Chapter 2.
32. K. J. Laidler, S. Glasstone, and H. Eyring, *J. Chem. Phys.*, **8**, 659 (1940).
33. E. A. Gulbransen and K. F. Andrew, *Ind. Eng. Chem.*, **41**, 2762 (1949).

Chapter 11

The Measurement of the Adsorption and Desorption of Gases from Solid Surfaces

A. W. CZANDERNA

Department of Physics and Institute of Colloid and Surface Science
Clarkson College of Technology, Potsdam, New York

I. INTRODUCTION

Adsorption is the accumulation of a surface excess at the interface of two immiscible phases. In this chapter, the measurement of the adsorption of a gas on a solid surface by gravimetric techniques will be discussed. Although physical and chemical adsorption measurements have been carried out by gravimetric techniques for over forty years, the usual quantitative relationships for the rate and equilibrium of adsorption are presented in terms of the volume of gas adsorbed (1). This is understandable since most adsorption studies were carried out by volumetric techniques prior to the advent of commercially available recording microbalances. The quantitative relationships for the rates of adsorption, of desorption, and of thermodesorption and for equilibrium at surfaces will be presented explicitly for gravimetric techniques.

The emphasis of this chapter will be on the application of the vacuum ultramicrobalance for comprehensive studies of chemisorption. As will be seen, the volumetric method, which is suitable for many studies of the physical adsorption of gases, is relatively inadequate for a systematic comprehensive *in situ* study of chemisorption. Some interesting comparisons are available (1–6) for the measurement of the amount of gas adsorbed by the volumetric method (7) and the gravimetric method (8). It is of special interest that the application of the microgravimetric technique to physical adsorption studies may lead to important new information about gas–solid interactions that is not obtainable by volumetric techniques (9).

The object of many studies of the chemisorption of gases on solids is to answer the fundamental questions: what are the adsorbed species, what is the relative population, and where are they located on the surface? The latter implies detailed knowledge of the surface structure. Thus, a detailed experimental study of chemisorption phenomena should allow the quantitative parameters of adsorption to be related to the structure of the adsorbent surface at various adsorbate coverages. This type of "ideal" study should greatly simplify the theoretical treatment of the nature of the interactions between the adsorbate and the adsorbent. At present, carrying out the "ideal" study is extremely difficult because of an either–or situation. As a broad generalization, the techniques of field ionization microscopy (FIM) and low energy electron diffraction (LEED) are dependable for specifying the structure of the surface on an atomic scale but are inadequate for studying adsorption equilibrium and kinetics. The vacuum ultramicrobalance can be used for the latter type of study but, in general, must utilize samples that present an inadequate definition of the surface structure. Auxiliary studies by electron diffraction and microscopy and/or simultaneous measurement of the mass and infrared absorption can provide a more complete description of the surface.

In principle, microgravimetric techniques have the potential for carrying out the "ideal" study by careful choice of the adsorbent material. For example, clean surfaces of single crystal wafers, thin films, or faceted powders may present a surface structure that is not only adequate for theoretical considerations but also exists in sufficient quantity for the reproducible measurement of the quantitative parameters of adsorption. On the other hand, the detailed surface arrangements of clean surfaces can vary widely and be quite complex, e.g., clean silicon surfaces have a variety of possible structures. In Table 11.1, there is presented a description of samples that have been (silver) or might be (nickel) used in an attempt to perform the "ideal" study with a microbalance. This presentation provides the reader considering microgravimetric methods for fundamental chemisorption studies with a perspective of the difficult compromise situations that may be required to gain another order of magnitude. Suggested possible methods of cleaning and preparing the surfaces and the requirements of the microbalance are also given. The data for (100) nickel were calculated assuming the experimental apparatus available would be the "ultimate microbalance" (Chapter 2, Section IV.F). The proposed size of the single crystal wafers and films were selected to provide a maximum surface with a minimum development of technology for their preparation. The problems of suspending many single crystals or supported films from the balance are indeed challenging. However, the potential gain in fundamental knowledge will make the effort worthwhile.

As has been discussed in previous chapters, the measurement of less than 1 μg of gas adsorbed at pressures exceeding 10^{-3} torr requires specialized techniques (10). It is precisely this pressure region where study of the quantitative parameters of adsorption must be carried out. Automated bakeable balances with the necessary capacity, long term stability, and a LPR (load to precision ratio)* of 10^8 are not available commercially. However, solutions to many of the problems of constructing acceptable specialized versions have been described (11) and are worthy of consideration in competition with other techniques.

The preparation of a clean surface is undoubtedly the most important aspect of any fundamental investigation of chemisorption. Techniques commonly employed to obtain a clean surface include sputtering, vacuum deposition. cleavage, flashing a filament, and outgassing and chemical reduction. In this chapter, results obtained during the study of the chemisorption of oxygen on silver powder surfaces cleaned by chemical reduction with carbon monoxide and/or hydrogen will be presented. By subjecting the sample repeatedly to a cycle of outgassing, oxygen adsorption, outgassing, and reduction (OAOR cycling), the usual problems with this method of surface preparation have been eliminated (12). The use of

* See Chapter 2, Section III for a discussion of LPR.

Table 11.1. Forms of Solid Surfaces Used or Suitable for Study with the Ultramicrobalance

	Nickel (100)			Silver[a]
	60 single crystal wafers, 0.1 mm thick and 2 cm in diam	250 single crystal films supported on inert substrates of density 2.25 g/cm^3, 0.1 mm thick, 2 cm diam	Spheres of powder predominantly with (100) facets developed by cyclic oxidation and reduction 68 μ diam 100 cm^2/g	Polycrystalline powder with a specific surface of 900 cm^2/g (spheres, 6.2 μ diam)
Mass (to be) used, g	16.8	17.8	18.6	1.5
Geometric surface area cm^2	377	1670	1860	1350
Surface preparation or cleaning	Sputtering	Vacuum deposition	OAOR cycling	OAOR cycling
Oxygen adsorbed at monolayer coverage,[b] μg	8.1	35.9	40	52

Minimum θ[c] detectable using a balance with a LPR of 1×10^8 and a 20 g load	0.021	0.005	0.004_7	0.0029[d]
Minimum θ detectable for a sensibility of 1 μg (high pressures)	0.123	0.028	0.025	0.019

[a] Typical data for silver powders used for chemisorption studies reported in Section IV.
[b] Calculated assuming one O atom adsorbs on each Ni surface atom where $A_0 = 3.52$ Å for the 100 Ni structure.
[c] θ is fraction of monolayer.
[d] For LPR of 1×10^7.

chemisorption data for evaluation of the quantitative adsorption and desorption parameters will be illustrated; pertinent data for typical samples of silver used in this study are appended to Table 11.1.

II. QUANTITATIVE RELATIONSHIPS

A. Adsorption Parameters

1. Introduction

The important quantitative parameters for an activated adsorption process are indicated in the potential energy diagram (Figure 11.1) where q is the isosteric heat of adsorption, E_A is the activation energy for adsorption, E_D is the activation energy for desorption, and r_0 is the equilibrium

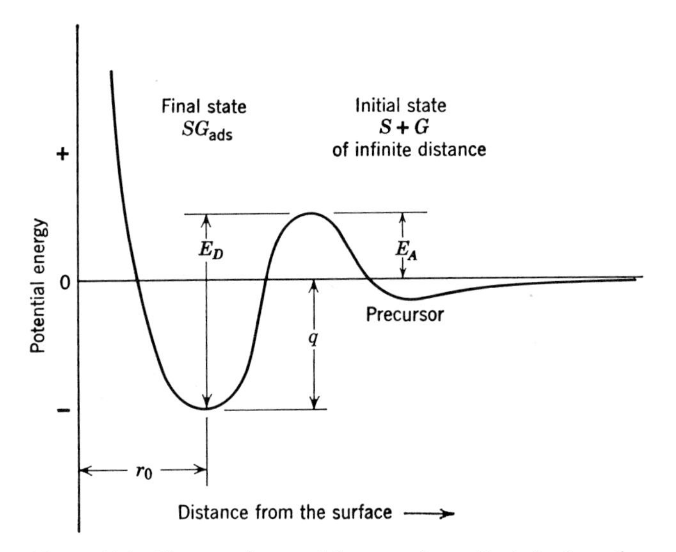

Figure 11.1. Change of potential energy for activated adsorption.

distance of the adsorbed gas from the surface. The potential energy reference of an adsorbate species in the gaseous state G relative to the solid surface S is defined as zero when the gas and solid have an infinite distance of separation.

2. Adsorption Equilibrium

At equilibrium between the gas and the solid, the mass of the adsorbate gas, m, adsorbed per unit mass of adsorbent is a function of temperature and pressure alone, viz. $m(P, T)$. Experimentally, a family of curves can be obtained by holding one of the variables, m, P, or T constant. To obtain a family of adsorption

1. isotherms, $m(P)$ is determined with T constant
2. isobars, $m(T)$ is determined with P constant
3. isosteres, $P(T)$ is determined with m constant

Obviously, adsorption isotherms and isobars may be determined readily by microgravimetric methods. Further, any family of curves may be transformed to another by replotting (1). When the data are presented as isotherms, the heat of adsorption, q, may be obtained at particular mass values, m, from the Clausius–Clapeyron relationship

$$(d \ln P/dT)_m = q_{\text{isos}}/RT^2 \qquad (11.1)$$

If the mass, m_m, corresponds to monolayer coverage of the surface, then θ, the fraction of a monolayer covered, is m/m_m. The slope of a plot of $\ln P$ versus $1/T$ corresponds to the magnitude of the isosteric heat, q. Since the latter usually depends on θ, viz., $q(\theta)$, the plots must be made at constant m or θ. The determination also incorporates the usual assumptions of the Clausius–Clapeyron equation, e.g., q is constant over the temperature range of the determination, etc.

3. Rate of Chemisorption

Chemisorption may occur at significantly different rates ranging from adsorption that is completed in microseconds to that which occurs for weeks. The rate of chemisorption, u, which corresponds to the time rate of mass gained dm/dt by a homogeneous adsorbent surface depends on four factors as given in equation 11.2.

$$u = [P/(2\pi m_g kT)^{1/2}] f(\theta)\sigma(\theta) \exp - [E_A(\theta)/RT] \qquad (11.2)$$

where $P/(2\pi m_g kT)^{1/2}$ is the number of impacts per unit time per unit area, m_g is the mass of the adsorbate species, k and R are the usual constants, the exponential is the fraction of gas atoms possessing the necessary activation energy, $f(\theta)$ is the probability the collision will occur on an unoccupied site, and $\sigma(\theta)$ is the probability that an adsorbate species with sufficient activation energy that hits an unoccupied site will adsorb. The latter term, which is frequently labeled the condensation coefficient and $f(\theta)$ are difficult to evaluate. However, if the rate is compared for the same values of θ at constant values of $PT^{-1/2}$ at a series of temperatures, the magnitude of the activation energy can be obtained from the slope of a plot of $\ln u$

versus $1/T$. The terms E_A and σ are not always a function of θ which leads to a simpler form of equation 11.2 but a heterogeneous surface leads to a much more complex expression.

4. Rate of Desorption and Thermodesorption

Desorption, which may occur at rates significantly different from those of adsorption, is always activated. For a uniform surface, where the variation in the desorption parameters results from surface interactions, the velocity of desorption, u', will occur from occupied sites according to

$$u' = K(\theta)f'(\theta) \exp - (E_D(\theta)/RT) \tag{11.3}$$

where $K(\theta)$ is the velocity constant, $f'(\theta)$ is the fraction of sites available for desorption, and the exponential is the fraction of adsorbed species with the necessary activation energy for desorption. From equation 11.3, it can be seen that in principle, the slope of a plot of ln u', the time rate of mass loss (dm/dt) by the surface, versus $1/T$ at a constant amount adsorbed will yield the magnitude of E_D. In practice, the true value of E_D is difficult to determine because readsorption of the desorbed species may occur especially for powders and films. The measurements thus yield an *apparent* activation energy of desorption. The interpretation of desorption data from powders must be handled carefully (1,13) to minimize the possibility of erroneous conclusions.

The activation energy of desorption also can be determined by the technique of thermodesorption (12). The latter refers to the desorption of gases from solids by a constantly increasing temperature. Theoretical expressions developed by Smith and Aranoff (14), Ehrlich (15), and Redhead (16) have been modified by Czanderna (12) to permit analysis of the ultramicromass loss on desorption. The expressions (12) are for first and second rate processes and a linear heating rate, β. For example, for a first order rate process the temperature, Tp, at which the maximum rate of desorption occurs, can be used to evaluate E_D from equation 11.4.

$$\frac{1}{X_p} = \frac{\nu I T_p}{\beta(1 - 2!/X_p + 3!/X_p{}^2 - 4!/X_p{}^3 + \cdots} \tag{11.4}$$

where ν is the frequency factor, X_p is E_D/RT_p, and I is an integral $I(X_p)$. Values of I have been published for X from 1 to 50 (14).

B. Surface Area

The determination of the surface area of a sample from a single adsorption isotherm is possible by utilizing the BET theory for multimolecular adsorption (17). The mass equivalent form of the usual two parameter BET equation is:

$$x/(1 - x)m = 1/m_m c + (c - 1)x/m_m c \qquad (11.5)$$

where $x = p/p_0$, p is the pressure of the adsorbate, p_0 is its vapor pressure at the temperature at which the isotherm is determined, m is the mass of the gas adsorbed at any pressure p, m_m is the mass adsorbed at monolayer coverage and c is a constant which depends on the nature of the adsorbate–adsorbent system (18). The derivation of the two parameter BET equation and an excellent discussion of the parameters c and v_m is available.[18] From equation 11.5, a plot of $x/(1 - x)m$ versus x is linear with slope S and intercept I from x of about 0.05–0.35 (17). Considerable care must be exercised when a linear plot is obtained in this region of x because this range is not valid for *all* materials (19). Thus,

$$m_m = (S + I)^{-1}$$

where S is $c - 1/\mathrm{c}m_m$ and I is $(\mathrm{c}m_m)^{-1}$. The specific surface, SA, is given in m^2/g by

$$SA = m_m K/W_A \qquad (11.7)$$

where W_A is the weight of the adsorbate and K is obtained from

$$K = A_m N/M \qquad (11.8)$$

where N is Avogadro's number, M is the molecular weight in grams, and A_m is the area occupied by one adsorbate molecule. For nitrogen, $A_m =$ 16.2 Å^2 and equation 11.7 becomes

$$SA_{N_2} = 3.48 m_m/W_A \qquad (11.9)$$

SA_{N_2} is in m^2/g when m_m is in μg and W_A is in mg (20).

An alternative method for estimating m_m from the isotherm is to use the point B method. Point B is where the curved and linear portions of a type II isotherm merge as illustrated in the plot of the adsorption of nitrogen in rutile in Figure 11.2. The region of mass–pressure boxed with dashed lines in the inset shows the part of the type II isotherm plotted on the larger scale. By introducing nitrogen in sufficiently small increments in the region of B, the departure of the isotherm from linearity can be estimated very accurately because of the ultrasensitivity of the microbalance.

The microbalance may be used for single point methods (19) or other variants of volumetric methods for obtaining approximate values of the specific surface, viz. SA from the measurement of m_m at one value of x. With presently available equipment completely automatic systems can be constructed for this purpose.

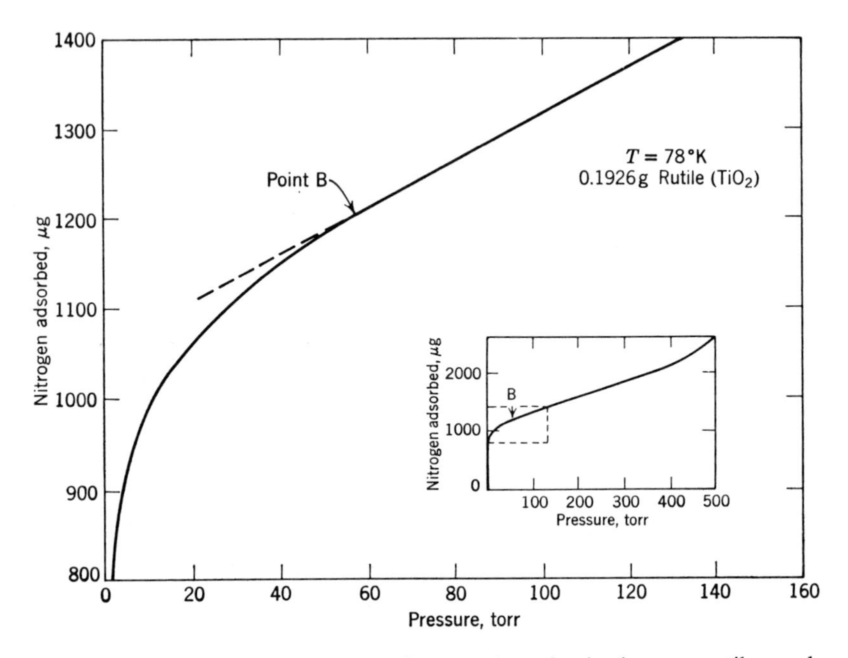

Figure 11.2. Location of point *B* on a nitrogen adsorption isotherm on rutile powder at 78°K; (inset after Czanderna and Honig, *J. Phys. Chem.*, **63**, 622 (1959)).

III. EXPERIMENTAL

A. Introduction

A beam type vacuum ultramicrobalance was used to study the adsorption of oxygen on silver powder. By utilizing furnaces and a vacuum system, it was possible to vary the ambient pressure and temperature over wide limits and to determine from the mass changes

1. the saturation uptake as a function of pressure and temperature
2. the rate of adsorption
3. the rate of desorption
4. the thermodesorption of adsorbed species
5. the adsorption isotherm
6. the constancy of the sample mass over long periods of time

From the saturation uptake as a function of pressure and temperature and equation 11.1, q may be obtained; from the rate of adsorption and equation 11.2, E_A may be obtained; from the rate of desorption and the thermodesorption of adsorbed species and equations 11.3 and 11.4, E_D may be obtained; from the adsorption isotherm and equations 11.5–11.9, the surface area may be obtained; and the constancy of the sample mass with time yields information essential for reliability and unambiguous interpretation of the data of (*1*) through (*5*) above (21).

B. Apparatus

The ultramicrobalance used for the study of the adsorption of oxygen on silver was a modified version of the pivotal beam balance described by Czanderna and Honig (22). The balance was calibrated by the buoyancy

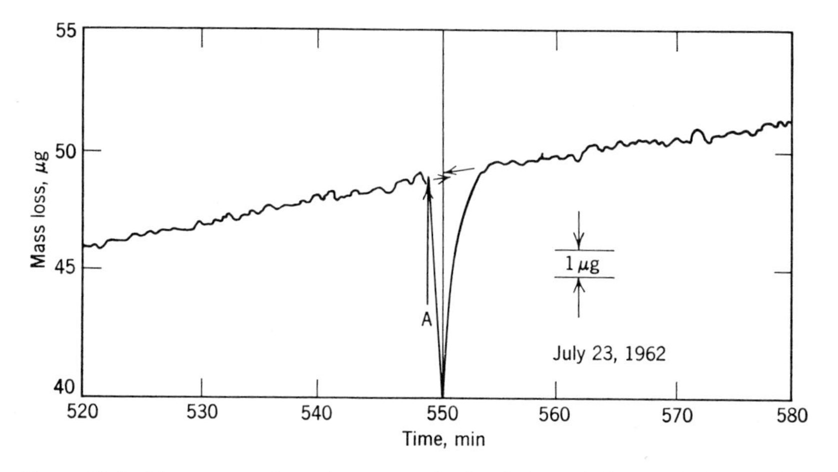

Figure 11.3. Mass versus time of automated microbalance before and after the beam was banged on the arrests.

technique (22) (see Chapter 2 Section V) but it was necessary to introduce an asymmetrical beam design to allow automation by the transducer-method described by Cochran (23). Details of the construction of the beam and automatic operation of the balance have been given (21,24). The limitations of the automated version (24,25) are much less restrictive than those of the original design (2,22). For example, zero shifts that occurred from the beam striking the arrest were eliminated by raising the beam arrests (11). Evidence for the zero stability is shown in Figure 11.3, which is a reproduction of a strip chart recording of the mass change being monitored by the balance. At point A, the beam was deliberately banged between the arrests by interrupting the current in the compensation solenoid. As can be seen, the mass reading before and after the event is not altered by more than 0.2 μg, the approximate error of the balance. Another advantage of the automated balance was the damping effect of the inductive transducer which enabled superior control of the beam during admission of gas. Thus, zero shifts were avoided, in general, which allowed most data to be taken at the full sensibility of the balance.

With the automated microbalance the rate of oxygen adsorption could be followed after 20 sec from the time of gas admission. Mass changes of ± 0.1 to 0.2 μg could be detected *in vacuo* or in the presence of the ambient

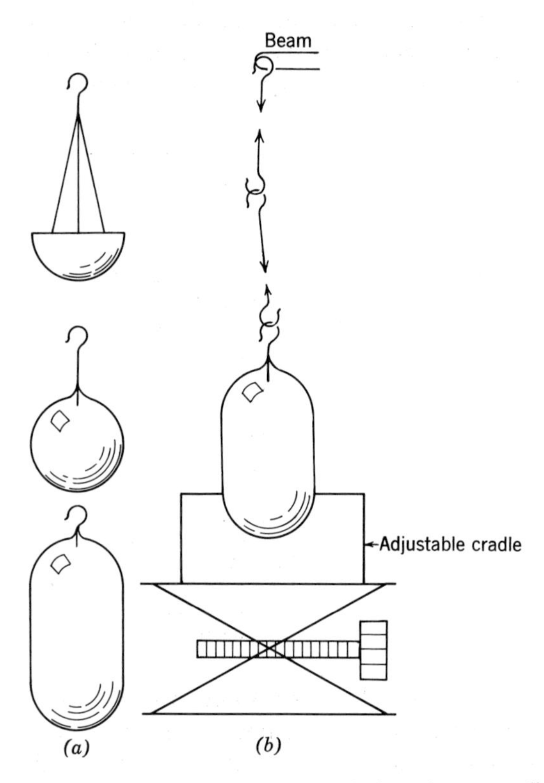

Figure 11.4. (*a*) Sample containers; (*b*) simple technique for suspending containers.

gas, respectively. Long-term changes in the total sample mass could be monitored to within 1 μg.

In the early work, vacuum in the system was produced with a two-stage mercury diffusion pump and mechanical forepump. The use of the appropriate pressure gauges, traps, and valves has been described (10,12,19, 26). In more recent work (27), an all-glass and metal valve system has been used in which vacuum was produced with a three stage CVC-25 oil diffusion pump and a mechanical forepump. The oil vapor was trapped with Linde Molecular Sieve 13X. The vacuum system was used to expose the sample alternately to outgassing conditions, to oxygen for adsorption, to a reducing gas for cleaning, and to nitrogen for surface area or buoyancy studies.

The calibration (2,21,28) and stability of the furnaces (21,23), the preparation of oxygen (28), nitrogen (28), carbon monoxide (21), hydrogen (21), and other gases (26,27), and the source and purity of the silver powder (12,21,26,27) have been reported. In all of the studies of the adsorption of gases on silver reported to date (10,12,13,21,24,26,27), aliquots of silver from the same preparation have been used.

Silver powder was loaded into a quartz bulb (or cylinder) similar to those shown in Figure 11.4*a*. Specimen holders of this type can be fabricated easily by a competent glassblower. The weighed bulb was cradled in lint free tissue and a clean glass rod with a flattened end was used to push the powder into the vial through the access hole. When this design is used, the size of the access hole may govern the pumping speed (29). The amount of sample placed in the bulb was determined from the weight of the bulb and sample. For suspending the bulb from the hangdown suspension of the balance, the bulb, supported in an adjustable cradle was hooked to the fiber without placing tension on the fiber. The cradle was then lowered slowly until all of the weight was taken by the suspension (Figure 11.4*b*). Use of this technique minimized breakage of fragile hooks.

C. Technique

1. General Comments on Buoyancy

For a gas obeying the equation of state $PV = nRT$, the mass, m, of the displaced fluid is given by (Chapter 2, Section VII.E)

$$m = (M\Delta V/RT)P \tag{11.10}$$

where P is the pressure in torr, M is the gram molecular weight of the gas, ΔV is in cm^3, T is the temperature in °K, and R is the gas constant (62,364 cm^3 torr mol^{-1} °K^{-1}). In addition to the assumed ideal gas behavior, equation 11.10 neglects second-order effects (9). The slope, S_b, of a plot of mass change versus pressure then is given by

$$S_b = M\Delta V/RT \tag{11.11}$$

In the standard symmetrical design for the hangdown tubes, shown in Figure 11.5, three net volumes can be considered because each may be subject to different temperatures. When a symmetrical design is employed, the temperature of the two chambers labeled T_{HC} is usually the same. If V_s is the volume of the hangdown suspension, sample container, and sample in the heated or cooled region T_{HC}, V_c is the volume of the hang-down suspension, counterweight container, and counterweight in the heated or cooled region T_{HC}, and V' is the net buoyancy volume at the beam

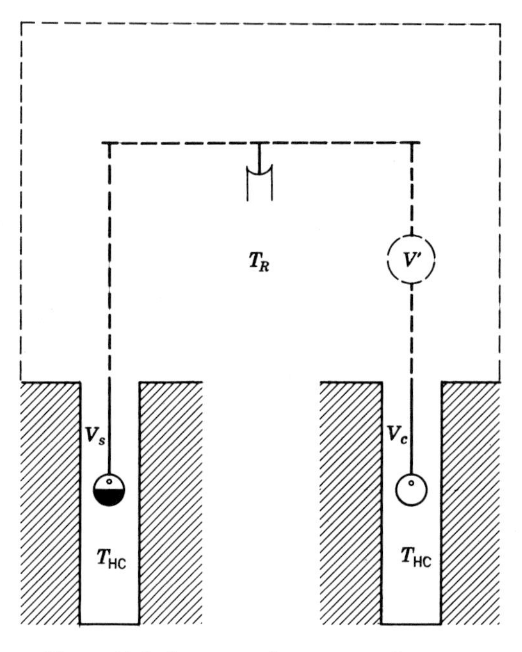

Figure 11.5. Important buoyancy volumes.

temperature, T_R, because of asymmetries in the beam, remaining suspension fibers, counterweights, and magnetic compensation wire enclosure, then

$$\Delta V = (V_s - V_c)_{\text{at } T_{HC}} - (V')_{\text{at } T_R} \tag{11.12}$$

When equations 11.11 and 11.12 are combined, the dependence of the slope on the surrounding temperatures is given by

$$S_b(T) = M(V_s - V_c)/RT_{HC} - MV'/RT_R \tag{11.13}$$

In any particular chemisorption study, the only variables in equation 13 are M and T_{HC} with V_s, V_c, and V' fixed prior to the study and T_R maintained a constant during the study. Thus, for S_b to be zero as a function of temperature, the relations, $V_s = V_c$ and $V' = 0$, must be satisfied. It is not easy to obtain these equalities in practice nor is it always desirable as will be shown. When an inert gas is used to obtain the slope, S_b, at various temperatures, the plotted data will resemble one of the sets of curves shown in Figure 11.6 depending on the relative contributions of V_s, V_c, and V' in

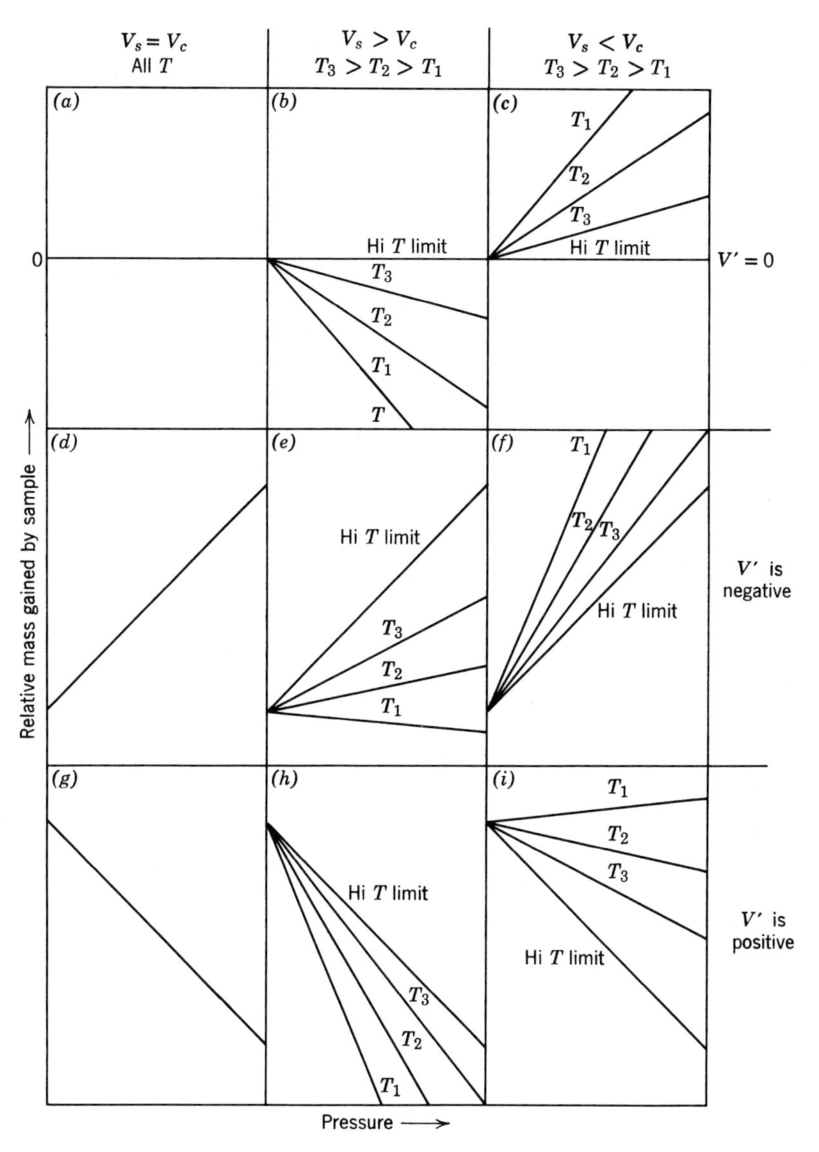

Figure 11.6. The effect of buoyancy on the apparent mass gained by the sample for possible value of V_s, V_c, and V'. The volume, V', is considered "negative" when an increase in pressure results in a buoyancy "mass loss" by the sample.

equation 11.13. Curves of this type are of considerable value when it is not feasible to adjust the volumes, V_s, V_c, and V' to allow $S_b(T)$ to be independent of temperature (Figure 11.6*a*).

When the volumes V_s and V_c are at a temperature $T_{HC} = T_2$, equation 11.13 becomes

$$S_{b2} = M(V_s - V_c)/RT_2 - MV'/RT_R \tag{11.13a}$$

and if they are at T_1,

$$S_{b1} = M(V_s - V_c)/RT_1 - MV'/RT_R \tag{11.14}$$

Equation 11.15 is obtained by subtracting equation 11.14 from equation 11.13a.

$$S_{b2} - S_{b1} = M(V_s - V_c)_{\text{at } T_2}/RT_2 - M(V_s - V_c)_{\text{at } T_1}/RT_1 \tag{11.15}$$

If the temperature dependence of the volume expansion of the solids comprising V_s and V_c is neglected, equation 11.15 becomes

$$S_{b2} - S_{b1} = [M(V_s - V_c)/R]\,(T_1 - T_2/T_1T_2) \tag{11.16}$$

Equation 11.16 is useful for calculating the slope at a second temperature, if the buoyancy has been determined at another temperature.

From inspection of equation 11.13, it is evident that a plot of the quantity $S_b(T) + V/RT_R$ versus $1/T$ will be linear with a slope of $M(V_s - V_c)/R$. This was verified experimentally by measuring the buoyancy with nitrogen in the balance system at −78, 0, 24, and 180–340°. The value of each slope was obtained from least squares analysis and plotted versus $1/T$ in Figure 11.7 as range bars along the heavy dashed line. From this straight line, the buoyancy slope at liquid nitrogen temperatures could be calculated for correcting nitrogen isotherm data. In Figure 11.7, the temperature dependence of the slope is also plotted for a series of assumed values of $V_s - V_c$ using nitrogen as the ambient gas. It is seen in Figure 11.7 that $V_s - V_c$ is 0.086 cm^3. The intercept of the ordinate, which is equivalent to the case $V_s = V_c$, can be used to determine V', if the components of V' are all at a known temperature. In this study, the temperature of V' was 34°; the intercept in Figure 11.7 is 0.4075 μg/torr; thus, equation 11.13 yields $V' = 0.278$ cm^3.

2. *Measurement of the Surface Area by the BET Method*

The following is the procedure for the determination of the surface area by the BET method from adsorption isotherms obtained at 78°K. After suitable outgassing, furnaces are replaced with dewars filled with liquid nitrogen. An inert gas such as helium may be used to obtain better control of the balance during the admission of nitrogen to the sample (28) but this is not necessary with automatic recording apparatus. Care must be exercised on each admission of nitrogen to allow for thermal equilibrium to be attained before continuing with further admissions. After about 15 min, no further mass change should be detectable unless pore volume filling, slow chemisorption, or some other slow process is occurring. The total mass gained by the sample is plotted as a function of the equilibrium pressures used. Measurements can be made as a function of both increasing and decreasing the pressure. After the appropriate corrections have

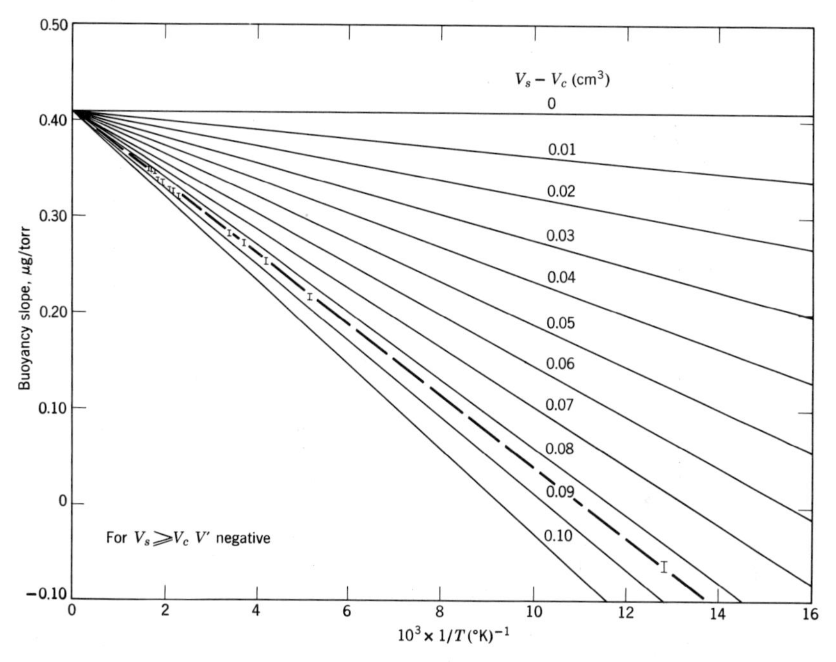

Figure 11.7. The temperature dependence of the buoyancy slope for V' negative and $V_s > V_c$ (cf. Figure 11.6*e*). Dashed line, experimental using nitrogen. Solid lines, calculated for nitrogen with assumed $V_s - V_c$, T.

been made for buoyancy, the mass data can be used for preparing a BET plot or evaluation of the specific surface from the point B in the isotherm. In Figure 11.8, the raw data obtained from the microbalance, the buoyancy slope, and the adsorption isotherm are plotted for the adsorption of nitrogen on a reduced silver surface during preparation for chemisorption studies. The isotherm may be obtained graphically as illustrated in Figure 11.8 or analytically from the raw data. Mass data corresponding to seven values of p/p_0 were used to construct the BET plot shown in Figure 11.9. The specific surface of the powder by the point B method was 0.111 ± 0.002 m^2/g; by the use of equations 11.5–11.9, S was 0.110 ± 0.004 m^2/g. The data plotted in Figure 11.9 are summarized in Table 11.2.

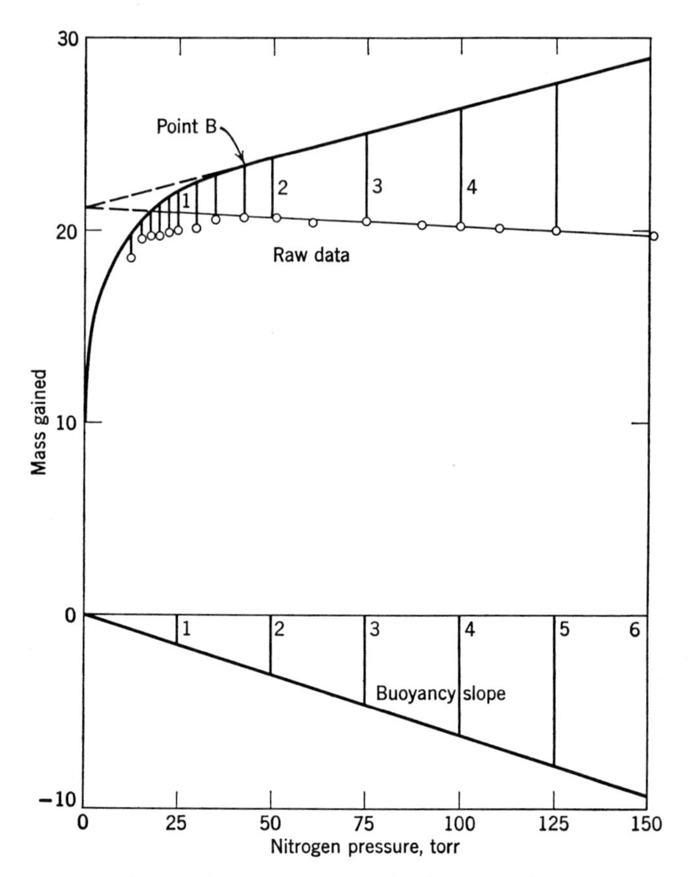

Figure 11.8. Graphical method for the determination of a nitrogen adsorption isotherm on silver powder at 77.8°K.

Table 11.2. Variables for BET Plot[a]

p	$x = p/p_0$	$m(\mu g)$[b]	$x/(1-x)$	$x/[1-x(m)]$
38	0.05	22.9	0.0527	0.00230
76	0.10	25.0	0.1111	0.00444
114	0.15	27.1	0.1766	0.00652
152	0.20	29.2	0.2500	0.00856
190	0.25	31.3	0.3333	0.01065
228	0.30	33.4	0.429	0.01285
266	0.35	35.5	0.539	0.01518

[a] $x/(1-x)m$ versus x is plotted in Figure 11.9.

[b] From nitrogen adsorption isotherm at 77.8°K, $p_0 = 760$ torr (Figure 11.8).

Difficulties from thermomolecular flow (TMF) (see Chapter 4) are not encountered for nitrogen adsorption because the lowest pressure corresponding to an x of 0.05 is above the TMF pressure region (2,20,22, 28,30–31). This is not true, however, for surface area studies using krypton (31).

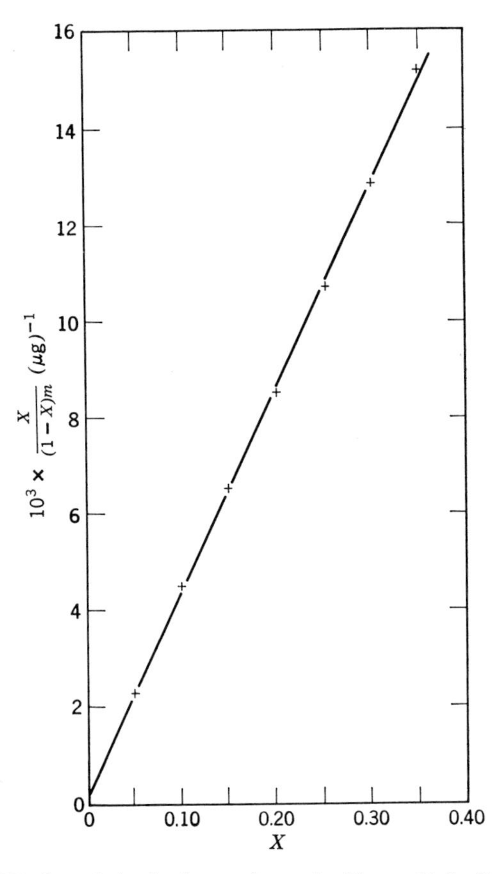

Figure 11.9. BET plot of the isotherm shown in Figure 11.8. $S = 0.04233\ \mu g^{-1}$; $I = 0.00020$; μg^{-1}; $M_m = 23.52\ \mu g$.

3. *Measurement of the Rate and Amount of Chemisorption*

After the sample surface has been prepared, determination of the amount of gas adsorbed at equilibrium may require specialized techniques adapted to the pressure region of interest and the particular gas–solid system being investigated. In the TMF region, Czanderna and Honig (28) and Katz and Gulbransen (32) developed graphical methods to correct for the effect of

TMF as illustrated in Figure 11.10. After the pressure effect is determined for an inert gas (nitrogen), the reacting gas (oxygen) is admitted in similar pressure increments. The mass gained by the sample shown in the inset is seen to be the difference in the mass reading between the two curves. Another method devised to increase the precision of measurement in the TMF region consists of diluting the reacting gas (oxygen) with an inert gas (nitrogen) (10). The success of the technique depends on maintaining

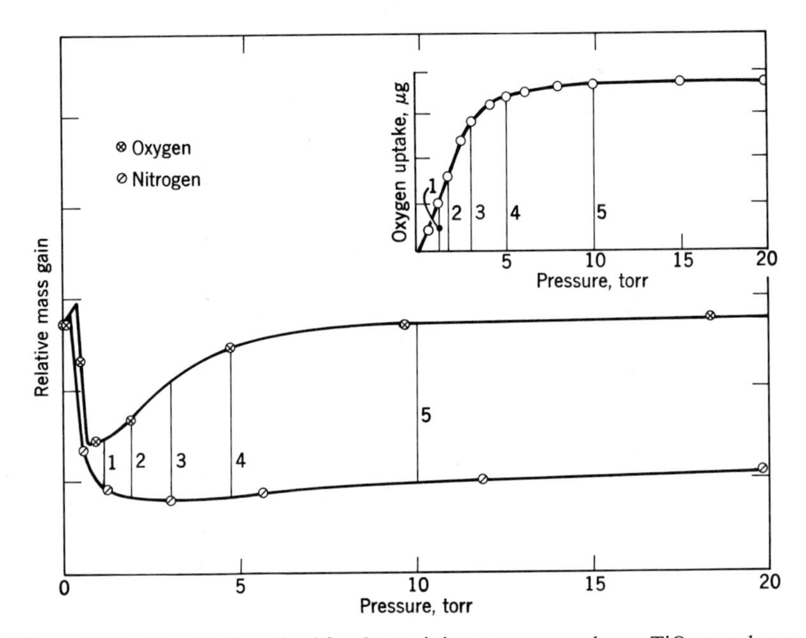

Figure 11.10. Graphical method for determining oxygen uptake on TiO_2 powders as a function of pressure. (After Czanderna and Honig, *J. Phys. Chem.*, **63**, 623 (1959).)

the total pressure in the system at pressures higher than the region of TMF. The techniques for systematic manipulation of the partial pressures of the reacting gas relative to the pressure of the inert gas to measure the equilibrium uptake through the entire TMF region have been described in detail (10).

For pressures greater than the region of TMF and for low area solids (9), the equilibrium amount of adsorption can be determined analytically. Thus, in Figure 11.11, the buoyancy in nitrogen was measured from 20 to 300 torr at the temperature for which the isotherm was to be determined. The slope was increased by the factor 32/28.016 for obtaining the buoyancy correction to the mass data obtained during the incremental admission of oxygen to the silver powder. The isotherm obtained is shown in the inset of

Figure 11.11. This method is simply a high pressure analytical variant of the method developed by Czanderna and Honig (28).

To measure the rate and amount of adsorption, the adsorbate gas is admitted when the sample is at the desired temperature. The mass change is monitored with the balance as a function of time. If the pressure remains constant during this period, the only correction that *should* be necessary is for the pressure effect arising from buoyancy and/ or TMF. The change from the zero in vacuum to the "zero" at some pressure can be determined using the techniques outlined in Figure 11.10 or 11.11.

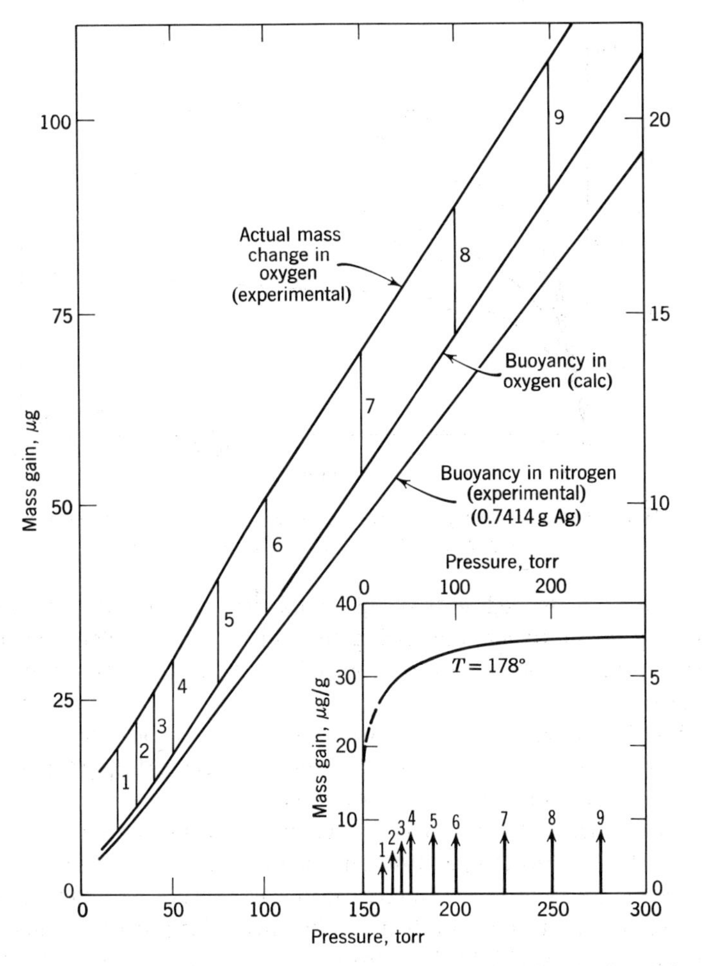

Figure 11.11. Graphical representation of analytical method for the determination of an adsorption isotherm at pressures greater than the region of TMF. (The experimental curve in oxygen was translated by $-10\ \mu g$ for convenience in plotting.)

4. Measurement of the Amount of Outgassing, Desorption, and Thermodesorption

A weighed quantity of the material to be outgassed is suspended from the microbalance in a suitable container. The balance is counterweighted to permit anticipated mass losses to be monitored over the outgassing period. Data taken in pressure decrements during the initial evacuation will allow precise determination of the amount of mass lost at room temperature by using a buoyancy slope determined in some inert gas such as nitrogen. While maintaining a continuous dynamic vacuum, the sample is

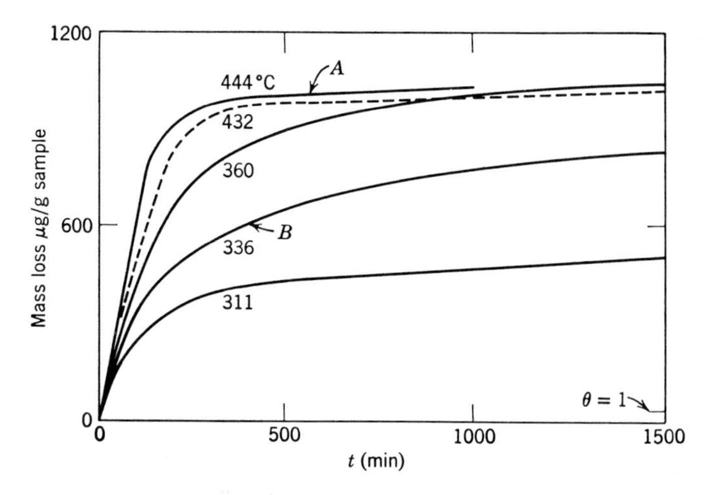

Figure 11.12. The outgassing of silver powders at various temperatures. (After Czanderna, *J. Phys. Chem.*, **70**, 2120 (1966).)

heated to any desired outgassing temperature, either directly or progressively. Some typical outgassing curves for silver powder are shown in Figure 11.12 in which the samples were heated directly to the temperatures indicated.

After the desired amount of gas has been adsorbed, the chamber is evacuated. The determination of the pressure effect is simply the reverse of the process described in the preceding section.

Prior to thermodesorption, it is customary to preadsorb a gas on the sample at a desired temperature and pressure and then to evacuate. The furnace temperature is then increased in a predetermined manner and the mass loss is monitored with the microbalance. This process is shown schematically in Figure 11.13. It is advantageous to carry out the evacuation following the preadsorption at a temperature where the desorption is too slow to be detectable. While this is not always possible, it simplifies the analysis of the data obtained. Thermodesorption data are most effectively presented as "desorption spectra" as shown in the inset of Figure 11.13. The raw data are plotted as the *rate of mass loss* versus temperature; the theoretical treatment of the peaks that appear on plots of this type has been discussed (13–16).

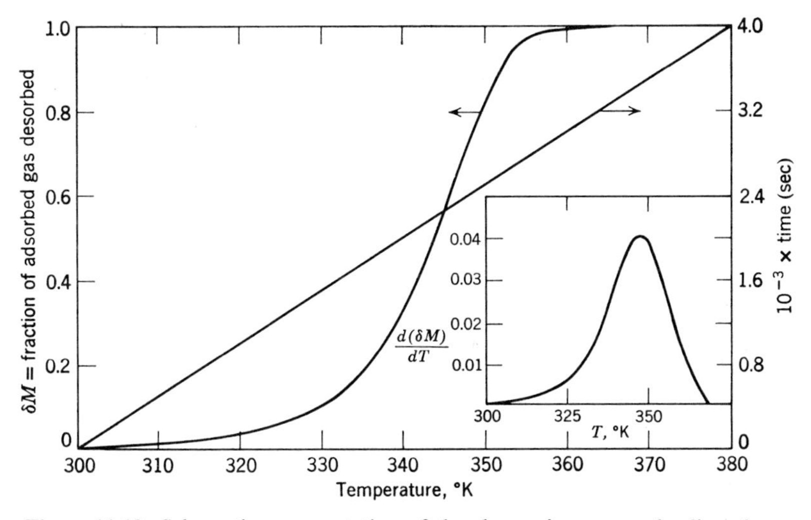

Figure 11.13. Schematic representation of the change in mass and a linearly programmed change in temperature during a complete adsorption–thermodesorption cycle. Inset: typical plot of the rate of desorption versus temperature. The data fit first order kinetics with $\beta = 0.02°/\text{sec}$, $r = 10^{13}\ \text{sec}^{-1}$, and $E_D = 25$ kcal/mol.

IV. SOME RESULTS OBTAINED DURING STUDY OF THE SILVER–OXYGEN SYSTEM

A. The Effect of Cyclic Oxygen Adsorption, Outgassing, and Reduction on the Adsorption Parameters

The purpose of the initial experiments was to establish that the usual objections to the use of the outgassing and reduction technique for surface cleaning could be dismissed. The possible problems associated with this method of surface cleaning are that the surface area may not become constant; the surface may be poisoned by diffusion from the bulk; the reducing gas and/or products may be chemisorbed on the surface; and finally, the "chemisorption behavior" of the cleaned surface may not be reproducible. Reproducibility of the rate of chemisorption and of the saturation uptake at the same temperature and pressure are included in the term "chemisorption behavior" and both must be the same if the sample surface reaches a stable configuration and is reproducibly cleaned by the reduction treatment.

Surface cleaning is a three-dimensional process; in the present study, after it was evident that solubility and incorporation effects could be dismissed, it was evident that there were three important parameters that could be following during OAOR cycling. These included the mass gained by the sample during adsorption and the mass lost by the sample during

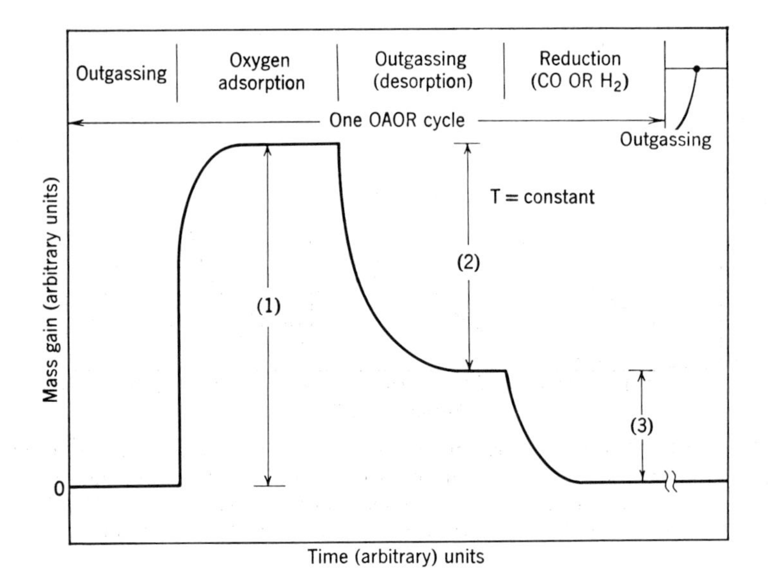

Figure 11.14. Schematic representation of mass changes by a silver powder in a typical OAOR cycle. (After Czanderna, *J. Phys. Chem.*, **70**, 2121 (1966).).)

desorption and reduction, as shown schematically by *1*, *2*, and *3* of Figure 11.14. In addition, the rates attending the mass changes of *1*, *2*, and *3* also could be studied, although for the adsorption and desorption steps, it was found preferable to use lower temperatures in the later OAOR cycling stages in order to detect the subtle changes that were occurring. Since a complete discussion of the resolution of the problems associated with the outgassing and reduction method has been published (12), only the changes in the amount of the parameters *1*, *2*, and *3* of Figure 11.14 and in the sample mass will be presented in this chapter. These are plotted for OAOR cycling at a temperature of 350° in Figure 11.15. Here, it can be seen that the sample mass, saturation oxygen uptake, amount of desorbable oxygen and amount of mass that could be removed from the surface by carbon

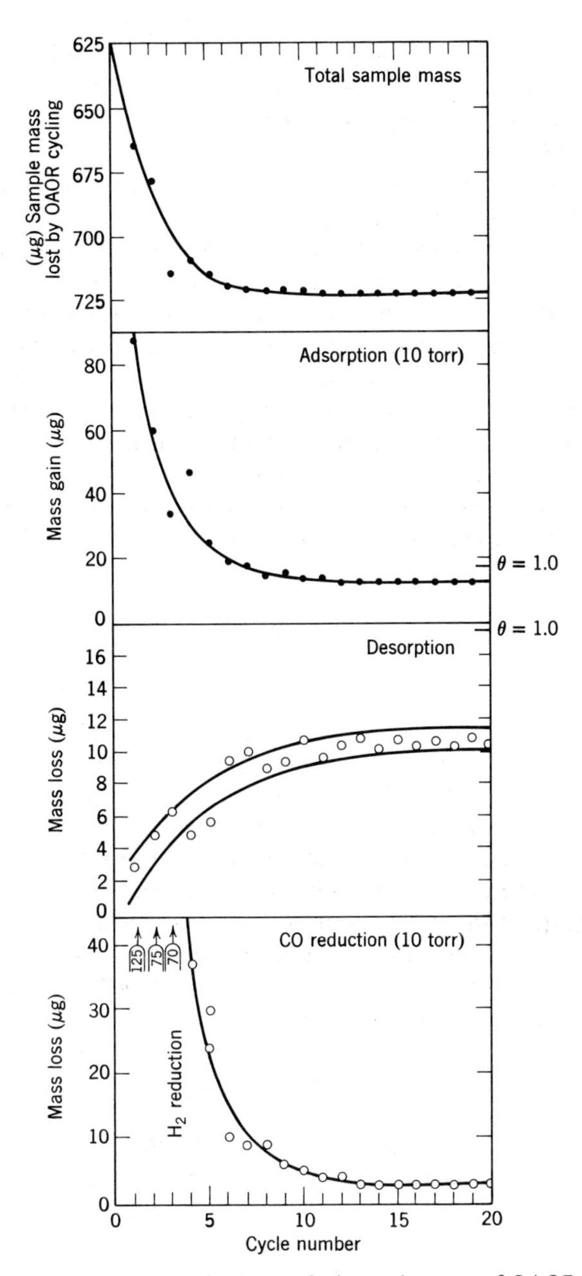

Figure 11.15. The mass change in oxygen during early stages of OAOR cycling. (After Czanderna, *J. Phys. Chem.*, **70**, 2121 (1966).

monoxide reduction change gradually with the number of OAOR cycles. When the parameters became constant with further cycling the chemisorption behavior was also reproducible at all lower temperatures and at any pressure employed for oxygen adsorption up to 300 torr. It is on this type of reproducible surface that meaningful kinetic and equilibrium studies might be made.

Consideration of the changes in the surface structure of the silver powder produced by OAOR cycling is of interest. Initially, the powder is probably polycrystalline with all possible index planes emerging at the surface because it was prepared by the decomposition of silver oxalate. Arguments that the reproducible surface consists of low index plane facets have been advanced (12) but have not been substantiated by an electron diffraction or microscopy investigation. However, the gravimetric adsorption data provide indirect evidence that the process occurring on the surface is consistent with that expected for a surface experiencing an ordering process to low index planes. An investigation is in progress to obtain direct evidence that faceting is occurring on the silver surface in a manner similar to that found in the copper oxide system (33).

B. Adsorption

1. Kinetics

After the silver powder reached a state of reproducible chemisorption behavior, the rate of oxygen adsorption was measured from -77 to 350° at various oxygen pressures. After an OAOR cycle was completed at 350°, the silver was cooled *in vacuo* to a lower temperature. The adsorption was measured and then the sample was subjected to one OAOR at 350° before another low temperature adsorption was determined. This treatment was adequate to ensure the silver was in the same cleaned state prior to each time dependent determination. A few curves obtained for the adsorption of oxygen on a silver sample with an area of 0.088 m^2/g are shown in Figure 11.16. The horizontal line indicates a coverage of $\theta = 0.73$ where the rate of adsorption was determined from each of the continuous experimental curves. Figure 11.17 is a plot of these slopes versus $1/T$ at $\theta = 0.73$.

The two activation energies obtained from the Arrhenius plot are 24.6 kcal/mole above 138° and 3.5 kcal/mole below 138°. The lower activation energy is attributed to the chemisorption of molecular oxygen; the higher one to surface diffusion of atomic oxygen on silver (21). A detailed account of the values of activation energies obtained at various coverages and temperatures using this technique has been reported (21).

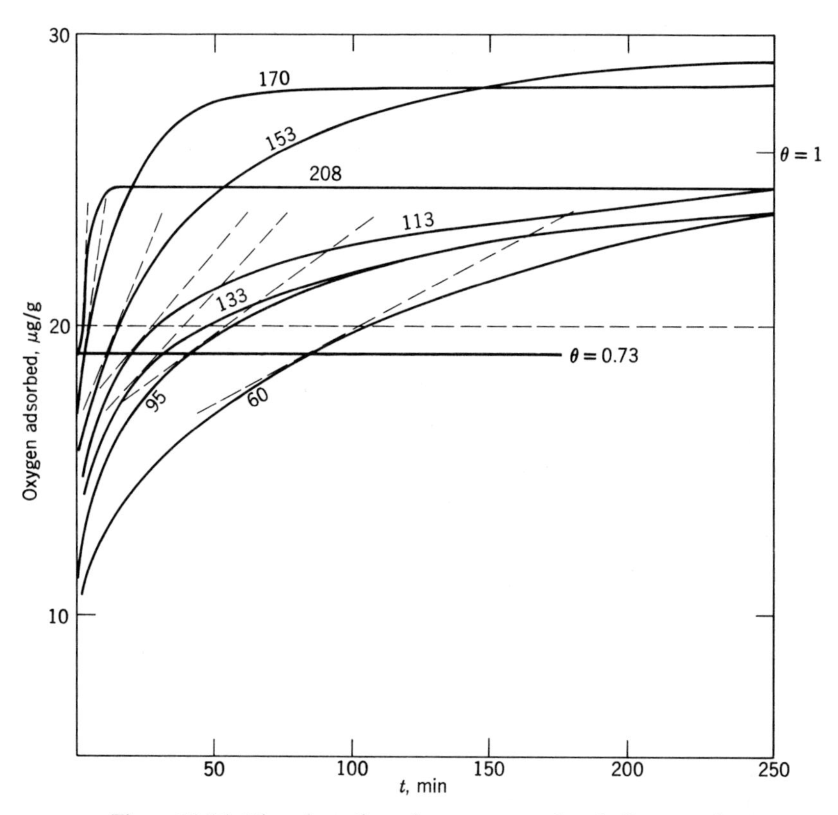

Figure 11.16. The adsorption of oxygen on reduced silver powder at various temperatures.

T(°C)	Po_2 (torr)
60	11.0 ± 0.2
95	10.5 ± 0.2
113	10.2 ± 0.2
133	10.0 ± 0.2
153	9.7_5 ± 0.2
170	9.5_5 ± 0.2
208	9.1_5 ± 0.2

2. Equilibrium

Adsorption isotherms obtained for oxygen on silver at several temperatures are shown in Figure 11.18. These data were gathered for silver reproducible to oxygen chemisorption behavior. The logarithm of the pressures at $\theta = 0.73$, the horizontal line in Figure 11.18, is plotted versus

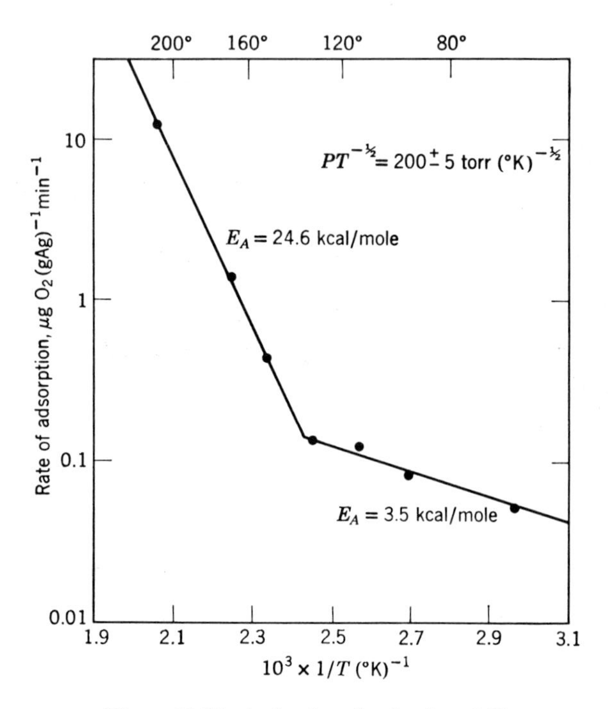

Figure 11.17. Arrhenius plot for $\theta = 0.73$.

the corresponding $1/T$ value in Figure 11.19. Thus, an isosteric heat of 17 kcal/mole was measured for this surface coverage. This value of q probably corresponds to molecularly adsorbed oxygen.

C. Desorption and Thermodesorption

After adsorbing oxygen at a series of temperatures below 350°, the system was evacuated. The mass loss measured during evacuation at each temperature is plotted in Figure 11.20. The rate of desorption below 160° and monolayer coverage is negligible, but above 200° it is quite rapid until low coverages are reached. The apparent activation energy for desorption was found to be about 25 kcal/mole for θ of 0.3–0.7. However, as was pointed out before, an activation energy calculated from the rate of desorption from powders cannot be exact.

For temperatures below 160°, thermodesorption studies were found to be particularly useful for identifying the coverage of various species adsorbed (21). A plot of the thermodesorption spectra yielded two peaks obtained after adsorption at 25° and 60° but only one after adsorption at

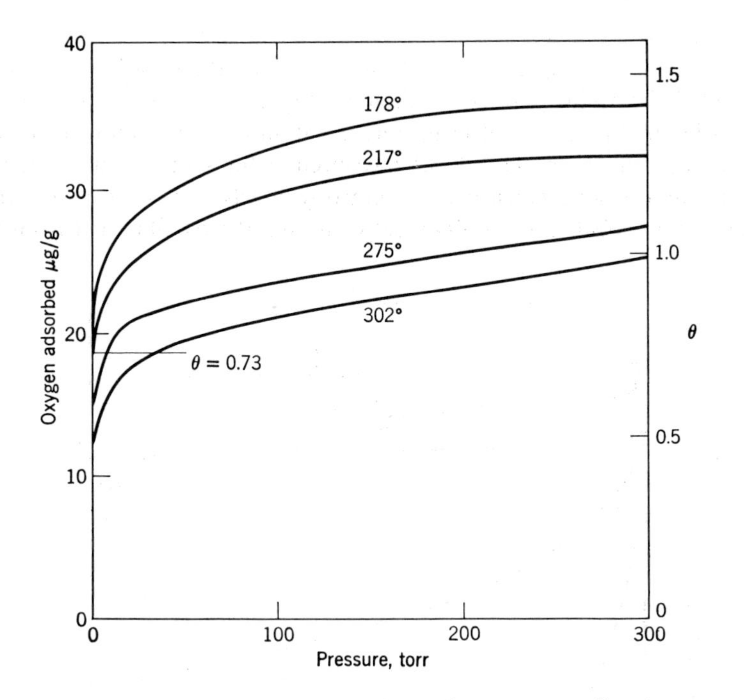

Figure 11.18. Adsorption isotherms for oxygen on silver.

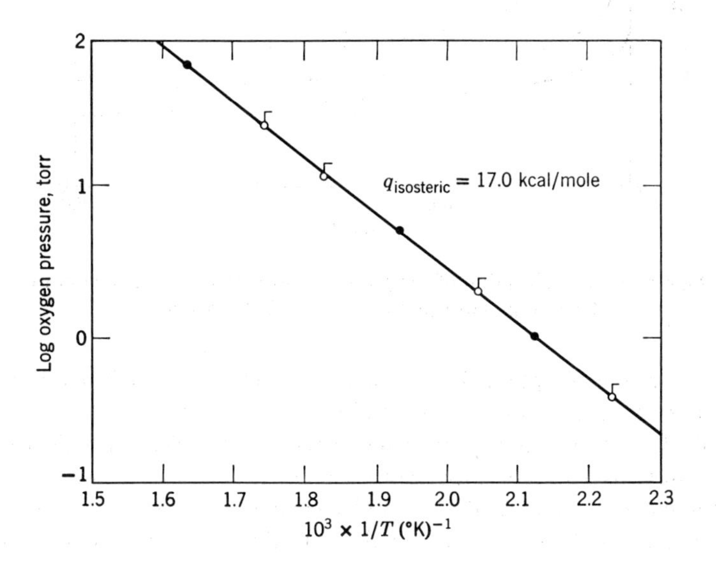

Figure 11.19. Plot of log p versus $1/T$ for oxygen on silver, $\theta = 0.73$. ⨽, from Figure 18; ●, $\theta = 0.73$; $q_{\text{isoteric}} = 17.0$ kcal/mol.

133°. The peaks correspond to activation energies of 26–30, and 37 kcal/mole. The highest value is significantly greater than the 25 kcal/mole estimated from the desorption data in Figure 20 in reference 21. At $\theta = 0.73$, an E_D of 37 ± 4 kcal/mole is within experimental error of the sum of q_{isos} and E_A of 41.5 ± 3.5 measured at the same coverage. If this simple comparison, which neglects entropy effects, is valid it shows that readsorption effects are not drastically altering the E_D obtained from the

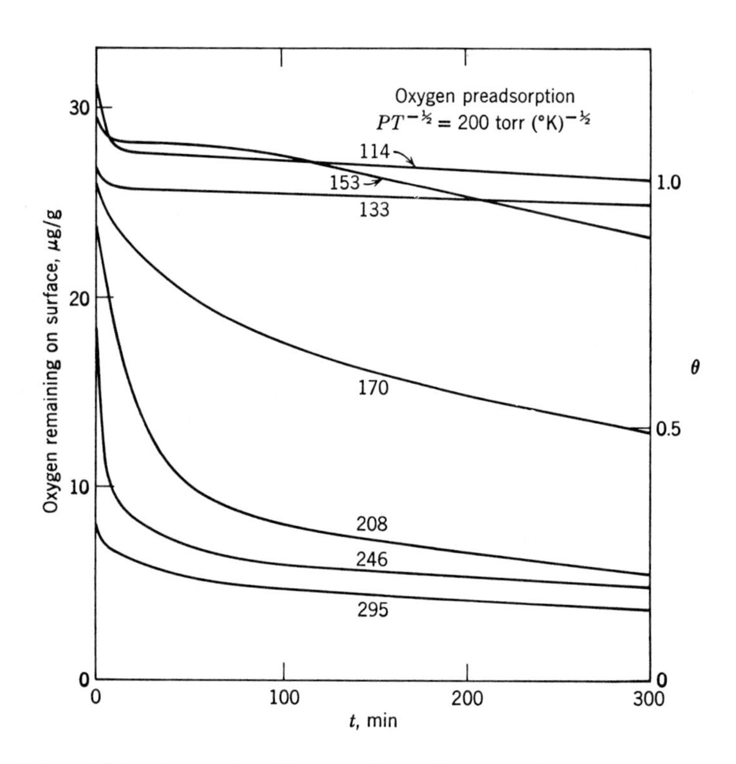

Figure 11.20. Desorption of oxygen from silver at various temperatures. Oxygen preadsorption: $PT^{-1/2} = 200$ torr (°K)$^{-1/2}$.

data of thermodesorption. This conclusion on readsorption is in agreement with a preliminary conclusion obtained by Czanderna (13) for carbon dioxide desorption from an oxygen covered silver surface. It is clear much more effort needs to be expended on thermodesorption studies (13) but the comparisons found to date are indeed encouraging; not only can values be obtained for E_D that are much closer to the true value but studies can be made in situ in a system where E_A and q are determined at the same value of θ.

V. CONCLUDING REMARKS

The vacuum ultramicrobalance has been found to be an extremely powerful tool for a comprehensive study of the chemisorption of oxygen on silver powder.

The technique of OAOR cycling is suitable for cleaning silver powder as a preparation for quantitative measurements. The fact that the cleaned surface does not adsorb carbon dioxide or gases with similar chemisorptive properties at any of the temperatures studied indicates that the extent of

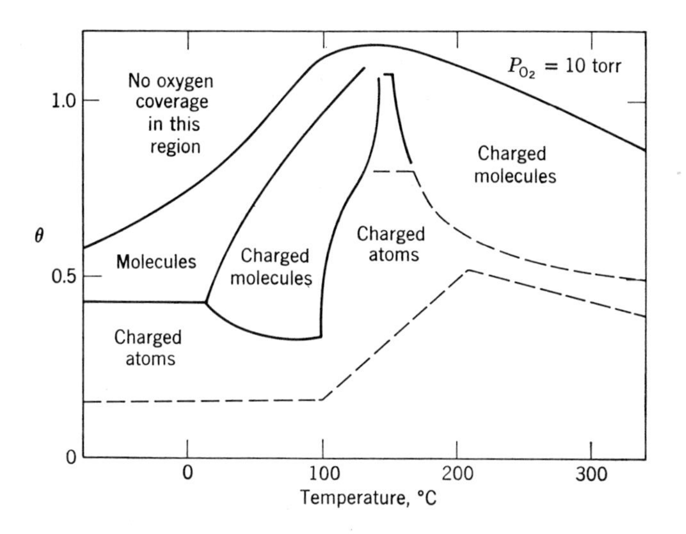

Figure 11.21. Species of oxygen probably adsorbed on silver at various coverages and temperatures. (After Czanderna, *J. Phys. Chem.*, **68**, 2770 (1964).)

surface cleaning is such that the amount of surface contamination remaining must either be very close to a θ of zero, or the surface contamination is not sensitive to chemisorption of water, carbon dioxide, etc. It seems unlikely that oxygen contamination of the surface could be present and not cause carbon dioxide chemisorption when oxygen introduced to the surface in measurable amounts results in the chemisorption of carbon dioxide.

The results of extensive kinetic and equilibrium studies of oxygen adsorption and desorption carried out on the powder have permitted an internally consistent interpretation to be reached concerning the coverage of silver with different species of oxygen. The plot in Figure 11.21 shows the various species of oxygen adsorbed on silver as a function of coverage and temperature. The major difficulty to date is that it is not possible to specify the structure of the surface of the material under study. With specialized balances developed recently, samples with a defined structure (Table 11.1) should be considered carefully for chemisorption studies on other gas–solid systems.

While the relative advantages of gravimetric and volumetric techniques for classical adsorption studies have been discussed in the literature, an unbiased comparison between the two methods is understandably difficult to find. Even the relative precision of the two methods for study of specific systems fluctuates dynamically as rather sophisticated instrumentation becomes available. As of now, the fact remains that fundamental studies of chemisorption must be carried out on low area materials and this is precisely where the volumetric method is least sensitive. The intrinsic advantages of the two techniques are tabulated in Table 11.3. It is the opinion

Table 11.3. A Comparison of the Gravimetric and Volumetric Methods for Chemisorption Studies on Low Area Materials

Experimental measurement	Derivative information	Method of study	
		Gravimetric	Volumetric
Physical adsorption isotherm	Specific surface	Yes	Yes
Rate of adsorption	E_A	Yes	Yes
Isotherm, isobar	q	Yes	Yes[a]
Thermodesorption	E_D	Yes	No
Desorption	E_D	Yes	No
Bulk solubility	Heat of solution	Yes	With difficulty[b]
Change in population of adsorbate species at constant amount adsorbed (surface diffusion)	E_A for surface mobility of adsorbate	Yes, with auxiliary techniques	No; no assurance of constant amount adsorbed

[a] The inability to detect long term mass changes in the sample *in situ* because of "sorption" is a serious limitation.

[b] Reversibility cannot be followed.

of the author that the gravimetric method is superior because of both the quantity and quality of the data obtainable. It can be anticipated that widespread adoption will gradually occur as the availability of the technology and the number of interested personnel increases.

ACKNOWLEDGMENTS

The author is indebted to Dr. S. P. Wolsky and Mr. W. Kollen for their helpful comments. Partial support of this work by the Division of Air Pollution Bureau of State Services, Public Health Service, under Research Grant 1 RO1 AP 00552-01 is gratefully acknowledged. Acknowledgment is made to the donors of the Petroleum Research Fund, administered by the American Chemical Society, for partial support of this research.

REFERENCES

1. D. O. Hayward and B. M. W. Trapnell, *Chemisorption*, 2nd ed., Butterworths, London, 1964.
2. P. A. Faeth, *Adsorption and Vacuum Technique*, Inst. Sci. Tech. Press, Univ. of Michigan, TR-66100-2-X (1962), Ann Arbor Michigan.
3. S. J. Gregg and K. S. W. Sing, *Adsorption, Surface Area, and Porosity*, Academic Press, New York, 1967.
4. A. W. Adamson, *Physical Chemistry of Surfaces*, 2nd ed., Interscience, New York, 1967.
5. P. W. M. Jacobs and F. C. Tompkins, in *Chemistry of the Solid State*, W. E. Garner, Ed., Butterworths, London, 1955.
6. D. L. Kantro, S. Brunauer, and L. E. Copeland, in *The Solid–Gas Interface*, Vol. I, E. A. Flood, ed., Dekker, New York, 1967, p. 413.
7. For example, S. Brunauer, *The Adsorption of Gases and Vapors*, Princeton Univ. Press, Princeton, N.J., 1943.
8. D. M. Young and A. Crowell, *Physical Adsorption of Gases*, Butterworths, London, 1962. For some more recent references, see A. W. Czanderna, Chapter 2, Section VIII.A, Table 2.3, this book.
9. R. A. Pierotti, *Vacuum Microbalance Tech.*, **6**, 1 (1967).
10. A. W. Czanderna, *Vacuum Microbalance Tech.*, **4**, 69 (1965).
11. A. W. Czanderna, reference 10 above, p. 175.
12. A. W. Czanderna, *J. Phys. Chem.*, **70**, 2120 (1966).
13. A. W. Czanderna, *Vacuum Microbalance Tech.*, A. W. Czanderna, Ed., **6**, 129 (1967), J. A. Poulis, discussion of paper.
14. A. W. Smith and S. Aranoff, *J. Phys. Chem.*, **62**, 684 (1958).
15. G. Ehrlich, *J. Appl. Phys.*, **32**, 4 (1961).
16. P. A. Redhead, *Vacuum*, **12**, 203 (1962).
17. S. Brunauer, P. H. Emmett, and E. Teller, *J. Am. Chem. Soc.*, **60**, 309 (1938).
18. S. Brunauer, L. E. Copeland, and D. L. Kantro, *The Solid–Gas Interface*, Vol. I, E. A. Flood, Ed., Dekker, N.Y., 1967, pp. 17, 77.
19. D. L. Kantro, S. Brunauer, and L. E. Copeland, reference 18 above, p. 413.
20. A. W. Czanderna, Ph.D. Thesis, Purdue University, W. Lafayette, Indiana, August 1957.
21. A. W. Czanderna, *J. Phys. Chem.*, **68**, 2765 (1964).
22. A. W. Czanderna and J. M. Honig, *Anal. Chem.*, **29**, 1206 (1957).
23. C. N. Cochran, *Rev. Sci. Instr.*, **29**, 1135 (1958); *Vacuum Microbalance Tech.*, **1**, 23 (1961).
24. A. W. Czanderna, *Vacuum Microbalance Tech.*, **4**, 57 (1965).

25. A. W. Czanderna and H. Wieder, *Vacuum Microbalance Tech.*, **2**, 147 (1962), and Appendix therein.
26. A. W. Czanderna, *J. Colloid. Interface Sci.*, **22**, 482 (1966).
27. A. W. Czanderna, *J. Colloid Interface Sci.*, **24**, 500 (1967).
28. A. W. Czanderna and J. M. Honig, *J. Phys. Chem*, **63**, 620 (1959).
29. W. Kollen and A. W. Czanderna, to be published.
30. E. L. Fuller, H. F. Holmes, and C. H. Secoy, *Vacuum Microbalance Tech.*, **4**, 109 (1965).
31. J. M. Thomas and B. R. Williams, reference 30 above, p. 209.
32. O. M. Katz and E. A. Gulbransen, *Vacuum Microbalance Tech.*, **1**, 111 (1961).
33. T. N. Rhodin, in *Advan. Catalysis*, **5**, 70 (1953).

Chapter 12

The Investigation of Sputtering Phenomena

S. P. WOLSKY AND E. J. ZDANUK

Laboratory for Physical Science, P. R. Mallory & Co. Inc.
Burlington, Massachusetts

I. INTRODUCTION

Sputtering, the ejection of atoms from a target surface as a result of ion bombardment, has become a subject of considerable practical importance (1). Although first observed more than 100 years ago (2), it is only comparatively recently that its utility has been widely recognized. The increased interest has resulted in a broad effort to understand the sputtering process.

A number of useful variations of the basic sputtering process have been developed. The more prominent of these are chemical or reactive sputtering (3,4), RF sputtering (5,6), and bias sputtering (7). Reactive sputtering arises from sputtering in an active gaseous ambient. RF sputtering is generally used to sputter insulator targets. The application of an RF rather than a dc potential eliminates the accumulation of charge at the target surface which otherwise would prevent the sputtering of the insulator. In bias sputtering a small dc potential placed on the substrate results in simultaneous bombardment of the substrate and the target. These innovations which have increased the practical utilization of sputtering have also generated a need for a basic understanding of the phenomenon.

This chapter is devoted to a discussion of the application of the vacuum microbalance to the investigation of sputtering phenomena. An introductory review of the nature of sputtering is included to provide the reader with background on the subject and to allow him to understand the virtues of the vacuum microbalance method as applied to sputtering studies.

II. SPUTTERING PHENOMENA

It is generally agreed that physical sputtering results from momentum transferred in collisions between the lattice atoms of the target and the bombarding ions, both of which, to a first approximation may be considered as hard spheres with the relative masses of the ion and target atom determining the energetics of the process. The energy transferred to the target atom, E_t may be determined from equation 12.1 where E_0 is the energy of the impinging ion, and M_1 and M_2 are the masses of the ion and target atom, respectively.

$$E_t = [4M_1M_2/(M_1 + M_2)^2]\, E_0 \tag{12.1}$$

Sputtering may be considered, therefore, as essentially a process of atomic billiards occurring primarily in the surface layers of the target.

The lack of reliable experimental data has generally limited the understanding of the sputtering process. Most of the early sputtering experiments were carried out in high pressure glow discharges (5,8); however, the interpretation of glow discharge data is complicated. In a glow, the energy and direction of incidence of the bombarding ion is poorly defined and the mean free path of the sputtered atoms is normally smaller than the distance from the target to a wall. The latter condition results in considerable back diffusion of sputtered material, thereby altering the original nature of the target surface.

The glow discharge studies did provide a valuable qualitative background for the more recent basic explorations performed at comparatively low bombarding gas pressures in which the mean free path of the sputtered atom is large compared to the target to wall distance. The development of ultrahigh vacuum techniques which significantly reduce residual gas contaminant concentrations and of sputtering sources utilizing hot filaments, magnetic fields, RF fields, or ion beams have made possible meaningful studies at gas pressures of 10^{-3} to 10^{-5} to 10^{-6} torr and at current densities ranging from a few microamps to several milliamps per square centimeter.

The sputtering yield, that is the number of atoms sputtered per impinging ion at a given ion energy, has been measured for many materials, both elemental and compound, at energies ranging from approximately 5 eV to several hundred keV. The optimum experimental method for sputtering yield measurement as a function of ion energy and angle of ion incidence is different in the various ion energy ranges. Figure 12.1 represents a generalized sputtering yield curve for a very broad range of ion energies. The sputtering yield generally increases very rapidly in the *low* energy regions (less than 5–10 keV), levels out in the *medium* energy range (5–25 keV), and finally decreases in the very *high* energy range (> 50 keV). The discussion in this chapter will deal primarily with the very low energy region where the vacuum microbalance method is most advantageously applied.

Experimental determination of the sputtering yield is particularly difficult in the low energy threshold region where an extremely small number of atoms are sputtered per impinging ion. The sputtering threshold energy, a characteristic parameter of a material approximately of the order of magnitude of the sublimation energy, is defined as the minimum energy

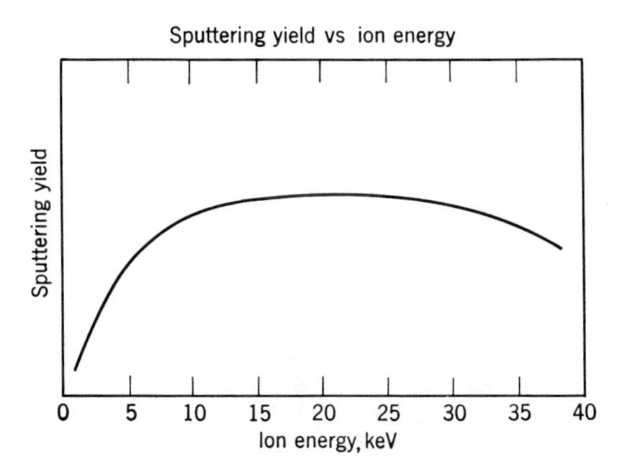

Figure 12.1. A generalized sputtering yield curve.

necessary to sputter an atom. In order to obtain reliable sputtering data in this region, the ion energy, the angle of incidence and the crystallographic orientation of the target must be carefully controlled, surface contamination must be minimized, and the measurement method must allow the detection of less than 10^{-4} sputtered atoms per impinging ion.

Considerable effort has been devoted to developing a theoretical understanding of the sputtering process. Present sputtering theory generally centers about detailed considerations of the mechanical collisions occurring between the impinging ion and the target atoms. The simplest theories (9–12) are based on the assumption that most sputtering results from the dissipation of the ion energy during early collisions in a near surface layer and that the sputtering yields are related to the collision mean free path. This approach has been particularly successful in the high energy region.

Systematic studies (13,14) of the sputtering of single crystal materials have revealed that sputtered atoms are ejected from a target preferentially in certain crystallographic directions. Figure 12.2 represents a typical spot pattern obtained from a 400 eV krypton ion bombardment of a (100) oriented copper crystal. As expected from a (100) surface of a cubic lattice, the spot pattern has fourfold symmetry. The copper target, suspended in the vacuum system from the balance, may be clearly seen in the middle of the pattern. It has generally been concluded that the sputtering spot patterns were direct experimental evidence of the focusing of momentum in

Figure 12.2. Sputtered atom ejection pattern for a Cu(100) target bombarded with 400 eV krypton ions.

the close packed directions of a crystal (15,16). Calculations based on computer simulation of the sputtering process indicate, however, that the spot patterns can be obtained without employing an extended collision sequence concept and that the sputtered atoms can yield deposition patterns as a result of random short range surface collisions (17). There is no widely accepted mechanism to account for the ejection patterns and considerable effort (18) is being devoted to determining the role of collision sequences in the sputtering process.

It has been well established that the sputtered atom is ejected from the target surface with an average energy of 3–10 eV (19,20). However, the range of sputtered particle energies may be considerably larger in the oblique ejection directions. The energy associated with the sputtered atom is, therefore, much greater than the thermal energy associated with an evaporated atom. Inherent to the widespread interest in the application of the sputtering process for film deposition has been the conviction that the *extra* energy associated with the sputtered atom contributes to film characteristics in a favorable manner such as through increased adhesion to the substrate or a lower epitaxial temperature. The epitaxial growth of sputtered metal films at liquid nitrogen temperature has been interpreted as resulting from the *extra* sputtered atom energy (21). However the formation and growth of sputtered films is a complex process in which it is difficult to separate the individual effects of the gaseous ambient, the deposition rate, and the energy of the sputtered atom. It will require a considerable amount of additional experimental effort to determine unequivocally the nature and magnitude of the effect of the sputtered atom energy upon film characteristics.

Sputtering yields at high ion energies have generally been determined gravimetrically. The usual method involves the weighing of a target on an ordinary analytical balance external to the sputtering system prior to and after the sputtering procedure. In the high energy region the weight changes are large enough to mask any errors introduced by handling procedures or by adsorption of gases onto the target surface. At very low energies, and especially in the threshold region where very small weight changes are involved, surface adsorption and handling can introduce a significant error in the weight change measurement. Removing the sample from the system for external weighing would not, therefore, be desirable.

The application of the quartz vacuum microbalance to sputtering studies by the authors arose almost incidently during experiments on the oxidation behavior of semiconductor surfaces (22,23). In the early days of semiconductor device development there was an intensive interest in many laboratories in the electrical and physical characteristics of the surfaces of germanium, the semiconductor material of prime concern at that time. In the course of studies of the adsorption of oxygen on germanium surfaces the microbalance was adapted to utilize the newly developed ion bombardment technique for the *in situ* preparation of clean sample surfaces (24). The recommended sputtering voltage, current and time for this method, although shown experimentally to be clearly adequate, were derived empirically since the amount of material sputtered per impinging ion, that is the sputtering yield, was unknown. To provide detailed information of the sputtering yield, the ion current and the weight change were recorded as a function of the applied voltage during the surface cleaning phase of the microbalance oxidation experiments. Beginning with this experience the vacuum microbalance has been developed into a valuable experimental tool for sputtering studies (25,26).

Until recently the sputtering data obtained with the microbalance has always been of a static nature in that because of thermal effects (discussed in greater detail later in this chapter) the balance readings could be taken only prior to and following, but not during, sputtering. Recent work (27, 28) has resulted in the development of a method which allows the balance to operate during the ion bombardment procedure. As a result it is now possible with the vacuum microbalance to make *in situ* absolute dynamic sputtering measurements in the low energy region.

There are a number of other methods which can be applied to obtain sputtering yield data in the low energy region. The most useful of these employ: (*1*) a quartz crystal oscillator (29) (see Chapter 5 for a complete discussion of crystal oscillators), or (*2*) spectroscopic techniques (30). In the oscillator method the sputtering yield is determined from the change in frequency resulting from the ion bombardment of a quartz crystal coated with the material under study. The method has been employed to detect changes of less than 10^{-9} g of the coated material. This process is limited by the following factors. (*1*) Single crystal studies are almost impossible since the material must be first deposited, either by plating, vacuum evaporation, or sputtering onto the quartz crystal thereby severely limiting the crystallograph perfection and orientation. (*2*) As noted in earlier chapters the quartz crystal oscillator can be utilized only with thin films. Since the sputtering of thin film material may differ substantially from that of the bulk material, the results could be misleading. (*3*) If the oscillator is used to measure sputtering yields by determining the amount of material sputtered from a target, there is always some uncertainty as to the exact fraction of the sputtered material collected. There is also the additional uncertainty concerning the sticking coefficient of the sputtered material on the quartz crystal.

Much of the sputtering data in the literature in the less than 100 eV ion energy region has been obtained by the spectroscopic technique developed by Stuart and Wehner (30). In this process the sputtered atoms are excited in the same discharge used to ionize the gas. The intensity of the light emitted by the excited sputtered atoms provides a measure of the flux of sputtered material and thus the sputtering yield. This method, however, is not absolute. In the spectroscopic method, it assumed that the excitation conditions and the sputtered atom emission and adsorption characteristics are constant throughout the low energy range being investigated and identical with those of higher energy reference points. The uncertainty of extrapolation of calibrations obtained at much higher energies is a serious limitation of the spectroscopic technique in the threshold energy region.

III. EXPERIMENTAL APPARATUS AND METHOD

A Gulbransen type microbalance (see Chapter 3) housed in an all-glass ultrahigh vacuum system has been used for sputtering yield measurements at energies as low as 10–20 eV (31). Schematics of the quartz microbalance and of the microbalance system are shown in Figures 12.3 and 12.4, respectively, and a photograph of the microbalance in Figure 12.5. The

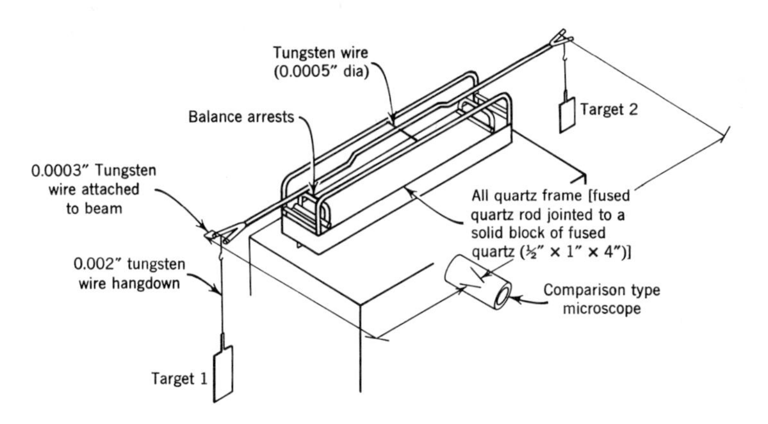

Figure 12.3. A schematic representation of the quartz microbalance.

precision made 0.075 in. diam quartz beam was supported on a 0.0005 in. annealed tungsten wire. The end support wires were 0.0003 in. annealed tungsten. The wires were joined to the beam and the quartz frame with fused silver chloride. Iron cores sealed in small quartz envelopes attached to the arrest allowed external manipulation of the balance by means of a small magnet. Static charge effects were eliminated through an external ground connection to a conductive coating of gold evaporated onto the balance. The gold film provided the additional advantage of minimizing thermal effects (32) (see Chapter 4). The stability of the balance zero point in terms of the actual mass change (Δm) over a period of 8 hr with an imposed temperature change (ΔT) of approximately 0.5–0.6°C is shown in Figure 12.6. The slight fluctuations were well within experimental error and probably can be attributed to Brownian motion (see Chapter 4). The balance tubes were coated on the outside with an aquadag film and grounded. Weight changes were followed by observing the deflection of the beam ends on a graduated fixed scale in the eyepiece of a specially designed comparison microscope. The balance was calibrated by measuring the deflection from an observed zero point caused by known weight differences. The capacity of the balance was approximately 1 g.

The characteristics of the simple quartz microbalance employed in these studies have been discussed in detail elsewhere (33). In general weight changes of approximately 2×10^{-8} g were reproducibly detected. With this sensibility the balance had a period of oscillation of approximately 20 secs. The balance was designed to have a constant sensitivity over the

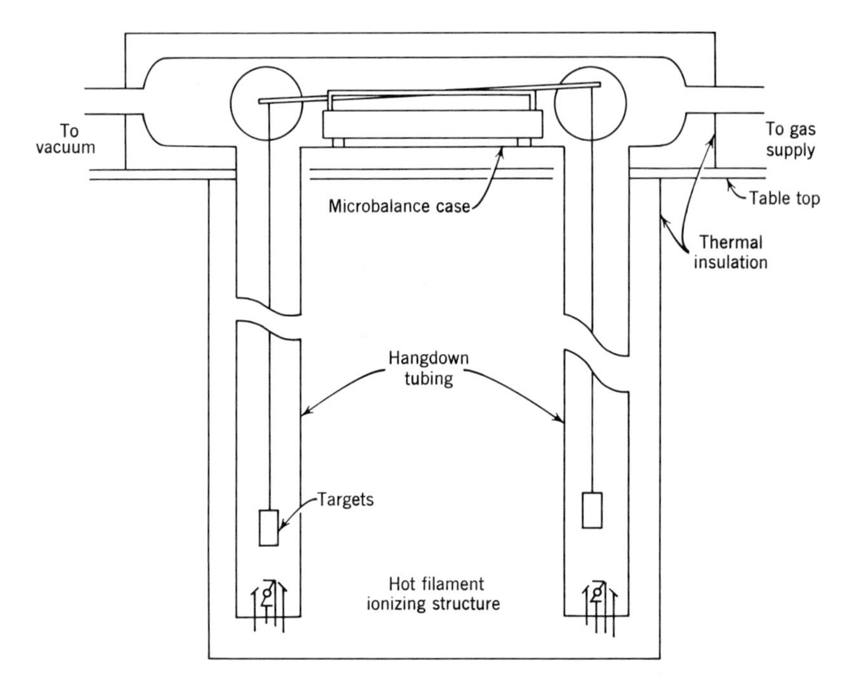

Figure 12.4. A schematic representation of the microbalance system.

relatively small weight change range involved in a single experimental sequence, but to vary with major changes (for example 0.1 g) in the load. In this manner the sensitivity could be controlled as desired, through a proper choice of load. Details of the fabrication of balance beams and of other experimental and theoretical factors affecting beam characteristics have been discussed in previous publications (26–28, 32, 33) and in other chapters in this book (see Chapters 2, 3, and 4).

The microbalance was readily adaptable for sputtering studies. The target materials were hung from each balance arm at a distance well below the balance itself (approximately 24 in.) The metal film on the balance provided a means for placing a potential on the target. The discharge was

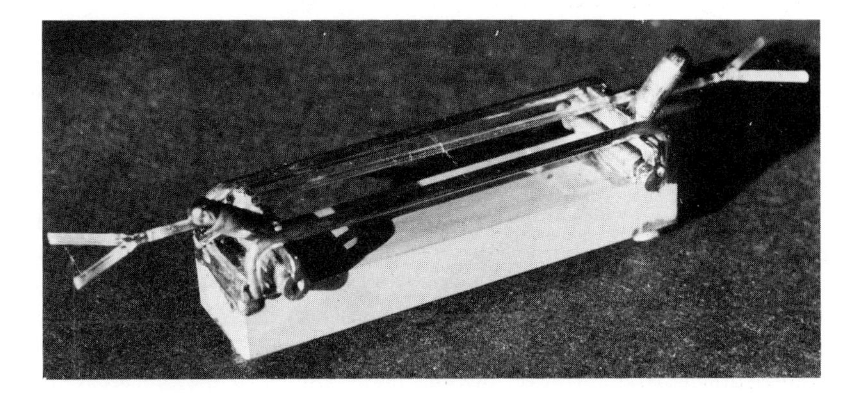

Figure 12.5. A photograph of the quartz microbalance.

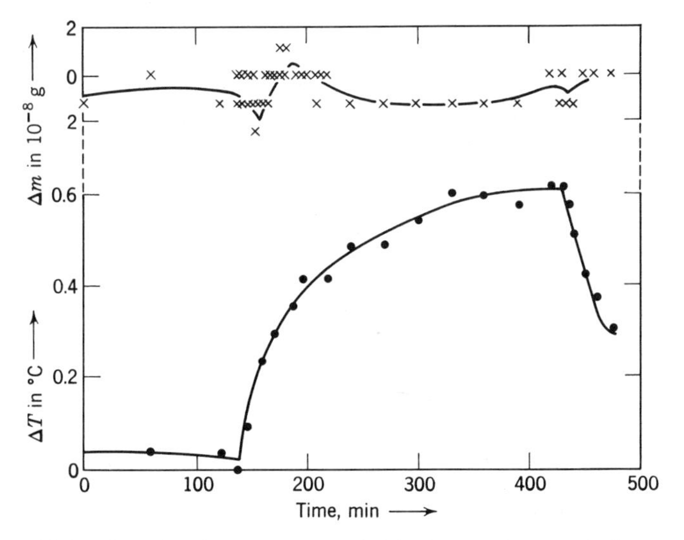

Figure 12.6. The microbalance zero point in terms of the actual mass change (Δm) as a function of time and temperature (ΔT) for a gold coated quartz balance beam.

obtained from a simple ionizing source, consisting of a tungsten filament and two molybdenum grid plates mounted on a standard tube base sealed into the bottom of the hangdown tube. A metal shield above the filament prevented evaporated tungsten from depositing onto the sample. The sputtering experiments were performed with spectroscopically pure inert gases at pressures of 10^{-3} to 10^{-4} torr. Ionizing potentials were 35–40 V to prevent the formation of double charged ions. The target was immersed in the manner of a Langmuir probe in a thermionically supported discharge. Under these bombardment conditions the ion energies were well defined and ion incidence was predominantly normal to the target surface (5). Carefully oriented single crystal materials were used in all experiments. The targets were given a clean up ion bombardment prior to detailed experimental measurements. Current densities generally were in the range of 1–25 $\mu A/cm^2$. Currents of up to 1 mA have been drawn through the balance for short periods of time and potentials of 3500 V have been applied without any apparent damage to the thin tungsten support wires. The bombarding current, however, was normally limited by the delicate nature of the fine beam support wires and the requirement to prevent overheating which would result in melting of the fused silver chloride sealing the wires to the beam. The ambient background pressures were normally 10^{-9} to 10^{-10} torr. With ion sources sealed into each hangdown tube the microbalance could be readily balanced through controlled sputtering of either one or both of the targets. The sensibility of the balance was unchanged after steady operation for more than one year.

Sputtering yields obtained statically provide valuable but limited information, since it is necessarily assumed that the yield was constant throughout the bombardment time. In actuality the instantaneous nature of the target surface can change significantly during bombardment and probably cannot be characterized accurately from static yield measurements. This contention has been verified experimentally from studies of the argon ion bombardment of copper and aluminium single crystal targets in the 20–100 eV range (31). These investigations indicated that the nature of the target surfaces varied about an average characteristic and was determined by such factors as the prebombardment cleanup treatment and the sputtering time. The value of the vacuum microbalance for the measurement of sputtering yields has been increased significantly by the more recent development of techniques which allow continuous weight change measurements during the sputtering process.

In both the continuous (dynamic) and static microbalance methods, the balance system is maintained symmetrical with carefully matched samples placed on each balance arm. Although ionizing sources are available in each balance tube, in the static procedure only one source is used (and therefore is hot) during a specific experiment. Under the normal static bombardment conditions of an inert gas pressure of a few millitorr, the temperature differential between the hot and cold balance arms created an unbalanced force resulting in very rapid and sometimes violent movement of the balance beam. (For a detailed discussion of thermal forces see Chapter 4.) In the static studies this problem was overcome by arresting the balance beam during the sputtering process. In order to make dynamic measurements it was necessary to balance the hot filament thermal effect sufficiently to establish an equilibrium or null condition in which the balance need not be arrested during the sputtering process. Both thermal and electromagnetic counterbalancing were considered as possible means of accomplishing this end.

It was experimentally observed that when one ion source filament was hot, the motion of the balance beam could be minimized by increasing the temperature of the other filament. Once at the equilibrium or null point, the displacement of the beam due to weight changes resulting from subsequent sputtering could be readily followed. Since thermal balancing could be so simply achieved without modification of the balance system, electromagnetic balancing was not seriously considered. Prior to general use of this method it was important to verify experimentally that the displacement of the beam from the equilibrium position due to the sputtering did not upset the null conditions and thereby introduce a variable error. Figure 12.7 shows the beam displacement at various initial deflection angles (from the horizontal) at a constant sputtering source filament voltage as a function of the filament voltage of the counterbalancing source. The range of angular deflections was considerably greater than that occurring during a normal sputtering experiment. The beam displacement appears to be a linear function of the counterbalancing filament voltage, or filament temperature. Excellent reproducibility was obtained as illustrated by the two sets of points for the 5° initial deflection taken several days apart. All data were collected with zero applied target voltage, zero grid (electron acceleration or ionizing) voltage and an argon pressure of 2 mtorr. These experiments showed that a well-defined reproducible relationship independent of the initial deflection angle existed between the balance rest point and the filament voltage indicating that the ion gun filaments could be used to provide the counterbalancing force required for continuous measurements.

Based on thermal balancing with the two ion source filaments, a simple experimental procedure was developed. The closely matched targets on each balance arm were carefully bombarded to adjust the beam position to a convenient deflection angle from its arrested position. The filament temperatures of each ion source were adjusted to establish an equilibrium position, that is, a reproducible zero reading for the unarrested balance beam, prior to sputtering. The bombardment was limited to the desired target by applying an ionizing grid potential only to that ion source in the hangdown tube with the target under immediate study. The balance readings were followed as a function of bombardment time and target ion current.

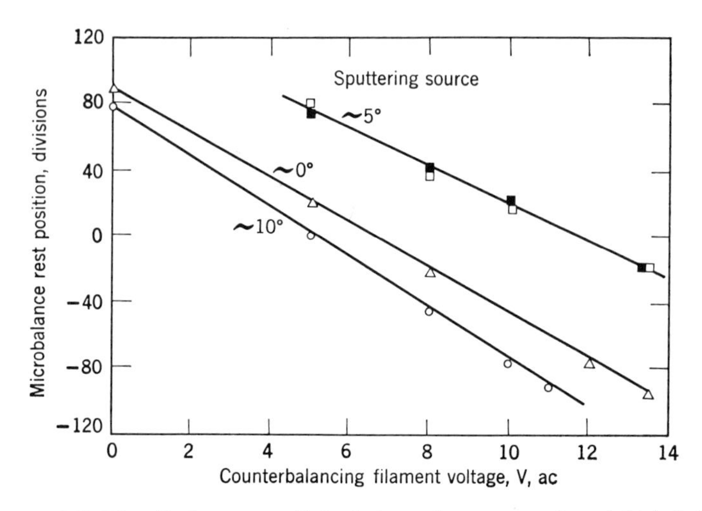

Figure 12.7. The displacement of the balance beam at various initial deflection angles (from the horizontal) as a function of the counterbalancing ion source filament voltage. Argon pressure, 2 mtorr. Sputtering source voltage, 9.5 V ac. Discharge voltage, 0V. target voltage, 0V.

In the static microbalance procedure the target ion current was changed by varying the ion source filament temperature. Deliberate large changes of the ion source filament temperature during the sputtering process in the dynamic method, however, were not possible without severely disturbing the established balancing conditions. The target ion current, therefore, was essentially fixed within very narrow limits by the presputtering equilibrium conditions of the filaments. Since the proper settings to obtain the desired ion currents could be determined through preliminary experimentation, this factor was not found to be a serious experimental limitation. The ion current was monitored with a recorder in order to follow variations resulting from small changes either in the gas pressure or the target potential.

In the dynamic method target voltages could be varied at will during the sputtering process. However, 5–15 min were required for return to the normal equilibrium conditions after sudden variations of more than 25 V. This problem was eliminated by changing the target voltage in increments of less than approximately 25 V or while the sputtering process had been interrupted through momentary removal of the discharge potentials.

IV. EXPERIMENTAL RESULTS AND DISCUSSION

The sputtering yields of single crystal copper and aluminium targets bombarded with argon ions in the 30–100 eV range have been determined both statically and dynamically with the vacuum microbalance (27). The ability to scan the sputtering yield curve rapidly *during* bombardment is demonstrated in Figure 12.8 for a Cu (111) target where the actual variation of microbalance readings at 1 min intervals is shown for sputtering voltages of 30, 50, 75, and 100 eV. The target ion current for each point is plotted in the lower region of Figure 12.8. The data could have been obtained at even shorter time intervals if desired. As expected with the constant filament voltage conditions, the ion current decreased with decreasing target voltage. The sudden 25 V target potential changes generally interrupted the measurement for no more than 5 min. The discontinuities associated with the voltage changes (Figure 12.8) have been shown by Poulis and Massen (34) to be of the order of magnitude of a static charge effect existing between the balance arm and the glass balance case. This capacitor-like effect, however, is constant at a given voltage and has been found experimentally not to affect the observed yield. The yields, S, were calculated from the constant slope established at each target voltage. Figure 12.9 is a plot of the sputtering yield versus target voltage for dynamic and static microbalance measurements using the same copper (100) and (111) targets. The solid lines represent the average of previous static measurements while the points are the dynamic data. The agreement between the two methods is good indicating that the static measurements were performed with carefully prepared surfaces. However, because of the continuous instantaneous nature of the measurements, the dynamic data are considered more reliable.

The value of continuous observation of sputtering yield as a function of time is illustrated in Figure 12.10 which is a plot of the actual weight changes observed during a sequence of argon ion bombardments of a single crystal Cu (100) sample as a function of bombardment time. The sputtering ion currents are also shown. This experiment involved the measurement of the yield even during what would have been the presputtering ion bombardment designed to clean the target surface prior to static measurements. The lack of weight change in the early part of the bombardment was typical of the initial sputtering of samples with no previous sputtering history and probably resulted from the presence of an oxide film. The increasing slope in the latter part of the experiment

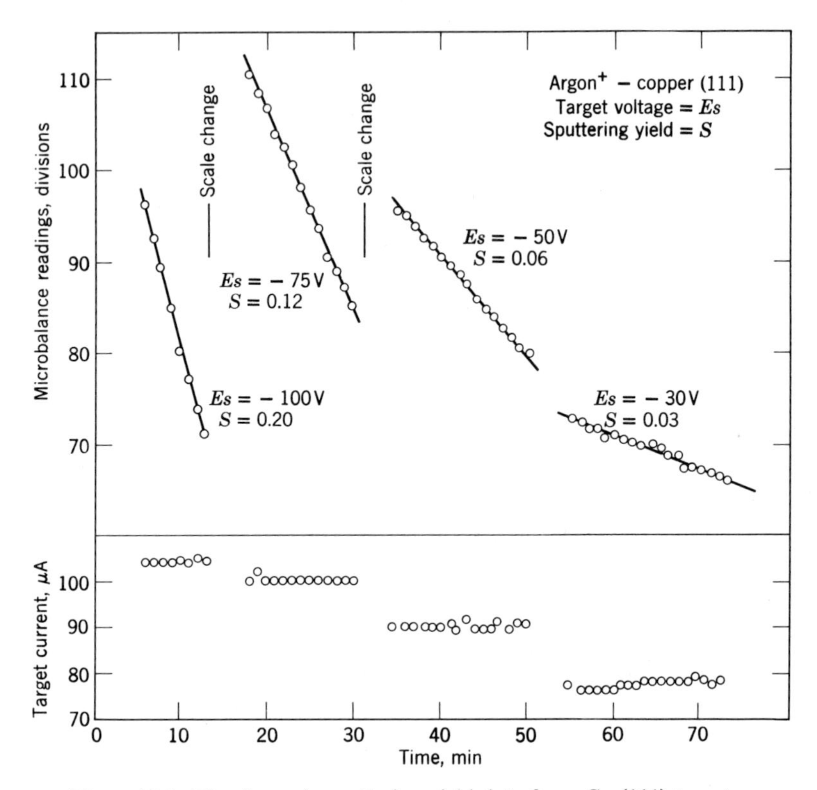

Figure 12.8. The dynamic sputtering yield data for a Cu (111) target.

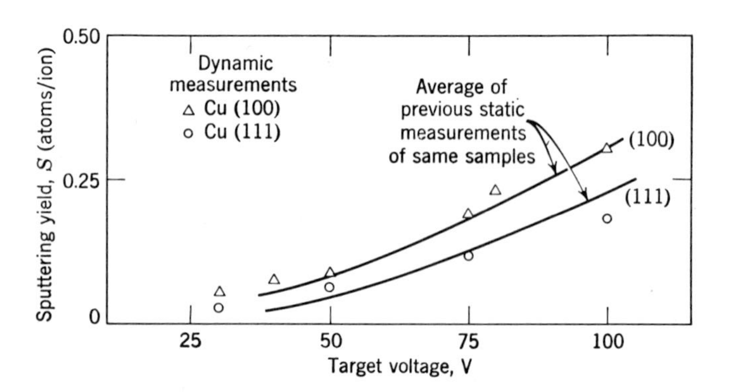

Figure 12.9. Static and dynamic yield data for Cu (100) and Cu (111) targets.

indicated the removal of the surface contamination and the approach to an equilibrium surface condition. Since it is assumed in the static microbalance method that the yield during the period of the bombardment is constant as a function of time, it is apparent that if experiment A had been performed statically, that is by observing the balance readings at the

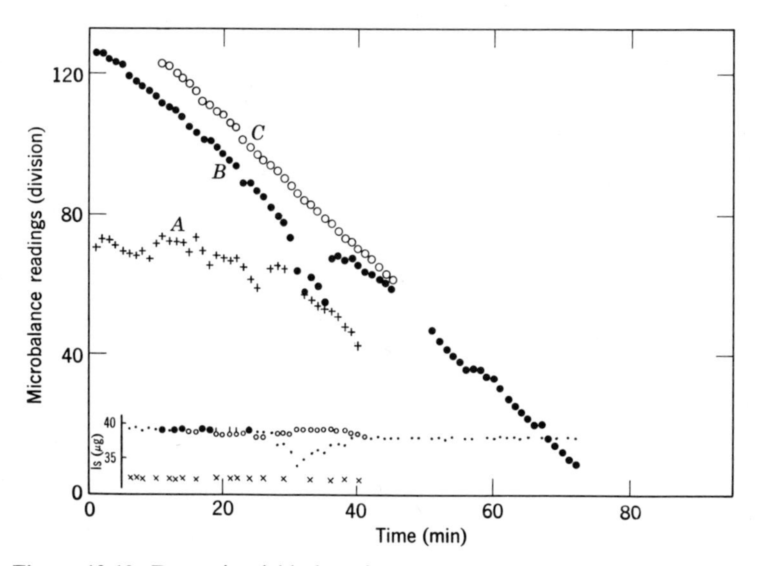

Figure 12.10. Dynamic yield data for a sequence of ion bombardments of a Cu (100) target. Target −80 V. Argon 2.0–2.2 mtorr. (A.) Initial bombardment after six weeks. (B) One day later. (C) Two hours later.

beginning and at the end of the bombardment, the calculated yield would have been less than that obtained from the slope of the last 10 min of the run. Experiment A indicates clearly the high dependence of the yield on the surface condition of the target and the greater reliability of dynamic measurements.

Experiment B was performed a day later and was followed within 2 hr by experiment C. In the interval between B and C the argon remained in the system and the filaments were maintained at the temperatures used for sputtering. The breaks observed in B near the middle of the experiment were associated with a sudden decrease in ion current arising from a power disturbance. The irregularities near the end of the experiment occurred when the system was disturbed by severe external vibrations. These irregularities illustrate the rapid recovery of the balance from a severe environmental disturbance. The reproducibility of the balance itself is evidenced by the fact that the slope at the end of run B, after the disturbances, was identical to that observed in run C. Experiment C is interesting in that the slope was essentially constant throughout and identical to that at the end of B indicating that significant contamination of the surface did not occur in the 2-hr interval between B and C, and that an equilibrium surface, that is a surface in which the sputtering yield is constant with time, had apparently been reached.

A rough estimate has been made of the number of layers of copper removed during the various stages of this series of sputtering experiments. The initial very low yield portion of run A represented the removal of approximately 3 atomic layers. The intermediate region in the remainder of A and through part of B, during which the sputtering yield increased towards a constant value, represented the removal of approximately 25 atomic layers. The results indicated, therefore, that approximately 28 atomic layers were removed prior to the establishment of an equilibrium surface. Interpretation of these measurements in terms of the depth of removal of material from a target surface is not completely valid since it is based on the assumption that sputtering occurs uniformly and non-selectively. These results represent, therefore, only the approximate order of magnitude of the effect. Other workers (35), however, have also concluded that the surface of a target may change with time in the early stages of sputtering until an equilibrium surface of essentially constant characteristics is obtained.

This work indicates the possibility of using sputtering as a surface probe. Careful experimentation with the sputtering of adsorbed gas layers or surface films formed in the microbalance system should provide valuable information on their detailed nature. Combining this information with kinetic observations of film formation and auxiliary studies with x-ray or similar methods can result in a greater understanding of solid surfaces than previously possible.

Some insight into sputtering mechanisms is obtained from an examination of the effect of crystallographic orientation on the sputtering yield. Figure 12.9 discussed earlier presents the results of microbalance sputtering yield measurements for the Cu (100) and Cu (111) planes, i.e, $S_{\mathrm{Cu}(100)}$ and $S_{\mathrm{Cu}(111)}$ respectively, in the 30–100 eV energy range. In this region where $S_{\mathrm{Cu}(100)} > S_{\mathrm{Cu}(111)}$ the major factor controlling the sputtering yield appears to be the binding energy of the surface atoms. The (111) plane having a greater binding energy than the (100) consequently has the lower yield. In Figure 12.11 microbalance data up to 500 eV are shown together with the results of other workers (12,36) above 500 eV. At higher energies the dominant influence on the sputtering process is the stopping power or transparency of the crystal plane. Since the mean free collision path of the bombarding ion is smaller in the more closely packed (111) direction than in the comparatively open (100) plane, the sputtering yield dependence is the reverse of that at low energies, i.e., $S_{Cu(111)} > S_{Cu(100)}$. Figure 12.11 also indicates that the (100) data above 500 eV will fit smoothly to the lower energy microbalance results, but that to match the (111) results, it is necessary for the (111) and (100) curves to cross. The crossover of the yield curves is indicative of a change in the factors controlling the sputtering process. Dynamic microbalance

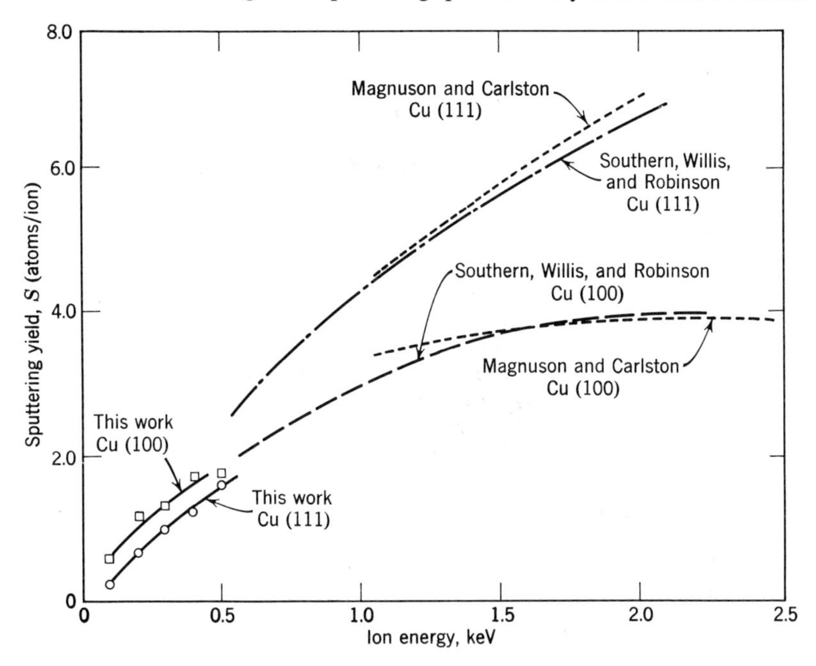

Figure 12.11. Static sputtering yield data for Cu (100) and Cu (111) targets up to 500 eV and that of other workers above 500 eV.

experiments with the (111) and (100) aluminium orientations show a similar orientation effect with the crossover of the yield curves occurring below 500 eV. The data of other workers (37,38) appears to support the thesis that the sputtering mechanism changes as a function of ion energy. The microbalance results suggest that the yield crossover point may be a characteristic parameter of fundamental significance, more easily measureable than the threshold energy, which merits further experimental and theoretical study.

V. CONCLUSION

The sensitivity, reliability and absolute nature of the vacuum microbalance make it an ideal tool for the investigation of sputtering phenomena in the threshold energy region. The potential is also considerable for use of the microbalance sputtering method for the study of the detailed nature of solid surfaces. This potential can be further heightened by the development of techniques for the simultaneous utilization of the microbalance and x-ray, infrared or low energy electron diffraction equipment. This latter area undoubtedly provides the greatest immediate challenge for the microbalance worker.

REFERENCES

1. *Symposium on the Deposition of Thin Films by Sputtering*, University of Rochester, Rochester, N.Y., 1966.
2. W. R. Grove, *Trans. Roy. Soc. (London)*, **142**, 87 (1852).
3. L. Holland and G. Siddall, *Vacuum*, **3**, 245 (1953).
4. N. Schwartz and R. Berry, "Thin Film Components and Circuits" in *Physics of Thin Films*, Vol. II, G. Hass, Academic Press, New York, 1964.
5. G. K. Wehner, *Advances in Electronics and Electron Physics*, Vol. 7, 1955, p. 239.
6. P. Davidse, *Solid State Technol.*, **9**, 30 (1966).
7. L. Maissell and P. Scheuble, *J. Appl. Phys.*, **36**, 237 (1965).
8. A. von Hipple, *Ann. Physik.*, **81**, 1043 (1926).
9. D. T. Goldman and A. Simon, *Phys. Rev.*, **111**, 383 (1958).
10. R. Pease, *R. C. Soc. Ital. Fis.*, **13**, 158 (1960).
11. P. Rol, J. Flint, and J. Kistemaker, *Physica*, **26**, 1009 (1960).
12. A. Southern, W. Willis, and M. Robinson, *J. Appl. Phys.*, **34**, 153 (1963).
13. G. Wehner, *J. Appl. Phys.*, **26**, 1056 (1955).
14. M. Koedam, *Proc. IV Intern. Conf. Ion Phen. Gases, Uppsala*, **1960.**
15. G. Anderson and G. Wehner, *J. Appl. Phys.*, **31**, 2305 (1960).
16. R. H. Silsbee, *J. Appl. Phys.*, **28**, 1246 (1957).
17. D. Harrison, J. Johnson, and N. Levy, *Appl. Phys. Letters*, **8**, 33 (1966).
18. R. Von Jan and R. Nelson, *Phil. Mag.*, **17**, 1017 (1968).
19. I. Veksler, *Zh. Exp. Theor. Phys.*, **38**, 324 (1960).
20. R. Stuart and G. Wehner, *Transactions of the 7th National Symposium on Vacuum Technology*, Pergamon Press, New York, 1960, 290.
21. K. L. Chopra and R. Randlett, *Appl. Phys. Letters*, **8**, 241 (1966).
22. S. Wolsky and A. Fowler, *Semiconductor Surface Physics*, Univ. of Pennsylvania Press, Philadelphia, Pa., 1957, p. 139.
23. S. Wolsky and E. Zdanuk, *J. Phys. Chem. Solids.*, **14**, 124 (1960).
24. R. Schlier and H. Farnsworth, *Semiconductor Surface Physics*, Univ. of Pennsylvania Press, Philadelphia, Pa., 1957, p. 3.
25. S. P. Wolsky, *Phys. Rev.*, **108**, 1131 (1957).
26. S. Wolsky and E. Zdanuk, *Vacuum Microbalance Techn.*, **2** (1962) p. 37.
27. E. Zdanuk and S. Wolsky, *Proc. 3rd Intern. Vacuum Congr.*, Pergamon Press, New York, 1965.
28. E. Zdanuk and S. Wolsky, *Vacuum Microbalance Tech.*, **5** (1966) p. 37.
29. D. McKeon, *Rev. Sci. Inst.*, **32**, 133 (1961).
30. R. Stuart and G. Wehner, *Phys. Rev. Letters*, **4**, 409 (1960).
31. E. Zdanuk and S. Wolsky, *J. Appl. Phys.*, **36**, 1683 (1965).

32. S. Wolsky, E. Zdanuk, C. Massen, and J. Poulis, *Vacuum Microbalance Techn.* **6** (1967) p. 111.
33. S. Wolsky and E. Zdanuk, *Vacuum Microbalance Techn.*, **1** (1961) p. 35.
34. C. Massen and J. Poulis, private communication.
35. J. Dillon and R. Oman, *Transactions of the 10th National Vacuum Symposium*, Macmillan, New York, 1963, p. 471.
36. G. Magnuson and C. Carlston, *J. Appl. Phys.*, **34**, 3267 (1963).
37. G. Ogilvie, J. Sanders, and A. Thomson, *J. Phys. Chem. Solids*, **24**, 247 (1963).
38. S. Wolsky and E. Zdanuk, *J. Appl. Phys.*, **37**, 3641 (1966).

Chapter 13

Some Unusual Applications of the Microbalance

THEODOR GAST

Technische Universität Berlin
Berlin, Germany

I. INTRODUCTION

The choice of method employed in the measurement of a property of a material is determined by the nature and magnitude of that property. As the sensitivity or resolution of a measuring instrument increases, new applications arise. This has occurred in the case of the development of reliable and sensitive electronic recording microbalances which have opened up new experimental areas in physics, chemistry and biology. For example, the rapid response of the microbalance allows the observation of dynamic surface tension effects, while the excellent long-term balance stability makes possible the measurement of the permeation of gases or vapors through membranes. Other new areas of application consist of the measurement of dielectric properties over large frequency ranges and the determination of the concentration and grain size distribution of dust and powders in gases.

In this chapter these rather unusual applications of the electronic microbalance will be briefly discussed.

II. PRECISION MEASUREMENT OF SURFACE TENSION

A. Basic Facts

A surface is a boundary which separates a condensed phase from a gas or its own vapor. To enlarge the surface by an increment dA, an amount of energy dW is necessary, which is proportional to the additional area.

$$dW = \sigma\, dA \tag{13.1}$$

The proportionality constant σ has the dimension of force per unit length and is measured in dyne/cm. It is called the surface tension and corresponds to a one-dimensional pressure.

B. Measurement of Surface Tension with the Aid of Wire Loops and Rings

The surface tension can be measured by the resistance to extension of a film of a liquid drawn from a flat surface by a body of exactly determined geometrical shape and dimensions. If a wire frame, which is bridged by a thin wire with a diameter of 0.005–0.01 cm is immersed in the liquid to be measured and slowly drawn out, a layer forms between the thin stretched wire and the surface of the liquid, and the surface tension of this layer exerts a force on the stretched wire from which σ can be calculated (1).

If r is the radius of the wire, l is its length, ρ is the density of the liquid, and g the acceleration of gravity, one obtains

$$\sigma = \frac{P_1 - P_2}{2l} - r(2\sigma\rho g)^{1/2} + \frac{r^2\pi\rho g}{4} + \frac{2\gamma\sigma}{l} \tag{13.1a}$$

where P_1 and P_2 are the measured forces with and without a layer in the frame for constant depth of immersion of the frame, P_1 being the maximum force observed with increasing length of the layer. Better accuracy can be obtained by using two frames with different lengths of stretched wire and evaluating according to the formula

$$\sigma_{\text{corr}} = \sigma_2 + (\sigma_2 - \sigma_1)l_1/(l_2 - l_1) \tag{13.2}$$

where l_1 is the corresponding length of the wire. This procedure eliminates the influence of the surface at the junction of frame and stretched wire.

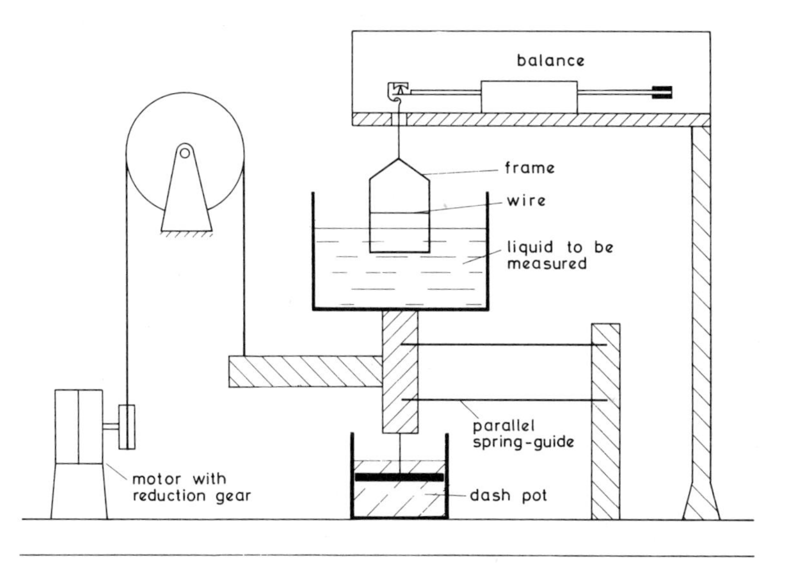

Figure 13.1. Measurement of surface tension with the aid of a stretched wire.

An electronic null-type microbalance is advantageously employed for this measurement since it allows one to record the force exerted as a function of the relative position of the wire and the plane surface of the liquid. Since the balance should not be disturbed by moving it up and down, it is necessary to raise and lower the surface of the liquid smoothly at a desired rate. This can be accomplished pneumatically or with the aid of an electric motor connected to a winch with the container for the liquid being guided by parallel springs. Figure 13.1 shows an example of such an assembly.

Instead of the stretched wire, a horizontal ring of metal can also be used in the measurement of surface tension. The measurement procedure is much the same as with the above-mentioned frame.

C. Measurement of Surface Tension with the Aid of an Immersed Solid Plate

If a thin solid plate is hooked to a balance and immersed in a liquid the force exerted onto it by surface tension is given by

$$P = 2l\sigma \cos \alpha \tag{13.3}$$

where l is the width of the plate measured along the surface of the liquid, α is the angle of contact between the plate and the liquid, and σ is the surface tension (2). By determining the force, the surface tension can be measured. The arrangement is shown in Figure 13.2. This method is

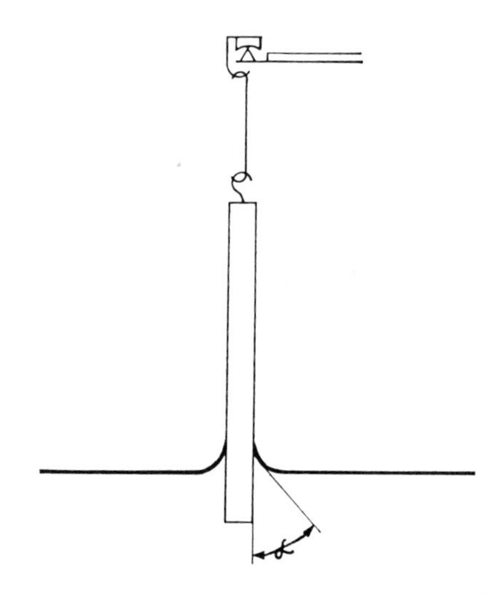

Figure 13.2. Method of Wilhelmy for the determination of surface tension.

especially useful if rapid variations of the surface tension are to be measured as is the case in certain biological applications (3).

As an example of the measurement of surface tension as a function of two-dimensional compression, the effect of a surfactant is observed, which has been extracted from an animal lung (4). The apparatus which is used is based upon the method of Wilhelmy (2). A phosphate buffer, $M/15$, pH 7, serves as hypophase. The surfactant is continuously fed to the surface. By means of glass barriers, which are driven back and fro with the aid of a gear, the surface film is periodically expanded and compressed. Thus, the area of the surface varies cyclically. The lateral displacement of the barriers is displayed on the x coordinate of an xy recorder, while the surface tension of the liquid between the barriers, as measured by the balance carrying the immersed plate, is recorded on the y coordinate. Figure 13.3 shows the hysteresis loop thus produced. The curve is somewhat broadened by fivefold repetition of the cycle.

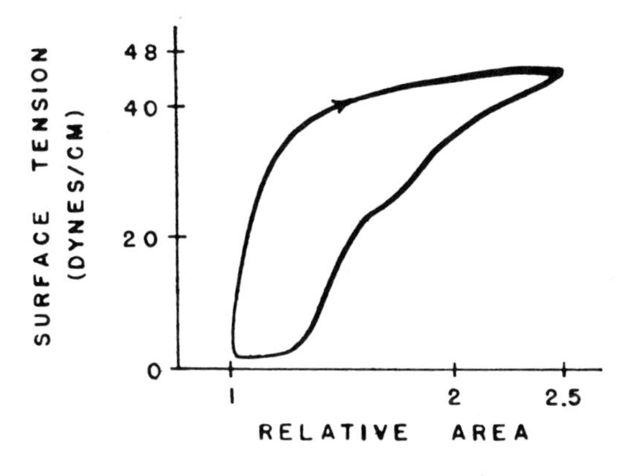

Figure 13.3. Hysteresis loop of surface tension by cyclic variation of area.

D. Measurement of Surface Tension with the Aid of a Cylinder

A method has been described by Seemann (4) for the precision measurement of surface tension, using a small cylinder of quartz, the mantle of which is finely ground to insure complete wetting. The cylinder is suspended from the balance and the balance is tared to zero. The liquid to be measured is contained in a vessel, which can be smoothly raised by a motor drive, until the surface of the liquid touches the lower plane surface of the cylinder. The liquid then jumps up to the cylinder, forming a band at its lower rim. The weight of this band is supported by the cylinder by means of surface tension and can be detected by the increase in weight at the balance. It is very important that the deflection of the beam is very close to zero so that the bottom of the cylinder rests at the level of the surface of the liquid. For zero deflection, there are no buoyancy effects. Small deflections of the balance can be easily corrected for, if the magnitude of the movement is known. The principle is illustrated by Figure 13.4.

III. PERMEATION OF WATER THROUGH PLASTIC MEMBRANES

A. Theory of Permeation

Henry and Fick's laws apply to the permeation of water through membranes, according to

$$C = Sp \tag{13.4}$$

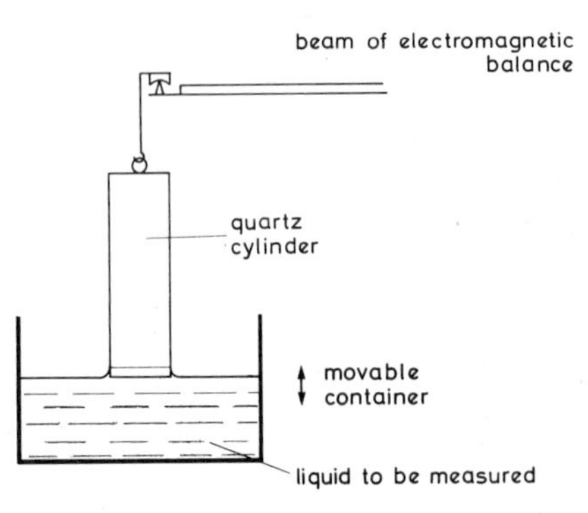

Figure 13.4. Measurement of surface tension according to Seemann.

where C is the concentration of the vapor and p is the pressure at equilibrium and S is the coefficient of solubility. Fick's law is expressed by the equation

$$q = DAt(C_1 - C_2)/d \tag{13.4a}$$

where q is the mass of water which has passed through the membrane, D is the diffusion constant, A is the area of the membrane, C_1 and C_2 are the respective concentrations at the two boundaries of the plastic, d is the thickness of the membrane, and t is time. By inserting equation 13.4 into the equation 13.4a we obtain

$$q = PAt(p_1 - p_2)/d \tag{13.5}$$

where p_1 and p_2 are the equilibrium pressures at both sides of the membrane and $P = DS$ is the coefficient of permeation. The permeability coefficient is very important, especially for sealants and diaphragms (5).

B. Measurement of Permeation

The permeation coefficient can be determined either by dynamic or stationary methods. The stationary methods can be either manometric or gravimetric. The following discussion will be concerned with the gravimetric stationary approach.

Figure 13.5 shows an arrangement, which has proved to be well suited for gravimetric determination of the permeation constant (6). The membrane to be measured is clamped between two cups. The first one is connected to a small container filled with water and held at a temperature corresponding to a desired vapor pressure of about 20 mm Hg. The other cup is connected to the housing of an electromagnetic vacuum balance. A small amount of a desiccating agent, e.g., phosphorus pentoxide, is attached to the beam of the balance. The P_2O_5 maintains an extremely low partial pressure of water vapor which may be neglected in comparison with the partial pressure of some 20 mm Hg at the other side. The increase in weight of the phosphorus pentoxide is recorded as a function of time.

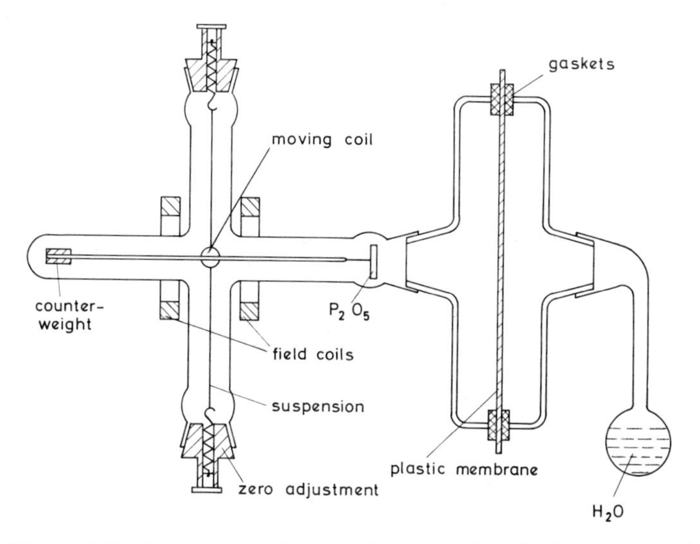

Figure 13.5. Measurement of permeation through a plastic membrane.

After an initial period of transitory permeation a straight line is recorded, the slope of which can be used to calculate the coefficient of permeation according to the equation.

$$P = \dot{q}d/A(p_1 - p_2) \tag{13.6}$$

where $\dot{q}$ is the slope of the line, A is area, d is the thickness, p_1 is the partial pressure of water on the water reservoir side, and p_2 is the negligible water vapor pressure over the P_2O_5, Figure 13.6 presents the data obtained with a disc of polystyrene with an area of 50 cm^2 and a thickness of 1.15 mm (6).

IV. MEASUREMENT OF PERMEATIVITY AND DIELECTRIC LOSS FACTOR

A. Theory of Electrostatic Forces on Dielectric Bodies

The usual methods for the determination of dielectric permittivity and loss factor are restricted to rather small bands of frequency. A well-known frequency-independent procedure is the Boltzmann method. Boltzmann determined the dielectric constant from the attraction of an electrically charged sphere by a rotational ellipsoid of the matter under study. If we

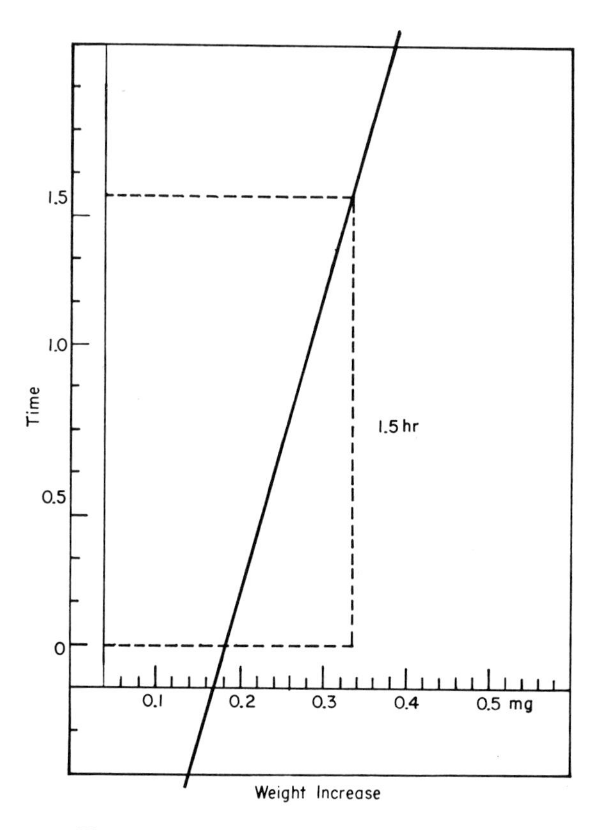

Figure 13.6. Evaluation of a permeation record.

suspend from a balance a small sphere of insulating material in a spherical condensor, the force K exerted on the sample follows the equation.

$$K = V\frac{\varepsilon - 1}{\varepsilon + 2}\varepsilon_0\frac{dE}{dr} \tag{13.7}$$

where V is the volume of the sphere, ε the dielectric constant, ε_0 the permittivity of vacuum, E the field strength, and r the distance of the sample from the center of the spherical condensor. The frequency does not enter this equation. Nevertheless, there are two limits existing:

1. The wavelengths of the corresponding electromagnetic wave should be much larger than the dimension of the condensor.

2. The period of the oscillation should be smaller than the time constant of the spherical sample, which can be calculated from electrical conductivity and dielectric permittivity.

From equation 13.7 it follows that

$$\varepsilon - 1 = \frac{K}{\varepsilon_0 EV(dE/d\gamma) - (K/3)} \tag{13.7a}$$

It is evident that the sensitivity of the method will be optimal for values of ε near 1 and will diminish rapidly for large permittivities. Thus, the method can be regarded as helpful, if a survey for a wide range of values for ε is desired. A method for ponderometric measurement of the dielectric loss factor *tg* has been worked out by Lertes (7). If a sample of a dielectric material with a certain dielectric loss factor $tg\delta$ is inserted into a rotating electric field, there exists the relationship between $tg\delta$ and torque M

$$tg\delta = \frac{9}{2}\frac{\varepsilon_0 V}{\varepsilon M} \pm \left[\left(\frac{9}{2}\frac{\varepsilon_0 VE}{\varepsilon M}\right) - 1 - \frac{4}{\varepsilon} - \frac{4}{\varepsilon^2}\right]^{1/2} \tag{13.8}$$

Of the two possible solutions of the equation, only the one where $tg\delta$ increases with growing M is useful.

B. Measurement of Dielectric Permittivity

Figure 13.7 shows an arrangement for the determination of the dielectric constant by gravimetric means (8). The system comprises a spherical condenser with electrodes which can be cooled to −50°C by liquid air sprayed into the jacket of the condenser, and heated up to 150°C by hot air. An ac voltage of about 1 kV with frequencies between 50 Hz and 100 MHz was applied. An electromagnetic balance of the type designed by Vieweg

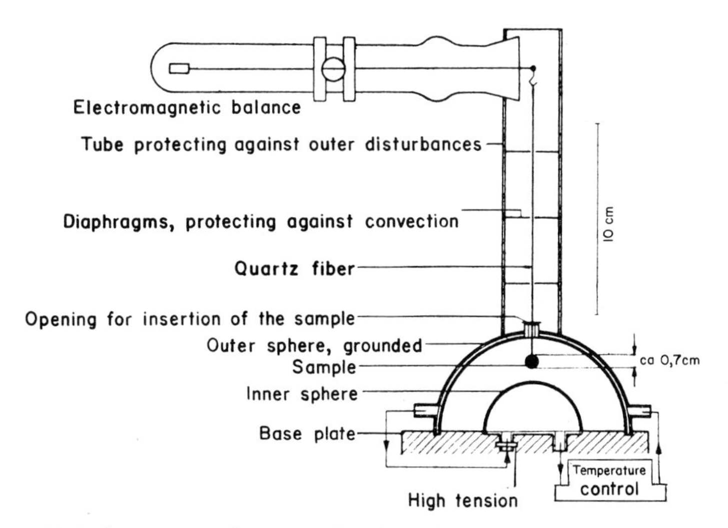

Figure 13.7. Arrangement for measuring the dielectric permittivity by weighing.

and Gast (6) was used. The system was airtight and the suspension fibre for the sample passed through narrow diaphragms to prevent thermal disturbances.

An accuracy of a few micrograms was attained. Figure 13.8 shows the anomalous dispersion of a certain plastic material (polyvinyl chloride with tricresylphosphate as plasticizer), as measured with this apparatus.

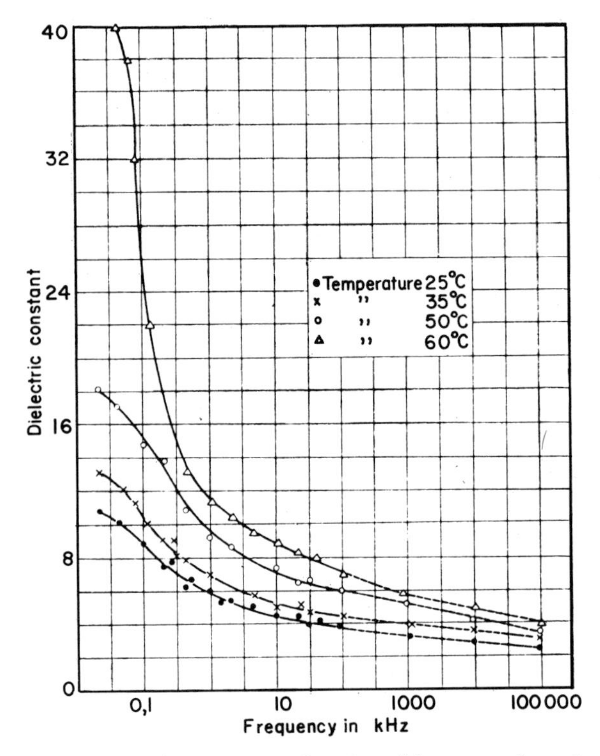

Figure 13.8. Dielectric constant as function of frequency for polyvinyl chloride with plasticizer.

C. Measurement of Dielectric Loss Factor

Figure 13.9 is a schematic drawing of the arrangement which has been used for the determination of the dielectric loss factor by measurement of the torque produced by a rotating electrical field.

A balance of the above-mentioned type, but provided with vertical torsion bands is connected with the disc-shaped sample. Four electrodes are mounted around the disc so that each opposing pair produces a rather homogeneous field in the center. Synchronous alternating current voltages are connected to the two pairs of electrodes. The voltages are produced by a tuneable oscillator which feeds two amplifiers provided with symmetrical outputs. One amplifier is coupled immediately to the oscillator. A phase

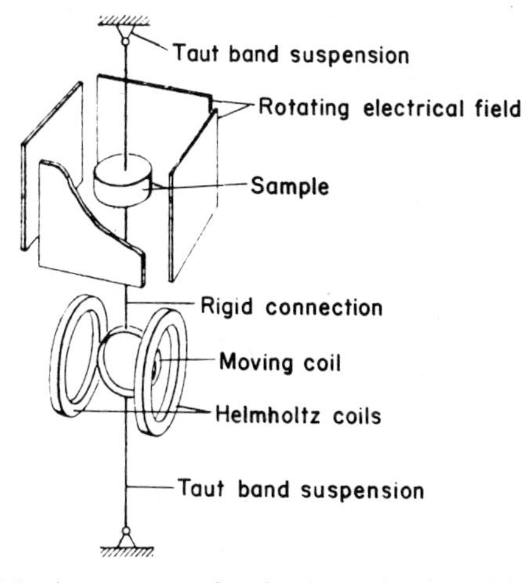

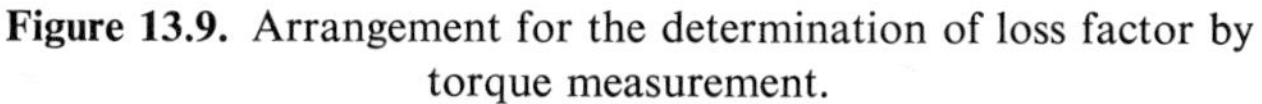

Figure 13.9. Arrangement for the determination of loss factor by torque measurement.

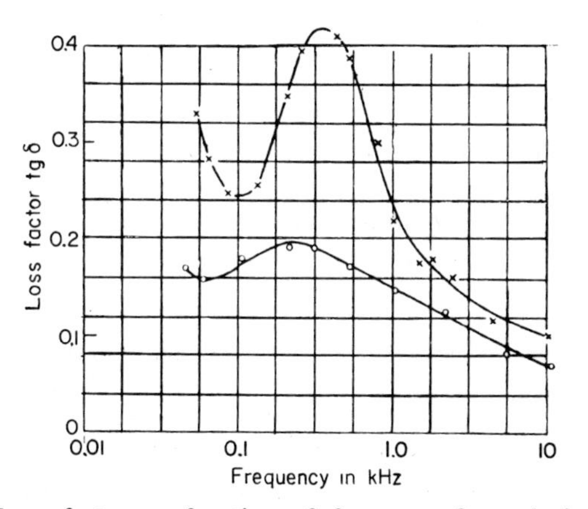

Figure 13.10. Loss factor as function of frequency for polyvinyl chloride with plasticizer. (○) Temperature 30°C; (×) Temperature 50°C.

shifter is inserted between the oscillator and the second amplifier which produces a 90° shift. Thus, a rotational electric field is produced in the center of the electrodes.

The curves in Figure 13.10 were obtained with plasticized polyvinyl chloride at 30 and 50°C. It can be seen that at low frequencies the loss factor tended to increase to infinity and the maximum of the curves was shifted to higher frequencies with increasing temperatures, as predicted by the theory of dielectric relaxation.

With the aid of the modern electromagnetic balances, the sensitivity of the measurement of forces and torques could easily be increased by a factor of 25. This would allow a reduction of the electric field strength by a factor of 5, thus either bringing the applied voltages down to a reasonable amplitude or improving the accuracy considerably. In this manner it could possibly compete with the accuracy attained by pure electrical measurements of permittivity.

V. MEASUREMENT OF DUST CONCENTRATION

A. The Basic Concept of the Dust Balance (9)

It has been shown by some experimental models that the concentration of dust in gases can be measured by precipitating the dust from a determined volume of air onto a surface attached to the beam of an electromagnetic balance. The measuring process is divided into four parts:

1. Precipitation of the dust onto the substrate
2. Weighing of the precipitated dust
3. Cleaning of the substrate
4. Check weighing to secure zero of the balance

The measured weights are recorded. The concentration, c, of the dust in the gas can be calculated using equation 13.9 from the volume of gas flowing through the apparatus in each cycle, $V(\mathrm{m}^3)$, and the difference of weight, $W(\mathrm{mg})$, in weighings *2* and *4*

$$c = W/V \ \mathrm{mg/m^3} \tag{13.9}$$

B. Combination of Electrostatic Precipitation and Electromagnetic Weighing

In Figure 13.11 the mode of operation of an automatic dust balance can be seen. At the right-hand side an electromagnetic balance, g, is shown, the beam being supported by torsion bands and provided with a counterweight and a disc of aluminium foil, d, at opposite ends. At one side of the disc there is an electrostatic precipitation chamber, b, which contains an electrode consisting of a pointed wire, c. The chamber has an opening in front of the disc where the polluted air enters. The electrode is connected to a potential of about 10,000 V. There is a sideward opening in the chamber, where the clean air is drawn out by an aspirator, a. This device produces a constant flow of air through the chamber during the period of precipitation. There is a slide, h, between the disc and the chamber, which is provided with a circular opening, somewhat smaller than the circular disc, the rim of the hole being fitted with a gasket. If this opening rests between the disc and the chamber and the cover plate, *e*, is pressed against *d*, an airtight connection is formed in such a way that the dust can be precipitated onto the substrate. In the next part of the cycle the cover plate is drawn back, and the slide is displaced by a fixed amount. Thus, the disc is shielded from the chamber. The high voltage is disconnected and the aspirator is stopped.

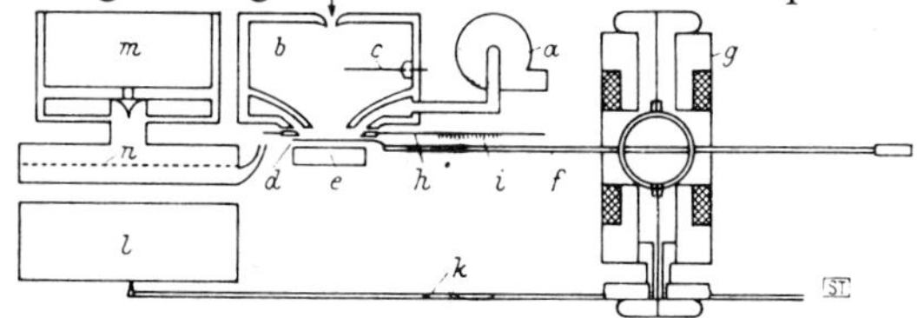

Figure 13.11. Schematic drawing of the recording dust balance.

With the apparatus in this condition, the dust on the surface of the disc is weighed. This can be done in two ways. First, the restoring current through the moving coil can be recorded or transmitted by conventional means. Second, this current can be fed to a very sensitive direct current motor, the rotation of which will twist one of the torsion bands of the balance thus compensating for the torque in the loaded balance. The latter method is applied in the described dust balance. In the drawing it is shown that the rotation of the torsion head of the balance is transmitted to a recording stylus, which engraves the measured value on waxed paper. The sensitivity is such that 10 μg of dust can be detected easily. Figure 13.12 presents a comparison between values of dust concentration recorded with a dust balance and determined by using a filter method (10).

C. Mass Determination by Vibrations of a Very Thin Band

It is well known that the thickness of evaporated layers on quartz crystals can be measured by the corresponding decrease in natural frequency of the crystal. If, however, this method of mass determination is applied to dust, the results are erroneous because the dust particles do not stick firmly to the surface of the crystal. The large acceleration at the boundary of the quartz by far exceeds the attracting molecular forces. Therefore, the coupling between the particles and the crystal is variable.

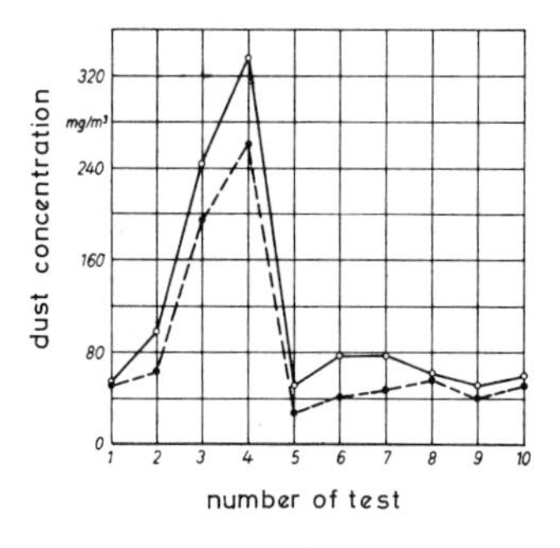

Figure 13.12. Comparison between dust balance (upper curve) and filter device (lower curve) with varying concentrations of dust.

In order to avoid the very high accelerations, the quartz crystal was replaced by a tautly stretched band of plastic. The band was a few micrometers thick, had a free length of about 10 cm, and showed a frequency of free oscillation of 1 kHz, when excited like a string. The measurements were performed in vacuum of about 0.01 torr to minimize air damping. The

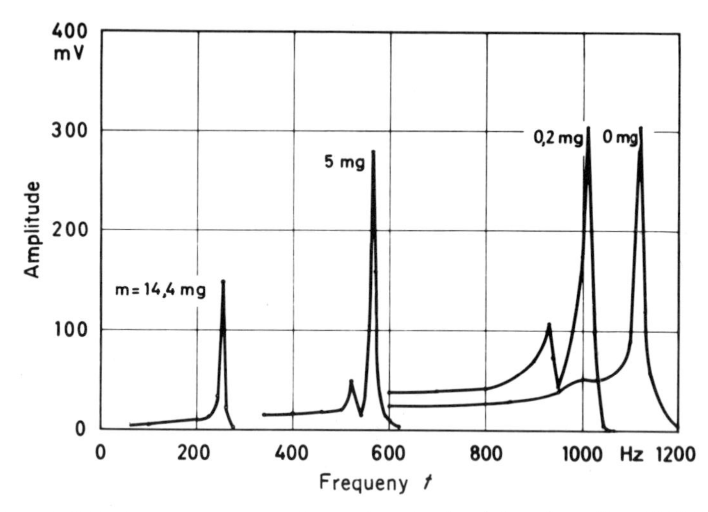

Figure 13.13. Resonance curves for various loads of dust ($p = 5 \times 10^{-3}$ torr).

band was excited by electrostatic forces and the deflections were detected by the fluctuation of voltage in a nearby electrode. The dust was electrostatically precipitated onto the metallized surface of the plastic. Figure 13.13 shows the shift in natural frequency, caused by different loads of dust (10). The method seems to hold promise for the continuous measurement of dust concentrations in air.

VI. DETERMINATION OF GRAIN SIZE DISTRIBUTIONS BY WEIGHING

A. Theory of Sedimentation Analysis

Several methods have been developed for the measurement of grain size distributions of powders and dusts with the aid of balances. They can be divided into classes according to the nature of the medium in which the particles are dispersed, i.e., gas, liquid or vacuum; there is still another classification according to the measurement procedure. It is common to most of these methods that the particles, collected on a pan after settling, are weighed and the curve of integral weight versus time is recorded.

Let us assume that all of the particles are randomly distributed over the volume of a freshly stirred container. The particles will then settle with different speeds, according to the Stokes law. According to Odén (12) the mass, $G_{(t)}$ which has settled on the bottom of the vessel after a certain time interval, t, is given by

$$G_{(t)} = g_{(t)} + t\frac{dG}{dt} \tag{13.10}$$

where $g_{(t)}$ is the mass of all grains, which have fallen through the full height of the liquid, h. The value of $g_{(t)}$ can be obtained by a tangent construction from the recorded curve of $G_{(t)}$ as in Figure 13.14. If $G_{(a)}$ is the total mass of the initially suspended grain, then the percentage R is given by

$$(g_{(t)}/Ga)100 = R \tag{13.11}$$

B. Grain Size Distribution by Sedimentation in Liquids

If a balance pan is suspended at a small distance above the bottom of the container, as it is shown in Figure 13.15, (12) the grains will settle partially upon this surface. The number of particles can be calculated from the ratio of the area of the pan and the cross section of the container.

Thus, the grain size distribution can be determined by recording the increase in weight of the balance pan. While Odén had used a balance with mechanically added balls as weight increments, Bachmann (13) has designed a system where the compensating increments are brought about by a stepping motor turning a torsion spring. It is of course, possible, to use the modern recording microbalances for this kind of sedimentation analysis.

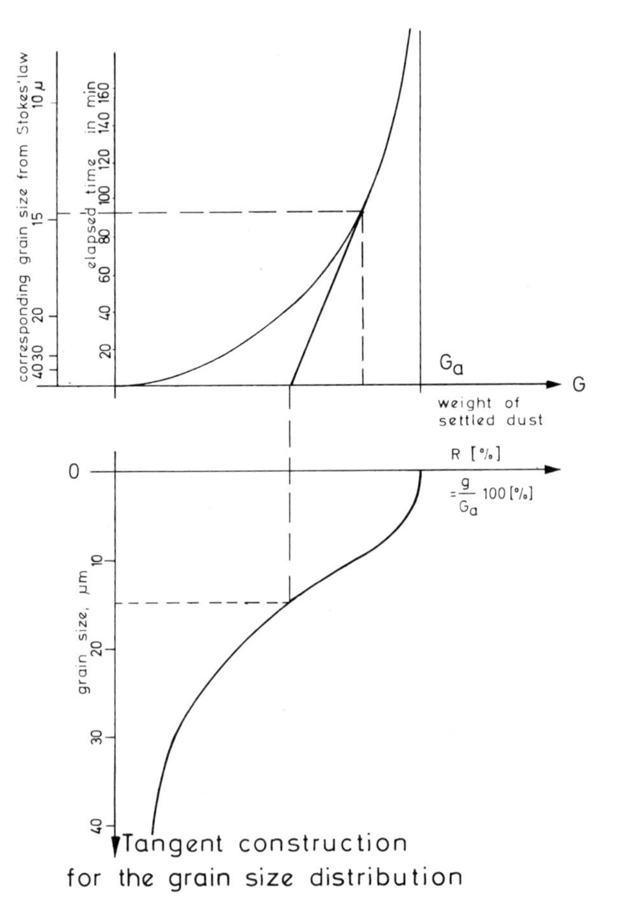

Figure 13.14. Tangent construction for the grain size distribution.

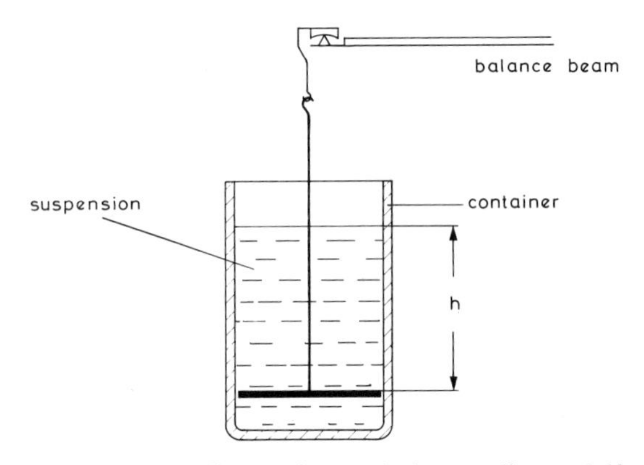

Figure 13.15. Sedimentation analysis according to Odén.

C. Grain Size Distribution by Sedimentation in Air

The time consumption of the sedimentation experiments depends on the viscosity of the suspension fluid and the size of the smallest grains of the suspension according to the equation

$$v = (1/18\eta)g(\rho_K - \rho_H)d^2 \qquad (13.12)$$

where v is the velocity of the particle, η the viscosity of the fluid, g the acceleration of gravity, ρ_K the density of the solid and ρ_H the density of the liquid, while d is the diameter of the grain.

It may therefore be useful, to use gases or partial vacuum for the sedimentation analysis, if the grains are very small. This principle is followed in the micromerograph by Sharples (14). There is an important difference between the micromerograph and the sedimentation balance described above, which results from the method of introduction of the solid substance. While the dust is initially homogeneously distributed throughout the sedimentation volume in the case of the sedimentation balances of Odén and Bachmann, it is suddenly introduced at the top of the cylinder by compressed air in the case of the Sharples apparatus. This latter mode of operation corresponds to a differential method and immediately leads to the cumulative curve of grain size distribution.

D. Grain Size Distribution by Sedimentation in Partial Vacua

In a similar way Götz and Gast (not published) have attempted to record the sedimentation of dust in partial vacuum. In this case Stokes law may need to be corrected for slip in a diluted gas because of the considerable mean free path of the particles. An arrangement for sedimentation analysis in diluted air is shown in Figure 13.16. With the aid of a vacuum pump, the sedimentation tube can be evacuated to any desired pressure between $\frac{1}{10}$ torr and 1 atm.

There is a sluice at the upper end of the tube, where the sample can be introduced into the evacuated system. Because the dust is under atmospheric pressure when the lower valve of the sluice is opened, there results an expansion of air from the sluice into the sedimentation tube. By the force of this expansion, the dust is thoroughly dispersed and distributed in the upper part of the tube, where the sedimentation starts.

The disadvantage of the last two methods results from the convection streams that form in the sedimentation tubes because of the unequal density distribution of the gaseous medium. These disturbances are avoided in the first method although the problem of unequal density distribution near the balance pan still exists. It is easily seen that even in the first method a region of lower density forms under the pan and creates convection streams around the edge of the pan. Thus, some of the dust particles which would otherwise settle on the pan, are deflected away. This error can be avoided by the special design of the sedimentation vessel used

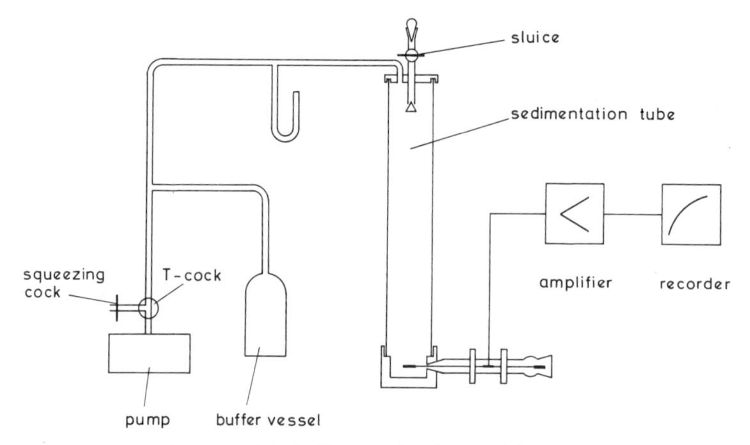

Figure 13.16. Sedimentation in partial vacuum.

by manufacturers of sedimentation balances. In this design, an outer vessel which is filled with the pure liquid surrounds a tube which contains the suspension. From the inner tube, the particles sedimentate onto the balance plate. The rim of this plate is bent upward, so as to form a sort of movable seal with the tube. By this means all of the dust is settled on the pan.

VII. CONCLUSION

A number of unusual applications of the recording electronic microbalance have been compiled in this chapter. It has been demonstrated that the microbalance can be a useful instrument for the measurement of a variety of different properties. Whether these properties are really unusual depends to a large degree on the viewpoint of the reader. For example, one of the first applications of the microbalance more than a quarter of a century ago was the measurement of the permeation of gases, while the measurement of surface tension with the balance could possibly develop into a standard method of testing surfactants.

REFERENCES

1. H. Umstätter, ATM V 9122–8 (Nov. 1951).
2. L. Wilhelmy, *Ann. Physik*, **119**, 177 (1863).

3. R. M. Mendenhall and A. L. Mendenhall, Jr., *Rev. Sci. Instr.*, **34**, 1350–1352 (1963).
4. F. W. Seemann, *Z. Instr.*, **72**, 82 (1964); *PTB-Mitteilungen*, **1**, 24 (1966).
5. G. J. van Amerongen, in *Nitsche Wolf Struktur und physikalisches Verhalten der Kunstoffe*, Springer-Verlag, Berlin, 1968.
6. R. Vieweg and Th. Gast, *Kunststoffe*, **34**, 117–119 (1944).
7. P. Lertes, *Z. Phys.*, **6**, 56 (1921).
8. Th. Gast and E. Alpers, *Z. Angew. Phys.*, **1**, 228–232 (1948).
9. Th. Gast, *Z. Staub*, **20**, 266–272 (1960).
10. D. Hasenclever, *Z. Staub*, **41**, 388–435 (1955).
11. D. Büker and Th. Gast, *Chem. Ing. Tech.*, **39**, 963–966 (1967).
12. S. Odén, *Kolloid Z.*, **18**, 25–31 (1916).
13. D. Bachmann and H. Gerstenberg, *Chem. Ing. Tech.*, **29**, 589–594 (1927).
14. F. S. Freddie and R. E. Payne, *Iron Age*, **174**, 99–102 (1954).

PART III

Commercial Apparatus

Chapter 14

The Availability of Commercial Microbalances and Quartz Crystal Oscillators

D. C. Fox and M. J. Katz

Institute for Exploratory Research
U.S. Army Electronics Command, Fort Monmouth, New Jersey

I. INTRODUCTION

The past decade has seen an ever-increasing interest in micro and sub-micro weighing techniques. To meet a growing need, manufacturers of conventional balances are marketing instruments with increased sensitivity and automatic recording capability. Quartz specialty companies have taken over from the experimenter the once tedious task of fabricating spring and torsion–fiber microbalances; vacuum and electronic specialty companies are producing monitors and controllers for thin film applications. Balances adequate to meet most weighing requirements are now generally available.

This chapter attempts to provide a compilation of pertinent information on the various commercial balance designs. The list, however, does not include all possible sources. In addition, the prices are subject to change and are included only to indicate the general order of costs. The information in this section will simplify the task of selecting and obtaining the proper commercial apparatus for a given need.

II. HELICAL SPRING BALANCES

The helical spring microbalance has been in use for a long time, but it was not until McBain (1) used this instrument for gas adsorption measurements that it came into its own. The spring or McBain balance (see Figure 14.1) consists of a simple helix, usually made of quartz fiber, hung vertically in a tube. The balance depends on fiber elasticity for its operation. Care must be exercised to select uniform fibers in order to achieve reliability. The design combines a good working length with simplicity. Springs of phosphor bronze and stainless steel have also been used. Some of the important features of this balance are: ease of construction, simplicity and speed of operation, and the ability to be used in controlled atmospheres. The balance can also be baked at elevated temperatures for long periods of time for outgassing.

A major disadvantage is that the ratio of load to sensitivity is low when compared with that of the torsion balance. Even with this limitation the balance can be extremely useful, and depending upon the load and method of determining weight changes, sensibilities in the 10^{-8} g region can be obtained. A more detailed discussion of the spring balance can be found in Chapters 2 and 3.

A. Microchemical Specialties Co. (Misco)*

A variety of balances and balance housings is available from Misco. Accessories or complete systems may be purchased depending upon the user's requirements.

There are four types of balance cases seen in Figure 14.2. *A* consists of a cap with a hook, a 30-mm o.d. mainbody, and a 35-mm o.d. bottom section. It is suitable for routine weighing of dry or wet samples, and the bottom section fits over the top section as a shield against drafts. *B* consists of a cap with a hook, a 32-mm o.d. mainbody, and a bottom section that extends 28 cm below the 34/45 taper joint. It is designed for weighing hygroscopic materials and a desiccant may be placed in the bottom of the tube. *C* has the same physical dimensions as *B* and is used in controlled atmosphere or vacuum at various temperatures. *D* has the same physical dimensions as *B* and is used in high temperature work in vacuum or controlled atmosphere, and has a water jacket to localize heat.

* 1825 Eastshore Highway, Berkeley, California 94710

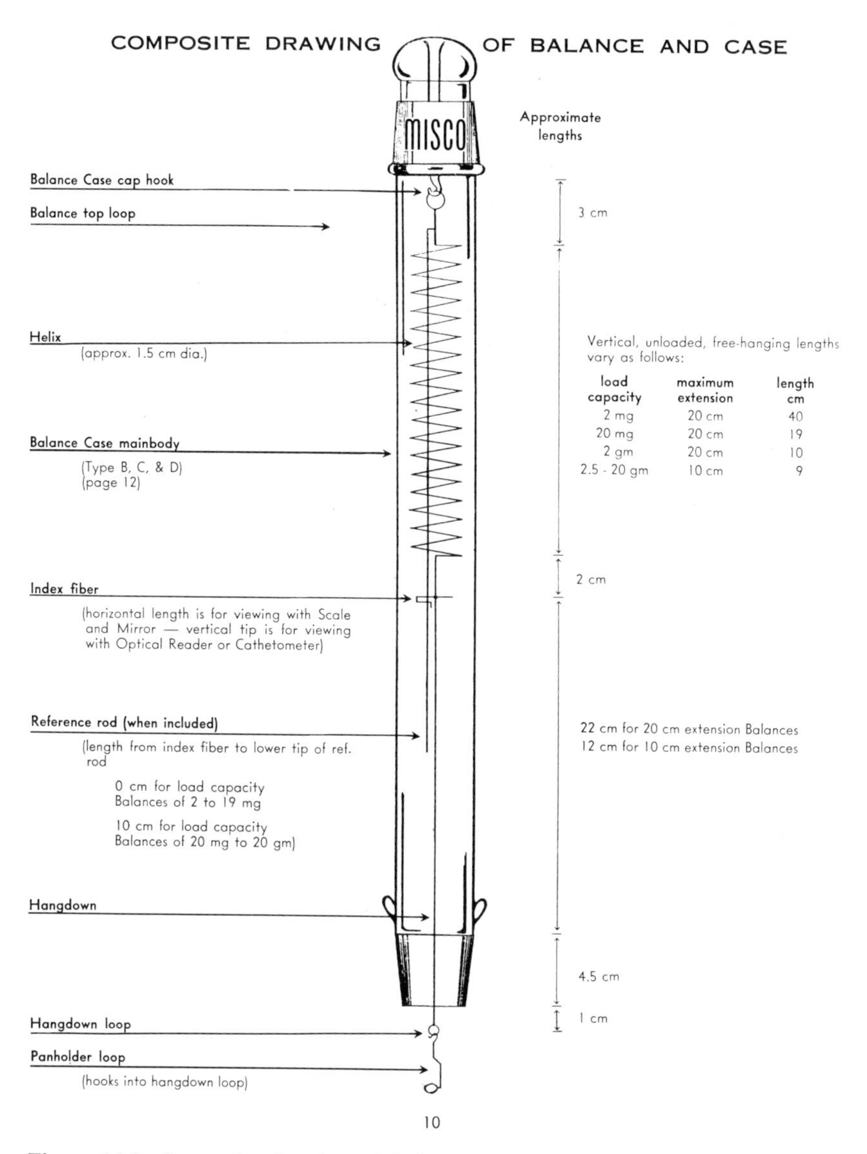

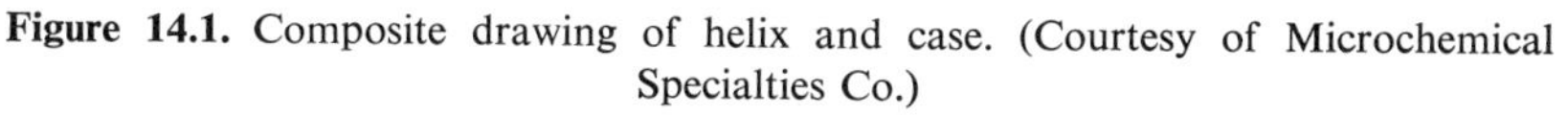
Figure 14.1. Composite drawing of helix and case. (Courtesy of Microchemical Specialties Co.)

Misco classifies their quartz helices by extension; either 10 or 20 cm for a given load capacity. The exact extension will be within 5% for all except the 2-mg helices; in this case it is 10% of the nominal extension. The indicated load capacity includes the sample, sample pan, and gain in

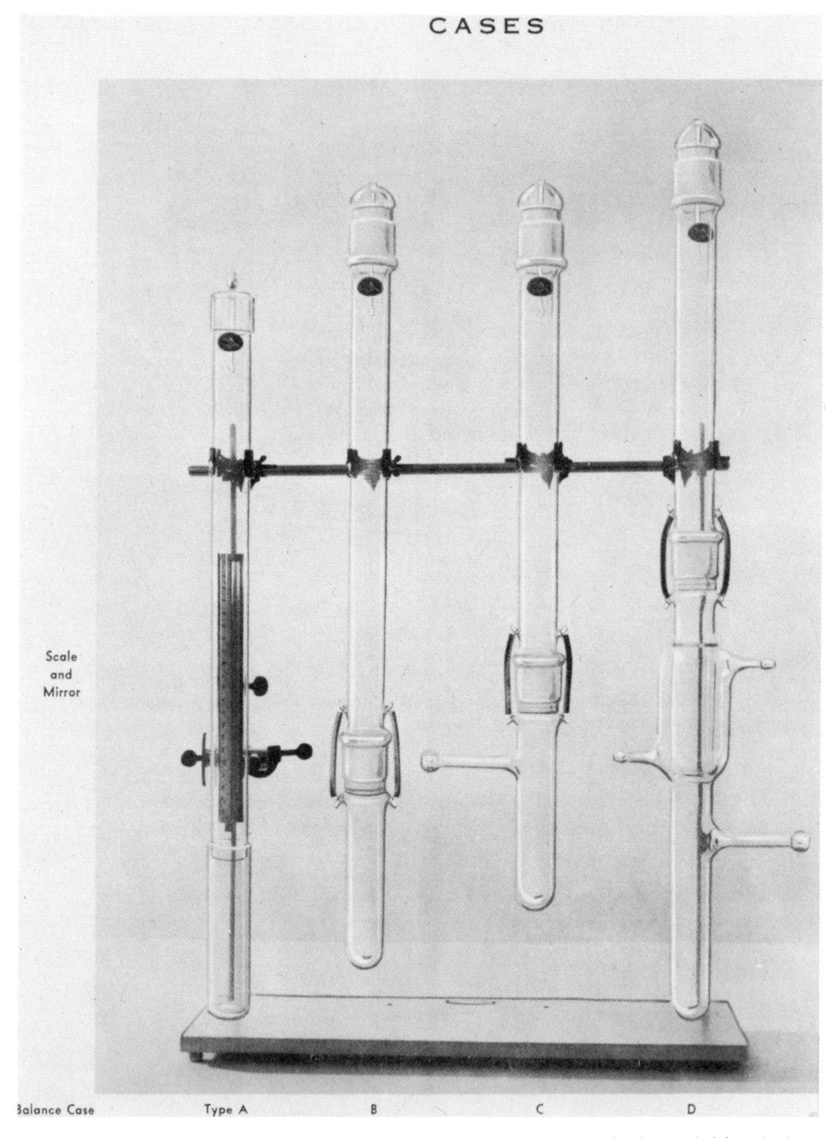

Figure 14.2. Types of balance cases. (Courtesy of Microchemical Specialties Co.)

sample weight. Misco allows a safety factor of 100% (50% for 40-g helices). However, this should not be added to the nominal maximum load or extension.

Performance characteristics and relative ranges for a 2-g capacity helix using different measuring devices are listed in Table 14.1.

Table 14.1.

	Detection of 2-g load capacity helices			
	10 cm max. extension		20 cm max. extension	
	Degree of detection	% Detection	Degree of detection	% Detection
Scale and Mirror: range, 20 cm to ±0.1 mm				
Sample weight (2 g, full load capacity)	±2 mg	0.2	±1 mg	0.1
Change in weight of sample	±2 mg	—[a]	±1 mg	—[a]
Optical Reader[b]: range, 5 mm[c] to 1 μ				
Small sample weight				
Up to 100 mg (equivalent to 5-mm extension of 10-cm helix)	±10 μg	0.02		
Up to 50 mg (equivalent to 5-mm extension of 20-cm helix)			±5 μg	0.02
Change in weight of sample				
Up to 100 mg (5% of max. load capacity)	±10 μg	0.02		
Up to 50 mg (2.5% of max. load capacity)			±5 μg	0.02
Cathetometer: range, 20 cm to 1 μ, for example				
Sample weight (2 g, full load capacity)	±10 μg	0.001	±5 μg	0.0005
Change in weight of sample	±10 μg	—[a]	±5 μg	—[a]

[a] Percent detection is dependent upon the actual change in weight.
[b] The scale and mirror and the optical reader complement one another in many applications.
[c] The optical reader makes its detection in a 5-mm field only, not over a 10 or 20 cm range.

A composite drawing of the quartz helix and case are shown in Figure 14.1. The quartz helix from top loop to hangdown loop are fused together thus preventing erratic rotation of the index fiber. The index fiber is 0.10 mm diameter and enables one to read the scale and mirror to ± 0.1 mm. The horizontal length is for viewing with a scale and mirror, and the vertical tip is for reading with an optical reader or cathetometer. Balances may be obtained with or without the reference rod. The rod is fused to the top loop and extends down inside the spring (see Figure 14.1). The reference rod is not needed with the scale and mirror or the optical reader, but is necessary with the cathetometer to detect changes in lengths less than 10 μ. The lower tip of the reference rod can serve as a check for the cathetometer especially if the cathetometer was moved during an experiment.

The spring balances are supplied with a nominal calibration. Additional calibrations must be made under normal operating conditions if an optical reader or cathetometer are to be used. Methods of exact calibration are also provided for various weight ranges.

The vertical, unloaded, free-hanging length of the helices vary, as shown in Figure 14.1. The diameter of the fiber in the coils is the most critical factor in determining the extension per unit weight, and the spring length. The helix extension varies inversely as the 4th power of the radius of the fiber. Available accessories include a scale and mirror, optical reader, sample pans of quartz, aluminium and platinum springs, pan holder loops, and quartz fibers. The scale is 20 cm long and graduated in millimeters. The mirror minimizes parallax; the index fiber is lined up with its image and read on the scale. With this arrangement extensions can be read to ± 0.1 mm. Thus, one obtains a 0.1% full-scale detection for the 20-cm extension balance, and 0.2% for the 10-cm extension balance. Figure 14.3 shows an assembly with a scale and mirror.

The optical reader is a microscope tube with a Bausch and Lomb objective and filar micrometer eyepiece, and is adapted vertically and horizontally by adjustments in the mounting. The objective, in combination with the tube length and eyepiece, has a working distance of 59 mm and a 5-mm diameter field. The eyepiece itself has a horizontal movable crossline which can divide the 5-mm field into 5000 equal parts, i.e., 0.02% or to ± 0.5 μ.

Sample pans are available in quartz, aluminum, or platinum in many sizes (diameters and weights) and are either flat or dish-shaped. The pan-holder loops are used only with a metal pan and must be compatible with other parts of the system.

Quartz suspension fibers are available in 18-in. lengths, two diameter ranges, 100–1000 μ, or 30–90 μ. Fibers are also available in 5-, 10-, and 20-μ diam in various lengths. Variation will be within ± 5%.

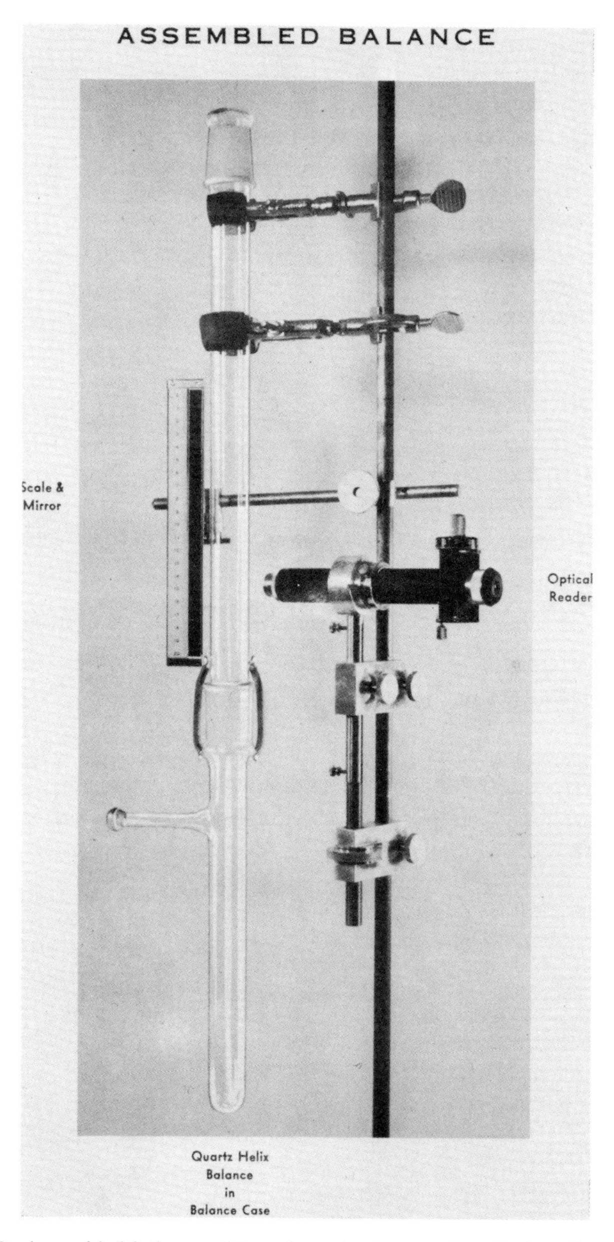

Figure 14.3. Assembled balance with scale and mirror and optical reader. (Courtesy of Microchemical Specialties Co.)

Misco offers helices with hooks on either end for customers who wish to make their own balances.

Balance, helix quartz, 10 or 20 cm extensions
Prices vary from $70.00 to $135.00 each depending upon load capacity and whether a reference rod is required.
Pans, platinum, quartz, and disposable aluminum
Pans have either a flat or a deep-dish shape and come in many sizes. Prices range from $2.00 to $7.00 each for a normal order. Special quartz pans range from $12.00 to $30.00 each.
Panholder loops for metal pans in various sizes are available at $4.00 each.
Balance case, Pyrex

	Each
Type A	$16.00
Type B	28.00
Type C	30.00
Type D	36.00

Spring, helix quartz 10 or 20 cm extensions
Prices vary from $25.00 to $70.00 each depending on load capacity.

B. Worden Quartz Products Inc.*

Worden Quartz Products, Inc. manufactures a complete line of quartz spring balances which can be purchased either individually or in an assembly ready for installation on the user's pumping station. As illustrated in Figure 14.4, the assembly consists of a Pyrex tube with a high vacuum joint at the top and two stopcocks for ambient control. Inside the tube is the quartz spring with reference mark, suspension hooks, and sample pan. Surrounding the portion of the Pyrex tube that contains the spring is a water jacket with a thermometer well that provides temperature control. The sample pan is at the base of the Pyrex tube and outside of the jacket, so that the temperature of the sample may be changed without changing the temperature of the spring. The entire assembly is supported by a tripod which also holds the micrometer-microscope in place.

Optional equipment which may be obtained includes sample heater, special sample holders, nonrotating fused quartz springs, and a tripod base and cathetometer which mounts on the tripod.

Fused quartz springs for this system may be obtained in almost any combination of sensitivity, extension, and load. They are normally supplied with hooks, but any changes can be made provided explicit instructions or drawings are supplied by the contractor.

* P. O. Box 36010, Houston, Texas 77036

Figure 14.4. Spring balance assembly. (Courtesy of Worden Quartz Products Inc.)

Table 14.2 Variation of Unloaded Spring Length with Maximum Load

Maximum load, g	Unloaded length, times maximum extension
To 1	0.1
2– 6	0.2
7– 12	0.3
13– 20	0.4
21– 50	0.5–0.7
41–100	0.7–1.0

The unloaded spring length varies with the maximum load (see Table 14.2).

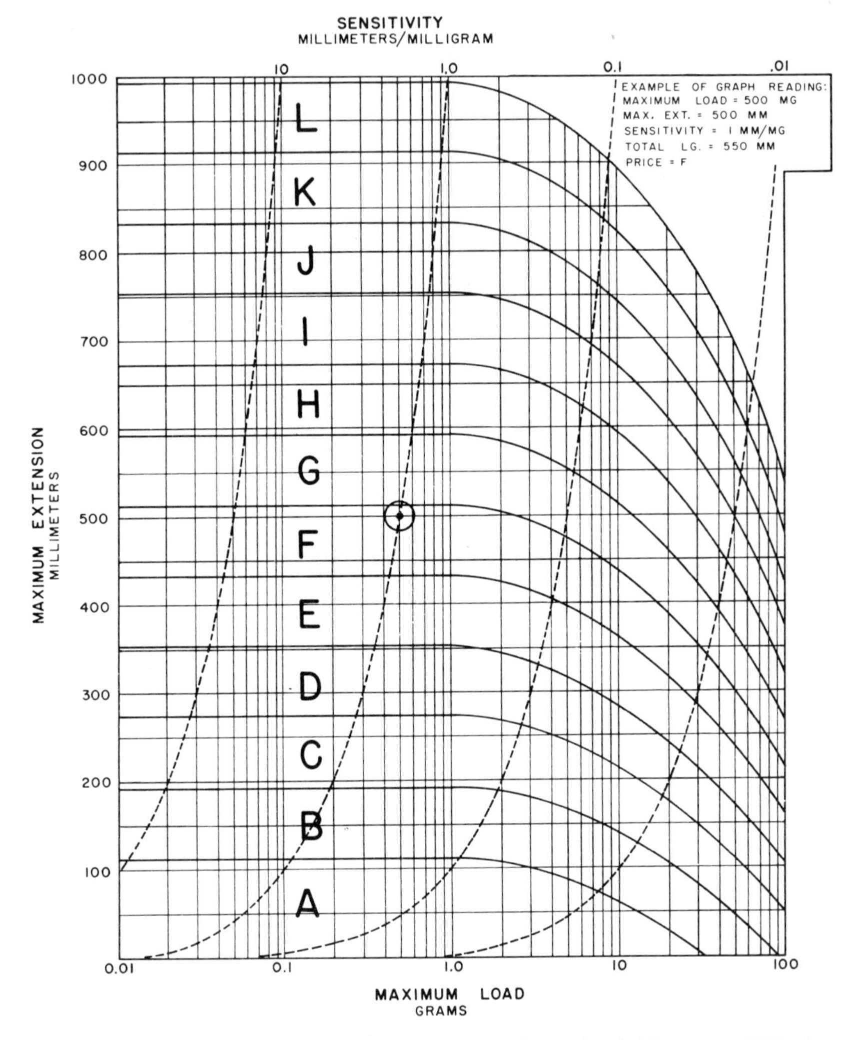

Figure 14.5. Plot of maximum extension versus maximum load. (Courtesy of Worden Quartz Products Inc.)

One may select quartz springs according to Figure 14.5, which plots maximum extension versus maximum load. The sensitivity is obtained by dividing the maximum extension by the maximum load. The letters in the alternate dark and light areas are price keys. As accessories, Worden offers fused quartz sample buckets ranging in weight from 5 to 500 mg and diameters of 5–50 mm; an assortment of quartz fibers ranging in diameter from 25 to 500 μ in 25 μ increments; and tweezers shaped for handling quartz fibers and sample pans. Also available is a fused quartz engineering kit that includes six quartz springs with maximum loads of 50, 100, and 500 mg, 1, 2, and 5 g with a maximum extension of all springs of approximately 75 mm, one flat spiral quartz spring with a torque constant of approximately 100 dyne-cm per radian, one hemispherical quartz sample pan 12–15 mm in diameter and weighing between 25 and 50 mg, an assortment of quartz fibers, and one pair of stainless tweezers.

A recent price list follows:

Spring balance assembly	$345.00
with tripod	495.00
with tripod and cathetometer	670.00
nonrotating fused quartz spring, standard system	20.00
Special sample holders	*
Sample heater	50.00
Tripod base	185.00
Cathetometer (for tripod mount only)	225.00
Cathetometer and tripod	410.00
Fused quartz pans (depending on size)	5.25 up
Tweezers	4.50
Fused quartz engineering kit	41.50
Quartz fiber kit	15.75
Fused quartz springs	10.50 to 120.00

(For nonrotary add 40% or $20.00, whichever is greater.)

* Price on request.

III. TORSION MICROBALANCES

Just prior to the turn of the century, Weber (2) used a fused quartz fiber as a microbalance beam. Since that time, people have used quartz fiber microbalances, making modifications to suit their needs. Rhodin (3) has outlined a brief history of quartz fiber microbalances. In Chapter 2, Czanderna gives an up-to-date review of the microbalance. Most balances used now are a modification of the Donau (4) balance. Figure 14.6 illustrates one type of torsion balance developed by Prof. T. Gray, Alfred University, N.Y. (5) and used by the authors.

The torsion balance, like the spring balance, is usually made of quartz. Metal beams are used occasionally, and some models have been made with fine tungsten wire as suspension fibers. The torsion balance is insensitive to small temperature changes and has a high zero stability. Care must be exercised in selecting the suspension fibers since they are critical in determining the sensitivity of the microbalance. These instruments are usually

Figure 14.6. Microbalance used by authors.

enclosed in vacuum housings and can be operated in controlled atmospheres. Generally they have greater sensitivity than spring balances and can be used either as deflection or null-type instruments. They are widely used in gas adsorption measurements, oxidation rate experiments, and high temperature reactions. The balance can be made quite sensitive. The authors have detected weight changes down to 10^{-9} g with this type of balance.

A detailed discussion of the theory of torsion microbalances is presented by Schwoebel in Chapter 3.

A. Worden Quartz Products Inc.*

Worden Quartz Products, Inc., in addition to the quartz spring balances, manufactures a fused quartz microbalance assembly. The apparatus, shown in Figure 14.7, consists of a Pyrex chamber with a flange and O-ring seal at the top for ready access. On the back just below the flange there is a small Pyrex tube for ambient control. At the bottom are two high vacuum joints; one for easy access to the sample, the other for the magnetic weight changer (optional). A flat window is provided for observing the pointer with the filar microscope. Inside the tube is the microbalance suspended from its frame which is attached to the metal flange; clamped to the outside on the flange is the filar microscope. The entire assembly is supported by a tripod.

* P. O. Box 36010, Houston, Texas 77036

Optional equipment which may be obtained includes magnetic damping, magnetic weight changer, special sample holders, extra long sample suspension fibers, sample heaters and a null reading device with electromagnetic balancing.

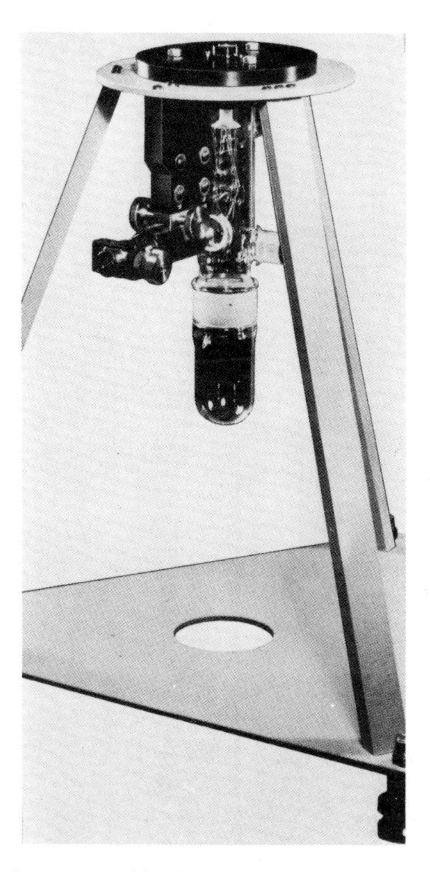

Figure 14.7. Microbalance assembly. (Courtesy of Worden Quartz Products Inc.)

Figure 14.8 shows the Worden microbalance which is constructed entirely of fused quartz. It is small, simple, relatively inexpensive, and easy to use. It can be baked at elevated temperature and used in high vacuum. If properly calibrated, it can be operated at high ambient temperatures. The sample loads are 100 mg to 1 g, 500 mg to 5 g, and 1 g to 10 mg with adjustable sensitivity depending on the model ordered. It has a precision of 0.5 μg when read with the filar microscope.The microscope is fitted with an 8-turn micrometer screw, with 100 dial divisions per turn.

The seals are either silicone or Viton. With silicone seals a temperature

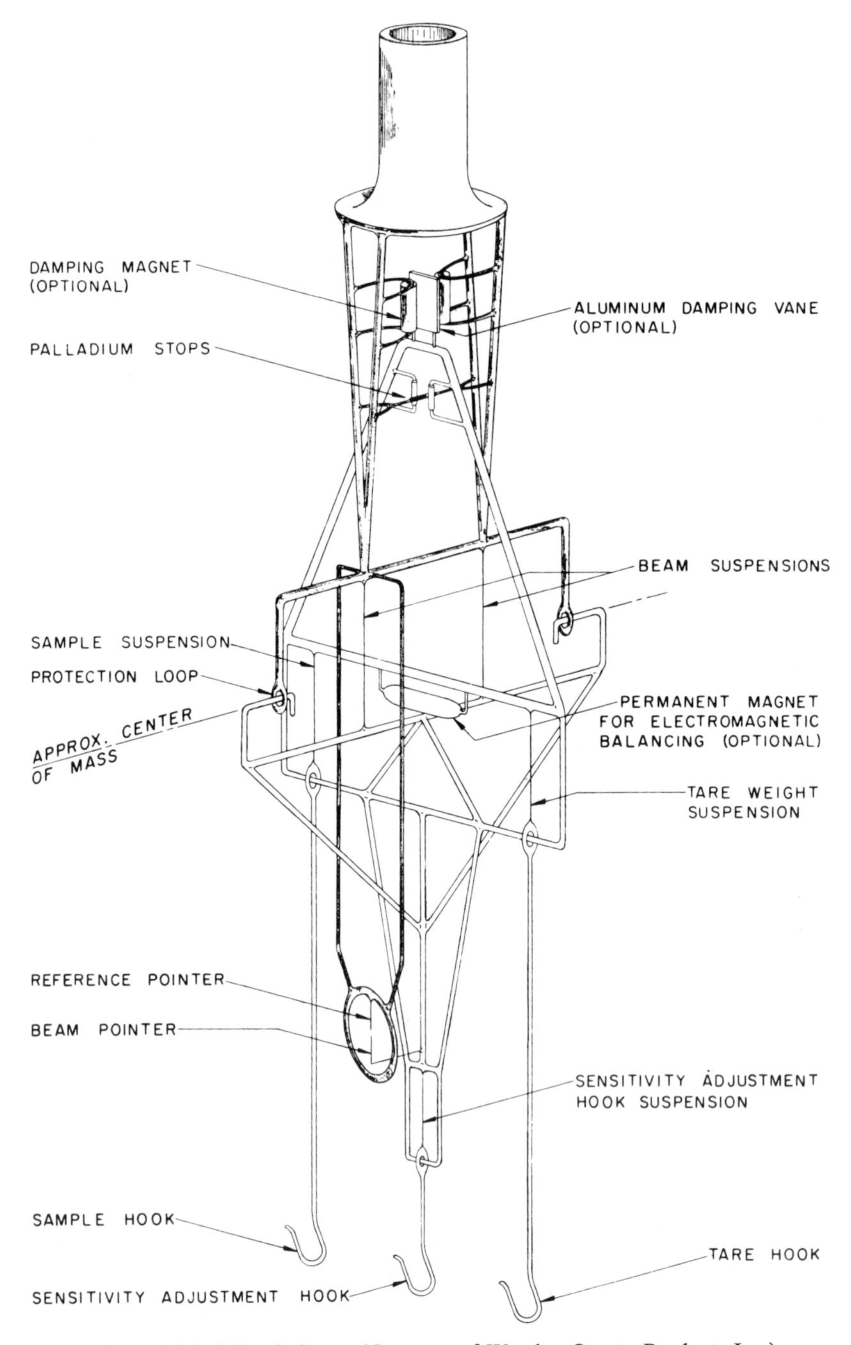

Figure 14.8. Microbalance. (Courtesy of Worden Quartz Products Inc.)

range of −300 to 500°F continuous may be expected; with Viton seals a temperature range of −300F to 400°F continuous to 700°F intermittent may be expected. The basic balance and tripod sells for $650. Optional equipment will increase the cost. Quotations may be obtained from the manufacturer.

Worden also produces an Auto-Null Micro-Balance which automatically maintains balance equilibrium. The balance and housing configuration are essentially as previously described. The balance is equipped with a solid

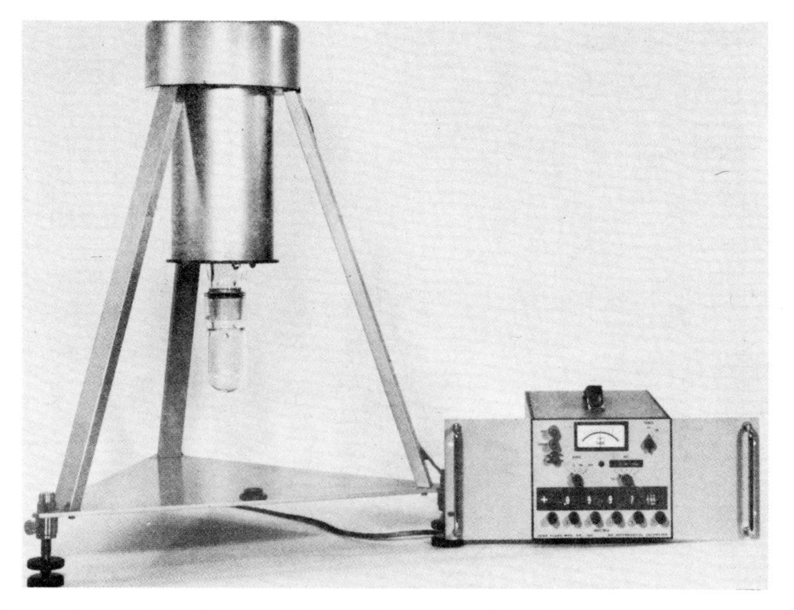

Figure 14.9. Auto-Null Micro-Balance. (Courtesy of Worden Quartz Products Inc.)

state electronic nulling circuit and has a two-position switch which allows manual or servo modes of operation. A microvolt per microgram readout is provided and it also has an output for a recorder. Figure 14.9 illustrates this apparatus. The sample weight has a maximum of 10 g and a minimum of 0.5 g. The apparatus has a precision of 0.5 μg and an adjustable sensitivity. This balance sells for approximately $4500.

B. Cahn Instrument Co.*

Cahn Instrument manufactures a variety of electrobalances which have sensitivities in the μg and sub-μg range. They are either of the recording or manual types and can be used in air, vacuum, or controlled atmospheres.

* 7500 Jefferson Street, Paramount, California 90723.

An example of the Cahn line, Figure 14.10, is the Recording Gram (RG) electrobalance shown in its vacuum bottle along with its controller. Materials used in the construction of the weighing mechanism are glass and alnico with small amounts of copper and nichrome, the balance itself being made of aluminum. The RG electrobalance is a null-type instrument

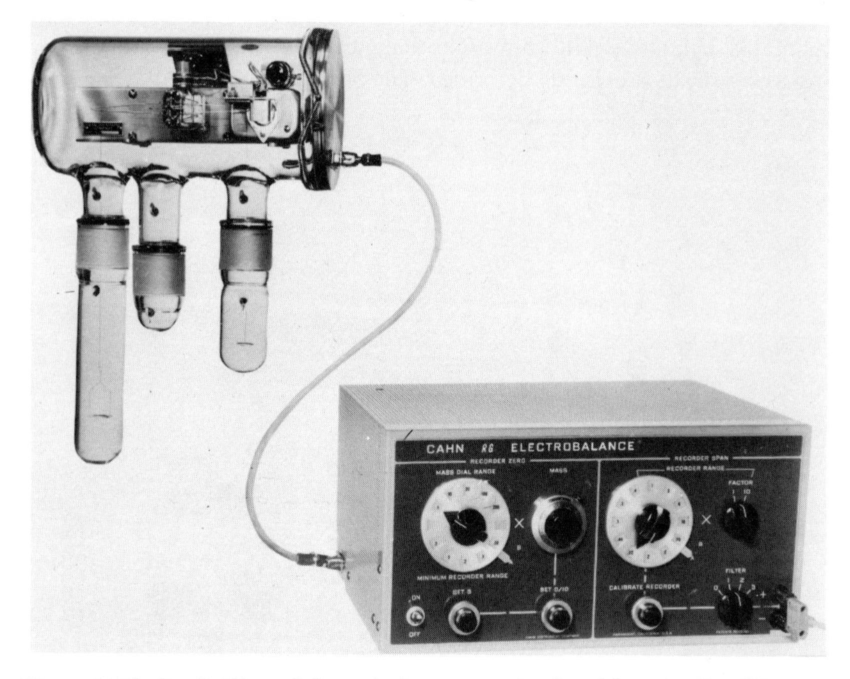

Figure 14.10. R. G. Electrobalance in its vacuum bottle with controller. (Courtesy of Cahn Instrument Co.)

the operating principle of which is indicated in Figure 14.11. A change in sample weight causes the beam to deflect. This deflection changes the phototube current, which is amplified and applied to the coil attached to the beam. The coil is in a magnetic field, so current through it produces a moment which restores the beam to balance. The current in the coil is a measure of sample weight.

The suspension ribbon eliminates friction, defines the axis of rotation, and is self-centering. It is rugged, hard to damage, and protected by stops when the balance is overloaded. Variations in rotational stiffness of the ribbon with temperature, age, and hysteresis do not affect balance readings because of the very small angle of rotation, and because readings are always made at the same angular position.

The RG electrobalance is a versatile instrument. It may be obtained in an aluminum case which permits use in air, or alternatively, the base of the case may be removed for weighing in another chamber. It may be placed on a platform for use in thermogravimetric analysis, e.g., or for recording surface tension. It may also be used in a vacuum system as shown in

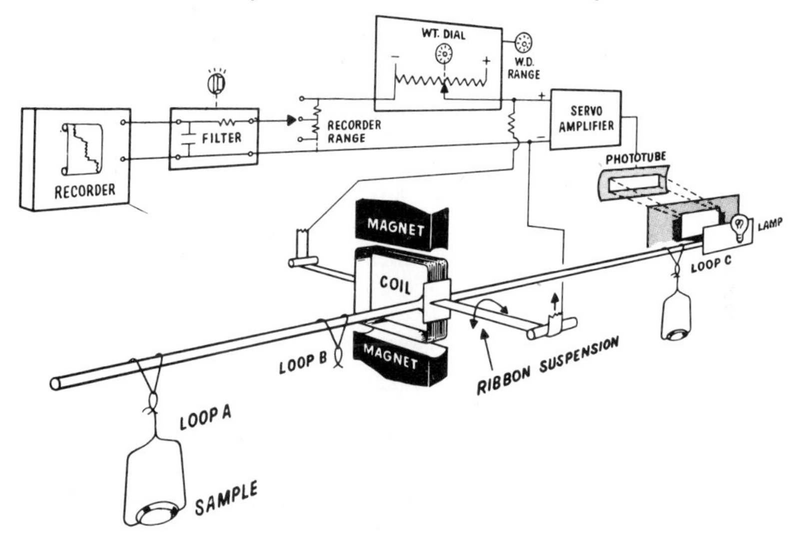

Figure 14.11. Operating principle of R. G. Electrobalance. (Courtesy of Cahn Instrument Co.)

Figure 14.10, or in a bell jar system for monitoring thin film evaporation. The electrobalance may also be obtained on a Varian Conflat Flange for operation in the ultrahigh vacuum region.

Specifications for the Cahn RG electrobalance are as follows:

Maximum capacity	2500 mg
Maximum weight change	1000 mg
Sensitivity, ultimate*	0.0001 mg
Suppression range	0–1 mg to 0–200 mg and 0–5 mg to 0–1 g selectable at will while recording
Suppression accuracy	To 0.0005 mg 0.05% of range (to 0.0001 mg and 0.01% of range on special order)
Sample temperature range	2540°C reported
Applications	Adsorption studies Thermogravimetry Magnetic susceptibility Surface tension

* As specified by the manufacturer.

Price	Complete with vacuum bottle for attaining ultrahigh vacuum but without recorder or other readout device. $2950.00

Cahn Instrument Co. manufactures a complete line of accessories for use with their electrobalances. A list of these accessories and their prices will be funished upon request.

C. Rodder Instrument*

For the ultimate in sensitivity in torsional fiber balances, the Rodder instrument stands alone. It is essentially custom-designed for a particular requirement. Since distribution is limited, only general specifications are available from the manufacturer. These should indicate, however, performance standards which may be attained in operating such balances in ultrahigh vacuum.

Figure 14.12. Microbalance assembly. (Courtesy of Rodder Instrument.)

The quartz microbalance assembly is shown in Figure 14.12. It is a null-type balance and consists of a fused quartz microbalance in a Pyrex housing with a metal flange on the sample hangdown tube. The microbalance is suspended by two long fibers connected at the ends of the pyrex tubes projecting from the sides of the main housing. The compensating mechanism is a magnet. The sample is positioned below the metal flange for easy access. The entire assembly may be baked for extended periods at 450°C in

* 775 Sunshine Drive, Los Altos, California

order to obtain ultrahigh vacuum. The detection unit, part of which may be seen in Figure 14.12, is a photocell.

The microbalance is small, simple, and easy to use. It has a load capacity of 200 mg with a differential load of 1 mg. The stability for 1 day is ± 0.04 μg.

This instrument is sold as a complete unit consisting of: microbalance, housing, flange, detection unit (less recorder), and installation for $8500.00.

IV. QUARTZ RESONATOR MICROBALANCES

Due to the growing importance of thin film evaporation, a number of companies have placed thin film thickness monitors on the market. These instruments are based on the principle that an oscillating quartz crystal will change frequency in direct proportion to the mass of any material deposited on the surface of the crystal. Although these are not microbalances in the conventional sense, they do measure a mass change. A more detailed treatment of this subject will be found in Chapters 5, 7 and 8.

A. Evaporation Apparatus Inc.*

Evaporation Apparatus Inc. has marketed a film thickness monitor and controller, Figure 14.13, designed for the precise measurement of deposited films. It may be used for routine work as well as for specialized research. The instruments will measure thin films during actual evaporation

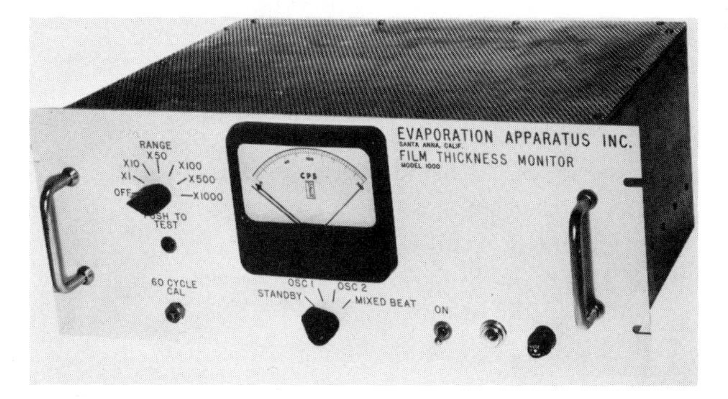

Figure 14.13. Monitor and controller. (Courtesy of Evaporation Apparatus Inc.)

by registering the change in frequency as material is deposited on the surface of the quartz crystal. If both the reference and monitor crystals are placed in the vacuum system, this unit will automatically compensate for variations in temperature, and ambient composition. Film thickness may be

* 2202 South Wright Street, Santa Ana, California 92705

controlled by a set-point indicator with automatic power cut off. Rates of evaporation can be calculated so that depositions may be duplicated. Film thickness corresponding to 10^{-8} g/cm^2 can be measured and controlled.

The equipment features self-compensation with respect to temperature, use in ion pumped systems, no watercooling, individual oscillator readout, crystals from 2 to 15 MHz, easy crystal changes, frequency shift range of ± 200 kHz, and film thickness control by adjustable cutoff relay.

Some specifications not already mentioned are as follows:

Range	± 200 kHz frequency shift
Crystal frequency	2–15 MHz (paired)
Crystal loading	For 10 MHz crystals:
	1% accuracy to ± 100 kHz (typical)
	0.05–2,600 Å (gold)
	0.37–18.500 Å (aluminum)
	2% accuracy to 200 kHz
	0.05–5,200 Å (gold)
	0.37–37,000 Å (aluminum)
Exposed crystal area	0.049 sq in.
Cost	$1,145.00, FOB Santa Ana, California

B. Ultek Corporation*

The thin film monitor and rate control system marketed by Ultek Corporation is shown in Figure 14.14. It monitors and controls the thickness of all types of vacuum deposited films during deposition and measures and regulates the rate of deposition on a substrate.

The apparatus consists of a water-cooled sensor head, accurately calibrated deposit thickness monitor, a deposit rate control, and an optional silicon controlled rectifier (SCR) power control module. The sensor head, connectors, cable, and feedthroughs are of 304 stainless steel, OFHC copper, or high density ceramic for high vacuum use.

The sensor head, when placed in vacuum, is connected to the vacuum side of the feedthrough by means of a copper-sheathed, bakeable, coaxial cable which extends to the feedthrough that is connected to the monitor by a cable. The head is contained in a welded 304 stainless steel enclosure controlled by an AT cut quartz crystal spring mounted on its face. The water lines from the feedthrough to the sensor head are $\frac{1}{4}$ in.

* Box 10920, Palo Alto, California

OFHC copper tubing that can be disconnected and changed when necessary.

The monitor unit is connected to the sensor head via the feedthrough by an RG-58U coaxial cable, and can be placed up to 56 ft away without weakening the signal. The sensor-head frequency output is compared to that of a stable tunable oscillator within the monitor. The difference between the two is amplified, shaped, and detected as a dc voltage. Other indications of frequency change include: An audible signal generated in an

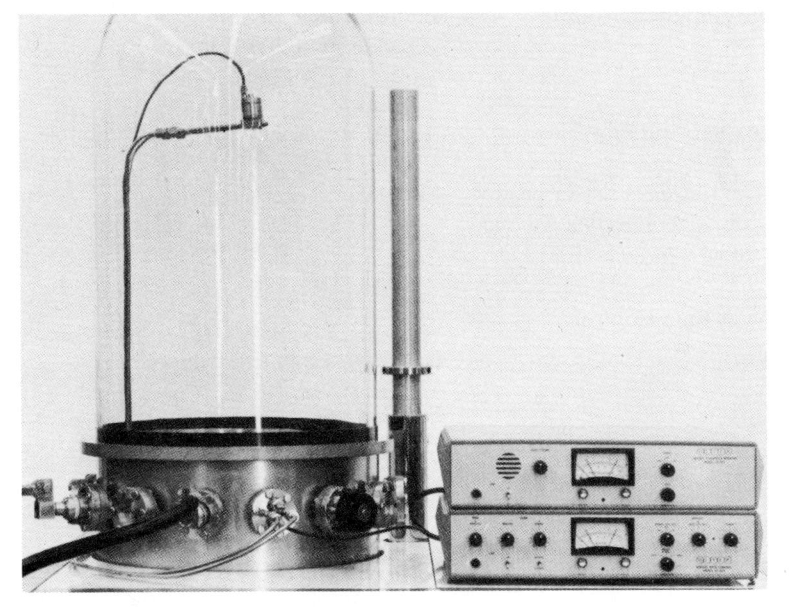

Figure 14.14. Thin film monitor and rate control system. (Courtesy of Ultek Corp.)

internal loud speaker; a pulse output which may be applied to an electronic counter; and a dc voltage output which may be fed directly into a recorder. The unit also has a double set point meter relay which allows automatic stop–start control of deposition by opening and closing a shutter mechanism.

The deposit rate control converts the frequency to a dc voltage proportional to the output of the crystal oscillator. This results in a voltage proportional to the deposition rate, that is compared to a preset rate. The difference, if any, produces a signal that is amplified as a 0 to 9 V dc output and fed to the accessory SCR module. The SCR module then modifies the power to the evaporation source until the rate of evaporation is the same as that desired.

Specifications are as follows:

Deposit Thickness Monitor

Range	100 kHz in 3, 10, or 30 kHz steps (30 kHz corresponds to 10,000 Å of aluminum)
Accuracy	Frequency 6 Hz to 30 kHz meter ±2% full scale
Stability	±10 ppm (1 hour)
Operating temperature	Indicator unit 10°C to 55°C pickup unit 200°C max.
Crystal	5 mHz AT cut, gold plated one side 0.75 in.^{2}
Auxiliary outputs	Audio, recorder, frequency, double set point relay
Price (includes sensor head, and all connecting cables, tubes, etc.)	\$1280.00

Deposit Rate Control

Range	0–300 Hz/sec in three stages 0–30 Hz/sec 0–100 Hz/sec 0–300 Hz/sec (useful minimum to 3 Hz/sec)
Control signal	0 to minimum 9 V dc for 2% full-scale rate change
Evaporation control	Preprogrammed rise and soak cycles and fully automatic deposit rate control
Auxiliary outputs	Double set point meter relay, recorder slave control
Price	\$1200.00

C. Sloan Instruments Corporation*

Sloan Instruments Corporation manufactures a line of thin film monitors which can be used for basic research or production control. The basic instrument consists of two units, a sensing head and an indicator, plus connecting cables. The sensing unit is a transitorized, crystal-controlled oscillator, housed in a stainless steel head. The indicator shown in Figure 14.15 features solid state circuitry. Information is displayed on a precision meter; an audio signal output has been added. Installation in a vacuum

* P. O. Box 4608, Santa Barbara, California

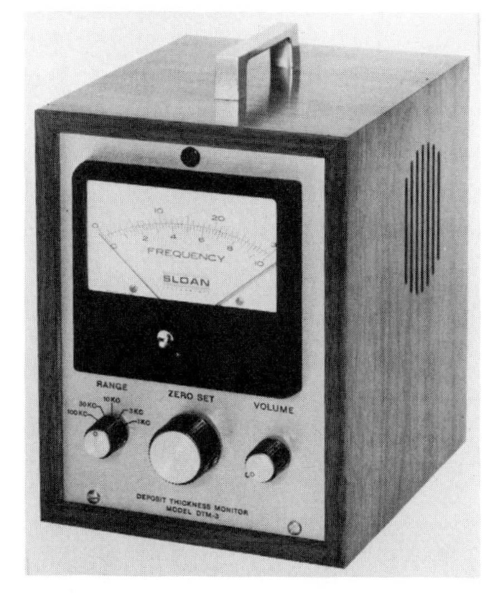

Figure 14.15. Deposit thickness monitor. (Courtesy of Sloan Instruments Corp.)

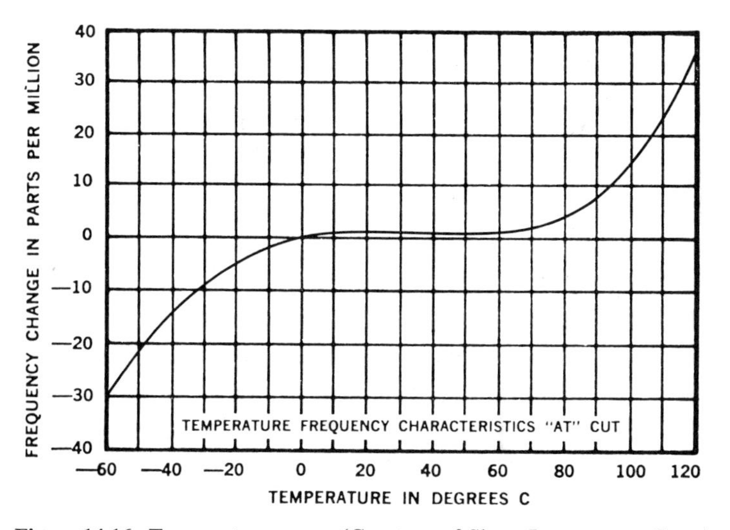

Figure 14.16. Temperature curve. (Courtesy of Sloan Instruments Corp.)

system requires only one electrical feedthrough and an associated ground. Sloan's method of mounting the crystal eliminates the need for a second temperature compensating crystal, by maintaining crystal temperature in the region of zero temperature response as shown in the curve in Figure 14.16.

Thickness is measured by beating the vacuum-crystal oscillator signal against a signal from a tunable reference oscillator. The demodulated beat frequency is filtered, amplified, and shaped. Frequency is converted to a dc signal and displayed. The meter relay can be preset to any frequency which automatically cuts off source power when the desired thickness is deposited.

Specifications for the deposit thickness monitor are as follows:

Oscillator tuning range	5 MHz tunable to +10 kHz and −140 kHz
Measurement range	0–100 kHz five ranges 1, 3, 10, 30, 100 kHz
Accuracy	±2% full scale
Stability (1 hr)	±30 ppm
Operating temperature range	See Figure 14.16
Crystal	5 MHz AT cut quartz, plated one side (0.75 in. × 0.75 in. × 0.015 in.)
Outputs	Indicators, 4½ in. precision panel meter and audio speaker Auxiliary, frequency: 0–100 kHz, 9 V peak to peak Recorder: 0 to −4½ V dc Relay: thickness limit, SPDT 5 A 125 V continuous
Price	$1200.00

Sloan Instruments is marketing a deposition rate control which is used in conjunction with their deposit thickness monitor, shown in Figure 14.17.

The deposit thickness monitor sends a signal to the deposition rate control. The frequency of this signal is directly related to the mass deposited on the sensing head of the monitor. This signal is amplified and converted to a dc voltage that is proportional to frequency. Differentiation of the dc voltage with respect to time gives the rate of frequency change and therefore, the rate of mass or thickness change. The deposition rate control compares this indicated rate with the desired rate and generates a signal that is proportional to the difference between the two. This signal is applied to the evaporator power supply and completes a control loop that will maintain the desired rate of deposition.

In addition the deposition rate control provides a source preheat warmup and hold cycle, automatic switches and relays for shuttering systems, and coordination with automatic vacuum system control.

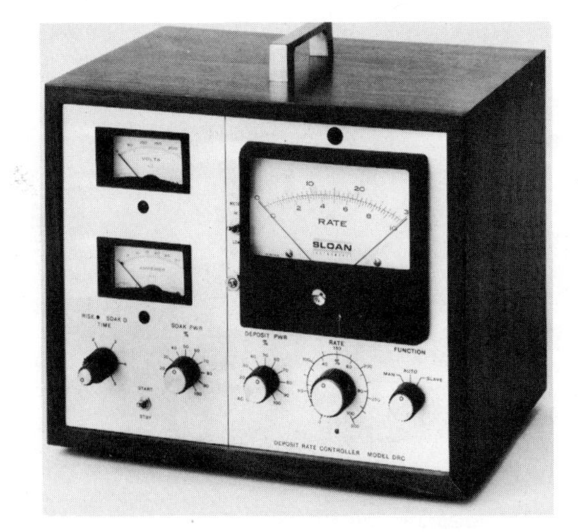

Figure 14.17. Deposit rate controller. (Courtesy of Sloan Instruments Corp.)

Specifications of the deposition rate control are as follows:

Metering	Rate: 0–100 Hz/sec and 0 to 300 Hz/sec $\pm 3\%$
	ac Volt: 0–250 $\pm 5\%$
	ac Amperes: 0–35 $\pm 5\%$ (with Sloan, SCR-P)
Inputs	Frequency: 0–30 kHz, 10 V peak to peak
	Slave: 0–9 V (0–300 Hz/sec)
	Start–stop: SPST
Outputs	Recorder: 0 to −9 V dc } 0–300 Hz/sec
	Slave: 0 to −9 V dc } 0–300 Hz/sec
Controls	Rise time = 5 sec to 5 min
	Soak time = 5 sec to 5 min
	Rate control: 10–300 Hz/sec
	Power control: 0 to −9 V dc
Price	$1200.00

The deposit thickness monitor and the deposition rate control may be obtained in one unit called OMNI-1, shown in Figure 14.18. The specifications for this unit are the same as those mentioned above. The price of OMNI-1 is $2400.00.

D. Heraeus-Engelhard Vacuum Inc.*

The Heraeus-Engelhard film thickness monitors have six frequency range multipliers (1X, 10X, 50X, 100X, 500X, 1000X) and can be operated with 2-, 5-, 10-, and 15- MHz crystals. Two 5- MHz crystals are furnished

* Seco Road, Monroeville, Pennsylvania 15146

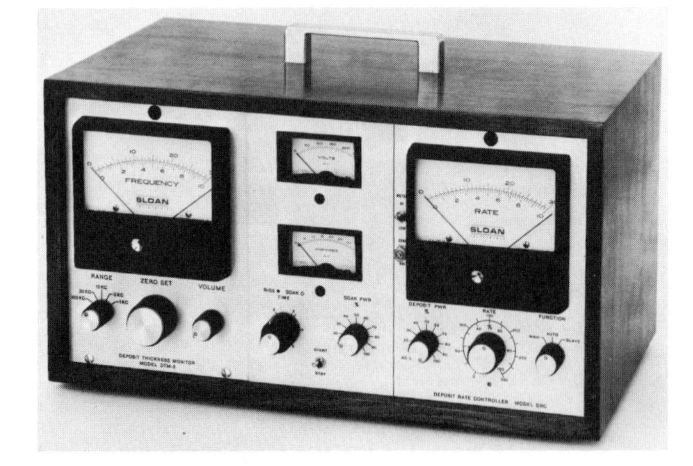

Figure 14.18. Deposit thickness monitor and deposit rate controller in one unit. (Courtesy of Sloan Instruments Corp.)

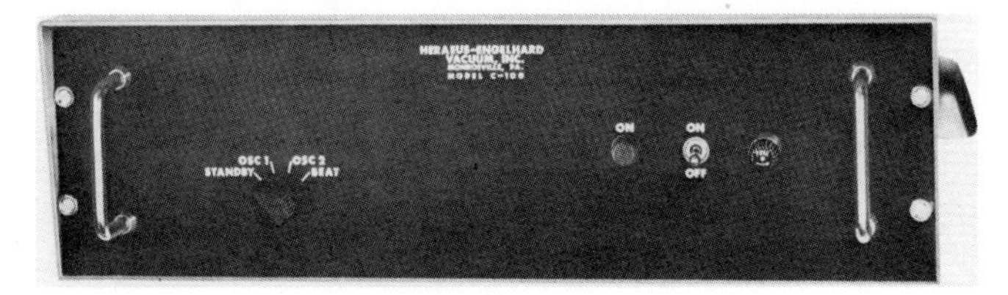

Figure 14.19. Film thickness monitor (basic unit). (Courtesy of Heraeus-Engelhard Vacuum Inc.)

with each unit. The measuring and the reference crystals are both inside the vacuum chamber and are not affected by low frequency sputtering; they can be changed quickly and easily.

During evaporation, the measuring crystal is exposed to the source while the reference crystal is shielded. The signals from the two crystals are passed into a mixer and the difference in frequency is fed to the readout.

Model C-100 is the basic unit (see Figure 14.19). It is rack mounted and has only one channel. A separate counter or frequency meter is required for operation.

Specifications are as follows:

Sensitivity	Greater than 10^{-9} g/cm^2
Range	200 kHz frequency shift
Crystal frequency	2–15 MHz in pairs
Crystal loading	Typical for 10 MHz crystals:
	1% accuracy to 100 kHz
	2% accuracy to 200 kHz

Exposed crystal area	0.049 sq in.
Crystal output	A periodic wave 1 V peak to peak, suited to all standard electronic counters (BNC connector on rear of panel)
Crystal input	Two 5-way binding posts
Oscillator test point	BNC connector on rear panel

Model C-110 is a dual channeled version of the C-100, having two monitors in one unit for alloying application. The specifications are the same as the C-100 except as follows:

Crystal output	Two BNC connectors on rear panel
Crystal input	Four 5-way binding posts
Oscillator test point	Two BNC connectors on rear panel

Model C-200 is a single channeled instrument which includes a frequency meter readout and relay cutoff circuit. The specifications are the same as the C-100 except as follows:

Relay control circuit	5 A at 115 V ac/dc inductive (NO) One output on rear panel
Meter	Reads in Hz; controls relay circuit with an increase or decrease in beat frequency. Range 0–200 kHz

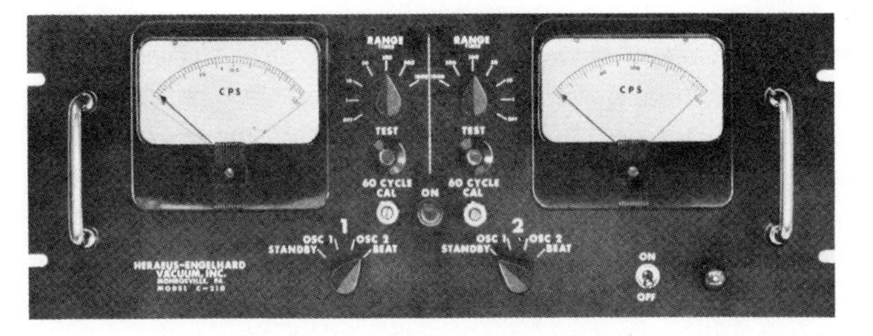

Figure 14.20. Film thickness monitor (two complete monitors in one unit). (Courtesy of Heraeus-Engelhard Vacuum Inc.)

Model C-210 shown in Figure 14.20 is a dual channel version of the C-200. It is two complete monitors in one unit for alloying applications. The specifications are the same as the C-100 except as follows:

Crystal output	Two BNC connectors on rear panel
Crystal input	Four 5-way binding posts
Oscillator test point	Two BNC connectors on rear panel
Relay control circuit	5 A 115 V ac/dc inductive (NO) Two outputs on rear panel

Model C-300 is a deluxe single thickness monitor with a built in electronic counter. It is designed to give extreme sensitivity for contaminant migration studies in ultrahigh vacuum environmental chambers, and for studies in surface physics. Specifications are available on request.

E. Edwards High Vacuum Inc.*

The Edwards High Vacuum thin film thickness monitor measures the thickness and rate of deposition in one unit. The monitor crystal is placed inside the vacuum chamber and the reference crystal is mounted in an oscillator unit outside the vacuum chamber. The difference between the two crystal frequencies is amplified within this unit and fed into a main unit where it is mixed with a signal from a variable oscillator to produce a final

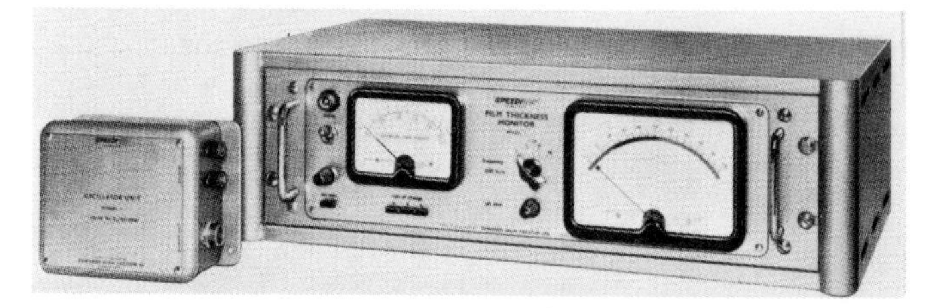

Figure 14.21. Thin film thickness monitor. (Courtesy of Edwards High Vacuum Inc.)

difference frequency of between 0 and 150 kHz. The deposited mass causes the change in the final difference frequency and this change is converted to a dc signal which actuates both the frequency shift meter and the rate meter.

Included in the thin film thickness monitor package (Figure 14.21) are the crystal support assembly, oscillator unit, five spare crystals (6 MHz AT cut), main control unit, and connecting cables.

A relay unit and an automatic evaporation control system for the thin film monitor unit are accessories which can be obtained.

The relay unit, Figure 14.22, enables the user to preset the film thickness so that when that thickness is reached the power to the source will be shut off.

The automatic evaporation control system (Figure 14.23) will automatically monitor the evaporation source degassing, rate of film growth, and film thickness. This system comprises a silicon controlled rectifier stack with drive circuit that has a source current stabilizer; a degas controller that includes separate switches for evaporant preheating and degas times; an evaporation controller which sets the evaporation rate; and a

* 3278 Grand Island Boulevard, Grand Island, New York 14072

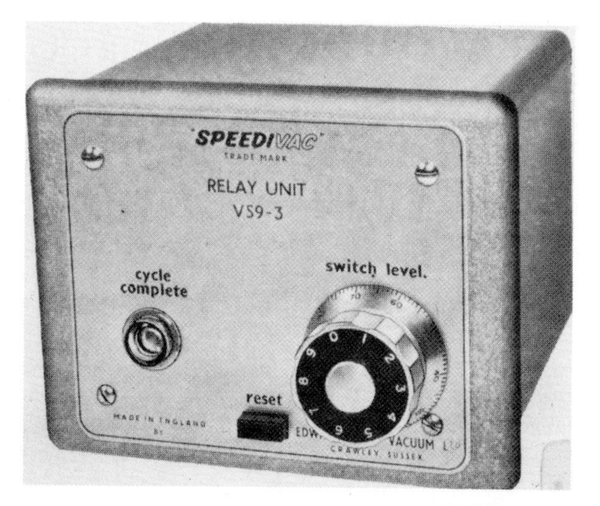

Figure 14.22. Relay unit. (Courtesy of Edwards High Vacuum Inc.)

Figure 14.23. Automatic evaporation control system. (Courtesy of Edwards High Vacuum Inc.)

process terminator which stops evaporation automatically at the required film thickness.

Specifications are as follows:

Thickness Measurement	
Min. detectable frequency shift	10 Hz
Max. frequency shift before resetting zero	
normal range	50 kHz

Crystal frequency shift before cleaning required	100 kHz approx. 150 kHz max.
Frequency shift tolerance between ranges	±3% of full scale
Frequency shift repeatable on same range	±1% of full scale deflection
Frequency shift drift over any 15-min period	less than Hz

Rate of Thickness Change Measurement

Max. rate of thickness change	10 frequency shift meter div./s/s
Rate of change tolerance for a linear frequency shift	±2% of full scale deflection on range A ±3% of full scale deflection on range B and C
Rate of change accuracy on any one range	±1% of full scale deflection
Price	$1150.00

Relay Unit

Price	$155.00

Automatic Evaporation Control System

Source current stabilizer: output excursion from 0–20 mA with an input source current of 5–0 A RMS
Degas controller: setting accuracy of both rise and total time within 6%

Evaporation Controller:

Frequency rate	1–5,000 Hz/sec
Counting rate	
Drift	1%/10°C change max.

Process terminator:

Stability	±2% of frequency meter setting
Price	$1600.00

V. EUROPEAN SOURCES

No detailed information is currently available on European products. However, readers may apply to the following companies as potential suppliers.

C. I. Electronics Ltd.	11 Greenclose Lane Wimborne, Dorset, U.K.

L. Oertling Ltd.	Cray Valley Works St. Mary Cray, Orpington Kent, U.K.
Research & Industrial Instruments Co.	17 Stannary Street London, S.E., U.K.
Stanton Instruments Ltd.	119 Oxford Street London, W.1, U.K.
Camlab (Glass) Ltd.	Cambridge
W. G. Pye & Co. Ltd.	P.O. Box 60, Scientific Instrument Center, York Street, Cambridge, U.K.
White Electrical Instrument Co. Ltd.	10 Amwell Street London, E.C.1, U.K.
Sermomex Controls Ltd.	Crowborough, Sussex, U.K.
Paul Bunge	2 Hamburg 22 Schellingstrasse 13, W. Germany
Colora Mess-Technik G.m.b.H.	7073 Lorch Barbarossa Str. 3, Germany
Mettler-Waagen Spoerhase AG	63 Giessen Marburger Str., W. Germany
Sartorius-Werke AG	34 Gottingen Weender Landstr., 96 W. Germany
Edwards High Vacuum International Ltd.	Manor Royal Crawley, Sussex, U.K.
W. C. Heraeaus, G.m.b.H.	6450 Hanau Heraeus Str. 12, Germany
Seavom	30 rue Raspail Argenteuil (S & O) France

ACKNOWLEDGEMENT

The Editors wish to thank the companies discussed in this chapter for providing information about their equipment.

REFERENCES

1. J. McBain and A. Baker, *J. Am. Chem. Soc.*, **48**, 690 (1920).
2. W. Weber, *Werke* (*Berlin*), **1**, 497 (1892).
3. T. N. Rhodin, *Advances in Catalysis*, Vol. V, Academic Press, New York, 1953, p. 40.
4. J. Donau, *Mikrochemie*, **9**, 1 (1931); **13**, 155 (1933).
5. T. J. Jennings, *The Defect Solid State*, T. J. Gray, Ed., Interscience, New York, 1957, p. 477.

AUTHOR INDEX

Numbers in parentheses are reference numbers and show that an author's work is referred to although his name is not mentioned in the text. Numbers in *italics* indicate the pages on which the full references appear.

C

D

H

I

J

K

L

M

N

O

T

Y

Z

SUBJECT INDEX